4.2 SIMPLEX METHOD

1. Determine the objective function.

2. Write all necessary constraints.

3. Convert each constraint into an equation by adding slack variables.

4. Set up the initial simplex tableau.

5. Locate the most negative indicator. If there are two such indicators, choose one.

6. Form the necessary quotients to find the pivot. Disregard any negative quotients or quotients with a 0 denominator. The smallest nonnegative quotient gives the location of the pivot. If all quotients must be disregarded, no maximum solution exists. If two quotients are equally the smallest, let either determine the pivot.

7. Use row operations to change the pivot to 1 and all other numbers in that column to 0.

8. If the indicators are all positive or 0, this is the final tableau. If not, go back to Step 5 above and repeat the process until a tableau with no negative indicators is obtained.

9. Read the solution from this final tableau. The maximum value of the objective function is the number in the lower right corner of the final tableau.

5.3 MULTIPLICATION PRINCIPLE

Suppose n independent choices must be made, with

m_1 ways to make choice 1,
m_2 ways to make choice 2,

and so on, with

m_n ways to make choice n.

Then there are

$$m_1 \cdot m_2 \cdots m_n$$

different ways to make the entire sequence of choices.

5.4 PERMUTATIONS AND COMBINATIONS

Permutations Different orderings or arrangements of the r objects are different permutations.

$$P(n, r) = \frac{n!}{(n - r)!}$$

Clue words: Arrangement, Schedule, Order

Combinations Each choice or subset of r objects gives one combination. Order within the group of r objects does not matter.

$$\binom{n}{r} = \frac{n!}{(n - r)!r!}$$

Clue words: Group, Committee, Sample

6.1 BASIC PROBABILITY PRINCIPLE

Let S be a sample space with n equally likely outcomes. Let event E contain m of these outcomes. Then the probability that event E occurs, written $P(E)$, is

$$P(E) = \frac{m}{n} = \frac{n(E)}{n(S)}.$$

FINITE MATHEMATICS
FOURTH EDITION.

FINITE MATHEMATICS
FOURTH EDITION

Margaret L. Lial
American River College

Charles D. Miller

SCOTT, FORESMAN AND COMPANY
Glenview, Illinois   London, England

TO THE STUDENT   If you want further help with this course, you may want to obtain a copy of the *Student's Solutions Manual* that accompanies this textbook. This manual provides detailed step-by-step solutions to the odd-numbered exercises in the textbook and can help you study and understand the course material. Your college bookstore either has this manual or can order it for you.

Cover art: Vasily Kandinsky, ''Extended,'' May-June 1926. Collection of the Solomon R. Guggenheim Museum, New York.
Photo: David Heald

Library of Congress Cataloging-in-Publication Data

Lial, Margaret L.
    Finite mathematics / Margaret L. Lial, Charles D. Miller—4th ed.
        p.   cm.
    Includes index.
1. Mathematics—1961–   I. Miller, Charles David.   II. Title.
QA37.2.L49   1989
510—dc19   88-38694
                                                                               CIP

ISBN 0-673-38253-2

Portions of this book appear in *Finite Mathematics and Calculus with Applications,* Third Edition, © 1989, 1985, 1980 Scott, Foresman and Company.

    3 4 5 6—VHJ—93 92 91 90 89

PREFACE

Finite Mathematics, Fourth Edition, is a solid, applications-oriented text covering the noncalculus portions of mathematics needed by students majoring in business, management, economics, or the life or social sciences. Explanations and examples are clear, direct, and to the point; exercises are carefully graded; and numerous examples are provided that directly correspond to the exercises. Highlights of the text are the Extended Applications and abundant applied problems that show students how mathematics is used in the careers they are exploring. The only prerequisite for this book is a previous course in algebra.

KEY FEATURES

FLEXIBILITY *Finite Mathematics* has been organized to be as flexible as possible. Some instructors may choose to begin the course with Chapter 1 and cover the chapters on linear programming. Some may choose to begin the course with Chapter 5 on sets and counting and continue with probability, statistics, and other chapters as desired. The book will facilitate either course organization.

APPLICATIONS A wide range of applications is included in examples and exercise sets to motivate student interest. An Index of Applications, grouped by discipline, includes those applications given in the exercises, examples, and Extended Applications, and is located after the table of contents.

EXAMPLES Nearly 400 worked-out examples clearly illustrate the techniques and concepts presented and prepare students for success with the exercises. For clarity, the end of each example is indicated with the symbol ▬.

EXERCISES More than 2500 exercises, including about 2025 drill problems and 475 applications, feature a wide range of difficulty from drill to challenging problems. A new format is used for exercises in this edition. Routine exercises are given first, followed by a clearly marked applications section arranged by discipline, with a descriptive heading for each application.

EXTENDED APPLICATIONS Special lengthy applications are included at the end of most chapters. Designed to help motivate students in the study of mathematics, the Extended Applications are derived from current literature in various fields and show how the course material can actually be applied.

FORMAT Important rules, definitions, theorems, and equations are enclosed in boxes and highlighted with a title in the margin, for ease of study and review. A second color is used to annotate equations, illuminate troublesome areas, and clarify concepts and processes. A new format in the exercise sets highlights the many diverse applications included in the book.

STUDY AIDS Each chapter ends with a list of key words presented in the chapter and an extensive set of review exercises. The key words are referenced to the section in which each appeared, for extra studying help.

▬ CONTENT FEATURES

The idea of **mathematical models** of real-world applications from business, economics, social sciences, and life sciences is emphasized throughout the book, beginning in Chapter 1. Both the strengths and the limitations of models are discussed.

Chapter 1 on linear functions is an **optional review chapter.** Well-prepared students can go directly to Chapter 2, on matrix theory, or to Chapter 5, which leads to probability. For many students, Chapter 1 will serve as a convenient reference.

The book has **two complete chapters on linear programming,** one on the graphical method and one on the simplex method. Mixed constraints are handled with the Phase I/Phase II method. This method also provides one way to solve minimizing problems. A section using the dual approach to minimizing is also included for those instructors who prefer that approach.

The **discussion of permutations and combinations has been expanded** to two sections (Sections 5.3 and 5.4) to give more emphasis to the multiplication rule and to the differences between permutations and combinations.

The **chapter on probability has been rearranged** to improve the pedagogical flow and the clarity of the presentation.

The **flexibility of the text** is indicated in the following chart of chapter prerequisites. As shown, the course could begin with either Chapter 1 or Chapter 5, and Chapter 10 on the mathematics of finance could be covered discretely.

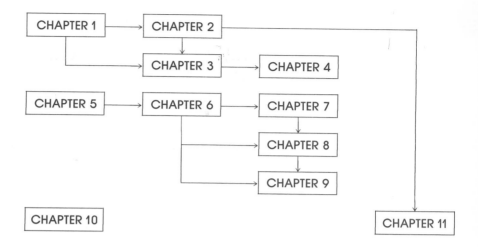

The **Instructor's Guide and Solutions Manual** gives a lengthy set of test questions for each chapter, organized by section, plus answers to all the test questions. It also provides complete solutions to all of the even-numbered text exercises.

The **Instructor's Answer Manual** gives the answers to every text exercise, collected in one convenient location and presented in an easy-to-use format.

The **Student's Solutions Manual,** available for purchase by students, provides detailed, worked-out solutions to all of the odd-numbered text exercises, plus a chapter test for each chapter. Answers to the chapter test questions are given at the back of the book.

The **Scott, Foresman Test Generator for Mathematics** enables instructors to select questions by section or chapter or to use a ready-made test for each chapter. Instructors may generate tests in multiple-choice or open-response formats, scramble the order of questions while printing, and produce multiple versions of each test (up to 9 with Apple, up to 25 with IBM). The system features a preview option that allows instructors to view questions before printing, to regenerate variables, and to replace or skip questions if desired. A Macintosh version of the Test Generator will be available.

Computer Applications for Finite Mathematics and Calculus by Donald R. Coscia is a softbound textbook packaged with two diskettes (in Apple II and IBM-PC versions) with programs and exercises keyed to the text. The programs allow students to solve meaningful problems without the difficulties of extensive computation. This book bridges the gap between the text and the computer by providing additional explanations and exercises for solution using a microcomputer. Appropriate exercises in *Finite Mathematics* are identified by the heading ''For the Computer.''

A set of two-color **overhead transparencies** showing charts, figures, and portions of examples is available and can be used to accompany a lecture.

■■ ACKNOWLEDGMENTS The following instructors reviewed the manuscript and made many helpful suggestions for improving this edition: Daniel D. Anderson, University of Iowa; Craig J. Benham, Mount Sinai School of Medicine; Richard A. Brualdi, University of Wisconsin–Madison; Mitzi Chaffer, Central Michigan University; Jay Graening, University of Arkansas; Cheryl M. Hawker, Eastern Illinois University; Joseph C. Hebert, University of Southwestern Louisiana; Richard D. Hobbs, Mission College; Jeffery T. McLean, College of Saint Thomas; Raymond Maruca, Delaware County Community College; Lenora Jane Mathis, Baylor University; William C. Ramaley, Fort Lewis College; James V. Rauff, College of Lake County; Burt M. Rosenbaum, Saint Leo College; and Diane Tischer, Metropolitan Technical Community College.

Thanks also to Bill Poole, Linda Youngman, and Janet Tilden, editors at Scott, Foresman, who contributed a great deal to the final book.

CONTENTS

Index of Applications   xv

1   LINEAR MODELS   1

1.1   Linear Equations and Inequalities in One Variable   2
1.2   Linear Functions   8
1.3   Slope and Equations of a Line   21
1.4   Linear Mathematical Models   32
1.5   Constructing Mathematical Models   (Optional)   44
Review Exercises   58
Extended Application: Estimating Seed Demands—The Upjohn Company   61
Extended Application: Marginal Cost—Booz, Allen & Hamilton   62

2   SYSTEMS OF LINEAR EQUATIONS AND MATRICES   65

2.1   Systems of Linear Equations; the Echelon Method   66
2.2   Solution of Linear Systems by the Gauss-Jordan Method   79
2.3   Basic Matrix Operations   90
2.4   Multiplication of Matrices   98
2.5   Matrix Inverses   108
2.6   Input-Output Models   119
Review Exercises   126

Extended Application: Routing   129

Extended Application: Contagion   131

Extended Application: Leontief's Model of the American Economy   132

3 LINEAR PROGRAMMING: THE GRAPHICAL METHOD   137

3.1 Graphing Linear Inequalities   138

3.2 Mathematical Models for Linear Programming   146

3.3 Solving Linear Programming Problems Graphically   154

Review Exercises   165

4 LINEAR PROGRAMMING: THE SIMPLEX METHOD   167

4.1 Slack Variables and the Pivot   168

4.2 Solving Maximization Problems   176

4.3 Mixed Constraints   187

4.4 Duality (Optional)   200

Review Exercises   210

Extended Application: Making Ice Cream   213

Extended Application: Merit Pay—The Upjohn Company   215

5 SETS AND COUNTING   217

5.1 Sets   218

5.2 Applications of Venn Diagrams   227

5.3 Permutations   236

5.4 Combinations   243

5.5 The Binomial Theorem (Optional)   253

Review Exercises   257

6 PROBABILITY 260

6.1 Basics of Probability 261

6.2 Further Probability Rules 270

6.3 Conditional Probability 283

6.4 Bayes' Formula 297

6.5 Applications of Counting 304

6.6 Bernoulli Trials 312

Review Exercises 322

Extended Application: Making a First Down 326

Extended Application: Medical Diagnosis 326

7 STATISTICS AND PROBABILITY DISTRIBUTIONS 328

7.1 Basic Properties of Probability Distributions 329

7.2 Expected Value 336

7.3 Variance and Standard Deviation 348

7.4 The Normal Distribution 358

7.5 The Normal Curve Approximation to the Binomial Distribution 370

Review Exercises 378

Extended Application: Optimal Inventory for a Service Truck 381

Extended Application: Bidding on a Potential Oil Field— Signal Oil 382

8 MARKOV CHAINS 385

8.1 Basic Properties of Markov Chains 386

8.2 Regular Markov Chains 396

8.3 Absorbing Markov Chains 404

Review Exercises 414

9 DECISION THEORY   417

9.1 Decision Making   418

9.2 Strictly Determined Games   423

9.3 Mixed Strategies   430

Review Exercises   441

Extended Application: Decision Making in Life Insurance   443

Extended Application: Decision Making in the Military   446

10 MATHEMATICS OF FINANCE   450

10.1 Simple Interest and Discount   451

10.2 Compound Interest   457

10.3 Sequences   465

10.4 Annuities   473

10.5 Present Value of an Annuity; Amortization   481

Review Exercises   488

Extended Application: A New Look at Athletes' Contracts   491

Extended Application: Present Value   493

11 DIGRAPHS AND NETWORKS   496

11.1 Graphs and Digraphs   497

11.2 Dominance Digraphs   501

11.3 Communication Digraphs   507

11.4 Networks   513

Review Exercises   520

▤ TABLES   523

Combinations   525

Area Under a Normal Curve   526

Compound Interest   528

Present Value   529

Amount of an Annuity   530

Present Value of an Annuity   531

Answers to Selected Exercises   533

Index   569

INDEX OF APPLICATIONS

BUSINESS AND
ECONOMICS

Advertising, 286, 314, 415, 439
Airline routing, 130
Airline usage, 302
Amortization, 483, 485, 486,
    487, 490
Annuities, 482, 486
Appliance reliability, 303
Athletes' contracts, 491
Average cost, 41
Bakery income, 184
Bakery orders, 117
Bank discount, 456, 490
Banking, use of computers in, 294
Bankruptcy, 380
Battery life, 356
Bidding on a potential oil
    field, 382
Blending milk, 152, 162
Blending seed, 198
Blending a soft drink, 199
Break-even analysis, 36, 42, 60
Break-even point, 41, 42
Breeding cats, income from, 176
Brewery repair costs and
    production levels, 54
Car payments, 483, 486
Car purchases, 319
Car rental, 7, 20
Charitable contributions, 183
Charitable trust, 487
Chicken farming, 233
Christmas card production, 419
Citrus farming, 418–19
Cola consumption, 235
Company earnings, 60
Company training programs,
    396, 414

Competing stores, 388, 394,
    399, 429, 430
Concert preparations, 421
Consumer credit, 303, 456
Consumer satisfaction, 320
Contractor bidding, 422
Correlation of sales and
    profits, 53
Correlation of stores and
    sales, 50
Cost analysis, 34, 102, 106, 166
Credit applications, 225
Credit cards, 415
Currency, 76
Defective items, production of, 300
Delinquent taxes, 456
Demand function, 15–16
Depreciation
    straight-line, 38, 41, 52,
      56, 60, 473
    sum-of-the-years'-digits,
      44, 46, 52, 56, 60
Doubling time, 463, 464, 465
Dry cleaning, 388, 394, 399
Effect of revenue on price, 57
Equilibrium demand, 17
Equilibrium price, 17
Equilibrium supply, 17
Farm profit, 170, 176, 187, 208
Fast food, consumption of, 319
Finance, 145, 152, 163, 197
Franchise fees, 21
Hotel occupancy, 346
Ice cream manufacturing, 213–14
Individual retirement accounts,
    464, 480
Inflation, 464
Inheritance income, 487
Input-output models, 119–24, 128
    closed, 122, 123, 125
    open, 122, 124

Insurance, 151, 161, 340,
    346, 394
Interest on investments, 4, 7
Inventory, 40, 91, 93, 95,
    381–82
Investment habits, 235
Investments
    comparing, 127, 212, 464
    and economic conditions,
      303, 420
    interest on, 4, 7, 79, 118, 198
IRAs, 464, 480
Job qualifications, 303
Labor relations, 441
Labor unions, 29
Land development, 395, 420
Land use, 395
Life insurance, 443–45
Light bulb life, 368
Linear cost function, 36, 59
Loan interest, 79, 456
Loan repayment, 453–55,
    456, 461, 490
Locating oil, 324
Machinery repairs, 421, 422
Manufacturing cost, 31
Marginal cost, 35, 40, 62
Market share, 394
Marketing a new product, 421
Maximizing profit, 146, 151,
    152, 161, 162, 166, 175,
    176, 184, 185, 186, 212
Maximizing revenue, 151–52,
    162, 175, 183
Merit pay, 215–16
Minimizing cost, 145, 146,
    161, 162, 192–93, 194,
    197, 518
Mining exploration, 319
Negative interest, 465
Net profit, 7

New car purchases, 319
Oil distribution, 128
Packaging, 90
Personnel screening, 320
Pet store stock, 295
Present value, 493
Price-to-earnings ratio, 224
Pricing, 439
Printing orders, 128
Process control, 353, 357
Production costs, 39, 197, 208
Production rates, 78
Production requirements, 118, 127
Production scheduling, 127,
   143, 145, 146
Profit, 59
  estimating, 346
  maximizing, 146, 151, 152,
   161, 162, 166, 175, 176,
   184, 185, 186, 212
  net, 7
  optimum, 208
Purchase orders, 439–40
Quality control, 295, 301,
   303, 306, 320, 322, 324,
   368, 374, 377, 380, 403
Rental fees, 19
Return on investment, 59, 79,
   118, 198
Revenue, 19
  maximizing, 151–52, 162,
   175, 183
Revenue/cost/profit, 7
Salary, 473
Sales, 10, 19, 31, 467
Sales accounts, rating, 346
Sales analysis, 33, 40, 98
Sales estimation, 42, 47, 50, 53
Sales management, 403
Sales promotion, 356
Savings annuity, 475, 479,
   480, 490
Savings deposits, 456, 464,
   470, 477
Seed demands, 61
Shipping costs, 89, 151, 152,
   161, 162, 192–93, 194,
   197, 198
Shipping decisions, 145, 146
Sinking fund, 477, 480, 490
Stock reports, 128
Stock returns, 379
Storage capacity, 148, 158
Straight-line depreciation, 38,
   41, 52, 56, 60, 473

Sum-of-the-years'-digits
   depreciation, 44, 46, 52,
   56, 60
Supply and demand, 15, 16,
   18, 20, 21, 59
Supply function, 59
Survey results, 322
Tax and interest, 7
Telephone service needs, 53
Television advertising, 303
Time management, 145, 166
Training programs, 396, 414
Transportation costs, 89, 151,
   152, 161, 162, 192–93,
   194, 197, 198
Unions, 29
Utility company management, 230
Worker error, 298

GENERAL

Chinese New Year
  celebrations, 233
Construction, 153, 164
Contests, 348
Family food expenditures, 21
Family income, 78
Football, 326, 429
Gambler's ruin, 406–7, 414
Grocery purchases, 97
I Ching, 321
Magazine readership, 229–30
Military decisions, 446–49
Music, 233
Native American ceremonies, 232
Postage costs, 61
Routing, 129–30
Tuition, 8
War games, 429

LIFE SCIENCES

Animal activity, 128
Animal growth, 97
Ant population growth, 43
Anti-smoking campaign, 422
Bacteria
  food requirements of, 90
  growth of, 43, 55
Blending nutrients, 186, 199, 206
Blood antigens, 234

Blood pressure, 346
Blood sugar and cholesterol
  levels, 61
Color blindness, 281, 296, 321
Contagion, 131–32, 413
Correlation of height and
  weight, 55
Dietetics, 79, 97, 107, 149,
   153, 163, 198, 209
Drugs
  effectiveness of, 320, 377
  prescribing, 438
  screening, 296
  sequencing, 243
  side effects of, 320
Fish
  effects of water pollution
   on, 31
  food requirements of, 26, 90
Flu inoculations, 321
Food web, 130
Genetics, 234, 245, 281, 282,
   291, 295, 393, 403, 416
Hazardous waste, 227
Health care, 153, 163, 198, 209
Height, estimating, 31
Hepatitis blood test, 303
Insect classification, 243
Insecticide spraying, 434
Medical diagnosis, 326
Medical research, 416
Plant foods, mixing, 90
Population growth, 473
Predator food requirements,
   153, 163
Radiation, effects of, 320
Rats
  diets of, 380
  weight gain of, 34
Sickle cell anemia, 325
Soil fertilization, 54
Surgery survival rate, 320
Thyroid problems, 227
Toxemia test, 304
Vitamin A deficiency, 320
Vitamin requirements, 369

PHYSICAL SCIENCES

Blending chemicals, 199, 210
Blending gasoline, 199, 210
Celsius and Fahrenheit
  temperatures, 56

Octane rating, 8
Seeding storms, 347
Weather, 404

SOCIAL SCIENCES

Border patrol, 429
Cities, distances between,
    515, 519
Class size, 44
Commuting times, 380
Education, 235, 369, 404, 423

Election results, 497, 501
Estimation of time, 43
High school and college GPAs, 56
Housing patterns, 395
Immigration, 32
Interpersonal relationships, 501
I.Q. scores, 380
Law enforcement, 440
Minority attendance at
    community colleges, 321
Political affiliation, 304
Politics, 186, 442
Preference tests, 506–7
Randomized response method, 325

Rat maze, 413
Reelection strategy, 423
Rumors, 404
Social class changes, 386
Stimulus effect, 43
Student retention, 412
Testing, 321
Tournament winners, 501,
    504, 506
Traffic control, 88
Transportation, 412
Voting, 31, 32, 304, 395
Work-related illness and
    injury, 32

1 Linear Models

1.1   Linear Equations and Inequalities in One Variable

1.2   Linear Functions

1.3   Slope and Equations of a Line

1.4   Linear Mathematical Models

1.5   Constructing Mathematical Models   (Optional)

Review Exercises

Extended Application
Estimating Seed Demands — The Upjohn Company

Extended Application
Marginal Cost — Booz, Allen & Hamilton

Before using mathematics to solve a real-world problem, we must usually set up a **mathematical model,** a mathematical description of the situation. Constructing such a model requires a solid understanding of the situation to be modeled, as well as familiarity with relevant mathematical ideas and techniques.

Much mathematical theory is available for building models, but the very richness and diversity of contemporary mathematics often prevents people in other fields from finding the mathematical tools they need. There are so many useful parts of mathematics that it can be hard to know which to choose.

To avoid this problem, it is helpful to have a thorough understanding of the most basic and useful mathematical tools that are available for constructing mathematical models. In this chapter we look at some mathematics of *linear* models, which are used for data whose graphs can be approximated by straight lines.

1.1 LINEAR EQUATIONS AND INEQUALITIES IN ONE VARIABLE

We begin this chapter by reviewing *linear equations,* because studying linear mathematical models often involves solving linear equations. A **linear equation in one variable** is an equation that can be simplified to the form $ax = b$, where a and b are real numbers, with $a \neq 0$. For example,

$$2x = 5, \qquad 3m + 7 = -12, \qquad \text{and} \qquad 6(a - 3) + 4 = 2a - 7$$

are linear equations.

The following properties are used in solving linear equations.

ADDITION AND MULTIPLICATION PROPERTIES OF EQUALITY

> For real numbers, a, b, and c,
>
> **1.** if $a = b$, then $a + c = b + c$;
> **2.** if $a = b$, then $ac = bc$ ($c \neq 0$).

By these properties any number can be added to both sides of an equation, and any nonzero number can be used to multiply both sides of an equation. Since subtraction is defined as adding the negative of a number, subtracting the same number from both sides of an equation is justified by the addition property. Also, division on both sides of an equation by the same nonzero number is justified by the multiplication property.

In solving a linear equation, the goal is to isolate the unknown variable on one side of the equation. To do this, we use the addition and multiplication properties as shown in the following examples.

▬▬ EXAMPLE 1

Solve each equation.

(a) $\frac{3}{2}x = -12$

Multiply both sides of the equation by 2/3.

$$\frac{2}{3} \cdot \frac{3}{2}x = \frac{2}{3}(-12)$$

$$x = -8$$

Check the solution, -8, by substitution in the original equation.

(b) $-6m + 2 = 14$

First subtract 2 on each side of the equation, then divide both sides by -6.

$$-6m + 2 = 14$$
$$-6m = 12$$
$$m = -2$$

The solution is -2. Remember to check the solution.

(c) $-3(k - 5) + 7 = 2(k + 1)$

Use the distributive property to clear parentheses.

$$-3(k - 5) + 7 = 2(k + 1)$$
$$-3k + 15 + 7 = 2k + 2$$

Now combine like terms on the left side of the equation to get

$$-3k + 22 = 2k + 2.$$

To get the variable terms on one side of the equation and the constants on the other side, subtract $2k$ on both sides of the equation, and then subtract 22 on both sides, which gives

$$-5k + 22 = 2$$
$$-5k = -20.$$

Finally, divide both sides by -5 to get

$$k = 4.$$

Check in the original equation to see that 4 is the correct solution.   ▬

Sometimes a linear equation involves more than one letter. As a general rule, the first letters of the alphabet—a, b, c, and so on—are used to represent constants, while letters near the end of the alphabet, such as x, y, and z, are used for variables.

▆ EXAMPLE 2
Solve each equation for x.

(a) $ax - b = c$

Here x is the variable, and the other letters are treated as constants. Add b to both sides of the equation, then divide both sides by a, assuming $a \neq 0$.

$$ax - b = c$$
$$ax = c + b$$
$$x = \frac{c + b}{a}$$

(b) $2x + y = 6(x - 3) + 12$

To solve for x, treat y as a constant. Get all terms with x on one side of the equation and all terms without x on other other side.

$$2x + y = 6x - 18 + 12 \qquad \text{Distributive property}$$
$$2x + y = 6x - 6 \qquad \text{Combine terms}$$
$$-4x + y = -6 \qquad \text{Subtract 6x}$$
$$-4x = -6 - y \qquad \text{Subtract y}$$
$$x = \frac{-6 - y}{-4} \qquad \text{Divide by } -4$$
$$x = \frac{6 + y}{4} \qquad \text{Simplify} \quad \blacksquare$$

The next example shows an application involving a linear equation.

▆ EXAMPLE 3
Richard Chinn receives a $14,000 bonus from his company. He invests part of the money in tax-free bonds at 6% and the remainder at 10%. He earns $1060 per year in interest from the investments. How much is invested at 6%?

Let x represent the amount Chinn invests at 6%; then $14,000 - x$ is the amount invested at 10%. Since interest is given by the product of principal, rate, and time, the interest earned in one year is as follows:

Interest at 6% $= x \cdot 6\% \cdot 1 = .06x;$

Interest at 10% $= (14,000 - x)(10\%)(1) = .10(14,000 - x).$

The total interest is $1060, so

$$.06x + .10(14,000 - x) = 1060.$$

Solve this equation.

$$.06x + 1400 - .10x = 1060$$
$$1400 - .04x = 1060$$
$$-.04x = -340$$
$$x = 8500$$

Check in the words of the original problem that $8500 of the bonus money was invested at 6%. ■

Linear Inequalities   Many mathematical models involve inequalities rather than equations, so it is useful to review the solution of linear inequalities. A **linear inequality in one variable** is an expression that can be simplified to the form $ax < b$. (Throughout our discussion of linear inequalities, $<$ can be replaced with $>$, $\le$, or $\ge$.) To solve a linear inequality, use the following properties, which are similar to the addition and multiplication properties of equations.

ADDITION AND MULTIPLICATION PROPERTIES OF INEQUALITY

For real numbers, a, b, and c:

1. If $a < b$, then $a + c < b + c$.
2. If $c > 0$ and $a < b$, then $ac < bc$.
3. If $c < 0$ and $a < b$, then $ac > bc$.

As with the corresponding properties for equality, the addition property can be used with subtraction, while the multiplication properties also apply to division. Note that when we multiply (or divide) both sides of an inequality by a *negative* number, the direction of the inequality symbol changes.

■ EXAMPLE 4

Solve the following inequalities.

(a) $3x - 5 < 7$

Use the properties as follows.

$$3x - 5 < 7$$
$$3x < 12 \qquad \text{Add 5}$$
$$x < 4 \qquad \text{Divide by 3}$$

The solution is $x < 4$.

(b) $4 - 3y \geq 7 + 2y$

Get the terms with y on one side and the terms without y on the other side.

$$4 - 3y \geq 7 + 2y$$
$$4 - 5y \geq 7 \qquad \text{Subtract } 2y$$
$$-5y \geq 3 \qquad \text{Subtract } 4$$
$$y \leq -\frac{3}{5} \qquad \text{Divide by } -5$$

Dividing by -5 required reversing the inequality from $\geq$ to $\leq$, so the solution is $y \leq -3/5$, not $y \geq -3/5$. ■

1.1 EXERCISES

Solve the following linear equations.

1. $4x - 1 = 15$

2. $-3y + 2 = 5$

3. $3m + 2 = -m + 7$

4. $-2k + 8 = 5k - 10$

5. $.2m - .5 = .1m + .7$

6. $.01p + 3.1 = 2.03p - 2.96$

7. $\frac{5}{6}k - 2k + \frac{1}{3} = \frac{2}{3}$

8. $\frac{3}{4} + \frac{1}{5}r - \frac{1}{2} = \frac{4}{5}r$

9. $2x - (x + 3) = 7 - x$

10. $4y + 3(1 - y) = 2 + y$

11. $3r + 2 - 5(r + 1) = 6r + 4$

12. $5(a + 3) + 4a - 5 = -(2a - 4)$

13. $5(3x - 2) = 7(x + 2)$

14. $3(2p + 5) = 5(p + 2)$

15. $\frac{x}{3} - 7 = x - \frac{3x}{4}$

16. $\frac{y}{3} + 1 = \frac{2y}{5} - 4$

Solve the following linear inequalities.

17. $2x - 5 \leq 15$

18. $-y + 10 \geq 18$

19. $6 - 4m \geq 12$

20. $8 + 3p \leq 20$

21. $5k + 2 < 2k - 3$

22. $6 - 4x > 14x + 10$

23. $9 - 3z > 8z + 12$

24. $2r - 12 < 5r + 10$

25. $\frac{4}{5}x + 3 \leq x - \frac{1}{5}$

26. $\frac{a}{3} - 4 \geq \frac{a}{4} + 7$

27. $3(t - 2) + 5 \geq t - 4$

28. $6 - 2(y + 1) \leq 6y + 1$

29. $5 - 3p + 2(p - 4) \leq 4p$

30. $-3 + 2(p - 1) + p \geq 4 - p$

31. $2(k - 5) + 3 < -(k + 1)$

32. $-x - (2x + 3) > 3x - 1$

Solve each equation for x.

33. $2(x - a) + b = 3x + a$

34. $5x - (2a + c) = a(x + 1)$

35. $ax + b = 3(x - a)$

36. $4a - ax = 3b + bx$

37. $x = a^2x - ax + 3a - 3$

38. $2a = ax - a - 6x + 6$

39. $a^2x + 3x = 2a^2$

40. $ax + b^2 = bx - a^2$

▤ APPLICATIONS

BUSINESS AND ECONOMICS

Interest on Investments   **41.** Weijen Luan invests $20,000 from an insurance settlement in two ways: some at 6% and some at 8%. Altogether, she makes $1360 per year interest. How much is invested at 8%?

Interest on Investments   **42.** Joe Gonzalves received $52,000 profit from the sale of some land. He invested part of the money at 10% interest and the rest at 8% interest. He earned a total of $4580 interest per year. How much did he invest at 10%?

Taxes and Interest   **43.** Matt Whitney won $100,000 in a state lottery. He paid income tax of 40% on the winnings. Of the rest, he invested some at 8½% and some at 6%, making $4950 interest per year. How much did he invest at 6%?

Taxes and Interest   **44.** Mary Collins earned $48,000 from royalties on her cookbook. She paid a 40% income tax on these royalties. Part of the balance was invested at 7½% and part at 10½%. The investments produced a total of $2550 interest income per year. Find the amount invested at 10½%.

Profit and Loss   **45.** Jane Martinelli bought two plots of land for a total of $120,000. When she sold the first plot, she made a profit of 15%. When she sold the second, she lost 10%. Her net profit was $5500. How much did she pay for the first piece of land?

Interest on Investments   **46.** Suppose $20,000 is invested at 9%. How much additional money must be invested at 6% to produce a yield of 7.2% on the entire amount invested?

Car Rental   **47.** Bill and Cheryl Bradkin went to Portland, Maine, for a week. They needed to rent a car, so they checked out two rental firms. Firm A wanted $28 per day, with no mileage fee. Firm B wanted $108 per week and 14¢ per mile. Let x represent the number of miles that the Bradkins would drive in one week.

(a) Write an expression for the cost to rent a car for one week from Firm A.

(b) Write an expression for the cost to rent a car for one week from Firm B.

(c) Write an inequality in which the cost to rent from Firm A is less than the cost to rent from Firm B; then solve the inequality to decide how many miles the Bradkins would have to drive before the car from Firm A was the better deal.

Revenue, Cost, and Profit   **48.** A company that produces videocassettes has found that revenue from the sales of the cassettes is $5 per cassette less sales costs of $100. Production costs are $125 plus $4 per cassette. Let x represent the number of cassettes produced and sold.

(a) Write an expression for the revenue from the sale of x cassettes.

(b) Write an expression for the cost to produce x cassettes.

(c) Write an inequality in which the revenue expression is greater than the cost expression and solve it to find the minimum production level at which the company can make a profit.

LIFE SCIENCES

Drug Dosage   **49.** A nurse must make sure that Ms. Kahali receives at least 30 units of a certain drug each day. This drug comes in red pills or green pills, each of which provides three units of the drug. The patient must have twice as many red pills as green pills. Find the smallest number of green pills that will satisfy the requirement.

PHYSICAL SCIENCES

Octane Rating

Exercises 50–51 involve the octane rating of gasoline, a measure of its antiknock qualities. The octane ratings of actual gasoline blends are compared with those of standard fuels. In one way to measure octane, a standard fuel is used that is made of only two ingredients, heptane and isooctane. For this fuel, the octane rating is its percentage of isooctane. For example, a gasoline with an octane rating of 98 has the same antiknock properties as a standard fuel that is 98% isooctane.

50. A service station has 92 octane and 98 octane gasoline. How many liters of 92 octane should be mixed with how many liters of 98 octane to provide 12 liters of 96 octane gasoline for a chemistry experiment?

51. How many liters of 94 octane gasoline should be mixed with 200 liters of 99 octane gasoline to get a mixture that is 97 octane?

GENERAL

Tuition

52. A business college charges annual tuition of $6440. Hong Le makes no more than $1610 per year in her summer job. What is the minimum number of summers that she must work in order to make enough money to pay for one year's tuition?

1.2   LINEAR FUNCTIONS

There are many everyday situations in which it is useful to be able to describe relationships between quantities. For example, we might want to express the relationship between the number of hours a student studies daily and the grade the student receives in a course. One way to do this is to set up a table showing the hours of study and the corresponding grades, as follows.

Hours of Study	Grade
3	A
2 1/2	B
2	C
1	D
0	F

In other relationships, a formula of some sort is used to describe how the value of one quantity depends on the value of another. For example, if a certain bank account pays 12% interest per year, then the interest I that a deposit of P dollars would earn in one year is given by

$$I = .12 \times P, \quad \text{or} \quad I = .12P.$$

The formula $I = .12P$ describes the relationship between interest and the amount of money deposited.

In this example, P, which represents the amount of money deposited, is called the **independent variable,** and I is called the **dependent variable.** (The amount of interest earned *depends* on the amount of money deposited.) When a specific number, say 2000, is substituted for P, then I takes on *one* specific value—here, $.12 \times 2000 = 240$. The variable I is said to be a *function* of P.

FUNCTION

> A **function** is a rule that assigns to each element from one set exactly one element from another set.

In almost every use of functions in this book, the "rule" mentioned above is given by an equation, such as $I = .12P$. When an equation is given for a function, we say that the equation *defines* the function. Whenever x and y are used in this book to define a function, x represents the independent variable and y the dependent variable.

The independent variable in a function can take on any value within a specified set of values called the *domain*.

DOMAIN AND RANGE

> The set of all possible values of the independent variable in a function is called the **domain** of the function, and the resulting set of possible values of the dependent variable is called the **range.**

The domain and range of a function may or may not be the same set.

EXAMPLE 1

Do the following equations define functions? Give the domain and range of each function.

(a) $y = -4x + 11$

For a given value of x, calculating $-4x + 11$ produces exactly one value of y. (For example, if $x = -7$, then $y = -4(-7) + 11 = 39$.) Since one value of the independent variable leads to exactly one value of the dependent variable, $y = -4x + 11$ defines a function. Both x and y may take on any real-number values at all, so the domain and range here are both the set of all real numbers.

(b) $y^2 = x$

Suppose $x = 36$. Then $y^2 = x$ becomes $y^2 = 36$, from which $y = 6$ or $y = -6$. Since one value of the independent variable can lead to two values of the dependent variable, $y^2 = x$ does not represent a function.

(c) $y = 2x + 7, \quad x \geq 0$

A given value of x produces exactly one value of y, making $y = 2x + 7$ a function. The independent variable x is restricted to values greater than or equal to 0, so the domain is the set $\{x | x \geq 0\}$. If $x \geq 0$, then

$$y = 2x + 7 \geq 2(0) + 7,$$

so that the range is $y \geq 7$. ∎

f(x) Notation   Letters such as f, g, or h often are used to name functions. For example, f might be used to name the function

$$y = 5 - 3x.$$

To show that this function is named f, and to also show that x is the independent variable, it is common to replace y with $f(x)$ (read "f of x") to get

$$f(x) = 5 - 3x.$$

By choosing 2 as a value of x, $f(x)$ becomes $5 - 3 \cdot 2 = 5 - 6 = -1$, written

$$f(2) = -1.$$

In a similar manner,

$$f(-4) = 5 - 3(-4) = 17, \qquad f(0) = 5, \qquad f(-6) = 23,$$

and so on.

■ EXAMPLE 2

Let $g(x) = x^2 - 4x + 5$. Find $g(3)$, $g(0)$, $g(a)$, and $g(b + 2)$.

To find $g(3)$, substitute 3 for x.

$$g(3) = 3^2 - 4(3) + 5 = 9 - 12 + 5 = 2$$

Find $g(0)$, $g(a)$, and $g(b + 2)$ in the same way:

$$g(0) = 0^2 - 4(0) + 5 = 5;$$

$$g(a) = a^2 - 4a + 5;$$

and

$$\begin{aligned} g(b + 2) &= (b + 2)^2 - 4(b + 2) + 5 \\ &= b^2 + 4b + 4 - 4b - 8 + 5 \\ &= b^2 + 1. \quad ■ \end{aligned}$$

■ EXAMPLE 3

Suppose the sales of a small company have been estimated to be

$$S(x) = 125 + 80x,$$

where $S(x)$ represents the total sales in thousands of dollars in year x, with $x = 0$ representing 1988. Estimate the sales in each of the following years.

(a) 1988

Since $x = 0$ corresponds to 1988, the sales for 1988 can be found from $S(0)$, that is, by substituting 0 for x.

$$\begin{aligned} S(0) &= 125 + 80(\mathbf{0}) \qquad \text{Let } x = 0 \\ &= 125 \end{aligned}$$

Since $S(x)$ represents sales in thousands of dollars, sales in 1988 would amount to 125×1000, or \$125,000.

(b) 1992

To estimate sales in 1992, let $x = 4$.

$$S(4) = 125 + 80(4) = 125 + 320 = 445,$$

so sales should be about \$445,000 in 1992. ■

Graphs Given a function $y = f(x)$, a given value in the domain of f produces a value for y. This pair of numbers, one for x and one for y, can be written as an **ordered pair** (x, y). For example, let $y = f(x) = 8 + x^2$. If $x = 1$, then $f(1) = 8 + 1^2 = 9$, producing the ordered pair $(1, 9)$. (Always write the value of the independent variable first.) If $x = -3$, then $f(-3) = 8 + (-3)^2 = 17$, giving $(-3, 17)$. Some other ordered pairs for this function are $(0, 8)$, $(-1, 9)$, and $(2, 12)$.

Ordered pairs are **graphed** with the perpendicular number lines of a **Cartesian coordinate system,** shown in Figure 1. The horizontal number line, or ***x*-axis,** represents the first components of the ordered pairs, while the vertical or ***y*-axis** represents the second components. The point where the number lines cross is the zero point on both lines; this point is called the **origin.**

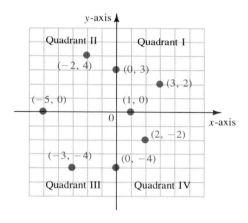

FIGURE 1

Each point on the xy-plane corresponds to an ordered pair of numbers, where the x-value is written first. From now on, we will refer to the point corresponding to the ordered pair (a, b) as "the point (a, b)."

Locate the point $(-2, 4)$ on the coordinate system by starting at the origin and counting 2 units to the left on the horizontal axis and 4 units upward, parallel to the vertical axis. This point is shown in Figure 1, along with several other sample points. The number -2 is the ***x*-coordinate** and the number 4 is the ***y*-coordinate** of the point $(-2, 4)$.

The x-axis and y-axis divide the graph into four parts or **quadrants.** For example, quadrant I includes all those points whose x- and y-coordinates are both positive. The quadrants are numbered as shown in Figure 1. The points on the axes themselves belong to no quadrant. The set of points corresponding to the ordered pairs of a function is the **graph** of the function.

In the graph of a function, each value of x in the domain of the function leads to exactly one value of y. Figure 2 shows a graph. For the value $x = x_1$,

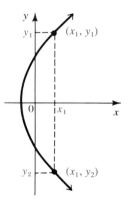

FIGURE 2

the graph gives the two values y_1 and y_2. Since the given value of x corresponds to two different values of y, this graph is not the graph of a function. The **vertical line test** for a function is based on this idea.

VERTICAL LINE TEST

> If a vertical line cuts a graph in more than one point, then the graph is not the graph of a function.

In this chapter, we will be particularly interested in *linear* functions, defined below.

LINEAR FUNCTION

> A function of the form
> $$f(x) = ax + b,$$
> for real numbers a and b, is a **linear function.**

Since $y = f(x)$, examples of linear functions include $y = 2x + 3$, $y = -5$, and $2x - 3y = 7$, which can be written as $y = (2/3)x - (7/3)$. A function such as $g(x) = x^2 - 4x + 5$ from Example 2 is *not* a linear function.

We can graph a linear function, such as $y = x + 1$, by finding several ordered pairs. For example, if $x = 2$, then $y = 2 + 1 = 3$, giving the ordered pair $(2, 3)$. Also, $(0, 1)$, $(4, 5)$, $(-2, -1)$, $(-5, -4)$, $(-3, -2)$, among many others, are ordered pairs that satisfy the equation.

To graph $y = x + 1$, begin by locating the ordered pairs obtained above, as shown in Figure 3(a). All the points of this graph appear to lie on a straight line, as in Figure 3(b). This straight line is the graph of $y = x + 1$. Since any vertical line will cut the graph in Figure 3(b) at only one point, the vertical line test verifies that $y = x + 1$ is a function.

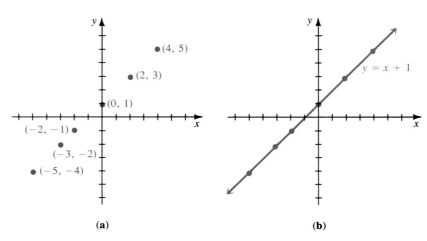

(a)   (b)

FIGURE 3

It can be shown that every linear function has a straight line as its graph. Although just two points are needed to determine a line, it is a good idea to plot a third point as a check when graphing a linear function.

▰ EXAMPLE 4

Use the equation $x + 2y = 6$ to complete the ordered pairs $(-6,\ )$, $(0,\ )$, and $(4,\ )$. Graph these points and then draw a straight line through them.

To complete the ordered pair $(-6,\ )$, let $x = -6$ in the equation $x + 2y = 6$.

$$x + 2y = 6$$
$$-6 + 2y = 6 \qquad \text{Let } x = -6$$
$$2y = 12 \qquad \text{Add 6 to both sides}$$
$$y = 6$$

The result is the ordered pair $(-6, 6)$.

In the same way, if $x = 0$, then $y = 3$, giving $(0, 3)$, and if $x = 4$, then $y = 1$, giving $(4, 1)$. These ordered pairs are graphed in Figure 4(a). A line is drawn through the points in Figure 4(b). ▰

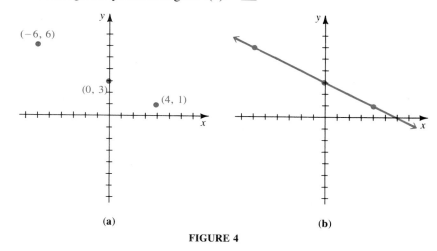

(a)   (b)

FIGURE 4

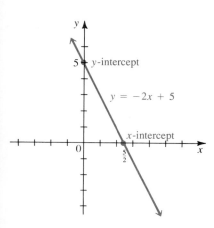

FIGURE 5

Intercepts   As mentioned above, since a straight line is completely determined by any two distinct points that it passes through, only two distinct points are needed for the graph. Two points that are often useful for this purpose are given by the *x*-intercept and the *y*-intercept. The **x-intercept** is the *x*-value (if one exists) where the graph of the equation crosses the *x*-axis. The **y-intercept** is the *y*-value (if one exists) at which the graph crosses the *y*-axis. At a point where the graph crosses the *y*-axis, $x = 0$. Also, $y = 0$ at an *x*-intercept. (See Figure 5.)

■ EXAMPLE 5

Use the intercepts to draw the graph of $y = -2x + 5$.

To find the y-intercept, the y-value at the point where the line crosses the y-axis, let $x = 0$.

$$y = -2x + 5$$
$$y = -2(0) + 5 \qquad \text{Let } x = 0$$
$$y = 5$$

The y-intercept is 5, leading to the ordered pair $(0, 5)$. In the same way, the x-intercept may be found by letting $y = 0$.

$$0 = -2x + 5 \qquad \text{Let } y = 0$$
$$-5 = -2x$$
$$\frac{5}{2} = x$$

The x-intercept is ⁵⁄₂, or 2½, with the graph going through (2½, 0).

Graph the points $(0, 5)$ and $(2½, 0)$ and connect them with a straight line to get the graph in Figure 5. To check these results, we can find a third point by choosing another value of x (or y) and finding the corresponding value of the other variable. Check that $(1, 3)$, $(2, 1)$, $(3, -1)$, and $(4, -3)$, among many other points, satisfy the equation $y = -2x + 5$ and lie on the line in Figure 5. ■

Not every line has two intercepts. In the discussion of intercepts given above, we added the phrase "if one exists" when talking about the place where a graph crosses an axis. The graph in the next example does not cross the x-axis, and thus has no x-intercept.

■ EXAMPLE 6

Graph $y = -3$.

The equation $y = -3$, or equivalently, $y = 0x - 3$, always gives the same y-value, -3, for any value of x. Therefore, no value of x will make $y = 0$, so the graph has no x-intercept. Since $y = -3$ is a linear function with a straight-line graph, and since the graph cannot cross the x-axis, the line must be parallel to the x-axis. For any value of x, the value of y is -3, making the graph the horizontal line parallel to the x-axis and with y-intercept -3, as shown in Figure 6. As the vertical line test shows, the graph is the graph of a function. In general, the graph of $y = k$, where k is a real number, is the horizontal line having y-intercept k. ■

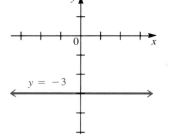

FIGURE 6

■ EXAMPLE 7

Graph $x = -1$.

To obtain the graph of $x = -1$, complete some ordered pairs using the equivalent form $x = 0y - 1$. For example, $(-1, 0)$, $(-1, 1)$, $(-1, 2)$, and $(-1, 4)$ are ordered pairs that satisfy $x = -1$. (The first coordinate of these

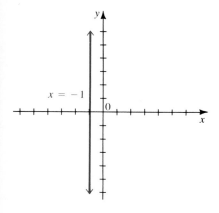

FIGURE 7

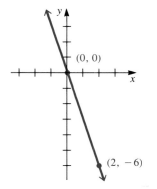

FIGURE 8

ordered pairs is always -1, which is what $x = -1$ means.) Here, more than one second coordinate corresponds to the same first coordinate, -1. As the graph in Figure 7 shows, a vertical line can cut this graph in more than one point. (In fact, a vertical line can cut the graph in an infinite number of points.) For this reason, $x = -1$ does not define a function. ■

The graphs in examples 6 and 7 were lines with only one intercept. Another type of linear function with only one intercept is graphed in Example 8.

■ EXAMPLE 8

Graph $y = -3x$.

Begin by looking for the x-intercept. If $y = 0$, then

$$y = -3x$$
$$0 = -3x \qquad \text{Let } y = 0$$
$$0 = x. \qquad \text{Divide both sides by } -3$$

We have the ordered pair $(0, 0)$. Starting with $x = 0$ gives exactly the same ordered pair, $(0, 0)$. Two points are needed to determine a straight line, and the intercepts have led to only one point. To get a second point, choose some other value of x (or y). For example, if $x = 2$, then

$$y = -3x = -3(2) = -6, \qquad \text{Let } x = 2$$

giving the ordered pair $(2, -6)$. These two ordered pairs, $(0, 0)$ and $(2, -6)$, were used to get the graph shown in Figure 8. ■

Linear functions can be very useful in setting up mathematical models for real-life situations. In almost every case, linear (or any other reasonably simple) functions provide only approximations to real-world situations. Nevertheless, these are often remarkably useful approximations.

Supply and Demand In particular, linear functions are often good choices for **supply and demand curves.** Typically, as the price of an item increases, the demand for the item decreases, while the supply increases. On the other hand, when demand for an item increases, so does the price causing the supply of the item to decrease.

■ EXAMPLE 9

Suppose that Greg Odjakjian, an economist, has studied the supply and demand for aluminum siding and has come up with the conclusion that price, p, and demand, x, in appropriate units and for an appropriate domain, are related by the linear function

$$p = 60 - \frac{3}{4}x.$$

(a) Find the demand at a price of $40.

$$p = 60 - \frac{3}{4}x$$

$$40 = 60 - \frac{3}{4}x \qquad \text{Let } p = 40$$

$$-20 = -\frac{3}{4}x$$

$$\frac{80}{3} = x$$

At a price of $40, 80/3 units will be demanded; this gives the ordered pair (80/3, 40). (It is customary to write the ordered pairs so that price comes second—that is, price is the dependent variable.)

(b) Find the price if the demand is 32 units.

$$p = 60 - \frac{3}{4}x$$

$$p = 60 - \frac{3}{4}(32) \qquad \text{Let } x = 32$$

$$p = 60 - 24$$
$$p = 36$$

When the demand is 32 units, the price is $36. This gives the ordered pair (32, 36).

(c) Graph $p = 60 - \frac{3}{4}x$.

Use the ordered pairs (80/3, 40) and (32, 36) to get the demand graph shown in Figure 9. Only the portion of the graph in quadrant I is shown, since this function is meaningful only for positive values of p and x. ▄▄

▄▄ EXAMPLE 10

Suppose that the economist in Example 9 concludes that the price and supply of siding are related by

$$p = \frac{3}{4}x,$$

where x now represents supply.

(a) Find the supply if the price is $60.

$$60 = \frac{3}{4}x \qquad \text{Let } p = 60$$

$$80 = x$$

If the price is $60, then 80 units will be supplied to the marketplace. This gives the ordered pair (80, 60).

(b) Find the price if the supply is 16 units.

$$p = \frac{3}{4}(16) = 12 \qquad \text{Let } x = 16$$

If the supply is 16 units, then the price is $12. This gives the ordered pair (16, 12).

(c) Graph $p = \frac{3}{4}x$.

Use the ordered pairs (80, 60) and (16, 12) to get the supply graph shown in Figure 9. ▬

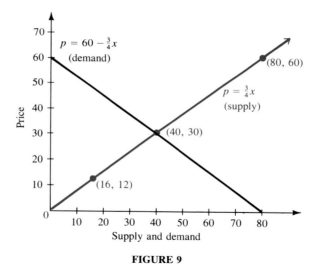

FIGURE 9

As shown in the graphs of Figure 9, both the supply and the demand graphs pass through the point (40, 30). If the price of the siding is more than $30, the supply will exceed the demand. At a price less than $30, the demand will exceed the supply. Only at a price of $30 will demand and supply be equal. For this reason, $30 is called the *equilibrium price*. When the price is $30, demand and supply both equal 40 units, the *equilibrium supply* or *equilibrium demand*. In general, the **equilibrium price** of a commodity is the price found at the point where the supply and demand graphs for that commodity cross. The **equilibrium demand** is the demand at that same point, and the **equilibrium supply** is the supply at that point.

■ EXAMPLE 11

Use algebra to find the equilibrium supply for the aluminum siding. (See Examples 9 and 10.)

The equilibrium supply is found when the prices from both supply and demand are equal. From Example 9, $p = 60 - (3/4)x$; in Example 10, $p = (3/4)x$. Set these two expressions for p equal to each other to get the following linear equation:

$$60 - \frac{3}{4}x = \frac{3}{4}x$$

$$240 - 3x = 3x \qquad \text{Multiply both sides by 4}$$

$$240 = 6x \qquad \text{Add } 3x \text{ to both sides}$$

$$40 = x.$$

The equilibrium supply is 40 units, the same answer found above. ■

In Example 11, the supply and demand functions were linear (since this chapter deals with linear functions). Supply and demand functions need not always be linear, however.

≡ 1.2 EXERCISES

Identify any of the following that are the graphs of functions.

1.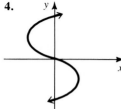

2.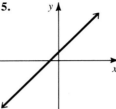

3.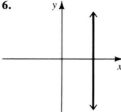

4.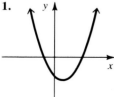

5.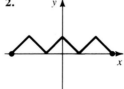

6.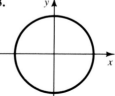

For each of the functions defined in Exercises 7–14, find **(a)** $f(4)$; **(b)** $f(-3)$; **(c)** $f(0)$; *and* **(d)** $f(a)$.

7. $f(x) = -2x - 4$

8. $f(x) = -3x + 7$

9. $f(x) = 6$

10. $f(x) = 0$

11. $f(x) = 2x^2 + 4x$

12. $f(x) = x^2 - 2x$

13. $f(x) = (x + 1)(x + 2)$

14. $f(x) = (x + 3)(x - 4)$

Graph each of the following equations.

15. $y = 2x + 1$ **16.** $y = 3x - 1$ **17.** $y = 4x$ **18.** $y = x + 5$

19. $3y + 4x = 12$ **20.** $4y + 5x = 10$ **21.** $y = -2$ **22.** $x = 4$

23. $x + 5 = 0$ **24.** $y - 4 = 0$ **25.** $8x + 3y = 10$ **26.** $9y - 4x = 12$

27. $y = 2x$ **28.** $y = -5x$ **29.** $x + 4y = 0$ **30.** $x - 3y = 0$

▆ APPLICATIONS

BUSINESS AND ECONOMICS

Sales   **31.** Suppose the sales of a small company that sells by mail are approximated by

$$S(t) = 1000 + 50(t + 1)$$

in thousands of dollars. Here t is time in years, with $t = 0$ representing the year 1989. Find the estimated sales in each of the following years.

   (a) 1989   **(b)** 1990   **(c)** 1992   **(d)** 1994

Revenue   **32.** Revenue for the company in Exercise 31 is approximated by

$$R(t) = 600t - 300$$

in thousands of dollars. Again, t is time in years, with $t = 0$ representing 1989. Find the revenue in each of the following years.

   (a) 1989   **(b)** 1990   **(c)** 1992   **(d)** 1994

Rental Fees   **33.** A chain-saw rental firm charges $7 per day or fraction of a day to rent a saw, plus a fixed fee of $4 for resharpening the blade. Let $S(x)$ represent the cost of renting a saw for x days.
Find each of the following.

   (a) $S\left(\dfrac{1}{2}\right)$ **(b)** $S(1)$ **(c)** $S\left(1\dfrac{1}{4}\right)$ **(d)** $S\left(3\dfrac{1}{2}\right)$

   (e) $S(4)$ **(f)** $S\left(4\dfrac{1}{10}\right)$ **(g)** $S\left(4\dfrac{9}{10}\right)$

   (h) A portion of the graph of $y = S(x)$ is shown here. Explain how the graph could be continued.

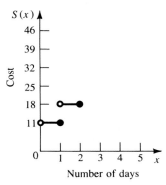

Car Rental   **34.** To rent a midsized car costs $40 per day or fraction of a day. If you pick up the car in Boston and drop it off in Utica, there is a fixed $40 charge. Let $C(x)$ represent the cost of renting the car for x days, taking it from Boston to Utica. Find each of the following.

(a) $C\left(\dfrac{3}{4}\right)$ (b) $C\left(\dfrac{9}{10}\right)$ (c) $C(1)$ (d) $C\left(1\dfrac{5}{8}\right)$ (e) $C\left(2\dfrac{1}{9}\right)$

(f) Graph the function $y = C(x)$.

Supply and Demand   **35.** Suppose that the demand and price for a certain model of electric can opener are related by

$$p = 16 - \frac{5}{4}x,$$

where p is price and x is demand, in appropriate units.
Find the price at each of the following levels of demand.

(a) 0 units   (b) 4 units   (c) 8 units

Find the demand for the electric can opener at each of the following prices.

(d) $6   (e) $11   (f) $16

(g) Graph $p = 16 - \dfrac{5}{4}x$.

Suppose the price and supply of the item above are related by

$$p = \frac{3}{4}x,$$

where x represents the supply and p the price.
Find the supply at each of the following prices.

(h) $0   (i) $10   (j) $20

(k) Graph $p = \dfrac{3}{4}x$ on the same axes used for part (g).

(l) Find the equilibrium supply.

(m) Find the equilibrium price.

Supply and Demand   **36.** Let the supply and demand functions for strawberry-flavored licorice be

$$\text{Supply: } p = \frac{3}{2}x \qquad \text{and} \qquad \text{Demand: } p = 81 - \frac{3}{4}x,$$

where x is the price in dollars.

(a) Graph these on the same axes.

(b) Find the equilibrium demand.

(c) Find the equilibrium price.

Supply and Demand   **37.** Let the supply and demand functions for butter pecan ice cream be given by

$$\text{Supply: } p = \frac{2}{5}x \qquad \text{and} \qquad \text{Demand: } p = 100 - \frac{2}{5}x,$$

where x is the price in dollars.

(a) Graph these on the same axes.

(b) Find the equilibrium demand.

(c) Find the equilibrium price.

Supply and Demand 38. Let the supply and demand functions for sugar be given by

$$\text{Supply: } p = 1.4x - .6 \quad \text{and} \quad \text{Demand: } p = -2x + 3.2.$$

(a) Graph these on the same axes.

(b) Find the equilibrium demand.

(c) Find the equilibrium price.

Franchise Fees 39. In a recent issue of *Business Week,* the president of Insta-Tune, a chain of franchised automobile tune-up shops, says that people who buy a franchise and open a shop pay a weekly fee of

$$y = .07x + \$135$$

to company headquarters. Here y is the fee and x is the total amount of money taken in during the week by the tune-up center.

Find the weekly fee for each of the following weekly revenues.

(a) \$0   (b) \$1000   (c) \$2000   (d) \$3000

(e) Graph the function.

GENERAL

Family Food Expenditures 40. In a recent issue of *The Wall Street Journal,* we are told that the relationship between the amount of money that an average family spends on food, x, and the amount of money it spends on eating out, y, is approximated by the model

$$y = .36x.$$

Find y for each of the following.

(a) $x = \$40$ (b) $x = \$80$ (c) $x = \$120$

(d) Graph the function.

1.3 SLOPE AND EQUATIONS OF A LINE

As mentioned in the previous section, we can graph a straight line if we know the coordinates of two different points on the line. The graph of a straight line also can be drawn based on knowledge of only *one* point on the line *if* we also know the "steepness" of the line. The number that represents the "steepness" of a line is called the *slope* of the line.

To see how slope is defined, look at the line in Figure 10 (on the next page). The line goes through the points $(x_1, y_1) = (-3, 5)$ and $(x_2, y_2) = (2, -4)$. The difference in the two x-values,

$$x_2 - x_1 = 2 - (-3) = 5$$

in this example, is called the **change in x.** The symbol Δx (read "delta x") is used to represent the change in x. In the same way, Δy represents the **change in y.** In our example,

$$\Delta y = y_2 - y_1 = -4 - 5 = -9.$$

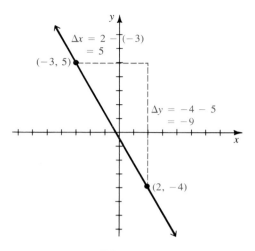

FIGURE 10

These symbols, Δx and Δy, are used in the following definition of slope.

SLOPE OF A LINE

> The **slope** of a line through the two different points (x_1, y_1) and (x_2, y_2) is defined as the change in y divided by the change in x, or
>
> $$\text{Slope} = \frac{\textbf{Change in } y}{\textbf{Change in } x} = \frac{\Delta y}{\Delta x} = \frac{y_2 - y_1}{x_2 - x_1},$$
>
> where $x_1 \neq x_2$.

By this definition, the slope of the line in Figure 10 is

$$\text{Slope} = \frac{\Delta y}{\Delta x} = \frac{-4 - 5}{2 - (-3)} = \frac{-9}{5}.$$

Using similar triangles, it can be shown that the slope of a line is independent of the choice of points on the line. That is, the same slope will be obtained for *any* choice of two different points on the line. (See Exercise 59 in this section.)

■ EXAMPLE 1

Find the slope of the line through the points $(-7, 6)$ and $(4, 5)$.

Let $(x_1, y_1) = (-7, 6)$; then $(x_2, y_2) = (4, 5)$. Use the definition of slope.

$$\text{Slope} = \frac{\Delta y}{\Delta x} = \frac{5 - 6}{4 - (-7)} = \frac{-1}{11} \quad ■$$

In finding the slope of the line in Example 1 we could have let $(x_1, y_1) = (4, 5)$ and $(x_2, y_2) = (-7, 6)$. In that case,

$$\text{Slope} = \frac{6 - 5}{-7 - 4} = \frac{1}{-11} = \frac{-1}{11},$$

the same answer. The order in which coordinates are subtracted does not matter, as long as it is done consistently.

The slope of a line is a measure of the steepness of the line. Figure 11 shows examples of lines with different slopes. Lines with positive slopes go up from left to right, while lines with negative slopes go down from left to right.

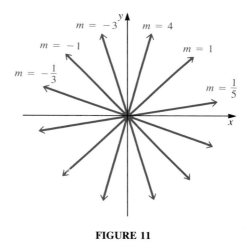

FIGURE 11

The quotient defining the slope of a line has a denominator of $x_2 - x_1$. If $x_1 = x_2$, this denominator is 0 and the quotient, and hence the slope, is undefined.

■ EXAMPLE 2

Find the slope of the line through $(2, -4)$ and $(2, 3)$.

Let $(x_1, y_1) = (2, -4)$ and $(x_2, y_2) = (2, 3)$. Then

$$\text{Slope} = \frac{3 - (-4)}{2 - 2} = \frac{7}{0},$$

which is undefined. Graphing the given points and drawing a line through them gives a vertical line. ■

Example 2 suggests the following statement.

> Slope is undefined for a vertical line.

▰ EXAMPLE 3

Find the slope of the line $3x - 4y = 12$.

Find the slope with two different points on the line. Here, use the points given by the intercepts. First let $x = 0$, and then let $y = 0$.

If $x = 0$,	If $y = 0$,
$3x - 4y = 12$	$3x - 4y = 12$
becomes $3(0) - 4y = 12$	becomes $3x - 4(0) = 12$
$-4y = 12$	$3x = 12$
so that $y = -3$.	so that $x = 4$.

This gives the ordered pairs $(0, -3)$ and $(4, 0)$. Now the slope can be found from the definition.

$$\text{Slope} = \frac{0 - (-3)}{4 - 0} = \frac{3}{4}$$ ▰

In the last section, we saw that the graph of a linear function is a line. The equation that defines a linear function may take many different forms. In the rest of this section we look at various forms of the equation of a line.

Slope-Intercept Form   A generalization of the method of Example 3 can be used to find the equation of a line given its y-intercept and slope. Assume that a line has y-intercept b, so that it goes through $(0, b)$. Let the slope of the line be represented by m. If (x, y) is any point on the line *other* than $(0, b)$, then the definition of slope can be used with the points $(0, b)$ and (x, y) to get

$$m = \frac{y - b}{x - 0}$$

$$m = \frac{y - b}{x}$$

$$mx = y - b$$

$$y = mx + b,$$

an equation of the line in linear function form. This result, also called the slope-intercept form of the equation of a line, is summarized as follows.

SLOPE-INTERCEPT FORM

> If a line has slope m and y-intercept b, then an equation of the line is
> $$y = mx + b,$$
> the **slope-intercept form** of the equation of a line.

▬ EXAMPLE 4

Find the slope-intercept form of the equation of the line having y-intercept 7/2 and slope $-5/2$.

Use the slope-intercept form with $b = 7/2$ and $m = -5/2$.

$$y = mx + b$$

$$y = -\frac{5}{2}x + \frac{7}{2}$$ ▬

We can find the slope of a line by solving its equation for y. Then the coefficient of x is the slope and the constant term is the y-intercept. (See Exercises 60 and 61.) For example, in Example 3 the slope of the line $3x - 4y = 12$ was found to be 3/4. This slope also could be found by solving the equation for y.

$$3x - 4y = 12$$

$$-4y = -3x + 12$$

$$y = \frac{3}{4}x - 3$$

The coefficient of x, 3/4, is the slope of the line.

▬ EXAMPLE 5

Find the slope and y-intercept for each of the following lines.

(a) $5x - 3y = 1$

Solve for y: $5x - 3y = 1$

$$-3y = -5x + 1$$

$$y = \frac{5}{3}x - \frac{1}{3}.$$

The slope is 5/3, and the y-intercept is $-1/3$.

(b) $-9x + 6y = 2$

Solve for y: $-9x + 6y = 2$

$$6y = 9x + 2$$

$$y = \frac{3}{2}x + \frac{1}{3}.$$

The slope is 3/2, and the y-intercept is 1/3. ▬

The slope and y-intercept of a line can be used to draw the graph of the line, as shown in the next example.

■ EXAMPLE 6

Use the slope and y-intercept to graph $3x - 2y = 2$.

Solve for y:

$$3x - 2y = 2$$
$$-2y = -3x + 2$$
$$y = \frac{3}{2}x - 1.$$

The slope is 3/2, and the y-intercept is -1.

To draw the graph, first graph the point $(0, -1)$ given by the y-intercept, as shown in Figure 12. To find a second point on the graph, use the slope. If m represents the slope, then

$$m = \frac{\Delta y}{\Delta x} = \frac{3}{2}$$

in this example. If x changes by 2 units ($\Delta x = 2$), then y will change by 3 units ($\Delta y = 3$). To find the second point, start at $(0, -1)$ in Figure 12 and move 3 units up and 2 units to the right. Once this second point is located, the line can be drawn through the two points. ■

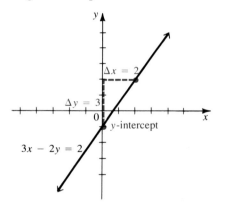

FIGURE 12

Point-Slope Form   The slope-intercept form of the equation of a line involves the slope and the y-intercept. Sometimes, however, the slope of a line is known, together with one point (perhaps *not* the y-intercept) that the line goes through. The *point-slope form* of the equation of a line is used to find the equation in this case. Let (x_1, y_1) be any fixed point on the line and let (x, y) represent any other point on the line. If m is the slope of the line, then by the definition of slope,

$$\frac{y - y_1}{x - x_1} = m,$$

or

$$y - y_1 = m(x - x_1).$$

POINT-SLOPE FORM

If a line has slope m and passes through the point (x_1, y_1), then an equation of the line is given by

$$y - y_1 = m(x - x_1),$$

the point-slope form of the equation of a line.

■ EXAMPLE 7

Find an equation of the line that passes through the given point and has the given slope.

(a) $(-4, 1)$, $m = -3$

We can use the point-slope form, since we know the coordinates of a point on the line as well as the slope of the line. Substitute the values $x_1 = -4$, $y_1 = 1$, and $m = -3$ into the point-slope form.

$$y - y_1 = m(x - x_1)$$
$$y - 1 = -3[x - (-4)]$$
$$y - 1 = -3(x + 4)$$
$$y - 1 = -3x - 12$$

This equation can be simplified by combining the constants to get

$$y = -3x - 11.$$

Of course, this equation could be written in other forms.

(b) $(3, -7)$, $m = 5/4$

Use the point-slope form.

$$y - y_1 = m(x - x_1)$$
$$y - (-7) = \frac{5}{4}(x - 3) \qquad \text{Let } y_1 = -7, \, m = 5/4, \, x_1 = 3$$
$$y + 7 = \frac{5}{4}(x - 3)$$
$$4y + 28 = 5(x - 3) \qquad \text{Multiply both sides by 4}$$
$$4y + 28 = 5x - 15$$
$$4y = 5x - 43 \qquad \text{Combine constants} \quad ■$$

As mentioned in Example 7(a), the equation of the same line can be given in many forms. To avoid confusion, the linear equations used in the rest of this section will be written in *standard form*.

STANDARD FORM

> The **standard form** of the equation of a line is
> $$ax + by = c.$$

The point-slope form also can be used to find an equation of a line if we know two different points that the line goes through. The procedure for doing this is shown in the next example.

■ EXAMPLE 8

Find an equation of the line through $(5, 4)$ and $(-10, -2)$.

Begin by using the definition of slope to find the slope of the line that passes through the given points.

$$\text{Slope} = m = \frac{-2 - 4}{-10 - 5} = \frac{-6}{-15} = \frac{2}{5}$$

Use $m = 2/5$ and either of the two points in the point-slope form. If $(x_1, y_1) = (5, 4)$, then

$$y - y_1 = m(x - x_1)$$

$$y - 4 = \frac{2}{5}(x - 5) \qquad \text{Let } y_1 = 4, m = \frac{2}{5}, x_1 = 5$$

$$5y - 20 = 2(x - 5) \qquad \text{Multiply both sides by 5}$$

$$5y - 20 = 2x - 10 \qquad \text{Distributive property}$$

$$5y = 2x + 10 \qquad \text{Combine constants}$$

$$-2x + 5y = 10. \qquad \text{Standard form}$$

Check that the same result is found if $(x_1, y_1) = (-10, -2)$. ■

■ EXAMPLE 9

Find an equation of the line through $(8, -4)$ and $(-2, -4)$.

Find the slope.

$$m = \frac{-4 - (-4)}{-2 - 8} = \frac{0}{-10} = 0$$

Choose, say, $(8, -4)$ as (x_1, y_1).

$$y - y_1 = m(x - x_1)$$

$$y - (-4) = 0(x - 8) \qquad \text{Let } y_1 = -4, m = 0, x_1 = 8$$

$$y + 4 = 0 \qquad 0(x - 8) = 0$$

$$y = -4 \quad ■$$

As shown in the previous section, $y = -4$ represents a horizontal line, with y-intercept -4. Generalizing from this example, every horizontal line has a slope of 0.

▰▰ EXAMPLE 10

Find an equation of the line through (4, 3) and (4, −6).

The slope of the line is

$$m = \frac{-6 - 3}{4 - 4} = \frac{-9}{0},$$

which is undefined. As mentioned above, the slope of a vertical line is undefined. It was shown in the last section that vertical lines have equations of the form $x = k$, where k can be any real number. Since the x-coordinate of the two ordered pairs given above is 4, the desired equation is

$$x = 4. \quad \blacksquare$$

The different forms of linear equations discussed in this section are summarized below.

EQUATIONS OF LINES

Equation	Description
$ax + by = c$	**Standard form:** if $a \neq 0$ and $b \neq 0$, line has x-intercept c/a and y-intercept c/b
$y = mx + b$	**Slope-intercept form:** slope m, y-intercept b
$y - y_1 = m(x - x_1)$	**Point-slope form:** slope m, line passes through (x_1, y_1)
$x = k$	**Vertical line:** x-intercept k, no y-intercept, undefined slope
$y = k$	**Horizontal line:** y-intercept k, no x-intercept, slope 0

Many real-world situations can be approximately described by a straight-line graph. One way to find the equation of such a straight line is to use two typical data points from the graph and the point-slope form of the equation of a line.

▰▰ EXAMPLE 11

In recent years, the percentage of the labor force in the United States that is unionized has steadily decreased. Assuming that the relationship between union membership and time is linear, find the equation of the line if the percentage of the labor force belonging to a union was 26% in 1965 and 15.5% in 1986. Let x represent time in years, with $x = 0$ for 1965, and let y represent the percent of workers unionized.

If 1965 corresponds to $x = 0$, then 1986 corresponds to $x = 86 - 65 = 21$. The ordered pairs representing the given information are (0, 26) and (21, 15.5). The slope of the line through these points is

$$m = \frac{15.5 - 26}{21 - 0} = \frac{-10.5}{21} = -.5.$$

Using $m = -.5$ and $(x_1, y_1) = (0, 26)$ in the point-slope form gives the required equation,

$$y - 26 = -.5(x - 0)$$
$$y = -.5x + 26. \quad \blacksquare$$

1.3 EXERCISES

Find the slope, if it exists, of the line going through each of the given pairs of points.

1. $(-8, 6), (2, 4)$ **2.** $(-3, 2), (5, 9)$ **3.** $(-1, 4), (2, 6)$ **4.** $(3, -8), (4, 1)$

5. The origin and $(-4, 6)$ **6.** The origin and $(8, -2)$ **7.** $(-2, 9), (-2, 11)$ **8.** $(7, 4), (7, 12)$

9. $(3, -6), (-5, -6)$ **10.** $(5, -11), (-9, -11)$

Find the slope and y-intercept of each line.

11. $y = 3x + 4$ **12.** $y = -3x + 2$ **13.** $y + 4x = 8$ **14.** $y - x = 3$

15. $3x + 4y = 5$ **16.** $2x - 5y = 8$ **17.** $3x + y = 0$ **18.** $y - 4x = 0$

19. $2x + 5y = 0$ **20.** $3x - 4y = 0$ **21.** $y = 8$ **22.** $y = -4$

23. $y + 2 = 0$ **24.** $y - 3 = 0$ **25.** $x = -8$ **26.** $x = 3$

Graph the line going through the given point and having the given slope.

27. $(-4, 2), m = 2/3$ **28.** $(3, -2), m = 3/4$ **29.** $(-5, -3), m = -2$

30. $(-1, 4), m = 2$ **31.** $(8, 2), m = 0$ **32.** $(2, -4), m = 0$

33. $(6, -5)$, undefined slope   **34.** $(-8, 9)$, undefined slope   **35.** $(0, -2), m = 3/4$

36. $(0, -3), m = 2/5$ **37.** $(5, 0), m = 1/4$ **38.** $(-9, 0), m = 5/2$

Find an equation in slope-intercept form for each line having the given y-intercept and slope.

39. $4, m = -3/4$ **40.** $-3, m = 2/3$ **41.** $-2, m = -1/2$

42. $3/2, m = 1/4$ **43.** $5/4, m = 3/2$ **44.** $-3/8, m = 3/4$

Write an equation in standard form for each line, based on the information given.

45. Through $(-4, 1), m = 2$ **46.** Through $(5, 1), m = -1$ **47.** Through $(0, 3), m = -3$

48. Through $(-2, 3), m = 3/2$ **49.** Through $(3, 2), m = 1/4$ **50.** Through $(0, 1), m = -2/3$

51. Through $(-1, 1)$ and $(2, 5)$ **52.** Through $(4, -2)$ and $(6, 8)$ **53.** Through $(9, -6)$ and $(12, -8)$

54. Through $(-5, 2)$ and $(7, 5)$ **55.** Through $(-8, 4)$ and $(-8, 6)$ **56.** Through $(2, -5)$ and $(4, -5)$

57. Through $(-1, 3)$ and $(0, 3)$ **58.** Through $(2, 9)$ and $(2, -9)$

59. Use similar triangles from geometry to show that the slope of a line is the same, no matter which two distinct points on the line are chosen to compute it.

60. Show that b is the y-intercept in the slope-intercept form $y = mx + b$.

61. Suppose that $(0, b)$ and (x_1, y_1) are distinct points on the line $y = mx + b$. Show that $(y_1 - b)/x_1$ is the slope of the line, so that $m = (y_1 - b)/x_1$.

▤ APPLICATIONS

BUSINESS AND ECONOMICS

In Exercises 62–63, assume that the data can be approximated by a straight line. Use the given information to find the slope of each line and to find an equation of the line.

Sales 62. The sales of a small company were $27,000 in its second year of operation and $63,000 in its fifth year. Let y represent sales in the xth year of operation.

Manufacturing Costs 63. A company finds that it can make a total of 20 solar heaters for $13,900, while 10 solar heaters cost $7500. Let y be the total cost to produce x solar heaters.

LIFE SCIENCES

Effects of Pollution 64. When a certain industrial pollutant is dumped into a river, the reproduction of catfish declines. In a given period of time, dumping three tons of the pollutant results in a fish population of 37,000. Also, 12 tons of pollutant produce a fish population of 28,000. Let y be the fish population when x tons of pollutant are dumped into the river.

(a) Assuming there is a linear relationship between the amount of pollution in the river and catfish reproduction, write an equation of the line based on the given data.

(b) Predict the fish population if 15 tons of pollution are dumped into the river.

Estimating Height 65. The radius bone goes from the wrist to the elbow. A female whose radius bone is 24 cm long would be 167 cm tall, while a radius bone of 26 cm corresponds to a height of 174 cm.

(a) Write a linear equation showing how the height h of a female corresponds to the length r of her radius bone.

(b) Estimate the heights of females with radius bones of length 23 cm and 27 cm.

(c) Estimate the length of a radius bone for a height of 170 cm.

Estimating Height 66. A person's tibia bone goes from ankle to knee. A male with a tibia 40 cm in length will have a height of 177 cm, while a tibia 43 cm in length corresponds to a height of 185 cm.

(a) Write a linear equation showing how the height h of a male relates to the length of his tibia.

(b) Estimate the heights of males having tibia bones of length 38 cm and 45 cm.

(c) Estimate the length of the tibia for a height of 190 cm.

SOCIAL SCIENCES

Voting Analysis 67. According to research done by the political scientist James March, if the Democrats win 45% of the two-party vote for the House of Representatives, they win 42.5% of the seats. If the Democrats win 55% of the vote, they win 67.5% of the seats. Let y be the percent of seats won and x the percent of the two-party vote.

(a) Write a linear equation satisfying this data.

(b) Use your equation to predict the number of seats the Democrats win if they get 50% of the vote.

Voting Analysis   **68.** If the Republicans win 45% of the two-party vote, they win 32.5% of the seats (see Exercise 67). If they win 60% of the vote, they get 70% of the seats. Let y represent the percent of the seats and x the percent of the vote.

(a) Write a linear equation satisfying this data.

(b) Use your equation to predict the percent of Republican seats won if the Republicans get 50% of the vote.

Immigration   **69.** In 1974, 86,821 people from other countries immigrated to the state of California. In 1984, the number of immigrants was 140,289.

(a) If the change in foreign immigration to California is considered to be linear, write an equation expressing the number of immigrants, y, in terms of the number of years after 1974, x.

(b) Use your result in part (a) to predict the foreign immigration to California in the year 2001.

Work-Related Illness   **70.** In 1943, the number of officially reported work-related illnesses and injuries in the
and Injury   state of California was 152,000, and in 1982 it was 330,870.

(a) If the change in the number of work-related illnesses and injuries is thought to be linear, write an equation expressing the number of illnesses and injuries, y, in terms of the number of years after 1943, x.

(b) Use your result in part (a) to predict the number of work-related illnesses and injuries in California in the year 2000.

1.4 LINEAR MATHEMATICAL MODELS

In this section, we will discuss mathematical modeling in more detail. We will develop some mathematical models using the linear equations discussed in the last two sections.

If the principles causing a certain event to happen are completely understood, then the mathematical model describing that event can be very accurate. As a rule, the mathematical models constructed in the physical sciences are excellent at predicting events. For example, if a body falls in a vacuum, then d, the distance in feet that the body will fall in t seconds, is given by

$$d = \frac{1}{2}gt^2,$$

where g is a fixed number representing gravity. (As an approximation, $g = 32$ feet per second per second.) Using this equation, we can predict an exact value of d for a known value of t.

The situation is different in the fields of management and social science. Mathematical models in these fields tend to be less accurate approximations, or even gross approximations. So many variables come into play that no mathematical model can ever hope to produce results comparable to those produced in physical science.

In spite of the limitations of mathematical models, they have found a large and increasing acceptance in management and economic decision making. There is one main reason for this: mathematical models produce very useful results, as will be shown in the next few examples.

Sales Analysis   It is common to compare the sales performances of two companies by comparing the rates at which sales change. If the sales of the two companies can be approximated by linear functions, the work of the last section can be used to find rates of change.

▬ EXAMPLE 1

The chart below shows sales (in dollars) in two different years for two different companies.

Company	Sales in 1984	Sales in 1987
A	10,000	16,000
B	5,000	14,000

A study of past records suggests that the sales of both companies have increased linearly (that is, the sales can be closely approximated by a linear function).

(a) Find a linear equation describing the sales of Company A.

To find a linear equation describing the sales, let $x = 0$ represent 1984, so that 1987 corresponds to $x = 3$. Then, by the chart above, the line representing the sales of Company A goes through the points $(0, 10,000)$ and $(3, 16,000)$. The slope of the line through these points is

$$m = \frac{16,000 - 10,000}{3 - 0} = 2000.$$

Using the point-slope form of the equation of a line,

$$y - 10,000 = 2000(x - 0)$$

or
$$y = 2000x + 10,000$$

gives a linear equation describing the sales of Company A.

(b) Find a linear equation describing the sales of Company B.

Since the sales of Company B have also increased linearly, its sales can be described by a line through $(0, 5000)$ and $(3, 14,000)$, leading to

$$y = 3000x + 5000$$

as the linear equation describing the sales of Company B.   ▬

Average Rate of Change   Notice that the sales for Company A in Example 1 increased from $10,000 to $16,000 over the 3-year period, making the average rate of change of sales

$$\frac{16,000 - 10,000}{3} = \frac{6000}{3} = 2000,$$

or $2000 per year. This is the same as the slope of the equation found in part (a) of the example. Verify that the annual rate of change in sales for Company B over the 3-year period also agrees with the slope found in part (b). Management needs to watch the rate of change of sales closely, because it indicates a trend. If the rate of change is decreasing, then the increase in sales is slowing down, and this trend may require some response. As these examples suggest, the average rate of change is the same as the slope of the line. This is always true for data that can be modeled with a linear function.

■■■ EXAMPLE 2

Suppose that a researcher has concluded that a dosage of x grams of a certain stimulant causes a rat to gain

$$y = 2x + 50$$

grams of weight, for appropriate values of x. If the researcher administers 30 grams of the stimulant, how much weight will the rat gain? What is the average rate of change in weight?

Let $x = 30$. The rat will gain

$$y = 2(30) + 50 = 110,$$

or 110 grams of weight.

The average rate of change of weight with respect to the amount of stimulant is given by the slope of the line. The slope of $y = 2x + 50$ is 2, so the change in weight is 2 grams when the dose is varied by 1 gram.   ■■■

Cost Analysis   In manufacturing, the cost of manufacturing an item often consists of two parts. One part is a *fixed cost* for designing the product, establishing a factory, training workers, and so on. Within broad limits, the fixed cost is constant for a particular product and does not change as more items are made. The second part of the cost is a *variable cost* per item for labor, materials, packing, shipping, and so on. The variable cost may well be the same per item, with total variable cost increasing as the number of items increases.

■■■ EXAMPLE 3

Suppose that the cost of producing clock-radios can be approximated by the linear model

$$C(x) = 12x + 100,$$

where $C(x)$ is the cost in dollars to produce x radios. The cost to produce 0 radios is

$$C(0) = 12(0) + 100 = 100,$$

or $100. This amount, $100, is the fixed cost.

Once the company has invested the fixed cost into the clock-radio project, what will be the additional cost per radio? To find out, first find the cost of manufacturing a total of 5 radios:

$$C(5) = 12(5) + 100 = 160,$$

or $160. The cost of making 6 radios is

$$C(6) = 12(6) + 100 = 172,$$

or $172.

Producing the sixth radio thus costs $172 - $160 = $12. In the same way, the 81st radio costs $C(81) - C(80) = $1072 - $1060 = $12 to produce. In fact, the $(n + 1)$st radio costs

$$C(n + 1) - C(n) = [12(n + 1) + 100] - [12n + 100] = 12,$$

or $12, to produce. Since each additional radio costs $12 to produce, $12 is the variable cost per radio. The value 12 is also the slope of the line with equation $C(x) = 12x + 100$. ■

In Example 3, the equation for cost could be written as

$$C(x) = (\text{Variable Cost})x + \text{Fixed Cost},$$

where x is the number of items produced.

In economics, the approximate cost of producing an additional item (the variable cost) is called the **marginal cost** of that item. In the clock-radio example, the marginal cost of each radio is $12.

COST

> If cost is defined by a linear function of the form $C(x) = mx + b$, then m represents the **variable cost** per item and b the **fixed cost.** Conversely, if the fixed cost of producing an item is b and the variable cost is m, then the **cost function** $C(x)$ for producing x items is given by $C(x) = mx + b$.

■ EXAMPLE 4

In a certain city, a taxi company charges riders $1.80 per mile plus a fixed fee of $1.50. Write a cost function $C(x)$ that is a mathematical model for a ride of x miles.

Here the fixed cost is $b = 1.50$ dollars and the variable cost is $m = 1.80$ dollars. The cost function is

$$C(x) = 1.80x + 1.50.$$

For example, a taxi ride of 4 miles will cost $C(4) = 1.80(4) + 1.50 = 8.70$, or $8.70. For each additional mile, the cost increases by $1.80. ▬

▬ **EXAMPLE 5**

The variable cost of raising a certain type of frog for laboratory study is $12 per unit of frogs, while the cost to produce 100 units is $1500. Find $C(x)$ if the cost is given by a linear function.

Since the cost function is linear, it can be expressed in the form $C(x) = mx + b$. The variable cost of $12 per unit gives the value for m in the model. The model can be written $C(x) = 12x + b$. To find b, use the fact that the cost of producing 100 units of frogs is $1500, or $C(100) = 1500$. Substituting $x = 100$ and $C(x) = 1500$ into $C(x) = 12x + b$ gives

$$C(x) = 12x + b$$
$$1500 = 12 \cdot 100 + b$$
$$1500 = 1200 + b$$
$$300 = b.$$

The desired model is given by $C(x) = 12x + 300$. The fixed cost of producing the frogs is $300. ▬

Break-Even Analysis The revenue $R(x)$ from selling x units of an item is the product of the price per unit p and the number of units sold (the demand) x, giving

$$R(x) = px.$$

The corresponding profit $P(x)$ is the difference between revenue $R(x)$ and cost $C(x)$, giving

$$P(x) = R(x) - C(x).$$

A company can make a profit only if the revenue received from its customers exceeds the cost of producing its goods and services. The point at which revenue just equals cost is the **break-even point.**

▬ **EXAMPLE 6**

A firm producing chicken feed finds that the total cost to produce x units is given by

$$C(x) = 20x + 100.$$

Management plans to charge $24 per unit for the feed.

(a) How many units must be sold for the firm to break even?

The firm will break even (no profit and no loss), as long as revenue just equals cost, or $R(x) = C(x)$. From the given information, since $R(x) = px$ and $p = 24,

$$R(x) = 24x.$$

Substituting for $R(x)$ and $C(x)$ in the equation $R(x) = C(x)$ gives

$$R(x) = C(x)$$
$$24x = 20x + 100 \qquad \text{Substitute for } R(x) \text{ and } C(x)$$
$$4x = 100$$
$$x = 25.$$

The firm will break even if it sells 25 units.

The graphs of $C(x) = 20x + 100$ and $R(x) = 24x$ are shown in Figure 13. The break-even point is shown on the graph. If the company produces more than 25 units (if $x > 25$) it makes a profit; if it produces less than 25 units it loses money.

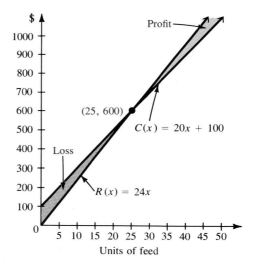

FIGURE 13

(b) What is the profit if 100 units of feed are sold?

Use the formula for profit.

$$P(x) = R(x) - C(x)$$
$$= 24x - (20x + 100)$$
$$= 4x - 100$$

Then $P(100) = 4(100) - 100 = 300$. The firm will make a profit of $300 from the sale of 100 units of feed.

(c) How many units must be sold to produce profit of $900?

Let $P(x) = 900$ in the equation $P(x) = 4x - 100$ and solve for x.

$$900 = 4x - 100$$
$$1000 = 4x$$
$$x = 250$$

Sales of 250 units will produce $900 profit. ▬

Depreciation   Because machines and equipment wear out or become obsolete over time, business firms must take into account the value that their equipment has lost during each year of its useful life. This lost value, called **depreciation,** may be calculated in several ways. The simplest way is to use **straight-line,** or **linear,** depreciation, in which an item having a useful life of *n* years is assumed to lose a constant $1/n$ of its value each year. For example, a typewriter with a 10-year life would be assumed to lose 1/10 of its value in each year and 4/10 of its value in 4 years.

A machine may have some **scrap value** at the end of its useful life. For this reason, depreciation is calculated on **net cost**—the difference between purchase price and scrap value. To find the annual straight-line depreciation on an item having a net cost of *x* dollars and a useful life of *n* years, multiply the net cost by the fraction of the value lost each year, $1/n$. The annual straight-line depreciation $D(x)$ is then

$$D(x) = \frac{1}{n} x.$$

■ EXAMPLE 7

An asset has a purchase price of $100,000 and a scrap value of $40,000. The useful life of the asset is 10 years. Find each of the following for this asset.

(a) Net cost

Since the net cost is the difference between purchase price and scrap value,

Net Cost $= x = \$100{,}000 - \$40{,}000 = \$60{,}000.$

(b) Annual depreciation

The useful life of the asset is 10 years. Therefore, 1/10 of the net cost is depreciated each year. The annual depreciation by the straight-line method is

$$D(x) = \frac{1}{10}x = \frac{1}{10}(\$60{,}000) = \$6000.$$

(c) Undepreciated balance after 4 years

The total amount that will be depreciated over the life of the asset is $60,000. In 4 years, the depreciation will be

$$4(\$6000) = \$24{,}000,$$

and the undepreciated balance will be

$$\$60{,}000 - \$24{,}000 = \$36{,}000. \quad ■$$

Straight-line depreciation is the easiest method of depreciation to use, but it often does not accurately reflect the rate at which assets actually lose value. Some assets, such as new cars, lose more value annually at the beginning of their useful life than at the end. For this reason, two other methods of depreciation, the *sum-of-the-years'-digits* method, discussed in Section 1.5, and the *double declining balance* method (a nonlinear method) often are used.

1.4 EXERCISES

Write a cost function for each of the following. Identify all variables used.

1. A chain saw rental firm charges $12 plus $1 per hour.

2. A trailer-hauling service charges $45 plus $2 per mile.

3. A parking garage charges 50¢ plus 35¢ per half-hour.

4. For a one-day rental, a car rental firm charges $44 plus 28¢ per mile.

Assume that each of the following can be expressed as a linear cost function. Find the cost function in each case.

5. Fixed cost: $100; 50 items cost $1600 to produce.

6. Fixed cost: $400; 10 items cost $650 to produce.

7. Fixed cost: $1000; 40 items cost $2000 to produce.

8. Fixed cost: $8500; 75 items cost $11,875 to produce.

9. Variable cost: $50; 80 items cost $4500 to produce.

10. Variable cost: $120; 100 items cost $15,800 to produce.

11. Variable cost: $90; 150 items cost $16,000 to produce.

12. Variable cost: $120; 700 items cost $96,500 to produce.

For each of the assets in Exercises 13–18 find the straight-line depreciation in year 4, and find the amount undepreciated after 4 years.

13. Cost: $50,000; scrap value: $10,000; life: 20 years

14. Cost: $120,000; scrap value: $0; life: 10 years

15. Cost: $80,000; scrap value: $20,000; life: 30 years

16. Cost: $720,000; scrap value: $240,000; life: 12 years

17. Cost: $1,400,000; scrap value: $200,000; life: 8 years

18. Cost: $2,200,000; scrap value: $400,000; life: 12 years

APPLICATIONS

BUSINESS AND ECONOMICS

Cost **19.** Yoshi Yamamura sells silk-screened T-shirts at community festivals and crafts fairs. Her variable cost to produce one T-shirt is $3.50. Her total cost to produce 60 T-shirts is $300, and she sells them for $9 each.

(a) Find the linear cost function for Yoshi's T-shirt production.

(b) How many T-shirts must she produce and sell in order to break even?

(c) How many T-shirts must she produce and sell to make a profit of $500?

Cost **20.** Enrique Gonzales owns a small publishing house specializing in Latin American poetry. His fixed cost to produce a typical poetry volume is $525, and his total cost to produce 1000 copies of the book is $2675. His books sell for $4.95 each.

(a) Find the linear cost function for Enrique's book production.

(b) How many poetry books must he produce and sell in order to break even?

(c) How many books must he produce and sell to make a profit of $1000?

Sales Analysis   **21.** Suppose the sales, in dollars, of a particular brand of electric guitar satisfy the relationship

$$S(x) = 300x + 2000,$$

where $S(x)$ represents the number of guitars sold in year x, with $x = 0$ corresponding to 1987.

Find the sales in each of the following years.

(a) 1989   (b) 1990   (c) 1991   (d) 1987

(e) Find the annual rate of change of the sales.

Sales Analysis   **22.** Assume that the sales of a certain automobile parts company are approximated by a linear function. Suppose that sales were $200,000 in 1981 and $1,000,000 in 1988. Let $x = 0$ represent 1981 and $x = 7$ represent 1988.

(a) Find the equation giving the company's yearly sales.

(b) Find the approximate sales in 1983.

(c) Estimate the sales in 1990.

Sales Analysis   **23.** Assume that the sales of a certain appliance dealer are approximated by a linear function. Suppose that sales were $850,000 in 1982 and $1,262,500 in 1987. Let $x = 0$ represent 1982.

(a) Find the equation giving the dealer's yearly sales.

(b) What were the dealer's sales in 1985?

(c) Estimate sales in 1990.

Inventory   **24.** Suppose the number of bottles $V(x)$ of a vitamin on hand at the beginning of the day in a health food store is given by

$$V(x) = 600 - 20x,$$

where $x = 1$ corresponds to June 1, and x is measured in days. The store is open every day of the month.

Find the number of bottles on hand at the beginning of each of the following days.

(a) June 6   (b) June 12   (c) June 24

(d) When will the last bottle from this stock be sold?

(e) What is the daily rate of change of this stock?

Marginal Cost   **25.** The manager of a restaurant found that his cost function for producing coffee was $C(x) = .097x$, where $C(x)$ was the total cost in dollars of producing x cups. (He ignored the cost of the coffee maker and the cost of labor.)

Find the total cost of producing the following numbers of cups.

(a) 1000 cups   (b) 1001 cups

(c) Find the marginal cost of the 1001st cup.

(d) What is the marginal cost for *any* cup?

Marginal Cost   **26.** In deciding whether to set up a new manufacturing plant, company analysts have decided that a reasonable function for the total cost $C(x)$ in dollars to produce x items is

$$C(x) = 500,000 + 4.75x.$$

(a) Find the total cost to produce 100,000 items.

(b) Find the marginal cost of the items to be produced in this plant.

Average Cost   *Let $C(x)$ be the total cost to manufacture x items. Then the quotient $(C(x))/x$ is the average cost per item. Use this definition in Exercises 27 and 28.*

27. $C(x) = 800 + 20x$; find the average cost per item at each of the following levels of production.

    (a) $x = 10$ **(b)** $x = 50$ **(c)** $x = 200$

28. $C(x) = 500{,}000 + 4.75x$; find the average cost per item at each of the following levels of production.

    (a) $x = 1000$ **(b)** $x = 5000$ **(c)** $x = 10{,}000$

Straight-Line Depreciation   29. Suppose an asset has a net cost of \$80,000 and a 4-year life.

    (a) Find the straight-line depreciation in each of years 1, 2, 3, and 4 of the item's life.

    (b) Find the sum of all depreciation for the 4-year life.

Straight-Line Depreciation   30. A forklift truck has a net cost of \$12,000, with a useful life of 5 years.

    (a) Find the straight-line depreciation in each of years 1, 2, 3, 4, and 5 of the forklift's life.

    (b) Find the sum of all depreciation for the 5-year life.

Depreciation   *Complete depreciation tables like the following for Exercises 31–34.*

Year	Depreciation	Accumulated Depreciation at End of Year
1		
2		
.		
.		
.		

31. An asset that costs \$28,000, has scrap value of \$16,000, and has a 6-year life

32. An asset that costs \$125,000, has scrap value of \$20,000, and has a 10-year life

33. An asset that costs \$15,000, has no scrap value, and has a 4-year life

34. An asset that costs \$9000, has no scrap value, and has a 5-year life

Break-Even Point   35. Producing x units of tacos costs $C(x) = 5x + 20$, for revenue of $R(x) = 15x$, where $C(x)$ and $R(x)$ are in dollars.

    (a) What is the break-even point?

    (b) What is the profit from 100 units?

    (c) How many units will produce a profit of \$500?

Break-Even Point   36. To produce x units of a religious medal costs $C(x) = 12x + 39$. The revenue is $R(x) = 25x$. Both $C(x)$ and $R(x)$ are in dollars.

    (a) Find the break-even point.

    (b) Find the profit from 250 units.

    (c) Find the number of units that must be produced for a profit of \$130.

Break-Even Point   37. The cost in dollars to produce x units of wire is $C(x) = 50x + 5000$, while the revenue in dollars is $R(x) = 60x$.

    (a) Find the break-even point and the revenue at the break-even point.

(b) Find the profit if 800 units are produced.

(c) How many units should be produced for a profit of $5000?

Break-Even Point   **38.** The cost in dollars to produce x units of squash is $C(x) = 100x + 6000$, while the revenue in dollars is $R(x) = 500x$.

(a) Find the break-even point.

(b) Find the profit from 100 units.

(c) How many units must be produced to earn a profit of $2000?

Break-Even Analysis   *You are the manager of a firm. You are considering the manufacture of a new product, so you ask the accounting department for cost estimates and the sales department for sales estimates. After you receive the data, you must decide whether to go ahead with production of the new product. Analyze the data in Exercises 39–42 (find a break-even point) and then decide what you would do in each case.*

39. $C(x) = 85x + 900$; $R(x) = 105x$; no more than 38 units can be sold.

40. $C(x) = 105x + 6000$; $R(x) = 250x$; no more than 400 units can be sold.

41. $C(x) = 70x + 500$; $R(x) = 60x$ (*Hint:* What does a negative break-even point mean?)

42. $C(x) = 1000x + 5000$; $R(x) = 900x$

Estimating Sales   **43.** The sales of a certain furniture company in thousands of dollars are shown in the chart below.

Year (x)	Sales (y)
0	48
1	59
2	66
3	75
4	80
5	90

(a) Graph this data, plotting years on the x-axis and sales on the y-axis. (Note that the data points can be closely approximated by a straight line.)

(b) Draw a line through the points $(2, 66)$ and $(5, 90)$. The other four points should be close to this line. (These two points were selected as "best" representing the line that could be drawn through the data points.)

(c) Use the two points from part (b) to find an equation for the line that approximates the data.

(d) Complete the following chart.

Year	Actual Sales	Predicted Sales (from equation in part (c))	Difference, Actual Minus Predicted
0			
1			
2			
3			
4			
5			

(We will obtain a formula for the "best" line through the points in Section 1.5.)

(e) Use the result of part (c) to predict sales in year 7.

(f) Do the same for year 9.

LIFE SCIENCES

Ant Population **44.** Suppose the population of ants in an anthill satisfies the relationship

$$A(x) = 1000x + 6000,$$

where $A(x)$ represents the number of ants present at the end of month x, and $x = 0$ represents June.

Find the number of ants present at the end of each of the following months.

(a) June   **(b)** July   **(c)** August   **(d)** December

(e) What is the monthly rate of change of the number of ants?

Bacterial Growth **45.** Let $N(x) = -5x + 100$ represent the number of bacteria in thousands present in a certain tissue culture at time x in hours after an antibacterial spray is introduced into the environment.

Find the number of bacteria present at each of the following times.

(a) $x = 0$ **(b)** $x = 6$ **(c)** $x = 20$

(d) What is the hourly rate of change in the number of bacteria? Interpret the negative sign in the answer.

SOCIAL SCIENCES

Stimulus Effect **46.** In psychology, the just-noticeable-difference (JND) for some stimulus is defined as the amount by which the stimulus must be increased so that a person will perceive it as having just barely been increased. For example, suppose a research study indicates that a line 40 cm in length must be increased to 42 cm before a subject thinks that it is longer. In this case, the JND would be $42 - 40 = 2$ cm. In a particular experiment, the JND is given by

$$y = .03x,$$

where x represents the original length of the line and y the JND.

Find the JNDs for lines having the following lengths.

(a) 10 cm   **(b)** 20 cm   **(c)** 50 cm   **(d)** 100 cm

(e) Find the rate of change in the JND with respect to the original length of the line.

Estimation of Time **47.** Most people are not very good at estimating the passage of time. Some people's estimations are too fast, and those of others are too slow. One psychologist has constructed a mathematical model for actual time as a function of estimated time: if y represents actual time and x estimated time, then

$$y = mx + b,$$

where m and b are constants that must be determined experimentally for each person. Suppose that for a particular person, $m = 1.25$ and $b = -5$. Find y for each of the following.

(a) $x = 30$ min   **(b)** $x = 60$ min   **(c)** $x = 120$ min   **(d)** $x = 180$ min

Suppose that for another person, $m = .85$ and $b = 1.2$. Find y for each of the following.

(e) $x = 15$ min   **(f)** $x = 30$ min   **(g)** $x = 60$ min   **(h)** $x = 120$ min

For the second person, find x for each of the following.

(i) $y = 60$ min   **(j)** $y = 90$ min

Class Size   **48.** Let $R(x) = -8x + 240$ represent the number of students present in a large business mathematics class, where x represents the number of hours of study required weekly.

Find the number of students present at each of the following levels of required study.

(a) $x = 0$ **(b)** $x = 5$ **(c)** $x = 10$

(d) What is the rate of change of the number of students in the class with respect to the number of hours of study? Interpret the negative sign in the answer.

≡ 1.5 CONSTRUCTING MATHEMATICAL MODELS (OPTIONAL)

We will construct two different mathematical models in this section: one for sum-of-the-years'-digits depreciation and one showing the relationship between the number of stores owned by a large company and sales of shoes in the company's stores over the last few years.

As these two models are developed, notice the fundamental difference between them. To develop the model for depreciation, we start with the basic definitions that apply to depreciation, use some of the mathematics learned in this course, and come up with a model that gives precise values for depreciation of an item. On the other hand, the model developed for the company's sales cannot go back to basic principles. (What are the basic principles for shoe sales? How do we write equations for them?) Rather, we construct a mathematical model for shoe sales by gathering data on past sales and using it to predict future sales. Such a method cannot give exact answers. The best to be expected is an approximation; if we are careful and lucky it will be a good approximation.

Sum-of-the-Years'-Digits Depreciation   As mentioned in the last section, when a business buys an asset (such as a machine or building) it does not treat the total cost of the asset as an expense immediately. Instead, it *depreciates* the cost of the asset over the lifetime of the asset. For example, a machine costing $10,000 and having a useful life of 8 years, after which time it is worthless, might be depreciated at the rate of $10,000/8 = $1250 per year. As noted in the previous section, this method of depreciation, called straight-line depreciation, assumes that the asset loses an equal amount of value during each year of its life.

This assumption of equal loss of value annually is not valid for many assets, such as new cars. A new car may lose 30% of its value during the first year. For assets that lose value quickly at first and then less rapidly in later years, the Internal Revenue Service permits the use of alternate methods of depreciation.

One common method of calculating depreciation is called **sum-of-the years'-digits-depreciation.** With this method, a successively smaller fraction is applied each year to the net cost of the asset. (Recall that the net cost is the original cost less any scrap value.) The denominator of the fraction, which remains constant, is the "sum of the years' digits" that gives the method its name. For example, if the asset has a 5-year life, the denominator is $5 + 4 + 3 + 2 + 1 = 15$. The numerator of the fraction, which changes each year, is the number of years of life that remain. For the first year the numerator is 5, for the second year it is 4, and so on. For example, for an asset with a net cost of $15,000 and a 5-year life, the depreciation in each year is shown in the following table.

Year	Fraction	Depreciation	Accumulated Depreciation
1	5/15	$5000	$ 5,000
2	4/15	$4000	$ 9,000
3	3/15	$3000	$12,000
4	2/15	$2000	$14,000
5	1/15	$1000	$15,000

Now, let us develop a model (or formula) for the depreciation D in year j of an asset with a net cost of x dollars and a life of n years. The constant denominator will be the sum

$$n + (n - 1) + (n - 2) + \ldots + 2 + 1.$$

Using the formula for the sum of an arithmetic sequence,* the sum can be written as

$$n + (n - 1) + \ldots + 2 + 1 = \frac{n(n + 1)}{2}.$$

Since the numerator is the number of years of life that remain, in the first year (when $j = 1$), the numerator is n; when $j = 2$, the numerator is $n - 1$; when $j = 3$, the numerator is $n - 2$; and so on. In each case, the value of j plus the numerator equals $n + 1$, so the numerator can be written as $n + 1 - j$. Thus, the fraction for the year j is

$$\frac{n + 1 - j}{\frac{(n)(n + 1)}{2}} = \frac{2(n + 1 - j)}{n(n + 1)},$$

and the depreciation in year j is

$$D = \frac{2(n + 1 - j)}{n(n + 1)}x.$$

---

*See the chapter entitled "Mathematics of Finance."

■ EXAMPLE 1

A copier costs $4500 and has a useful life of 3 years. Find the depreciation each year using the sum-of-the-years'-digits method.

By the formula above, depreciation in year 1 is

$$D = \frac{2(n + 1 - j)}{n(n + 1)}x$$

$$= \frac{2(3 + 1 - 1)}{3(3 + 1)}(4500) \qquad \text{Let } n = 3, j = 1, \text{ and } x = 4500$$

$$= \frac{6}{12}(4500)$$

$$= 2250,$$

or $2250; depreciation in year 2 is

$$D = \frac{2(3 + 1 - 2)}{3(3 + 1)}(4500)$$

$$= \frac{4}{12}(4500)$$

$$= 1500,$$

or $1500; and depreciation in year 3 is

$$D = \frac{2(3 + 1 - 3)}{3(3 + 1)}(4500)$$

$$= \frac{2}{12}(4500)$$

$$= 750,$$

or $750. The total depreciation is thus $2250 + $1500 + $750 = $4500, as required. ■

■ EXAMPLE 2

Oxford Typo, Inc., buys a new printing press for $42,000. The press has a life of 6 years. Find the depreciation in years 1, 2, and 3 using the sum-of-the-years'-digits method.

Use the mathematical model given above for D; replace j in turn by 1, 2, and 3. In year 1,

$$D = \frac{2(6 + 1 - 1)}{6(7)}(42{,}000) = 12{,}000,$$

or $12,000; in year 2,

$$D = \frac{2(6 + 1 - 2)}{6(7)}(42{,}000) = 10{,}000,$$

or $10,000; and in year 3,

$$D = \frac{2(6 + 1 - 3)}{6(7)}(42{,}000) = 8000,$$

or $8000. ▬

Curve Fitting—The Least Squares Method   Suppose a large company that owns many different shoe stores has increased the number of its stores rapidly over the last few years. This growth has naturally led to increased sales. In the rest of this section, we develop a mathematical model showing the relationship between the number of stores and sales. To begin, we gather data on the number of stores and the sales for each of the past five years, as shown in the following chart. (The numbers are rounded for simplicity.)

Number of Stores	Sales (billions of $)
900	.75
1000	.90
1100	.95
1300	1.05
1350	1.20

Clearly, we cannot use the available data to obtain a precise, exact mathematical model, such as the one for depreciation. There are too many variables here; the state of the economy, shoe prices, and other factors affect shoe sales. For example, from the chart, when the number of stores increased from 1100 to 1300, an increase of 200, sales increased by .1 billion dollars, or $100,000,000. However, an increase of only 50 stores, from 1300 to 1350, corresponded to increased sales of .15 billion, or $150,000,000.

We want an equation that describes the relationship between the number of stores and the sales as closely as possible. Figure 14 shows a graph of the data from the chart. Since we want to predict sales from the number of stores, let sales be the dependent variable, y, and the number of stores be the independent variable, x. The points graphed in Figure 14 appear to lie approximately along a line, which means that a linear function may give a good approximation to the data. Many lines could be drawn on the graph that would be "close" to all the points of the graph. How do we decide on the "best" possible line? The line used most often in applications is that in which the sum of the squares of the vertical distances from the data points to the line is as small as possible. Such a line is called the **least squares line.**

FIGURE 14

In Figure 15 a line has been drawn that seems to lie close to the points corresponding to the data. The distances from these points to the line are indicated by d_1, d_2, d_3, d_4, and d_5 (read "d-sub-one, d-sub-two, d-sub-three," and so on). For n points, corresponding to the n pairs of data, the least squares line is found by minimizing the sum $(d_1)^2 + (d_2)^2 + (d_3)^2 + \ldots + (d_n)^2$.

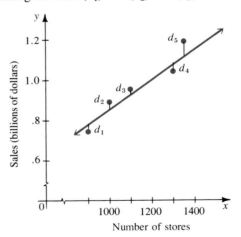

FIGURE 15

For the points (x_1, y_1), (x_2, y_2), ..., (x_n, y_n), if the equation of the desired line is $y' = mx + b$, then

$$d_1 = y'_1 - y_1 = mx_1 + b - y_1,$$
$$d_2 = y'_2 - y_2 = mx_2 + b - y_2,$$

and so on. The sum to be minimized becomes

$$(mx_1 + b - y_1)^2 + (mx_2 + b - y_2)^2 + \ldots + (mx_n + b - y_n)^2,$$

where (x_1, y_1), (x_2, y_2), . . ., (x_n, y_n) are known and m and b are to be found.

The method of minimizing this sum requires calculus and is not given here. The result gives equations for finding the slope and y-intercept of the least squares line. The symbol y' was used instead of y in the equation of the line to distinguish the predicted y-values on the line from the y-values of the given pairs of data points.

LEAST SQUARES LINE

The least squares line $y' = mx + b$ that gives the best fit to the data points (x_1, y_1), (x_2, y_2), . . ., (x_n, y_n) has

$$\text{Slope } m = \frac{n(\Sigma xy) - (\Sigma x)(\Sigma y)}{n(\Sigma x^2) - (\Sigma x)^2}$$

and

$$y'\text{-intercept } b = \frac{\Sigma y - m(\Sigma x)}{n}.$$

The symbol Σ indicates "the sum of:" Σxy means the sum $x_1 y_1 + x_2 y_2 + \ldots + x_n y_n$, Σx means $x_1 + x_2 + \ldots + x_n$, and so on. Note that n is the number of data points.

To find the least squares line for the given data, first find the required sums. Let x represent the number of stores (in thousands) and y represent the sales (in billions of dollars), and prepare a table, as follows.

x	y	xy	x^2	y^2
.90	.75	.6750	.8100	.5625
1.00	.90	.9000	1.0000	.8100
1.10	.95	1.0450	1.2100	.9025
1.30	1.05	1.3650	1.6900	1.1025
1.35	1.20	1.6200	1.8225	1.4400
5.65	4.85	5.6050	6.5325	4.8175

(The column headed y^2 will be used later.) Now calculate m and b; here $n = 5$.

$$m = \frac{n(\Sigma xy) - (\Sigma x)(\Sigma y)}{n(\Sigma x^2) - (\Sigma x)^2} \qquad b = \frac{\Sigma y - m(\Sigma x)}{n}$$

$$= \frac{5(5.605) - (5.65)(4.85)}{5(6.5325) - (5.65)^2} \qquad = \frac{4.85 - (.84)(5.65)}{5}$$

$$= .84 \quad \text{(rounded)} \qquad\qquad = .02 \quad \text{(rounded)}$$

Substitute *m* and *b* into the least squares line equation, $y' = mx + b$; the least squares line that best fits the five data points has equation $y' = .84x + .02$. This is the mathematical model of the relationship between number of stores and sales. The equation can be used to predict *y* from a given value of *x*, as shown in Example 3; however, it is not a good idea to use the least squares equation to predict data points that are not close to those points on which the equation was modeled.

▬ EXAMPLE 3

Suppose the company opens 50 new stores for a total of 1400 or 1.4 thousand stores. Estimate the total sales.

Use the least squares line equation given above with $x = 1.4$.

$$y' = .84x + .02$$
$$= .84(1.4) + .02$$
$$= 1.196$$

Sales for the 1400 stores will be about $1.2 billion.   ▬

▬ EXAMPLE 4

How many stores would be needed to increase sales to $2 billion?

Let $y' = 2$ in the equation above and solve for *x*.

$$2 = .84x + .02$$
$$1.98 = .84x$$
$$x = 2.36$$

The number of stores must increase to about 2360 to produce sales of $2 billion.   ▬

Correlation   Once an equation is found for the least squares line, we might well ask, "Just how good is this equation for prediction purposes?" If the points from the data fit the line quite closely, then we can expect future data pairs to do so. Also, if the points are widely scattered about even the "best-fitting" line, then predictions are not likely to be accurate.

In order to have a quantitative basis for confidence in our predictions, we need a measure of the "goodness of fit" of the original data to the prediction line. One such measure is called the **coefficient of correlation,** denoted *r*.

COEFFICIENT OF CORRELATION

$$r = \frac{n(\Sigma xy) - (\Sigma x)(\Sigma y)}{\sqrt{n(\Sigma x^2) - (\Sigma x)^2} \cdot \sqrt{n(\Sigma y^2) - (\Sigma y)^2}}$$

The coefficient of correlation *r* is always equal to or between 1 and −1. Values of exactly 1 or −1 indicate that the data points lie *exactly* on the least squares line. If $r = 1$, the least squares line has a positive slope; $r = -1$ gives a negative slope. If $r = 0$, there is no linear correlation between the data points (but some *nonlinear* function might provide an excellent fit for the data). Some graphs that correspond to these values of r are shown in Figure 16.

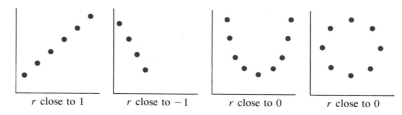

| r close to 1 | r close to -1 | r close to 0 | r close to 0 |

FIGURE 16

▬ EXAMPLE 5

Find r for the shoe sales data given in the table.

From the table, $\Sigma x = 5.65$, $\Sigma y = 4.85$, $\Sigma xy = 5.6050$, $\Sigma x^2 = 6.5325$, and $\Sigma y^2 = 4.8175$. Also, $n = 5$. Substituting these values into the formula for r gives

$$r = \frac{5(5.6050) - (5.65)(4.85)}{\sqrt{5(6.5325) - (5.65)^2} \cdot \sqrt{5(4.8175) - (4.85)^2}}$$

$$= \frac{.6225}{\sqrt{.74} \cdot \sqrt{.565}}$$

$$= .96.$$

This is a very high correlation, as one would expect. Increasing the number of stores clearly would increase sales, so it is not surprising that the coefficient of correlation indicates a close relationship. ▬

▬ 1.5 EXERCISES

For each of the assets in Exercises 1–4, use the sum-of-the-years'-digits method to find the depreciation in year 1 and year 4.

1. Cost: $36,500; scrap value: $8500; life: 10 years

2. Cost: $6250; scrap value: $250; life: 5 years

3. Cost: $18,500; scrap value: $3900; life: 6 years

4. Cost: $275,000; scrap value: $25,000; life: 20 years

▰ APPLICATIONS

BUSINESS AND ECONOMICS

Depreciation
5. For a certain asset, $n = 4$ and $x = \$10,000$.

(a) Use the sum-of-the-years'-digits method to find the depreciation for each of the 4 years covering the useful life of the asset.

(b) What would be the annual depreciation by the straight-line method?

Depreciation
6. A machine tool costs $105,000 and has a scrap value of $25,000, with a useful life of 4 years.

(a) Use the sum-of-the-years'-digits method to find the depreciation in each of the 4 years of the machine tool's life.

(b) Find the total depreciation by this method.

Depreciation
7. An apartment building costs $210,000. The owners decide to depreciate it over 6 years (it isn't built very well). Use the sum-of-the-years'-digits method to find the depreciation in each of the following years.

(a) Year 1   (b) Year 2   (c) Year 3   (d) Year 4

Depreciation
8. A new airplane costs $660,000 and has a useful life of 10 years. Use the sum-of-the-years'-digits method to find the depreciation in each of the following years.

(a) Year 1   (b) Year 2   (c) Year 3   (d) Year 4

Depreciation Comparisons
Complete the tables in Exercises 9 and 10, which compare the two methods of depreciation discussed in this chapter.

9. Cost: $1400; life: 3 years; scrap value: $275

Year	Straight-Line Depreciation	Sum-of-the-Years'-Digits Depreciation
1		
2		
3		
Totals		

10. Cost: $55,000; life: 5 years; scrap value: $0

Year	Straight-Line Depreciation	Sum-of-the-Years'-Digits Depreciation
1		
2		
3		
4		
5		
Totals		

Correlation of Sales and Profits

11. The following table shows profit as a percentage of sales for a typical fast-food restaurant in a recent year.

Annual Store Sales in Thousands (x)	250	375	450	500	600	650
Percent Pretax Profit (y)	9.3	14.8	14.2	15.9	19.2	21.0

(a) Find the least squares line for this data.

(b) Use the results of part (a) to estimate the percentage of profit for sales of $400,000.

(c) Use the results of part (a) to determine what level of sales would produce a profit of 15%.

(d) Find the coefficient of correlation.

Estimating Sales

12. Records show that the annual sales of the EZ Life Company in 5-year periods for the last 20 years were as follows.

Year (x)	Sales in Millions (y)
1969	1.0
1974	1.3
1979	1.7
1984	1.9
1989	2.1

The company wishes to estimate sales from these records for the next few years. Code the years so that $1969 = 0$, $1970 = 1$, and so on.

(a) Plot the 5 points on a graph.

(b) Find the equation of the least squares line, and graph it on the graph from part (a).

(c) Predict the company's sales for 1993 and 1994.

(d) Compute the correlation coefficient.

Estimating Phone Service Needs

13. U.S. West is the largest regional holding company created by the AT&T divestiture. The company includes Mountain Bell, Northwestern Bell, and Pacific Northwest Bell. The table shows the number of customer lines (in millions) serviced by U.S. West over a 6-year period.

Year (x)	1983	1984	1985	1986	1987	1988
Number of Lines (y)	5	7.8	8.9	9.8	9.6	10.4

Let 1983 represent year 1, 1984 represent year 2, and so on.

(a) Use the data to find a least squares line.

(b) Use the results of (a) to predict the number of customer lines that will be serviced by U.S. West in 1990.

(c) Use the results to determine in what year there will be 15 million customer lines.

(d) Find the coefficient of correlation.

Repair Costs and Production Levels

14. The following data furnished by a major brewery were used to determine whether there is a relationship between repair costs and barrels of beer produced. The data in thousands are given for a 10-month period.

Month	Barrels of Beer (x)	Repairs (y)
January	369	299
February	379	280
March	482	393
April	493	388
May	496	385
June	567	423
July	521	374
August	482	357
September	391	106
October	417	332

(a) Find the equation of the least squares line.

(b) Find the coefficient of correlation.

(c) If 500,000 barrels of beer are produced, what will the equation from part (a) give as the predicted repair costs?

LIFE SCIENCES

Effect of Fertilizer on Growth

15. In a study to determine the linear relationship between the size (in decimeters) of an ear of corn (y) and the amount (in tons per acre) of fertilizer used (x), the following data were collected.

$$n = 10$$
$$\Sigma x = 30$$
$$\Sigma y = 24$$
$$\Sigma xy = 75$$
$$\Sigma x^2 = 100$$
$$\Sigma y^2 = 80$$

(a) Find an equation for the least squares line.

(b) Find the coefficient of correlation.

(c) If 3 tons of fertilizer are used per acre, what length (in decimeters) would the equation in (a) predict for an ear of corn?

Correlation of Height and Weight

16. A sample of 10 adult men gave the following data on their heights and weights.

Height in Inches (x)	62	62	63	65	66	67	68	68	70	72
Weight in Pounds (y)	120	140	130	150	142	130	135	175	149	168

(a) Find the equation of the least squares line.

Use the results of part (a) to predict the weights of men of the following heights.

(b) 60 inches   **(c)** 70 inches

(d) Compute the coefficient of correlation.

Bacterial Growth

17. (This problem is appropriate only for those who have studied common logarithms.) Sometimes the scatter diagram of data does not have a linear pattern. This is particularly true in biological applications. In these applications, however, the scatter diagram of the *logarithms* of the data often has a linear pattern. A least squares line can be used to predict the logarithm of any desired value, from which the value itself can be found. Suppose that a certain kind of bacterium grows in number as shown in Table A. The actual number of bacteria present at each time period is replaced with the common logarithm of that number (Table B).

Table A

Time in Hours	0	1	2	3	4	5
Number of Bacteria	1000	1649	2718	4482	7389	12182

Table B

Time (x)	0	1	2	3	4	5
Log (y)	3.0000	3.2172	3.4343	3.6515	3.8686	4.0857

We can now find a least squares line that will predict y, given x.

(a) Plot the original pairs of numbers. The pattern should be nonlinear.

(b) Plot the log values against the time values. The pattern should be almost linear.

(c) Find the equation of the least squares line. (First round off the log values to the nearest hundredth.)

(d) Predict the log value for a time of 7 hours. Find the number whose logarithm is your answer. This will be the predicted number of bacteria.

PHYSICAL SCIENCES

Celsius and Fahrenheit Temperatures

18. In an experiment to determine the linear relationship between temperatures on the Celsius scale (y) and on the Fahrenheit scale (x), a student got the following results.

$$n = 5 \qquad \Sigma xy = 28,050$$
$$\Sigma x = 376 \qquad \Sigma x^2 = 62,522$$
$$\Sigma y = 120 \qquad \Sigma y^2 = 13,450$$

(a) Find an equation for the least squares line.

(b) Find the reading on the Celsius scale that corresponds to a reading of 120° Fahrenheit, using the equation from part (a).

(c) Find the coefficient of correlation.

SOCIAL SCIENCES

High School and College GPAs

19. To determine whether high school grade-point averages and college grade-point averages can be closely approximated by a linear function, a college registrar randomly selected 10 students, finding the averages shown in the chart.

High School GPA (x)	2.5	2.8	2.9	3.0	3.2	3.3	3.3	3.4	3.6	3.9
College GPA (y)	2.0	2.5	2.5	2.8	3.0	3.0	3.5	3.3	3.4	3.6

(a) Find the equation of the least squares line for this data.

Using the results of (a), estimate the college grade-point average of students with the following high school grade-point averages.

(b) 2.7   **(c)** 3.5

(d) What high school grade-point average does the least squares equation require for a college grade-point average of 2.5?

(e) Find the coefficient of correlation.

FOR THE COMPUTER

Depreciation

Find the depreciation by the straight-line and sum-of-the-years'-digits methods for items with the following characteristics.

20. (a) Cost: $12,482; life: 10 yr   **(b)** Cost: $29,700; life: 12 yr

21. (a) Cost: $145,000; life: 30 yr   **(b)** Cost: $258,000; life: 15 yr

It is sometimes possible to get a better prediction for a variable by considering its relationship with more than one other variable. For example, one should be able to predict college GPAs more precisely if both high school GPAs and scores on the ACT are considered. To do this, we alter the equation used to find a least squares line by adding a term for the new variable as follows. If y represents college GPAs, x_1 high school GPAs, and x_2 ACT scores, then y', the predicted GPA, is given by

$$y' = ax_1 + bx_2 + c.$$

*This equation represents a **least squares plane**. The equations for the constants a, b,* *and c are more complicated than those given in the text for m and b, so calculating a least squares equation for three variables is more likely to require the aid of a computer.*

Effect of Revenue on Price
22. Alcoa* used a least squares plane with two independent variables, x_1 and x_2, to predict the effect on revenue of the price of aluminum forged truck wheels, as follows.

$$x_1 = \text{average price in dollars per wheel}$$
$$x_2 = \text{production of large trucks in thousands}$$
$$y = \text{sales of aluminum forged truck wheels in thousands}$$

Using data for an 11-year period, the company found the equation of the least squares line to be

$$y' = 49.2755 - 1.1924x_1 + .1631x_2,$$

for which the correlation coefficient was .902. The following figures were then forecast for truck production.

Year	1	2	3	4	5	6
Production	160.0	165.0	170.0	175.0	180.0	185.0

Three possible price levels per wheel were considered: $42, $45, and $48.

(a) Use the least squares plane equation given above to find the estimated sales of wheels (y') for year 3 at each of the three price levels.

(b) Repeat part (a) for year 6.

(c) For which price level, on the basis of the figures for years 3 and 6, are total estimated sales greatest?

(By comparing total estimated sales for the years 1 through 6 at each of the three price levels, the company found that the selling price of $42 per wheel would generate the greatest sales volume over the 6-year period.)

Find the least squares line and the coefficient of correlation for the sets of data pairs in Exercises 23 and 24.

23.

x	2.1	3.2	4.6	4.9	5.3	5.8
y	10.3	12.7	11.9	13.2	14.0	14.8

24.

x	143	152	169	175	178	185	190	202
y	43	58	55	62	69	71	78	83

*This example was supplied by John H. Van Denender, Public Relations Department, Aluminum Company of America.

≡ **KEY WORDS** ≡   *To understand the concepts presented in this chapter, you should know the meaning and use of the following words. For easy reference, the section in the chapter where a word (or expression) was first used is given with each item.*

	mathematical model		supply curve
1.1	linear equation		demand curve
	linear inequality		equilibrium price
1.2	independent variable	1.3	slope
	dependent variable		slope-intercept form
	function		point-slope form
	domain		standard form
	range	1.4	fixed cost
	ordered pair		variable cost
	Cartesian coordinate system		marginal cost
	axes		break-even point
	origin		straight-line depreciation
	coordinates		scrap value
	quadrant		net cost
	graph	1.5	sum-of-the-years'-digits
	vertical line test		depreciation
	linear function		least squares line
	intercepts		coefficient of correlation

≡ **CHAPTER 1**   **REVIEW EXERCISES**

Solve each of the following linear equations.

1. $3x + 2 = 8$ **2.** $-m + 7 = 12$ **3.** $2k - 3 = 6 - k$

4. $4p - 2p + 3 = 3p$ **5.** $5n - (n + 1) = 3(n + 2)$ **6.** $4(y + 3) - 2y = -(y - 1)$

7. $\dfrac{3}{2} + 3z = 5 + \dfrac{2z}{4}$ **8.** $\dfrac{x}{5} - \dfrac{3}{5} = 2x + 1$

Solve each of the following linear inequalities.

9. $4m + 2 \le 12$ **10.** $3z + 5 \le 8$

11. $5 - 2m > 7$ **12.** $9 - 5p > 12$

13. $2k + 3 \le 5k - 4$ **14.** $4 - 3z \le 2z + 1$

15. $\dfrac{2}{3}x + 4 \ge x - \dfrac{1}{3}$ **16.** $\dfrac{a}{5} - 3 \ge 2a + \dfrac{2}{5}$

17. $2(k + 5) - 3k < k + 4$ **18.** $5 - 4b < 2(b + 3) - b$

For each of the functions in Exercises 19–22, find **(a)** $f(6)$; **(b)** $f(-2)$;
(c) $f(-4)$; *and* **(d)** $f(r + 1)$.

19. $f(x) = 4x - 1$ **20.** $f(x) = 3 - 4x$

21. $f(x) = -x^2 + 2x - 4$ **22.** $f(x) = 8 - x - x^2$

Graph each linear function defined as follows.

23. $y = 4x + 3$ **24.** $y = 6 - 2x$ **25.** $3x - 5y = 15$ **26.** $2x + 7y = 14$

27. $x + 2 = 0$ **28.** $y = 1$ **29.** $y = 2x$ **30.** $x + 3y = 0$

Find the slope for each of the lines in Exercises 31–38 that has a slope.

31. Through $(-2, 5)$ and $(4, 7)$

32. Through $(4, -1)$ and $(3, -3)$

33. Through the origin and $(11, -2)$

34. Through the origin and $(0, 7)$

35. $2x + 3y = 15$

36. $4x - y = 7$

37. $x + 4 = 9$

38. $3y - 1 = 14$

Find an equation for each of the lines described below.

39. Through $(5, -1)$, slope $= 2/3$

40. Through $(8, 0)$, slope $= -1/4$

41. Through $(5, -2)$ and $(1, 3)$

42. Through $(2, -3)$ and $(-3, 4)$

43. Through $(-1, 4)$, undefined slope

44. Through $(-2, 5)$, slope $= 0$

▤ APPLICATIONS

BUSINESS AND ECONOMICS

Return on Investment   **45.** Virginia Wallace invested an inheritance of $30,000 in two ways: part at 8½% interest and the rest at 10%. Her total annual interest on the money is $2820. How much is invested at 10%?

Profit   **46.** To manufacture x thousand computer chips requires fixed expenditures of $352 plus $42 per thousand chips. Receipts from the sale of x thousand chips amount to $130 per thousand.

(a) Write an expression for expenditures.

(b) Write an expression for receipts.

(c) For profit to be made, receipts must be greater than expenditures. How many chips must be sold to produce a profit?

Supply and Demand   **47.** The supply and demand for a certain commodity are related by

$$\text{Supply: } p = 6x + 3$$

and

$$\text{Demand: } p = 19 - 2x,$$

where p represents the price at a supply or demand, respectively, of x units.

Find the supply and the demand at the following prices.

(a) $10   (b) $15   (c) $18

(d) Graph both the supply and the demand functions on the same axes.

(e) Find the equilibrium price.

(f) Find the equilibrium supply and the equilibrium demand.

Supply   **48.** For a particular product, 72 units will be supplied at a price of $6, while 104 units will be supplied at a price of $10. Write a supply function for this product.

Cost   *In Exercises 49–52, write a linear cost function for each situation.*

49. Eight units cost $300; fixed cost is $60.

50. Fixed cost is $2000; 36 units cost $8480.

51. Twelve units cost $445; 50 units cost $1585.

52. Thirty units cost $1500; 120 units cost $5640.

Break-Even Analysis   **53.** The cost of producing x units of a product is $C(x)$, where
$$C(x) = 20x + 100.$$
The product sells for $40 per unit.

(a) Find the break-even point.

(b) What revenue will the company receive if it sells just that number of units?

Depreciation   *Work the following problems involving depreciation.*

54. An asset costs $21,000 and has a 6-year life with no scrap value. Find the first-year depreciation for this asset by each of the following methods.

(a) Straight-line depreciation

(b) Sum-of-the-years' digits depreciation

55. An asset with a purchase price of $21,000 has a scrap value of $2000 and a 7-year life. Find the third-year depreciation for this asset using the following methods.

(a) Straight-line depreciation

(b) Sum-of-the-years'-digits depreciation

56. Find the first- and fourth-year depreciation by both straight-line and sum-of-the-years'-digits methods for an asset that costs $12,000, has no scrap value, and has a 5-year life.

57. Complete the following table for an asset that costs $79,000, has a scrap value of $11,000, and has a 4-year life.

Year	Straight-Line Depreciation	Sum-of-the-Years'-Digits Depreciation
1		
2		
3		
4		

58. Complete a table similar to the one in Exercise 57 for an asset that costs $430,000, has a scrap value of $70,000, and has a 5-year life.

59. Complete a table similar to the one in Exercise 57 for an asset that costs $18,000, has no scrap value, and has a 3-year life.

60. Repeat Exercise 59 for an asset that costs $24,000, has no scrap value, and has a 4-year life.

Earnings   **61.** Find the least squares line for the following data that give the earnings y in ten-thousands for a certain company after x years.

x	3	5	7	8
y	4	11	20	23

62. Use your equation from Exercise 61 to predict the company's earnings after 6 years.

63. Find the coefficient of correlation for the data in Exercise 61.

LIFE SCIENCES

Blood Sugar and Cholesterol Levels

64. The following data show the connection between blood sugar levels and cholesterol levels for 8 different patients.

Patient	1	2	3	4	5	6	7	8
Blood Sugar Level (x)	130	138	142	159	165	200	210	250
Cholesterol Level (y)	170	160	173	181	201	192	240	290

For this data, $\Sigma x = 1394$, $\Sigma y = 1607$, $\Sigma xy = 291{,}990$, $\Sigma x^2 = 255{,}214$, and $\Sigma y^2 = 336{,}155$.

(a) Find the equation of the least squares line, $y' = mx + b$.

(b) Predict the cholesterol level for a person whose blood sugar level is 190.

(c) Find r.

GENERAL

Postage Costs

65. Assume that it costs $.30 to mail a letter weighing 1 oz or less, with each additional ounce, or portion of an ounce, costing $.27. Let $C(x)$ represent the cost in dollars to mail a letter weighing x oz.

Find the cost of mailing a letter of each of the following weights.

(a) 3.4 oz   **(b)** 1.02 oz   **(c)** 5.9 oz   **(d)** 10 oz

(e) Graph C.

(f) Give the domain and range for C.

EXTENDED APPLICATION

ESTIMATING SEED DEMANDS—THE UPJOHN COMPANY

The Upjohn Company has a subsidiary that buys seeds from farmers and then resells them. Each spring the firm contracts with farmers to grow the seeds. The firm must decide on the number of acres that it will contract for. The problem faced by the company is that the demand for seed is not constant, but fluctuates from year to year. Also, the number of tons of seed produced per acre varies, depending on weather and other factors. In an attempt to decide the number of acres that should be planted in order to maximize profits, a company mathematician created a model of the variables involved in determining the number of acres to plant.*

The analysis of this model required advanced methods that we will not go into. We can, however, give the conclusion: the number of acres that will maximize profit in the long run is found by solving the equation

$$F(AX + Q) = \frac{(S - C_p) X - C_A}{(S - C_p + C_c) X} \tag{1}$$

*Based on work by David P. Rutten, Senior Mathematician, The Upjohn Company, Kalamazoo, Michigan.

for *A*. The function $F(z)$ represents the chances that z tons of seed will be demanded by the marketplace. The variables in the equation are

A = number of acres of land contracted by the company,
X = quantity of seed produced per acre of land,
Q = quantity of seed in inventory from previous years,
S = selling price per ton of seed,
C_p = variable cost (production, marketing, etc.) per ton of seed,
C_c = cost to carry over one ton of seed from previous year,
C_A = variable cost per acre of land.

To advise management of the number of acres of seed to contract for, the mathematician studied past records to find the values of the various variables. From these records and from predictions of future trends, it was concluded that S = $10,000 per ton, X = .1 ton per acre (on the average), Q = 200 tons, C_p = $5000 per ton, C_A = $100 per acre, C_c = $3000 per ton.

The function $F(z)$ is found by the same process to be approximately

$$F(z) = \frac{z}{1000} - \frac{1}{2}, \text{ if } 500 \le z \le 1500 \text{ tons.} \tag{2}$$

EXERCISES

1. Evaluate $AX + Q$ using the values of X and Q given above.
2. Find $F(AX + Q)$, using equation (2) and your results from Exercise 1.
3. Solve equation (1) for A.
4. How many acres should be planted?
5. How many tons of seed will be produced?
6. Find the total revenue that will be received from the sale of the seeds.

EXTENDED APPLICATION

MARGINAL COST—BOOZ, ALLEN & HAMILTON

Booz, Allen & Hamilton Inc. is a large management consulting firm.* One of the services it provides to client companies is the preparation of profitability studies that show ways in which the client can increase profit levels. The client company requesting the analysis presented in this case is a large producer of a staple food. The company buys from farmers and then processes the food in its mills, resulting in a finished product. The company sells both at retail under its own brands, and in bulk to other companies who use the product in the manufacture of convenience foods.

The client company has been reasonably profitable in recent years, but the management retained Booz, Allen & Hamilton to see whether its consultants could suggest ways of increasing company profits. The management of the company had long operated with the philosophy of trying to process and sell as much of its product as possible, because they felt this would lower the average processing cost per unit sold. The consultants found, however, that the client's fixed mill costs were quite low and that processing extra units made the cost per unit start to increase. (There are several reasons for this: the company must run three shifts, machines break down more often, and so on.)

---

*This case was supplied by John R. Dowdle of the Chicago office of Booz, Allen & Hamilton Inc.

In this case, we shall discuss the marginal cost of two of the company's products. The marginal cost of production (cost of producing an extra unit) of product A was found by the consultants to be approximated by the linear function

$$y = .133x + 10.09,$$

where x is the number of units produced (in millions) and y is the marginal cost.

For example, at a level of production of 3.1 million units, producing an additional unit of product A would cost about

$$y = .133(3.1) + 10.09 \approx \$10.50.†$$

At a level of production of 5.7 million units, producing an extra unit would cost $10.85. Figure 17 shows a graph of the marginal cost function from $x = 3.1$ to $x = 5.7$, the domain over which the function above was found to apply.

Since the selling price for product A is $10.73 per unit, the company was losing money on many units of the product that it sold (see Figure 17). Since the selling price could not be raised if the company were to remain competitive, the consultants recommended that production of product A be cut.

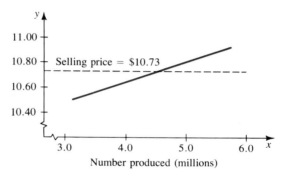

FIGURE 17

For product B, the Booz, Allen & Hamilton consultants found a marginal cost function given by

$$y = .0667x + 10.29,$$

with x and y as defined above. Verify that at a production level of 3.1 million units, the marginal cost is about $10.50, while at a production level of 5.7 million units, the marginal cost is about $10.67. Since the selling price of this product is $9.65, the consultants again recommended a cutback in production.

The consultants ran similar cost analyses of other products made by the company and then issued their recommendation: the company should reduce total production by 2.1 million units. The analysts predicted that this would raise profits for the products under discussion from $8.3 million annually to $9.6 million—which is very close to what actually happened when the client took the advice.

---

†The symbol $\approx$ means *approximately equal to.*

EXERCISES

1. At what level of production x was the marginal cost of a unit of product A equal to the selling price?

2. Graph the marginal cost function for product B from $x = 3.1$ million units to $x = 5.7$ million units.

3. Find the number of units for which marginal cost equals the selling price for product B.

4. For product C, the marginal cost of production is

$$y = .133x + 9.46.$$

(a) Find the marginal cost at each of the following production levels: 3.1 million units and 5.7 million units.

(b) Graph the marginal cost function.

(c) For a selling price of $9.57, find the production level at which the cost equals the selling price.

ms pg 81

2

Systems of Linear Equations and Matrices

2.1   Systems of Linear Equations; the Echelon Method

2.2   Solution of Linear Systems by the Gauss -Jordan Method

2.3   Basic Matrix Operations

2.4   Multiplication of Matrices

2.5   Matrix Inverses

2.6   Input - Output Models

Review Exercises

Extended Application
Routing

Extended Application
Contagion

Extended Application
Leontief's Model of the American Economy

Many mathematical models require finding the solutions of two or more equations. The solutions must satisfy *all* of the equations in the model. A set of equations related in this way is called a **system of equations.** In this chapter we will discuss systems of equations, introduce the idea of a *matrix,* and then show how matrices are used to solve systems of equations.

2.1 SYSTEMS OF LINEAR EQUATIONS; THE ECHELON METHOD

Suppose that an animal feed is made from three ingredients: corn, soybeans, and cottonseed. One unit of each ingredient provides the number of units of protein, fat, and fiber shown in the table. For example, the entries in the first column, .25, .4, and .3, indicate that one unit of corn provides twenty-five hundredths (one-fourth) of a unit of protein, four-tenths of a unit of fat, and three-tenths of a unit of fiber.

	Corn	*Soybeans*	*Cottonseed*
Protein	.25	.4	.2
Fat	.4	.2	.3
Fiber	.3	.2	.1

Now suppose we need to know the number of units of each ingredient that should be used to make a feed that contains 22 units of protein, 28 units of fat, and 18 units of fiber. To find out, we let x represent the required number of units of corn, y the number of units of soybeans, and z the number of units of cottonseed. Since the total amount of protein is to be 22 units,

$$.25x + .4y + .2z = 22.$$

The feed must supply 28 units of fat, so

$$.4x + .2y + .3z = 28,$$

and 18 units of fiber, so

$$.3x + .2y + .1z = 18.$$

To solve this problem, we must find values of x, y, and z that satisfy this system of equations. Verify that $x = 40$, $y = 15$, and $z = 30$ is a solution of the system, since these numbers satisfy all three equations. In fact, this is the only solution of this system. Many practical problems lead to such systems of *first-degree equations.*

A **first-degree equation** in n unknowns is any equation of the form

$$a_1x_1 + a_2x_2 + \cdots + a_nx_n = k$$

where a_1, a_2, $\cdot \cdot \cdot$, a_n and k are all real numbers.* Each of the three equations from the animal feed problem is a first-degree equation in three unknowns. A **solution** of the first-degree equation

$$a_1x_1 + a_2x_2 + \cdot \cdot \cdot + a_nx_n = k$$

is a sequence of numbers s_1, s_2, $\cdot \cdot \cdot$, s_n such that

$$a_1s_1 + a_2s_2 + \cdot \cdot \cdot + a_ns_n = k.$$

A solution of an equation is usually written between parentheses as $(s_1, s_2, \cdot \cdot \cdot, s_n)$. For example, $(1, 6, 2)$ is a solution of the equation $3x_1 + 2x_2 - 4x_3 = 7$, since $3(1) + 2(6) - 4(2) = 7$. This is an extension of the idea of an ordered pair, which was introduced in Chapter 1. A solution of a first-degree equation in two unknowns is an ordered pair, and the graph of the equation is a straight line. For this reason all first-degree equations are also called linear equations. In this section we develop a method for solving a system of first-degree equations. Although the discussion will be confined to equations with only a few variables, the method of solution can be extended to systems with many variables.

Because the graph of a linear equation in two unknowns is a straight line, there are three possibilities for the solutions of a system of two linear equations in two unknowns.

1. The two graphs are lines intersecting at a single point. The system has one solution, and it is given by the coordinates of this point. [See Figure 1(a).]

2. The graphs are distinct parallel lines. When this is the case, the system is **inconsistent;** that is, there is no solution common to both equations. [See Figure 1(b).]

3. The graphs are the same line. In this case, the equations are said to be **dependent,** since any solution of one equation is also a solution of the other. There are infinitely many solutions. [See Figure 1(c).]

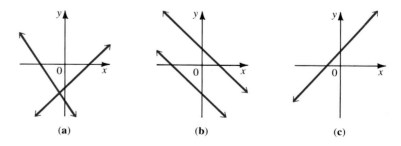

(a)   (b)   (c)

FIGURE 1

In larger systems, with more equations and more variables, there also may be exactly one solution, no solutions, or infinitely many solutions. If no solution satisfies every equation in the system, the system is inconsistent, and if there are infinitely many solutions that satisfy all the equations in the system, the equations are dependent.

Transformations   To solve a linear system of equations, we use properties of algebra to change, or transform, the system into a simpler *equivalent* system. An **equivalent system** is one that has the same solutions as the given system. Algebraic properties are the basis of the transformations listed below.

TRANSFORMATIONS OF A SYSTEM

> The following transformations can be applied to a system of equations to get an equivalent system:
>
> **1.** exchanging any two equations;
> **2.** multiplying both sides of an equation by any nonzero real number;
> **3.** adding to any equation a multiple of some other equation.

Use of these transformations leads to an equivalent system because each transformation can be reversed or "undone," allowing a return to the original system.

The Echelon Method   A systematic approach for solving systems of equations using the three transformations is called the **echelon method.** This method, using matrix notation (discussed later in this chapter), makes it possible to solve systems of equations by computer. The echelon method involves transforming a system of equations to a special form, as shown in the following examples and summarized later in this section.

■ EXAMPLE 1

Solve the system

$$2x + 3y = 12 \qquad (1)$$
$$3x - 4y = 1. \qquad (2)$$

The goal of the echelon method is to transform the system to a simpler equivalent system. The first equation in the new system should have a coefficient of 1 for x. Multiply equation (1) on both sides by 1/2 to get

$$x + \frac{3}{2}y = 6. \qquad (3)$$

The new system is

$$x + \frac{3}{2}y = 6 \qquad (3)$$
$$3x - 4y = 1. \qquad (2)$$

Now transform equation (2) so that the coefficient of x is 0. Using transformation 3, multiply equation (3) by -3 and add the results to equation (2), getting

$$-3\left(x + \frac{3}{2}y\right) + 3x - 4y = -3(6) + 1$$

$$-3x + 3x + \frac{-9}{2}y - 4y = -17$$

$$\frac{-17}{2}y = -17.$$

The result of this step is the equivalent system

$$x + \frac{3}{2}y = 6 \qquad\qquad\text{(3)}$$

$$\frac{-17}{2}y = -17. \qquad\qquad\text{(4)}$$

Next, using transformation 2, multiply both sides of equation (4) by $-2/17$ to get a new system with the coefficient of y equal to 1.

$$x + \frac{3}{2}y = 6 \qquad\qquad\text{(3)}$$

$$y = 2 \qquad\qquad\text{(5)}$$

Equation (5) gives the value of y. Substitute 2 for y in equation (3) to get $x = 3$. Thus, the solution of the system is (3, 2). The graphs of the two equations in Figure 2 suggest that (3, 2) indeed satisfies both equations in the system. ■

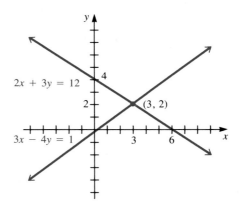

FIGURE 2

■ EXAMPLE 2

Solve the system

$$3x - 2y = 4 \qquad \textbf{(1)}$$
$$-6x + 4y = 7. \qquad \textbf{(2)}$$

Multiply equation (1) by 1/3 so that x has a coefficient of 1. The new system is

$$x - \frac{2}{3}y = \frac{4}{3} \qquad \textbf{(3)}$$
$$-6x + 4y = 7. \qquad \textbf{(2)}$$

Multiply equation (3) by 6 and add the result to equation (2), so the coefficient of x in equation (2) becomes 0.

$$6\left(x - \frac{2}{3}y\right) + (-6x + 4y) = 6\left(\frac{4}{3}\right) + 7$$
$$6x - 6x - 4y + 4y = 8 + 7$$
$$0 = 15$$

The new system is

$$x - \frac{2}{3}y = \frac{4}{3}$$
$$0 = 15.$$

In the second equation, both variables have been eliminated, leaving a false statement. This is a signal that these two equations have no common solution. This system is inconsistent and has no solution. As Figure 3 shows, the graph of the system is made up of two distinct parallel lines. ■

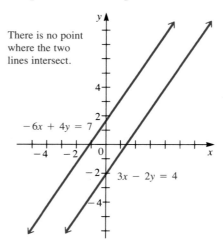

FIGURE 3

▬ EXAMPLE 3

Solve the system

$$-4x + y = 2 \tag{1}$$

$$8x - 2y = -4. \tag{2}$$

Multiply both sides of equation (1) by $-1/4$ (so that the coefficient of x is 1) to get the new system

$$x - \frac{1}{4}y = -\frac{1}{2} \tag{3}$$

$$8x - 2y = -4. \tag{2}$$

To make the coefficient of x in equation (2) equal to 0, multiply equation (1) on both sides by -8 and add to equation (2).

$$-8(x - \frac{1}{4}y) + (8x - 2y) = -8\left(-\frac{1}{2}\right) + (-4)$$

$$-8x + 8x + 2y - 2y = 4 - 4$$

$$0 = 0$$

The system becomes

$$x - \frac{1}{4}y = -\frac{1}{2} \tag{3}$$

$$0 = 0. \tag{4}$$

The true statement in equation (4) indicates that the two equations have the same graph, which means that there is an infinite number of solutions for the system. In this case, all the ordered pairs that satisfy the equation $-4x + y = 2$ (or $8x - 2y = -4$) are solutions. If $x = a$, then $y = 2 + 4a$, and all ordered pairs of the form $(a, 2 + 4a)$ are solutions. For example, some solutions are $(1, 6)$, $(0, 2)$, and $(-3, -10)$. See Figure 4. ▬

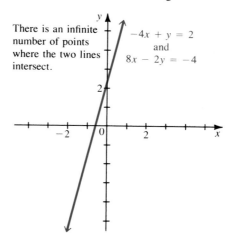

There is an infinite number of points where the two lines intersect.

$-4x + y = 2$
and
$8x - 2y = -4$

FIGURE 4

Additional steps are needed to solve a system with three equations in three unknowns by the echelon method.

▬▬ EXAMPLE 4

Solve the system

$$2x + y - z = 2 \tag{1}$$
$$x + 3y + 2z = 1 \tag{2}$$
$$x + y + z = 2. \tag{3}$$

As in previous examples, our goal is to transform the given system into an equivalent system that is easier to solve. In the new system, the coefficient of x in the first equation should be 1. Use the first transformation to exchange equations (1) and (2). (Alternately, we could have multiplied both sides of equation (1) by 1/2.) The new system is

$$x + 3y + 2z = 1 \tag{2}$$
$$2x + y - z = 2 \tag{1}$$
$$x + y + z = 2. \tag{3}$$

To get a new second equation without x, multiply each term of equation (2) by -2 and add the results to equation (1). This gives

$$-5y - 5z = 0. \tag{4}$$

Multiply both sides of equation (4) by $-1/5$ to make the coefficient of y equal to 1.

$$y + z = 0 \tag{5}$$

We now have the equivalent system

$$x + 3y + 2z = 1 \tag{2}$$
$$y + z = 0 \tag{5}$$
$$x + y + z = 2. \tag{3}$$

Get a new third equation without x by multiplying both sides of equation (2) by -1 and adding the results to equation (3). The resulting equation is

$$-2y - z = 1. \tag{6}$$

To eliminate y in equation (6) multiply both sides of equation (5) by 2 and add to equation (6) to get

$$z = 1. \tag{7}$$

The original system has now led to the equivalent system

$$x + 3y + 2z = 1 \tag{2}$$
$$y + z = 0 \tag{5}$$
$$z = 1. \tag{7}$$

Note the triangular form of this last system. This is the typical echelon form. From equation (7), $z = 1$. Substitute 1 for z in equation (5) to get $y = -1$. Finally, substitute 1 for z and -1 for y in equation (2) to get $x = 2$. The solution of the system is $(2, -1, 1)$. ▬

In summary, to solve a linear system in n variables by the echelon method, perform the following steps using the three transformations given earlier.

THE ECHELON METHOD OF SOLVING A LINEAR SYSTEM

1. If possible, arrange the equations so that there is an x_1 term in the first equation, an x_2 term in the second equation, and so on.
2. Make the coefficient of the first variable equal to 1 in the first equation and 0 in the other equations.
3. Make the coefficient of the second variable equal to 1 in the second equation and 0 in all remaining equations.
4. Make the coefficient of the third variable equal to 1 in the third equation and 0 in all remaining equations.
5. Continue in this way until the last equation is of the form $x_n = k$, where k is a constant.

Parameters   The systems of equations discussed so far have had the same number of equations as variables. Such systems have either one solution, no solution, or infinitely many solutions. Some mathematical models lead to systems of equations with fewer equations than variables, however. Such systems have either infinitely many solutions or no solution.

▬▬ **EXAMPLE 5**

Solve the system

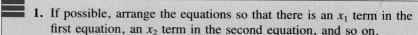

$$2x + 3y + 4z = 6 \tag{1}$$
$$x - 2y + z = 9 \tag{2}$$
$$3x - 6y + 3z = 27. \tag{3}$$

Since the third equation here is a multiple of the second equation, this system really has only the two equations

$$2x + 3y + 4z = 6 \tag{1}$$
$$x - 2y + z = 9. \tag{2}$$

Use the steps of the echelon method as far as possible. Exchange the two equations so that the coefficient of x is 1 in the first equation of the system. This gives

$$x - 2y + z = 9 \tag{2}$$
$$2x + 3y + 4z = 6. \tag{1}$$

To get a new equation without x, multiply both sides of equation (2) by -2 and add to equation (1). The result is

$$7y + 2z = -12.$$

Multiply both sides of this equation by $1/7$ so that the coefficient of y is 1. The new system is

$$x - 2y + z = 9 \qquad\qquad (2)$$

$$y + \frac{2}{7}z = \frac{-12}{7}. \qquad\qquad (3)$$

Since there are only two equations, it is not possible to continue with the echelon method. To complete the solution, solve equation (3) for y.

$$y = \frac{-2}{7}z - \frac{12}{7}$$

Now substitute the result for y in equation (2), and solve for x.

$$x - 2y + z = 9 \qquad\qquad (2)$$

$$x - 2\left(\frac{-2}{7}z - \frac{12}{7}\right) + z = 9$$

$$x + \frac{4}{7}z + \frac{24}{7} + z = 9$$

$$7x + 4z + 24 + 7z = 63$$

$$7x + 11z = 39$$

$$7x = -11z + 39$$

$$x = \frac{-11}{7}z + \frac{39}{7}$$

Each choice of a value for z leads to values for x and y. For example,

if $z = 1$,

then $x = -\dfrac{11}{7} \cdot 1 + \dfrac{39}{7} = 4$ and $y = -\dfrac{2}{7} \cdot 1 - \dfrac{12}{7} = -2$;

if $z = -6$,

then $x = 15$ and $y = 0$;

if $z = 0$,

then $x = \dfrac{39}{7}$ and $y = \dfrac{-12}{7}$.

To write a general solution, let $z = a$. Then

$$x = -\frac{11}{7}a + \frac{39}{7} \qquad \text{and} \qquad y = -\frac{2}{7}a - \frac{12}{7}.$$

There are infinitely many solutions for the original system, since a can take on infinitely many values. The set of solutions may be written

$$\left\{ \left(-\frac{11}{7}a + \frac{39}{7}, \quad -\frac{2}{7}a - \frac{12}{7}, \quad a \right) \right\}. \quad \blacksquare$$

Since both x and y in Example 5 were expressed in terms of z, the variable z is called a **parameter.** If we solved the system in a different way, x or y could be the parameter.

The system in Example 5 had one more variable than equations. If there are two more variables than equations, there usually will be two parameters, and so on. What if there are fewer variables than equations? For example, consider the system

$$2x + 3y = 8 \qquad \qquad \textbf{(1)}$$
$$x - y = 4 \qquad \qquad \textbf{(2)}$$
$$5x + y = 7, \qquad \qquad \textbf{(3)}$$

which has three equations in two variables. Since each of these equations has a line as its graph, there are various possibilities: three lines that intersect at a common point, three lines that cross at three different points, three lines of which two are the same line so that the intersection would be a point, three lines of which two are parallel so that the intersection would be two different points, three lines that are all parallel so that there would be no intersection, and so on. As in the case of n equations with n variables, the possibilities result in a unique solution, no solution, or infinitely many solutions.

To solve the system using the echelon method, begin by exchanging equations (1) and (2) to get a 1 for the coefficient of x in the first equation.

$$x - y = 4 \qquad \qquad \textbf{(2)}$$
$$2x + 3y = 8 \qquad \qquad \textbf{(1)}$$
$$5x + y = 7 \qquad \qquad \textbf{(3)}$$

Now multiply both sides of equation (2) by -2 and add the result to equation (1) to get $5y = 0$. The new system is

$$x - y = 4 \qquad \qquad \textbf{(2)}$$
$$5y = 0 \qquad \qquad \textbf{(4)}$$
$$5x + y = 7. \qquad \qquad \textbf{(3)}$$

Multiply both sides of equation (4) by 1/5 to get $y = 0$. Substituting $y = 0$ leads to $x = 4$ in equation (2) and $x = 7/5$ in equation (3), a contradiction. This contradiction shows that the system has no solution.

Applications   The mathematical techniques in this text will be useful to you only if you are able to apply them to practical problems. To do this, always begin by reading the problem carefully. Next, identify what must be found.

Let each unknown quantity be represented by a variable. (It is a good idea to *write down* exactly what each variable represents.) Now reread the problem, looking for all necessary data. Write that down, too. Finally, look for one or more sentences that lead to equations or inequalities. The next example illustrates these steps.

▰ EXAMPLE 6

A bank teller has a total of 70 bills in five-, ten- and twenty-dollar denominations. The number of fives is three times the number of tens, while the total value of the money is $960. Find the number of each type of bill.

Let x be the number of fives, y the number of tens, and z the number of twenties. Then, since the total number of bills is 70,

$$x + y + z = 70.$$

The number of fives, x, is 3 times the number of tens, y.

$$x = 3y$$

Finally, the total value is $960. The value of the fives is $5x$, the value of the tens is $10y$, and the value of the twenties is $20z$, so that

$$5x + 10y + 20z = 960.$$

Rewriting the second equation as $x - 3y = 0$ gives the system

$$x + \quad y + \quad z = 70 \qquad \textbf{(1)}$$
$$x - \quad 3y \qquad\quad = 0 \qquad \textbf{(2)}$$
$$5x + 10y + 20z = 960. \qquad \textbf{(3)}$$

To solve by the echelon method, first eliminate x from equation (2). Do this by multiplying both sides of equation (1) by -1 and adding to equation (2), then multiplying both sides of the result by $-1/4$. The new system is

$$x + \quad y + \quad z = 70 \qquad \textbf{(1)}$$
$$y + \frac{1}{4}z = \frac{35}{2} \qquad \textbf{(4)}$$
$$5x + 10y + 20z = 960. \qquad \textbf{(3)}$$

Multiply equation (1) by -5 and add to equation (3) to eliminate x from equation (3), getting the system

$$x + y + \quad z = 70 \qquad \textbf{(1)}$$
$$y + \frac{1}{4}z = \frac{35}{2} \qquad \textbf{(4)}$$
$$5y + 15z = 610. \qquad \textbf{(5)}$$

To eliminate y from equation (5), multiply both sides of equation (4) by -5 and add to equation (5). The final equivalent system is

$$x + y + z = 70 \tag{1}$$

$$y + \frac{1}{4}z = \frac{35}{2} \tag{5}$$

$$\frac{55}{4}z = \frac{1045}{2}. \tag{6}$$

Equation (6) gives $z = 38$. Substituting this result into equation (5) gives $y = 8$. Finally, from equation (1), $x = 24$. The teller had 24 fives, 8 tens, and 38 twenties. Check this answer in the words of the original problem.   ▬

2.1 EXERCISES

Use the echelon method to solve each system of two equations in two unknowns. Check your answers.

1. $x + y = 9$
$2x - y = 0$

2. $4x + y = 9$
$3x - y = 5$

3. $5x + 3y = 7$
$7x - 3y = -19$

4. $2x + 7y = -8$
$-2x + 3y = -12$

5. $3x + 2y = -6$
$5x - 2y = -10$

6. $-6x + 2y = 8$
$5x - 2y = -8$

7. $2x - 3y = -7$
$5x + 4y = 17$

8. $4m + 3n = -1$
$2m + 5n = 3$

9. $5p + 7q = 6$
$10p - 3q = 46$

10. $12s - 5t = 9$
$3s - 8t = -18$

11. $6x + 7y = -2$
$7x - 6y = 26$

12. $2a + 9b = 3$
$5a + 7b = -8$

13. $3x + 2y = 5$
$6x + 4y = 8$

14. $9x - 5y = 1$
$-18x + 10y = 1$

15. $4x - y = 9$
$-8x + 2y = -18$

16. $3x + 5y + 2 = 0$
$9x + 15y + 6 = 0$

Use the echelon method to solve each system. Check your answers.

17. $\dfrac{x}{2} + \dfrac{y}{3} = 8$
$\dfrac{2x}{3} + \dfrac{3y}{2} = 17$

18. $\dfrac{x}{5} + 3y = 31$
$2x - \dfrac{y}{5} = 8$

19. $\dfrac{x}{2} + y = \dfrac{3}{2}$
$\dfrac{x}{3} + y = \dfrac{1}{3}$

20. $x + \dfrac{y}{3} = -6$
$\dfrac{x}{5} + \dfrac{y}{4} = -\dfrac{7}{4}$

Use the echelon method to solve each system of three equations in three unknowns. Check your answers.

21. $x + y + z = 2$
$2x + y - z = 5$
$x - y + z = -2$

22. $2x + y + z = 9$
$-x - y + z = 1$
$3x - y + z = 9$

23. $x + 3y + 4z = 14$
$2x - 3y + 2z = 10$
$3x - y + z = 9$

24. $4x - y + 3z = -2$
$3x + 5y - z = 15$
$-2x + y + 4z = 14$

25. $x + 2y + 3z = 8$
$3x - y + 2z = 5$
$-2x - 4y - 6z = 5$

26. $3x - 2y - 8z = 1$
$9x - 6y + 24z = -2$
$x - y + z = 1$

27. $2x - 4y + z = -4$
$x + 2y - z = 0$
$-x + y + z = 6$

28. $4x - 3y + z = 9$
$3x + 2y - 2z = 4$
$x - y + 3z = 5$

29. $x + 4y - z = 6$
$2x - y + z = 3$
$3x + 2y + 3z = 16$

30. $3x + y - z = 7$
$2x - 3y + z = -7$
$x - 4y + 3z = -6$

31. $5m + n - 3p = -6$
$2m + 3n + p = 5$
$-3m - 2n + 4p = 3$

32. $2r - 5s + 4t = -35$
$5r + 3s - t = 1$
$r + s + t = 1$

33. $a - 3b - 2c = -3$
$3a + 2b - c = 12$
$-a - b + 4c = 3$

34. $2x + 2y + 2z = 6$
$3x - 3y - 4z = -1$
$x + y + 3z = 11$

Solve each of the following systems of equations. Let z be the parameter.

35. $5x + 3y + 4z = 19$
$3x - y + z = -4$

36. $3x + y - z = 0$
$2x - y + 3z = -7$

37. $x + 2y + 3z = 11$
$2x - y + z = 2$

38. $-x + y - z = -7$
$2x + 3y + z = 7$

39. $x + y - z + 2w = -20$
$2x - y + z + w = 11$
$3x - 2y + z - 2w = 27$

40. $4x + 3y + z + 2w = 1$
$-2x - y + 2z + 3w = 0$
$x + 4y + z - w = 12$

Solve the following systems of equations.

41. $5x + 2y = 7$
$-2x + y = -10$
$x - 3y = 15$

42. $9x - 2y = -14$
$3x + y = -4$
$-6x - 2y = 8$

43. $x + 7y = 5$
$4x - 3y = 2$
$-x + 2y = 10$

44. $-3x - 2y = 11$
$x + 2y = -14$
$5x + y = -9$

45. $x + y = 2$
$y + z = 4$
$x + z = 3$
$y - z = 8$

46. $2x + y = 7$
$x + 3z = 5$
$y - 2z = 6$
$x + 4y = 10$

In Exercises 47 and 48, find the value of k for which each system has a single solution, and then find the solution of the system.

47. $4x + 8z = 12$
$2y - z = -2$
$3x + y + z = -1$
$x + y - kz = -7$

48. $2x + y + z = 4$
$3y + z = -2$
$x + y - z = -3$
$4x + kz = 8$

▬▬ APPLICATIONS

GENERAL

Income **49.** A working couple earned a total of $4352. The wife earned $64 per day; the husband earned $8 per day less. Find the number of days each worked if the total number of days worked by both was 72.

BUSINESS AND ECONOMICS

Production **50.** Midtown Manufacturing Company makes two products, plastic plates and plastic cups. Both require time on two machines: a batch of plates takes 1 hr on machine *A* and 2 hr on machine *B;* a batch of cups takes 3 hr on machine *A* and 1 hr on machine *B*. Both machines operate 15 hr a day. How many batches of each product can be produced in a day under these conditions?

Production **51.** A company produces two models of bicycles, model 201 and model 301. Model 201 requires 2 hr of assembly time, and model 301 requires 3 hr of asembly time. The parts for model 201 cost $25 per bike, and the parts for model 301 cost $30 per bike. If the company has a total of 34 hr of assembly time and $365 available per day for these two models, how many of each can be made in a day?

Investments **52.** Juanita invests $10,000, received from her grandmother, in three ways. With one part, she buys mutual funds that offer a return of 8% per year. She uses the second part, which amounts to twice the first, to buy government bonds at 9% per year. She puts the rest in the bank at 5% annual interest. The first year her investments bring a return of $830. How much did she invest in each way?

Loans **53.** To get the necessary funds for a planned expansion, a small company took out three loans totaling $25,000. The company was able to borrow some of the money at 16%. They borrowed $2000 more than one-half the amount of the 16% loan at 20%, and the rest at 18%. The total annual interest on the loans was $4440. How much did they borrow at each rate?

LIFE SCIENCES

Dietetics **54.** A hospital dietician is planning a special diet for a certain patient. The total amount per meal of food groups A, B, and C must equal 400 grams. The diet should include one-third as much of group A as of group B, and the sum of the amounts of group A and group C should equal twice the amount of group B. How many grams of each food group should be included?

2.2 SOLUTION OF LINEAR SYSTEMS BY THE GAUSS-JORDAN METHOD

In the last section we used the echelon method to solve linear systems of equations. Since the variables are always the same, we really need to keep track of just the coefficients and the constants. For example, look at the system solved in Example 4 of the previous section.

$$2x + y - z = 2$$
$$x + 3y + 2z = 1$$
$$x + y + z = 2.$$

This system can be written in an abbreviated form as

$$\begin{bmatrix} 2 & 1 & -1 & 2 \\ 1 & 3 & 2 & 1 \\ 1 & 1 & 1 & 2 \end{bmatrix}.$$

Such a rectangular array of numbers enclosed by brackets is called a **matrix** (plural: **matrices**). Each number in the array is an **element** or **entry.** To separate the constants in the last column of the matrix from the coefficients of the variables, use a vertical line, producing the following **augmented matrix.**

$$\begin{bmatrix} 2 & 1 & -1 & | & 2 \\ 1 & 3 & 2 & | & 1 \\ 1 & 1 & 1 & | & 2 \end{bmatrix}$$

The rows of the augmented matrix can be transformed in the same way as the equations of the system, since the matrix is just a shortened form of the system. The following **row operations** on the augmented matrix correspond to the transformations of systems of equations given earlier.

ROW OPERATIONS

> For any augmented matrix of a system of equations, the following operations produce the augmented matrix of an equivalent system:
>
> **1.** interchanging any two rows;
> **2.** multiplying the elements of a row by any nonzero real number;
> **3.** adding a multiple of the elements of one row to the corresponding elements of some other row.

Row operations, like the transformation of systems of equations, are reversible. If they are used to change matrix A to matrix B, then it is possible to use row operations to transform B back into A. In addition to their use in solving equations, row operations are very important in the simplex method to be described in Chapter 4.

In the examples in this section, we will use a shortened notation in color to show the row operation used. The notation R_1 indicates row 1 of the previous matrix, for example, and $-3R_1 + R_2$ means that row 1 is multiplied by -3 and added to row 2.

By the first row operation, the matrix

$$\begin{bmatrix} 1 & 3 & 5 & 6 \\ 0 & 1 & 2 & 3 \\ 2 & 1 & -2 & -5 \end{bmatrix} \quad \text{becomes} \quad \begin{bmatrix} 0 & 1 & 2 & 3 \\ 1 & 3 & 5 & 6 \\ 2 & 1 & -2 & -5 \end{bmatrix} \quad \begin{matrix} \text{Interchange} \\ R_1 \text{ and } R_2 \end{matrix}$$

by interchanging the first two rows. Row 3 is left unchanged.

The second row operation allows us to change

$$\begin{bmatrix} 1 & 3 & 5 & 6 \\ 0 & 1 & 2 & 3 \\ 2 & 1 & -2 & -5 \end{bmatrix} \quad \text{to} \quad \begin{bmatrix} -2 & -6 & -10 & -12 \\ 0 & 1 & 2 & 3 \\ 2 & 1 & -2 & -5 \end{bmatrix} \quad -2R_1$$

by multiplying the elements of row 1 of the original matrix by -2. Note that rows 2 and 3 are left unchanged.

Using the third row operation, we change

$$\begin{bmatrix} 1 & 3 & 5 & 6 \\ 0 & 1 & 2 & 3 \\ 2 & 1 & -2 & -5 \end{bmatrix} \quad \text{to} \quad \begin{bmatrix} -1 & 2 & 7 & 11 \\ 0 & 1 & 2 & 3 \\ 2 & 1 & -2 & -5 \end{bmatrix} \quad -1R_3 + R_1$$

by first multiplying each element in row 3 of the original matrix by -1 and then adding the results to the corresponding elements in the first row of that matrix. Work as follows.

$$\begin{bmatrix} 1 + 2(-1) & 3 + 1(-1) & 5 + (-2)(-1) & 6 + (-5)(-1) \\ 0 & 1 & 2 & 3 \\ 2 & 1 & -2 & -5 \end{bmatrix} = \begin{bmatrix} -1 & 2 & 7 & 11 \\ 0 & 1 & 2 & 3 \\ 2 & 1 & -2 & -5 \end{bmatrix}$$

Again rows 2 and 3 are left unchanged, *even though the elements of row 3 were used to transform row 1.*

The Gauss-Jordan Method   The *Gauss-Jordan method* is an extension of the echelon method of solving systems. Before the Gauss-Jordan method can be used, the system must be in proper form: the terms with variables should be on the left and the constants on the right in each equation, with the variables in the same order in each equation. The following example illustrates the use of the Gauss-Jordan method to solve a system of equations.

■■■ EXAMPLE 1
Solve the system

$$3x - 4y = 1 \qquad \textbf{(1)}$$
$$5x + 2y = 19. \qquad \textbf{(2)}$$

The system is already in the proper form for use of the Gauss-Jordan method. The solution procedure is parallel to the echelon method of Section 2.1, except for the last step. The echelon method is shown on the left below. The Gauss-Jordan method, shown on the right, uses the augmented matrix for the system.

Echelon Method	*Gauss-Jordan Method*

$3x - 4y = 1$ **(1)** $\begin{bmatrix} 3 & -4 & \bigm| & 1 \\ 5 & 2 & \bigm| & 19 \end{bmatrix}$
$5x + 2y = 19$ **(2)**

Multiply both sides of equation (1) by 1/3 so that x has a coefficient of 1.   Multiply each element of row 1 by 1/3.

$x - \dfrac{4}{3}y = \dfrac{1}{3}$ **(3)** $\begin{bmatrix} 1 & -\frac{4}{3} & \bigm| & \frac{1}{3} \\ 5 & 2 & \bigm| & 19 \end{bmatrix}$ $\frac{1}{3}R_1$

$5x + 2y = 19$ **(2)**

Eliminate x from equation (2) by adding -5 times equation (3) to equation (2).   Add -5 times the elements of row 1 to the elements of row 2.

$x - \dfrac{4}{3}y = \dfrac{1}{3}$ **(3)** $\begin{bmatrix} 1 & -\frac{4}{3} & \bigm| & \frac{1}{3} \\ 0 & \frac{26}{3} & \bigm| & \frac{52}{3} \end{bmatrix}$ $-5R_1 + R_2$

$\dfrac{26}{3}y = \dfrac{52}{3}$ **(4)**

Multiply both sides of equation (4) by 3/26 to get $y = 2$.   Multiply the elements of row 2 by 3/26.

$x - \dfrac{4}{3}y = \dfrac{1}{3}$ **(3)** $\begin{bmatrix} 1 & -\frac{4}{3} & \bigm| & \frac{1}{3} \\ 0 & 1 & \bigm| & 2 \end{bmatrix}$ $\frac{3}{26}R_2$

$y = 2$ **(5)**

Substitute $y = 2$ into equation (3) and solve for x to get $x = 3$.

$$x = 3 \qquad \textbf{(6)}$$
$$y = 2 \qquad \textbf{(5)}$$

The solution of the system is $(3, 2)$.

Multiply the elements of row 2 by 4/3 and add to the elements of row 1.

$$\begin{bmatrix} 1 & 0 & | & 3 \\ 0 & 1 & | & 2 \end{bmatrix} \qquad \tfrac{4}{3}R_2 + R_1$$

The solution of the system, $(3, 2)$, can be read directly from the last column of the final matrix.

In both cases, check the solution by substitution in the equations of the original system. ▬

When the Gauss-Jordan method is used to solve a system, the final matrix always will have zeros above and below the diagonal of 1's on the left of the vertical bar. To transform the matrix, it is best to work column by column from left to right. For each column, first get a 1 in the proper position, and then perform the steps that give zeros in the remainder of the column. With dependent or inconsistent systems, it will not be possible to get the complete diagonal of 1's.

▬▬ **EXAMPLE 2**

Use the Gauss-Jordan method to solve the system

$$x + 5z = -6 + y$$
$$3x + 3y = 10 + z$$
$$x + 3y + 2z = 5.$$

First, rewrite the system in proper form as follows.

$$x - y + 5z = -6$$
$$3x + 3y - z = 10$$
$$x + 3y + 2z = 5$$

Begin the solution by writing the augmented matrix of the linear system.

$$\begin{bmatrix} 1 & -1 & 5 & | & -6 \\ 3 & 3 & -1 & | & 10 \\ 1 & 3 & 2 & | & 5 \end{bmatrix}$$

Row transformations will be used to rewrite this matrix in the form

$$\begin{bmatrix} 1 & 0 & 0 & | & m \\ 0 & 1 & 0 & | & n \\ 0 & 0 & 1 & | & p \end{bmatrix},$$

where m, n, and p are real numbers. From this final form of the matrix, the solution can be read: $x = m$, $y = n$, $z = p$, or (m, n, p).

In the matrix of the linear system above, there is already a 1 for the first element in column 1. To get 0 for the second element in column 1, multiply each element in row 1 by -3 and add the results to the corresponding elements in row 2 (using the third row operation.

$$\begin{bmatrix} 1 & -1 & 5 & | & -6 \\ 0 & 6 & -16 & | & 28 \\ 1 & 3 & 2 & | & 5 \end{bmatrix} \qquad -3R_1 + R_2$$

To change the last element in column 1 to 0, multiply each element in row 1 by -1 and add each result to the corresponding elements of row 3 (again using the third row operation).

$$\begin{bmatrix} 1 & -1 & 5 & | & -6 \\ 0 & 6 & -16 & | & 28 \\ 0 & 4 & -3 & | & 11 \end{bmatrix} \qquad -1R_1 + R_3$$

This transforms the first column. Transform the second and third columns in a similar manner.

$$\begin{bmatrix} 1 & -1 & 5 & | & -6 \\ 0 & 1 & -\frac{8}{3} & | & \frac{14}{3} \\ 0 & 4 & -3 & | & 11 \end{bmatrix} \qquad \frac{1}{6}R_2$$

$$\begin{bmatrix} 1 & 0 & \frac{7}{3} & | & -\frac{4}{3} \\ 0 & 1 & -\frac{8}{3} & | & \frac{14}{3} \\ 0 & 4 & -3 & | & 11 \end{bmatrix} \qquad R_2 + R_1$$

$$\begin{bmatrix} 1 & 0 & \frac{7}{3} & | & -\frac{4}{3} \\ 0 & 1 & -\frac{8}{3} & | & \frac{14}{3} \\ 0 & 0 & \frac{23}{3} & | & -\frac{23}{3} \end{bmatrix} \qquad -4R_2 + R_3$$

$$\begin{bmatrix} 1 & 0 & \frac{7}{3} & | & -\frac{4}{3} \\ 0 & 1 & -\frac{8}{3} & | & \frac{14}{3} \\ 0 & 0 & 1 & | & -1 \end{bmatrix} \qquad \frac{3}{23}R_3$$

$$\begin{bmatrix} 1 & 0 & 0 & | & 1 \\ 0 & 1 & -\frac{8}{3} & | & \frac{14}{3} \\ 0 & 0 & 1 & | & -1 \end{bmatrix} \qquad -\frac{7}{3}R_3 + R_1$$

$$\begin{bmatrix} 1 & 0 & 0 & | & 1 \\ 0 & 1 & 0 & | & 2 \\ 0 & 0 & 1 & | & -1 \end{bmatrix} \qquad \frac{8}{3}R_3 + R_2$$

The linear system associated with the final augmented matrix is

$$\begin{aligned} x \quad\quad &= 1 \\ y \quad &= 2 \\ z &= -1, \end{aligned}$$

and the solution is $(1, 2, -1)$. ∎

In summary, the Gauss-Jordan method of solving a linear system requires the following steps.

THE GAUSS-JORDAN
METHOD OF SOLVING A
LINEAR SYSTEM

1. Write each equation so that variable terms are in the same order on the left side of the equals sign and constants are on the right.
2. Write the augmented matrix that corresponds to the system.
3. Use row operations to transform the first column so that the first element is 1 and the remaining elements are 0.
4. Use row operations to transform the second column so that the second element is 1 and the remaining elements are 0.
5. Use row operations to transform the third column so that the third element is 1 and the remaining elements are 0.
6. Continue in this way until the last row is written in the form
$$[0 \quad 0 \quad 0 \ldots 0 \quad 1 \mid k],$$
where k is a constant.

If some of the given steps cannot be performed, the solution cannot be completed with the Gauss-Jordan method. The next examples illustrate how such cases may be handled.

 EXAMPLE 3

Use the Gauss-Jordan method to solve the system

$$x + y = 2$$
$$2x + 2y = 5.$$

Begin by writing the augmented matrix.

$$\begin{bmatrix} 1 & 1 & \mid & 2 \\ 2 & 2 & \mid & 5 \end{bmatrix}$$

The first element in column 1 is already 1. To get a 0 for the second element in column 1, multiply the numbers in row 1 by -2 and add the results to the corresponding elements in row 2.

$$\begin{bmatrix} 1 & 1 & \mid & 2 \\ 0 & 0 & \mid & 1 \end{bmatrix} \qquad -2R_1 + R_2$$

The next step is to get a 1 for the second element in column 2. Since this is impossible, we cannot go further with the method. This matrix corresponds to the system

$$x + y = 2$$
$$0x + 0y = 1.$$

Since the second equation is $0 = 1$, the system is inconsistent and therefore has no solution. The row $[0 \quad 0 \mid 1]$ is a signal that the given system is inconsistent. ▬

▬ EXAMPLE 4

Use the Gauss-Jordan method to solve the system

$$x + 2y - z = 0$$
$$3x - y + z = 6$$
$$-2x - 4y + 2z = 0.$$

The augmented matrix is

$$\begin{bmatrix} 1 & 2 & -1 & | & 0 \\ 3 & -1 & 1 & | & 6 \\ -2 & -4 & 2 & | & 0 \end{bmatrix}.$$

The first element in column 1 is 1. Use row operations to get zeros in the rest of column 1.

$$\begin{bmatrix} 1 & 2 & -1 & | & 0 \\ 0 & -7 & 4 & | & 6 \\ -2 & -4 & 2 & | & 0 \end{bmatrix} \qquad -3R_1 + R_2$$

$$\begin{bmatrix} 1 & 2 & -1 & | & 0 \\ 0 & -7 & 4 & | & 6 \\ 0 & 0 & 0 & | & 0 \end{bmatrix} \qquad 2R_1 + R_3$$

The row of zeros in the last matrix is a signal that two of the equations (the first and last) are dependent. To continue, multiply row 2 by $-1/7$.

$$\begin{bmatrix} 1 & 2 & -1 & | & 0 \\ 0 & 1 & -\frac{4}{7} & | & -\frac{6}{7} \\ 0 & 0 & 0 & | & 0 \end{bmatrix} \qquad -\frac{1}{7}R_2$$

Finally, add -2 times row 2 to row 1.

$$\begin{bmatrix} 1 & 0 & \frac{1}{7} & | & \frac{12}{7} \\ 0 & 1 & -\frac{4}{7} & | & -\frac{6}{7} \\ 0 & 0 & 0 & | & 0 \end{bmatrix} \qquad -2R_2 + R_1$$

This is as far as we can go with the Gauss-Jordan method. To complete the solution, write the equations that correspond to the first two lines of the matrix.

$$x \quad + \frac{1}{7}z = \frac{12}{7}$$

$$y - \frac{4}{7}z = -\frac{6}{7}$$

Since both equations involve z, solve the first equation for x and the second equation for y to get

$$x = -\frac{1}{7}z + \frac{12}{7} \qquad \text{and} \qquad y = \frac{4}{7}z - \frac{6}{7}.$$

As shown in the previous section, for any real number a, let $z = a$ to get the general solution

$$\left\{ \left(-\frac{1}{7}a + \frac{12}{7}, \ \frac{4}{7}a - \frac{6}{7}, \ a \right) \right\}. \quad \blacksquare$$

Although the examples have used only systems with two equations in two unknowns or three equations in three unknowns, the Gauss-Jordan method can be used for any system with n equations and n unknowns. In fact, it can be used with n equations and m unknowns, as in Example 4. The system in Example 4 actually had just two equations with three unknowns. The method does become tedious even with three equations in three unknowns; however, it is very suitable for use by computers. A computer can produce the solution to a fairly large system very quickly.*

2.2 EXERCISES

Write the augmented matrix for each of the following systems. **Do not solve.**

1. $2x + 3y = 11$
$\quad\ x + 2y = \ \ 8$

2. $3x + 5y = -13$
$\quad\ 2x + 3y = \ \ -9$

3. $x \qquad = 6 - 5y$
$\qquad\quad y = 1$

4. $\qquad\quad 7y = \ \ 1 - 2x$
$\qquad 5x \qquad = -15$

5. $2x + \ y + \ z = \ \ \ 3$
$\quad\ 3x - 4y + 2z = -7$
$\qquad\ x + \ y + \ z = \ \ \ 2$

6. $4x - 2y + 3z = 4$
$\quad\ 3x + 5y + \ z = 7$
$\quad\ 5x - \ y + 4z = 7$

7. $\ \ y = \ \ 2 - x$
$\ 2y = -4 - z$
$\ \ z = \ \ 2$

8. $x = 6$
$\ y = 2 - 2z$
$\ x = 6 + 3z$

9. $x = \ \ 5$
$\ y = -2$
$\ z = \ \ 3$

10. $\qquad x = 8$
$\quad\ y + z = 6$
$\qquad\quad z = 2$

Write the system of equations associated with each of the following augmented matrices. **Do not solve.**

11. $\begin{bmatrix} 1 & 0 & 2 \\ 0 & 1 & 3 \end{bmatrix}$

12. $\begin{bmatrix} 1 & 0 & 5 \\ 0 & 1 & -3 \end{bmatrix}$

13. $\begin{bmatrix} 2 & 1 & 1 \\ 3 & -2 & -9 \end{bmatrix}$

14. $\begin{bmatrix} 1 & -5 & -18 \\ 6 & 2 & 20 \end{bmatrix}$

15. $\begin{bmatrix} 1 & 0 & 0 & 2 \\ 0 & 1 & 0 & 3 \\ 0 & 0 & 1 & -2 \end{bmatrix}$

16. $\begin{bmatrix} 1 & 0 & 0 & 4 \\ 0 & 1 & 0 & 2 \\ 0 & 0 & 1 & 3 \end{bmatrix}$

---

*See D. R. Coscia, *Computer Applications for Finite Mathematics and Calculus,* Scott, Foresman and Company, 1986.

17. $\begin{bmatrix} 1 & 4 & | & 10 \\ 0 & 0 & | & 0 \end{bmatrix}$

18. $\begin{bmatrix} 1 & 0 & | & 9 \\ 0 & 0 & | & 6 \end{bmatrix}$

19. $\begin{bmatrix} 1 & 0 & 5 & | & 15 \\ 0 & 1 & 3 & | & 12 \\ 0 & 0 & 0 & | & 0 \end{bmatrix}$

20. $\begin{bmatrix} 1 & 0 & 2 & | & 10 \\ 0 & 1 & 1 & | & 8 \\ 0 & 0 & 0 & | & 0 \end{bmatrix}$

Use the Gauss-Jordan method to solve the following systems of equations.

21. $x + y = 5$
$x - y = -1$

22. $x + 2y = 5$
$2x + y = -2$

23. $3x + 3y = -9$
$2x - 5y = -6$

24. $3x - 2y = 4$
$3x + y = -2$

25. $2x = 10 + 3y$
$2y = 5 - 2x$

26. $y = 5 - 4x$
$2x = 3 - y$

27. $2x - 5y = 10$
$4x - 5y = 15$

28. $4x - 2y = 3$
$-2x + 3y = 1$

29. $2x - 3y = 2$
$4x - 6y = 1$

30. $x + 2y = 1$
$2x + 4y = 3$

31. $6x - 3y = 1$
$-12x + 6y = -2$

32. $x - y = 1$
$-x + y = -1$

33. $2x + 2y = -2$
$y + z = 4$
$x + z = 1$

34. $x - z = -3$
$y + z = 9$
$x + z = 7$

35. $4x + 4y - 4z = 24$
$2x - y + z = -9$
$x - 2y + 3z = 1$

36. $x + 3y - 6z = 7$
$2x - y + 2z = 0$
$x + y + 2z = -1$

37. $y = x - 1$
$y = 6 + z$
$z = -1 - x$

38. $x = 1 - y$
$2x = z$
$2z = -2 - y$

39. $3x - 6y + 3z = 15$
$2x + y - z = 2$
$-2x + 4y - 2z = 2$

40. $3x + 5y - z = 0$
$4x - y + 2z = 1$
$-6x - 10y + 2z = 0$

41. $2x + 3y + z = 9$
$4x - y + 3z = -1$
$6x + 2y - 4z = -8$

42. $3x + 2y - z = -16$
$6x - 4y + 3z = 12$
$3x + 3y + z = -11$

43. $5x - 4y + 2z = 4$
$10x + 3y - z = 27$
$15x - 5y + 3z = 25$

44. $4x - 2y - 3z = -23$
$-4x + 3y + z = 11$
$8x - 5y + 4z = 6$

45. $x + 2y - w = 3$
$2x + 4z + 2w = -6$
$x + 2y - z = 6$
$2x - y + z + w = -3$

46. $x + 3y - 2z - w = 9$
$2x + 4y + 2w = 10$
$-3x - 5y + 2z - w = -15$
$x - y - 3z + 2w = 6$

::: APPLICATIONS

SOCIAL SCIENCES

Traffic Control   **47.** At rush hours, substantial traffic congestion is encountered at the traffic intersections shown in the figure. (The streets are one-way, as shown by the arrows.)

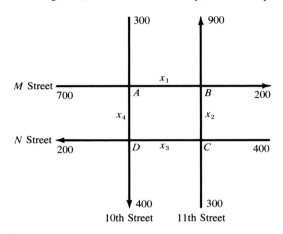

The city wishes to improve the signals at these corners so as to speed the flow of traffic. The traffic engineers first gather data. As the figure shows, 700 cars per hour come down M Street to intersection A, and 300 cars per hour come down 10th Street to intersection A. A total of x_1 of these cars leave A on M Street, and x_4 cars leave A on 10th Street. The number of cars entering A must equal the number leaving, so that

$$x_1 + x_4 = 700 + 300$$

or

$$x_1 + x_4 = 1000.$$

For intersection B, x_1 cars enter on M Street and x_2 on 11th Street. The figure shows that 900 cars leave B on 11th and 200 on M. We have

$$x_1 + x_2 = 900 + 200$$
$$x_1 + x_2 = 1100.$$

At intersection C, 400 cars enter on N Street and 300 on 11th Street, while x_2 leave on 11th Street and x_3 on N Street. This gives

$$x_2 + x_3 = 400 + 300$$
$$x_2 + x_3 = 700.$$

Finally, intersection D has x_3 cars entering on N and x_4 on 10th. There are 400 leaving D on 10th and 200 on N, so that

$$x_3 + x_4 = 400 + 200$$
$$x_3 + x_4 = 600.$$

(a) Use the four equations to set up an augmented matrix, and then use the Gauss-Jordan method to solve it. (*Hint:* Keep going until you get a row of all zeros.)

(b) Since you got a row of all zeros, the system of equations does not have a unique solution. Write three equations, corresponding to the three nonzero rows of the matrix.

(c) Solve each of the equations for x_4.

(d) One of your equations should have been $x_4 = 1000 - x_1$. What is the largest possible value of x_1 so that x_4 is not negative? What is the largest value of x_4 so that x_1 is not negative?

(e) Your second equation should have been $x_4 = x_2 - 100$. Find the smallest possible value of x_2 so that x_4 is not negative.

(f) For the third equation, $x_4 = 600 - x_3$, find the largest possible values of x_3 and x_4 so that neither variable is negative.

(g) Look at your answers for parts (d)–(f). What is the maximum value of x_4 so that all the equations are satisfied and all variables are nonnegative? Of x_3? Of x_2? Of x_1?

BUSINESS AND ECONOMICS

Transportation **48.** A manufacturer purchases a part for use at both of its plants—one at Roseville, California, the other at Akron, Ohio. The part is available in limited quantities from two suppliers. Each supplier has 75 units available. The Roseville plant needs 40 units, and the Akron plant requires 75 units. The first supplier charges $70 per unit delivered to Roseville and $90 per unit delivered to Akron. Corresponding costs from the second supplier are $80 and $120. The manufacturer wants to order a total of 75 units from the first, less expensive, supplier, with the remaining 40 units to come from the second supplier. If the company spends $10,750 to purchase the required number of units for the two plants, find the number of units that should be purchased from each supplier for each plant as follows:

(a) Assign variables to the four unknowns.

(b) Write a system of five equations with the four variables. (Not all equations will involve all four variables.)

(c) Use the Gauss-Jordan method to solve the system of equations.

Transportation **49.** An auto manufacturer sends cars from two plants, I and II, to dealerships A and B located in a midwestern city. Plant I has a total of 28 cars to send, and plant II has 8. Dealer A needs 20 cars, and dealer B needs 16. Transportation costs based on the distance of each dealership from each plant are $220 from I to A, $300 from I to B, $400 from II to A, and $180 from II to B. The manufacturer wants to limit transportation costs to $10,640. How many cars should be sent from each plant to each of the two dealerships? Use the Gauss-Jordan method to find the solution.

Transportation **50.** A knitting shop ordered yarn from three suppliers, I, II, and III. One month the shop ordered a total of 100 units of yarn from these suppliers. The delivery costs were $80, $50, and $65 per unit for the orders from suppliers I, II, and III, respectively, with total delivery costs of $5990. The shop ordered the same amount from suppliers I and III. How many units were ordered from each supplier? Use the Gauss-Jordan method to find the solution.

FOR THE COMPUTER

Solve the linear systems in Exercises 51–54.

51. $2.1x + 3.5y + 9.4z = 15.6$
$6.8x - 1.5y + 7.5z = 26.4$
$3.7x + 2.5y - 6.1z = 18.7$

52. $9.03x - 5.91y + 2.68z = 29.5$
$3.94x + 6.82y + 1.53z = 35.4$
$2.79x + 1.68y - 6.23z = 12.1$

53. $10.47x + 3.52y + 2.58z - 6.42w = 218.65$
$8.62x - 4.93y - 1.75z + 2.83w = 157.03$
$4.92x + 6.83y - 2.97z + 2.65w = 462.3$
$2.86x + 19.1\ y - 6.24z - 8.73w = 398.4$

54. $28.6x + 94.5y + 16.0z - 2.94w = 198.3$
$16.7x + 44.3y - 27.3z + 8.9w = 254.7$
$12.5x - 38.7y + 92.5z + 22.4w = 562.7$
$40.1x - 28.3y + 17.5z - 10.2w = 375.4$

In Exercises 55–58, write a system of equations and then solve it.

Mixing Plant Foods   **55.** Natural Brand plant food is made from three chemicals. The mix must include 10.8% of the first chemical, and the other two chemicals must be in a ratio of 4 to 3 as measured by weight. How much of each chemical is required to make 750 kg of the plant food?

Bacterial Food Requirements   **56.** Three species of bacteria are fed three foods, I, II, and III. A bacterium of the first species consumes 1.3 units each of foods I and II and 2.3 units of food III each day. A bacterium of the second species consumes 1.1 units of food I, 2.4 units of food II, and 3.7 units of food III each day. A bacterium of the third species consumes 8.1 units of I, 2.9 units of II, and 5.1 units of III each day. If 16,000 units of I, 28,000 units of II, and 44,000 units of III are supplied each day, how many of each species can be maintained in this environment?

Fish Food Requirements   **57.** A lake is stocked each spring with three species of fish, A, B, and C. Three foods, I, II, and III, are available in the lake. Each fish of species A requires an average of 1.32 units of food I, 2.9 units of food II, and 1.75 units of food III each day. Species B fish each require 2.1 units of food I, .95 unit of food II, and .6 unit of food III daily. Species C fish require .86, 1.52, and 2.01 units of I, II, and III per day, respectively. If 490 units of food I, 897 units of food II, and 653 units of food III are available daily, how many of each species should be stocked?

Packaging   **58.** A company produces three combinations of mixed vegetables that sell in 1-kg packages. Italian style combines .3 kg of zucchini, .3 of broccoli, and .4 of carrots. French style combines .6 kg of broccoli and .4 of carrots. Oriental style combines .2 kg of zucchini, .5 of broccoli, and .3 of carrots. The company has a stock of 16,200 kg of zucchini, 41,400 kg of broccoli, and 29,400 kg of carrots. How many packages of each style should it prepare to use up existing supplies?

2.3 BASIC MATRIX OPERATIONS

As shown in Section 2.2, matrices can be used to represent systems of linear equations. The study of matrices has been of interest to mathematicians for some time. Recently the use of matrices has gained increasing importance in the fields of management, natural science, and social science because matrices provide a convenient way to organize data, as Example 1 demonstrates.

■■■ EXAMPLE 1

The EZ Life Company manufactures sofas and armchairs in three models, A, B, and C. The company has regional warehouses in New York, Chicago, and San Francisco. In its August shipment, the company sends 10 model-A sofas, 12 model-B sofas, 5 model-C sofas, 15 model-A chairs, 20 model-B chairs, and 8 model-C chairs to each warehouse.

To organize this data, we might first list it as follows.

Sofas	10 model-A	12 model-B	5 model-C
Chairs	15 model-A	20 model-B	8 model-C

Alternatively, we might tabulate the data in a chart.

		Model		
		A	B	C
Furniture Type	Sofas	10	12	5
	Chairs	15	20	8

With the understanding that the numbers in each row refer to the furniture type (sofa, chair) and the numbers in each column refer to the model (A, B, C), the same information can be given by a matrix, as follows.

$$M = \begin{bmatrix} 10 & 12 & 5 \\ 15 & 20 & 8 \end{bmatrix}$$ ■■

Matrices often are named with capital letters, as in Example 1. A matrix is classified by its **order** (or **dimension**), that is, by the number of rows and columns that it contains. For example, matrix M above has two rows and three columns. This matrix is of order 2×3 (read "2 by 3") or dimension 2×3. By definition, a matrix with m rows and n columns is of order $m \times n$. The number of rows is always given first.

■■■ EXAMPLE 2

(a) The matrix $\begin{bmatrix} 6 & 5 \\ 3 & 4 \\ 5 & -1 \end{bmatrix}$ is of order 3×2.

(b) $\begin{bmatrix} 5 & 8 & 9 \\ 0 & 5 & -3 \\ -4 & 0 & 5 \end{bmatrix}$ is of order 3×3.

(c) $[1 \quad 6 \quad 5 \quad -2 \quad 5]$ is of order 1×5.

(d) $\begin{bmatrix} 3 \\ -5 \\ 0 \\ 2 \end{bmatrix}$ is of order 4 × 1. ▄

A matrix with the same number of rows as columns is called a **square matrix.** The matrix in Example 2(b) is a square matrix.

A matrix containing only one row is called a **row matrix** or a **row vector.** The matrix in Example 2(c) is a row matrix, as are

$$[5 \quad 8], \qquad [6 \quad -9 \quad 2], \qquad \text{and} \qquad [-4 \quad 0 \quad 0 \quad 0].$$

A matrix of only one column, as in Example 2(d), is a **column matrix** or a **column vector.**

Two matrices are **equal** if they are of the same order and if each pair of corresponding elements is equal. By this definition, the matrices

$$\begin{bmatrix} 2 & 1 \\ 3 & -5 \end{bmatrix} \qquad \text{and} \qquad \begin{bmatrix} 1 & 2 \\ -5 & 3 \end{bmatrix}$$

are not equal (even though they contain the same elements and are of the same order) since the corresponding elements differ.

▄▄ EXAMPLE 3

(a) From the definition of matrix equality given above, the only way that the statement

$$\begin{bmatrix} 2 & 1 \\ p & q \end{bmatrix} = \begin{bmatrix} x & y \\ -1 & 0 \end{bmatrix}$$

can be true is if $2 = x$, $1 = y$, $p = -1$, and $q = 0$.

(b) The statement

$$\begin{bmatrix} x \\ y \end{bmatrix} = \begin{bmatrix} 1 \\ 4 \\ 0 \end{bmatrix}$$

can never be true, since the two matrices are of different order. (One is 2 × 1 and the other is 3 × 1.) ▄

Addition   The matrix given in Example 1,

$$M = \begin{bmatrix} 10 & 12 & 5 \\ 15 & 20 & 8 \end{bmatrix},$$

shows the August shipment from the EZ Life plant to each of its warehouses.

If matrix N below gives the September shipment to the New York warehouse, what is the total shipment of each item of furniture to the New York warehouse for these two months?

$$N = \begin{bmatrix} 45 & 35 & 20 \\ 65 & 40 & 35 \end{bmatrix}$$

If 10 model-A sofas were shipped in August and 45 in September, then altogether $10 + 45 = 55$ model-A sofas were shipped in the two months. The other corresponding entries can be added in a similar way to get a new matrix, Q, that represents the total shipment for the two months.

$$Q = \begin{bmatrix} 55 & 47 & 25 \\ 80 & 60 & 43 \end{bmatrix}$$

It is convenient to refer to Q as the sum of M and N.

The way these two matrices were added illustrates the following definition of addition of matrices.

ADDITION OF MATRICES

> The **sum** of two $m \times n$ matrices X and Y is the $m \times n$ matrix $X + Y$ in which each element is the sum of the corresponding elements of X and Y.

It is important to remember that only matrices with the same order or dimension can be added.

■ EXAMPLE 4

Find each sum if possible.

(a) $\begin{bmatrix} 5 & -6 \\ 8 & 9 \end{bmatrix} + \begin{bmatrix} -4 & 6 \\ 8 & -3 \end{bmatrix} = \begin{bmatrix} 5 + (-4) & -6 + 6 \\ 8 + 8 & 9 + (-3) \end{bmatrix} = \begin{bmatrix} 1 & 0 \\ 16 & 6 \end{bmatrix}$

(b) The matrices

$$A = \begin{bmatrix} 5 & 8 \\ 6 & 2 \end{bmatrix} \quad \text{and} \quad B = \begin{bmatrix} 3 & 9 & 1 \\ 4 & 2 & 5 \end{bmatrix}$$

are of different orders. Therefore, the sum $A + B$ does not exist. ■

■ EXAMPLE 5

The September shipments from the EZ Life Company to the New York, San Francisco, and Chicago warehouses are given in matrices N, S, and C below.

$$N = \begin{bmatrix} 45 & 35 & 20 \\ 65 & 40 & 35 \end{bmatrix}, \quad S = \begin{bmatrix} 30 & 32 & 28 \\ 43 & 47 & 30 \end{bmatrix}, \quad C = \begin{bmatrix} 22 & 25 & 38 \\ 31 & 34 & 35 \end{bmatrix}$$

What was the total amount shipped to the three warehouses in September?

The total of the September shipments is represented by the sum of the three matrices N, S, and C.

$$N + S + C = \begin{bmatrix} 45 & 35 & 20 \\ 65 & 40 & 35 \end{bmatrix} + \begin{bmatrix} 30 & 32 & 28 \\ 43 & 47 & 30 \end{bmatrix} + \begin{bmatrix} 22 & 25 & 38 \\ 31 & 34 & 35 \end{bmatrix}$$

$$= \begin{bmatrix} 97 & 92 & 86 \\ 139 & 121 & 100 \end{bmatrix}.$$

For example, this sum shows that the total number of model-C sofas shipped to the three warehouses in September was 86. ■

The **additive inverse** (or **negative**) of a matrix X is the matrix $-X$ in which each element is the additive inverse of the corresponding element of X. If

$$A = \begin{bmatrix} 1 & 2 & 3 \\ 0 & -1 & 5 \end{bmatrix} \quad \text{and} \quad B = \begin{bmatrix} -2 & 3 & 0 \\ 1 & -7 & 2 \end{bmatrix},$$

then by the definition of the additive inverse of a matrix,

$$-A = \begin{bmatrix} -1 & -2 & -3 \\ 0 & 1 & -5 \end{bmatrix} \quad \text{and} \quad -B = \begin{bmatrix} 2 & -3 & 0 \\ -1 & 7 & -2 \end{bmatrix}.$$

By the definition of matrix addition, for each matrix X the sum $X + (-X)$ is a **zero matrix,** O, whose elements are all zeros. There is an $m \times n$ zero matrix for each pair of values of m and n. Such a matrix serves as an $m \times n$ **additive identity,** similar to the additive identity 0 for any real number. Zero matrices have the following identity property: if O is an $m \times n$ zero matrix, and A is any $m \times n$ matrix, then

$$A + O = O + A = A.$$

Compare this with the identity property for real numbers: for any real number a, $a + 0 = 0 + a = a$. Exercises 33–37 give other properties of matrices that are parallel to the properties of real numbers.

Subtraction   The **subtraction** of matrices is defined in a manner comparable to subtraction of real numbers.

SUBTRACTION OF MATRICES

For two $m \times n$ matrices X and Y, the **difference** of X and Y, or $X - Y$, is the matrix defined by

$$X - Y = X + (-Y).$$

With A, B, and $-B$ as defined above,

$$A - B = A + (-B) = \begin{bmatrix} 1 & 2 & 3 \\ 0 & -1 & 5 \end{bmatrix} + \begin{bmatrix} 2 & -3 & 0 \\ -1 & 7 & -2 \end{bmatrix}$$

$$= \begin{bmatrix} 3 & -1 & 3 \\ -1 & 6 & 3 \end{bmatrix}.$$

According to this definition, matrix subtraction can be performed by subtracting corresponding elements.

■ EXAMPLE 6

(a) $[8 \quad 6 \quad -4] - [3 \quad 5 \quad -8] = [5 \quad 1 \quad 4]$

(b) The matrices

$$\begin{bmatrix} -2 & 5 \\ 0 & 1 \end{bmatrix} \quad \text{and} \quad \begin{bmatrix} 3 \\ 5 \end{bmatrix}$$

have different orders and cannot be subtracted. ■

■ EXAMPLE 7

During September the Chicago warehouse of the EZ Life Company shipped out the following numbers of each model.

$$K = \begin{bmatrix} 5 & 10 & 8 \\ 11 & 14 & 15 \end{bmatrix}$$

What was the Chicago warehouse inventory on October 1, taking into account only the number of items received and sent out during the month?

The number of each kind of item received during September is given by matrix C from Example 5; the number of each model sent out during September is given by matrix K. The October 1 inventory will be represented by the matrix $C - K$:

$$\begin{bmatrix} 22 & 25 & 38 \\ 31 & 34 & 35 \end{bmatrix} - \begin{bmatrix} 5 & 10 & 8 \\ 11 & 14 & 15 \end{bmatrix} = \begin{bmatrix} 17 & 15 & 30 \\ 20 & 20 & 20 \end{bmatrix}. \quad ■$$

■ 2.3 EXERCISES

Mark each statement as true *or* false. *If false, tell why.*

1. $\begin{bmatrix} 1 & 3 \\ 5 & 7 \end{bmatrix} = \begin{bmatrix} 1 & 5 \\ 3 & 7 \end{bmatrix}$

2. $\begin{bmatrix} 1 \\ 2 \\ 3 \end{bmatrix} = [1 \quad 2 \quad 3]$

3. $\begin{bmatrix} x \\ y \end{bmatrix} = \begin{bmatrix} 3 \\ 5 \end{bmatrix}$ if $x = 3$ and $y = 5$.

4. $\begin{bmatrix} 3 & 5 & 2 & 8 \\ 1 & -1 & 4 & 0 \end{bmatrix}$ is a 4 × 2 matrix.

5. $\begin{bmatrix} 1 & 9 & -4 \\ 3 & 7 & 2 \\ -1 & 1 & 0 \end{bmatrix}$ is a square matrix.

6. $\begin{bmatrix} 2 & 4 & -1 \\ 3 & 7 & 5 \\ 0 & 0 & 0 \end{bmatrix} = \begin{bmatrix} 2 & 4 & -1 \\ 3 & 7 & 5 \end{bmatrix}$

Find the order of each matrix. Identify any square, column, or row matrices.

7. $\begin{bmatrix} -4 & 8 \\ 2 & 3 \end{bmatrix}$

8. $\begin{bmatrix} -9 & 6 & 2 \\ 4 & 1 & 8 \end{bmatrix}$

9. $\begin{bmatrix} -6 & 8 & 0 & 0 \\ 4 & 1 & 9 & 2 \\ 3 & -5 & 7 & 1 \end{bmatrix}$

10. $[8 \quad -2 \quad 4 \quad 6 \quad 3]$

11. $\begin{bmatrix} 2 \\ 4 \end{bmatrix}$

12. $[-9]$

Find the values of the variables in each equation.

13. $\begin{bmatrix} 2 & 1 \\ 4 & 8 \end{bmatrix} = \begin{bmatrix} x & 1 \\ y & z \end{bmatrix}$

14. $\begin{bmatrix} -5 \\ y \end{bmatrix} = \begin{bmatrix} -5 \\ 8 \end{bmatrix}$

15. $\begin{bmatrix} x + 6 & y + 2 \\ 8 & 3 \end{bmatrix} = \begin{bmatrix} -9 & 7 \\ 8 & k \end{bmatrix}$

16. $\begin{bmatrix} 9 & 7 \\ r & 0 \end{bmatrix} = \begin{bmatrix} m - 3 & n + 5 \\ 8 & 0 \end{bmatrix}$

17. $\begin{bmatrix} -7 + z & 4r & 8s \\ 6p & 2 & 5 \end{bmatrix} + \begin{bmatrix} -9 & 8r & 3 \\ 2 & 5 & 4 \end{bmatrix} = \begin{bmatrix} 2 & 36 & 27 \\ 20 & 7 & 12a \end{bmatrix}$

18. $\begin{bmatrix} a + 2 & 3z + 1 & 5m \\ 4k & 0 & 3 \end{bmatrix} + \begin{bmatrix} 3a & 2z & 5m \\ 2k & 5 & 6 \end{bmatrix} = \begin{bmatrix} 10 & -14 & 80 \\ 10 & 5 & 9 \end{bmatrix}$

Perform the indicated operations where possible.

19. $\begin{bmatrix} 1 & 2 & 5 & -1 \\ 3 & 0 & 2 & -4 \end{bmatrix} + \begin{bmatrix} 8 & 10 & -5 & 3 \\ -2 & -1 & 0 & 0 \end{bmatrix}$

20. $\begin{bmatrix} 1 & 5 \\ 2 & -3 \\ 3 & 7 \end{bmatrix} + \begin{bmatrix} 2 & 3 \\ 8 & 5 \\ -1 & 9 \end{bmatrix}$

21. $\begin{bmatrix} 1 & 5 & 7 \\ 2 & 2 & 3 \end{bmatrix} + \begin{bmatrix} 4 & 8 & -7 \\ 1 & -1 & 5 \end{bmatrix}$

22. $\begin{bmatrix} 2 & 4 \\ -8 & 1 \end{bmatrix} + \begin{bmatrix} 9 & -3 \\ 8 & 5 \end{bmatrix}$

23. $\begin{bmatrix} 1 & 3 & -2 \\ 4 & 7 & 1 \end{bmatrix} + \begin{bmatrix} 3 & 0 \\ 6 & 4 \\ -5 & 2 \end{bmatrix}$

24. $\begin{bmatrix} 1 & 3 & -2 \\ 4 & 7 & 1 \end{bmatrix} - \begin{bmatrix} 3 & 6 & -5 \\ 0 & 4 & 2 \end{bmatrix}$

25. $\begin{bmatrix} 2 & 8 & 12 & 0 \\ 7 & 4 & -1 & 5 \\ 1 & 2 & 0 & 10 \end{bmatrix} - \begin{bmatrix} 1 & 3 & 6 & 9 \\ 2 & -3 & -3 & 4 \\ 8 & 0 & -2 & 17 \end{bmatrix}$

26. $\begin{bmatrix} 2 & 1 \\ 5 & -3 \\ -7 & 2 \\ 9 & 0 \end{bmatrix} + \begin{bmatrix} 1 & -8 & 0 \\ 5 & 3 & 2 \\ -6 & 7 & -5 \\ 2 & -1 & 0 \end{bmatrix}$

27. $\begin{bmatrix} 2 & 3 \\ -2 & 4 \end{bmatrix} + \begin{bmatrix} 4 & 3 \\ 7 & 8 \end{bmatrix} - \begin{bmatrix} 3 & 2 \\ 1 & 4 \end{bmatrix}$

28. $\begin{bmatrix} 4 & 3 \\ 1 & 2 \end{bmatrix} - \begin{bmatrix} 1 & 1 \\ 1 & 0 \end{bmatrix} + \begin{bmatrix} 1 & 1 \\ 1 & 4 \end{bmatrix}$

29. $\begin{bmatrix} 1 & 5 \\ -3 & 7 \end{bmatrix} - \begin{bmatrix} 6 & 3 \\ 2 & 4 \end{bmatrix} + \begin{bmatrix} 8 & 10 \\ -1 & 0 \end{bmatrix}$

30. $\begin{bmatrix} -1 & -1 \\ -1 & 0 \end{bmatrix} + \begin{bmatrix} 4 & 3 \\ 1 & 2 \end{bmatrix} + \begin{bmatrix} 1 & 1 \\ 1 & 4 \end{bmatrix}$

31. $\begin{bmatrix} -4x + 2y & -3x + y \\ 6x - 3y & 2x - 5y \end{bmatrix} + \begin{bmatrix} -8x + 6y & 2x \\ 3y - 5x & 6x + 4y \end{bmatrix}$

32. $\begin{bmatrix} 4k - 8y \\ 6z - 3x \\ 2k + 5a \\ -4m + 2n \end{bmatrix} - \begin{bmatrix} 5k + 6y \\ 2z + 5x \\ 4k + 6a \\ 4m - 2n \end{bmatrix}$

Using matrices $O = \begin{bmatrix} 0 & 0 \\ 0 & 0 \end{bmatrix}$, $P = \begin{bmatrix} m & n \\ p & q \end{bmatrix}$, $T = \begin{bmatrix} r & s \\ t & u \end{bmatrix}$, *and* $X = \begin{bmatrix} x & y \\ z & w \end{bmatrix}$,

verify the statements in Exercises 33–37.

33. $X + T$ is a 2 × 2 matrix   (closure property)

34. $X + T = T + X$ (commutative property of addition of matrices)

35. $X + (T + P) = (X + T) + P$ (associative property of addition of matrices)

36. $X + (-X) = 0$ (inverse property of addition of matrices)

37. $P + O = P$ (identity property of addition of matrices)

38. Which of the above properties are valid for matrices that are not square?

▦ APPLICATIONS

GENERAL

Inventory   **39.** When John inventoried his collection of bolts, he found that he had 7 flathead long bolts, 9 flathead medium, 8 flathead short, 2 roundhead long, no roundhead medium, and 6 roundhead short. Write this information first as a 3 × 2 matrix and then as a 2 × 3 matrix.

Grocery Purchases   **40.** At the grocery store, Miguel bought 4 quarts of milk, 2 loaves of bread, 4 chickens, and an apple. Mary bought 2 quarts of milk, a loaf of bread, 5 chickens, and 4 apples. Write this information first as a 2 × 4 matrix and then as a 4 × 2 matrix.

LIFE SCIENCES

Diet Analysis   **41.** A dietician prepares a diet specifying the amounts a patient should eat of four basic food groups: group I, meats; group II, fruits and vegetables; group III, breads and starches; group IV, milk products. Amounts are given in "exchanges" that represent 1 oz (meat), 1/2 cup (fruits and vegetables), 1 slice (bread), 8 oz (milk), or other suitable measurements.

(a) The number of "exchanges" for breakfast for each of the four food groups respectively are 2, 1, 2, and 1; for lunch, 3, 2, 2, and 1; and for dinner, 4, 3, 2, and 1. Write a 3 × 4 matrix using this information.

(b) The amounts of fat, carbohydrates, and protein (in appropriate units) in each food group respectively are as follows.

> Fat: 5, 0, 0, 10
> Carbohydrates: 0, 10, 15, 12
> Protein: 7, 1, 2, 8

Use this information to write a 4 × 3 matrix.

(c) There are 8 calories per exchange of fat, 4 calories per exchange of carbohydrates, and 5 calories per exchange of protein; summarize this data in a 3 × 1 matrix.

Animal Growth   **42.** At the beginning of a laboratory experiment, five baby rats measured 5.6, 6.4, 6.9, 7.6, and 6.1 centimeters in length, and weighed 144, 138, 149, 152, and 146 grams respectively.

(a) Write a 2 × 5 matrix using this information.

(b) At the end of two weeks, their lengths in centimeters were 10.2, 11.4, 11.4, 12.7, and 10.8, and their weights in grams were 196, 196, 225, 250, and 230. Write a 2 × 5 matrix with this information.

(c) Use matrix subtraction and the matrices found in parts (a) and (b) to write a matrix that gives the amount of change in length and weight for each rat.

(d) During the third week the rats gained as shown in the matrix below.

$$\begin{array}{l} \text{Length} \\ \text{Weight} \end{array} \begin{bmatrix} 1.8 & 1.5 & 2.3 & 1.8 & 2.0 \\ 25 & 22 & 29 & 33 & 20 \end{bmatrix}$$

What were their lengths and weights at the end of this week?

2.4 MULTIPLICATION OF MATRICES

In work with matrices, a real number is called a **scalar.**

PRODUCT OF A MATRIX AND A SCALAR

> The **product** of a scalar k and a matrix X is the matrix kX, each of whose elements is k times the corresponding element of X.

For example,

$$(-3)\begin{bmatrix} 2 & -5 \\ 1 & 7 \end{bmatrix} = \begin{bmatrix} -6 & 15 \\ -3 & -21 \end{bmatrix}.$$

Finding the product of two matrices is more involved, but such multiplication is important in solving practical problems. To understand the reasoning behind matrix multiplication, it may be helpful to consider another example concerning the EZ Life Company discussed in Section 2.3. Suppose sofas and chairs of the same model are often sold as sets. Matrix W shows the number of sets of each model in each warehouse.

$$\begin{array}{l} \text{New York} \\ \text{Chicago} \\ \text{San Francisco} \end{array} \begin{array}{ccc} A & B & C \\ \begin{bmatrix} 10 & 7 & 3 \\ 5 & 9 & 6 \\ 4 & 8 & 2 \end{bmatrix} \end{array} = W$$

If the selling price of a model-A set is $800, of a model-B set $1000, and of a model-C set $1200, the total value of the sets in the New York warehoue is found as follows.

Type	Number of Sets		Price of Set		Total
A	10	×	$800	=	$8000
B	7	×	$1000	=	$7000
C	3	×	$1200	=	$3600
					$18,600
					(Total for New York)

The total value of the three kinds of sets in New York is $18,600.

The work done in the table above is summarized as follows:

$$10(\$800) + 7(\$1000) + 3(\$1200) = \$18,600.$$

In the same way, we find that the Chicago sets have a total value of

$$5(\$800) + 9(\$1000) + 6(\$1200) = \$20,200,$$

and in San Francisco, the total value of the sets is

$$4(\$800) + 8(\$1000) + 2(\$1200) = \$13,600.$$

The selling prices can be written as a column matrix, P, and the total value in each location as another column matrix, V.

$$\begin{bmatrix} 800 \\ 1000 \\ 1200 \end{bmatrix} = P \qquad \begin{bmatrix} 18,600 \\ 20,200 \\ 13,600 \end{bmatrix} = V$$

Look at the elements of W and P below; multiplying the first, second, and third elements of the first row of W by the first, second, and third elements respectively of the column matrix P and then adding these products gives the first element in V. Doing the same thing with the second row of W gives the second element of V; the third row of W leads to the third element of V, suggesting that it is reasonable to write the product of matrices

$$W = \begin{bmatrix} 10 & 7 & 3 \\ 5 & 9 & 6 \\ 4 & 8 & 2 \end{bmatrix} \qquad \text{and} \qquad P = \begin{bmatrix} 800 \\ 1000 \\ 1200 \end{bmatrix}$$

as

$$WP = \begin{bmatrix} 10 & 7 & 3 \\ 5 & 9 & 6 \\ 4 & 8 & 2 \end{bmatrix} \begin{bmatrix} 800 \\ 1000 \\ 1200 \end{bmatrix} = \begin{bmatrix} 18,600 \\ 20,200 \\ 13,600 \end{bmatrix} = V.$$

The product was found by multiplying the elements of *rows* of the matrix on the left and the corresponding elements of the *column* of the matrix on the right, and then finding the sum of these separate products. Notice that the product of a 3×3 matrix and a 3×1 matrix is a 3×1 matrix.

The **product** AB of an $m \times n$ matrix A and an $n \times k$ matrix B is found as follows. Multiply each element of the *first row* of A by the corresponding element of the *first column* of B. The sum of these n products is the *first row, first column* element of AB. Similarly, the sum of the products found by multiplying the elements of the *first row* of A times the corresponding elements of the *second column* of B gives the *first-row, second-column* element of AB, and so on.

PRODUCT OF TWO MATRICES

> Let A be an $m \times n$ matrix and let B be an $n \times k$ matrix. To find the element in the ith row and jth column of the **product matrix** AB, multiply each element in the ith row of A by the corresponding element in the jth column of B, then add these products. The product matrix AB is of order $m \times k$.

▰▰ EXAMPLE 1

Find the product *AB* of matrices

$$A = \begin{bmatrix} 2 & 3 & -1 \\ 4 & 2 & 2 \end{bmatrix} \quad \text{and} \quad B = \begin{bmatrix} 1 \\ 8 \\ 6 \end{bmatrix}.$$

Step 1   Multiply the elements of the first row of *A* and the corresponding elements of the column of *B*.

$$\begin{bmatrix} 2 & 3 & -1 \\ 4 & 2 & 2 \end{bmatrix} \begin{bmatrix} 1 \\ 8 \\ 6 \end{bmatrix} \quad 2 \cdot 1 + 3 \cdot 8 + (-1) \cdot 6 = 20$$

Thus, 20 is the first-row entry of the product matrix *AB*.

Step 2   Multiply the elements of the second row of *A* and the corresponding elements of *B*.

$$\begin{bmatrix} 2 & 3 & -1 \\ 4 & 2 & 2 \end{bmatrix} \begin{bmatrix} 1 \\ 8 \\ 6 \end{bmatrix} \quad 4 \cdot 1 + 2 \cdot 8 + 2 \cdot 6 = 32$$

The second-row entry of the product matrix *AB* is 32.

Step 3   Write the product as a column matrix using the two entries found above.

$$AB = \begin{bmatrix} 2 & 3 & -1 \\ 4 & 2 & 2 \end{bmatrix} \begin{bmatrix} 1 \\ 8 \\ 6 \end{bmatrix} = \begin{bmatrix} 20 \\ 32 \end{bmatrix} \quad \blacksquare$$

▰▰ EXAMPLE 2

Find the product *CD* of matrices

$$C = \begin{bmatrix} -3 & 4 & 2 \\ 5 & 0 & 4 \end{bmatrix} \quad \text{and} \quad D = \begin{bmatrix} -6 & 4 \\ 2 & 3 \\ 3 & -2 \end{bmatrix}.$$

Step 1

$$\begin{bmatrix} -3 & 4 & 2 \\ 5 & 0 & 4 \end{bmatrix} \begin{bmatrix} -6 & 4 \\ 2 & 3 \\ 3 & -2 \end{bmatrix} \quad (-3) \cdot (-6) + 4 \cdot 2 + 2 \cdot 3 = 32$$

Step 2

$$\begin{bmatrix} -3 & 4 & 2 \\ 5 & 0 & 4 \end{bmatrix} \begin{bmatrix} -6 & 4 \\ 2 & 3 \\ 3 & -2 \end{bmatrix} \quad (-3) \cdot 4 + 4 \cdot 3 + 2 \cdot (-2) = -4$$

Step 3

$$\begin{bmatrix} -3 & 4 & 2 \\ 5 & 0 & 4 \end{bmatrix} \begin{bmatrix} -6 & 4 \\ 2 & 3 \\ 3 & -2 \end{bmatrix} \quad 5 \cdot (-6) + 0 \cdot 2 + 4 \cdot 3 = -18$$

Step 4

$$\begin{bmatrix} -3 & 4 & 2 \\ 5 & 0 & 4 \end{bmatrix} \begin{bmatrix} -6 & 4 \\ 2 & 3 \\ 3 & -2 \end{bmatrix} \quad 5 \cdot 4 + 0 \cdot 3 + 4 \cdot (-2) = 12$$

Step 5   The product is

$$CD = \begin{bmatrix} -3 & 4 & 2 \\ 5 & 0 & 4 \end{bmatrix} \begin{bmatrix} -6 & 4 \\ 2 & 3 \\ 3 & -2 \end{bmatrix} = \begin{bmatrix} 32 & -4 \\ -18 & 12 \end{bmatrix}.$$

Here the product of a 2×3 matrix and a 3×2 matrix is a 2×2 matrix. ▬

As the definition of matrix multiplication shows,

> **the product *AB* of two matrices *A* and *B* can be found only if the number of columns of *A* is the same as the number of rows of *B*.**

The final product will have as many rows as *A* and as many columns as *B*.

▬ EXAMPLE 3

Suppose matrix *A* is 2×2 and matrix *B* is 2×4. Can the product *AB* be calculated? What is the order of the product?

The following diagram helps decide the answers to these questions.

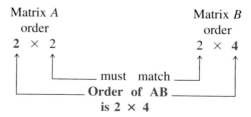

The product of *A* and *B* can be found because *A* has two columns and *B* has two rows. The order of the product is 2×4. ▬

■ EXAMPLE 4

Find *BA*, given

$$A = \begin{bmatrix} 1 & -3 \\ 7 & 2 \end{bmatrix} \quad \text{and} \quad B = \begin{bmatrix} 1 & 0 & -1 \\ 3 & 1 & 4 \end{bmatrix}.$$

Since *B* is a 2 × 3 matrix and *A* is a 2 × 2 matrix, the product *BA* cannot be found. ■

In Example 4, the product *BA* could not be found. The product *AB*, however, is given by

$$AB = \begin{bmatrix} 1 & -3 \\ 7 & 2 \end{bmatrix} \cdot \begin{bmatrix} 1 & 0 & -1 \\ 3 & 1 & 4 \end{bmatrix}$$

$$= \begin{bmatrix} -8 & -3 & -13 \\ 13 & 2 & 1 \end{bmatrix}.$$

This means that matrix multiplication is *not* commutative. Even if both *A* and *B* are square matrices, in general, matrices *AB* and *BA* are not equal. (See Exercise 31).

Matrix multiplication *is* associative, however. For example, if

$$C = \begin{bmatrix} 2 & 1 \\ 3 & 4 \\ 1 & 5 \end{bmatrix},$$

$(AB)C = A(BC)$, where *A* and *B* are the matrices given in Example 4. (Verify this.) Also, there is a distributive property of matrices such that, for appropriate matrices *A*, *B*, and *C*,

$$A(B + C) = AB + AC.$$

(See Exercises 32 and 33.) Other properties of matrix multiplication involving scalars are included in the exercises. Multiplicative inverses and multiplicative identities are defined in the next section.

■ EXAMPLE 5

A contractor builds three kinds of houses, models A, B, and C, with a choice of two styles, Spanish or contemporary. Matrix *P* shows the number of each kind of house planned for a new 100-home subdivision. The amounts for each of the exterior materials depend primarily on the style of the house. These amounts are shown in matrix *Q* (Concrete is in cubic yards, lumber in units of 1000 board feet, brick in 1000's, and shingles in units of 100 square feet.) Matrix *R* gives the cost in dollars for each kind of material.

$$\begin{array}{c} \\ \text{Model A} \\ \text{Model B} \\ \text{Model C} \end{array} \begin{array}{cc} \text{Spanish} & \text{Contemporary} \\ \begin{bmatrix} 0 & 30 \\ 10 & 20 \\ 20 & 20 \end{bmatrix} \end{array} = P$$

$$\begin{array}{c} \\ \text{Spanish} \\ \text{Contemporary} \end{array} \begin{array}{cccc} \text{Concrete} & \text{Lumber} & \text{Brick} & \text{Shingles} \\ \left[\begin{array}{cccc} 10 & 2 & 0 & 2 \\ 50 & 1 & 20 & 2 \end{array} \right] \end{array} = Q$$

$$\begin{array}{c} \\ \text{Concrete} \\ \text{Lumber} \\ \text{Brick} \\ \text{Shingles} \end{array} \begin{array}{c} \text{Cost per Unit} \\ \left[\begin{array}{c} 20 \\ 180 \\ 60 \\ 25 \end{array} \right] \end{array} = R$$

(a) What is the total cost of these materials for each model?

To find the cost for each model, first find PQ, which shows the amount of each material needed for each model.

$$PQ = \begin{bmatrix} 0 & 30 \\ 10 & 20 \\ 20 & 20 \end{bmatrix} \begin{bmatrix} 10 & 2 & 0 & 2 \\ 50 & 1 & 20 & 2 \end{bmatrix}$$

$$= \begin{array}{c} \begin{array}{cccc} \text{Concrete} & \text{Lumber} & \text{Brick} & \text{Shingles} \end{array} \\ \begin{bmatrix} 1500 & 30 & 600 & 60 \\ 1100 & 40 & 400 & 60 \\ 1200 & 60 & 400 & 80 \end{bmatrix} \begin{array}{c} \text{Model A} \\ \text{Model B} \\ \text{Model C} \end{array} \end{array}$$

Now multiply PQ and R, the cost matrix, to get the total cost of the exterior materials for each model.

$$\begin{bmatrix} 1500 & 30 & 600 & 60 \\ 1100 & 40 & 400 & 60 \\ 1200 & 60 & 400 & 80 \end{bmatrix} \begin{bmatrix} 20 \\ 180 \\ 60 \\ 25 \end{bmatrix} = \begin{array}{c} \text{Cost} \\ \begin{bmatrix} 72{,}900 \\ 54{,}700 \\ 60{,}800 \end{bmatrix} \begin{array}{c} \text{Model A} \\ \text{Model B} \\ \text{Model C} \end{array} \end{array}$$

(b) How much of each of the four kinds of material must be ordered?

The totals of the columns of matrix PQ will give a matrix whose elements represent the total amounts of each material needed for the subdivision. Call this matrix T, and write it as a row matrix.

$$T = [3800 \quad 130 \quad 1400 \quad 200]$$

(c) What is the total cost for exterior materials?

For the total cost of all the exterior materials, find the product of matrix T, the matrix showing the total amount of each material, and matrix R, the cost matrix. (To multiply these and get a 1×1 matrix representing total cost we need a 1×4 matrix multiplied by a 4×1 matrix. This is why T was written as a row matrix in (b) above.)

$$TR = [3800 \quad 130 \quad 1400 \quad 200] \begin{bmatrix} 20 \\ 180 \\ 60 \\ 25 \end{bmatrix} = [188{,}400].$$

(d) Suppose the contractor builds the same number of homes in five subdivisions. Calculate the total amount of each exterior material for each model for all five subdivisions.

Multiply PQ by the scalar 5, as follows.

$$5\begin{bmatrix} 1500 & 30 & 600 & 60 \\ 1100 & 40 & 400 & 60 \\ 1200 & 60 & 400 & 80 \end{bmatrix} = \begin{bmatrix} 7500 & 150 & 3000 & 300 \\ 5500 & 200 & 2000 & 300 \\ 6000 & 300 & 2000 & 400 \end{bmatrix} \blacksquare$$

It is helpful to use a notation that keeps track of the quantities a matrix represents. For example, let matrix P, from Example 5, represent models/styles, matrix Q represent styles/materials, and matrix R represent materials/cost. In each case, write the meaning of the rows first and the columns second. In the product PQ from Example 5, the rows of the matrix represented models and the columns represent materials. Therefore, the matrix product PQ represents models/materials. Note that the common quantity, styles, in both P and Q was eliminated in the product PQ. By this method, the product $(PQ)R$ represents models/cost.

In practical problems this notation helps us decide in which order to multiply matrices so that the results are meaningful. In Example 5(c) either RT or TR could have been found. Since T represents subdivisions/materials and R represents materials/cost, the product TR gives subdivisions/cost.

2.4 EXERCISES

In Exercises 1–8, the dimensions of two matrices A and B are given. Find the dimensions of the product AB and the product BA, whenever these products exist.

1. A is 2×2, and B is 2×2.

2. A is 3×3, and B is 3×3.

3. A is 4×2, and B is 2×4.

4. A is 3×1, and B is 1×3.

5. A is 3×5, and B is 5×2.

6. A is 4×3, and B is 3×6.

7. A is 4×2, and B is 3×4.

8. A is 7×3, and B is 2×7.

Let $A = \begin{bmatrix} -2 & 4 \\ 0 & 3 \end{bmatrix}$ and $B = \begin{bmatrix} -6 & 2 \\ 4 & 0 \end{bmatrix}$.

Find each of the following.

9. $2A$

10. $-3B$

11. $-4B$

12. $5A$

13. $-4A + 5B$

14. $3A - 10B$

Find each of the matrix products in Exercises 15–30 if possible.

15. $\begin{bmatrix} 1 & 2 \\ 3 & 4 \end{bmatrix}\begin{bmatrix} -1 \\ 7 \end{bmatrix}$

16. $\begin{bmatrix} -1 & 5 \\ 7 & 0 \end{bmatrix}\begin{bmatrix} 6 \\ 2 \end{bmatrix}$

17. $\begin{bmatrix} 1 & 5 & 3 \\ -1 & 2 & 7 \end{bmatrix}\begin{bmatrix} 4 \\ 2 \\ -3 \end{bmatrix}$

18. $\begin{bmatrix} 5 & 2 \\ 7 & 6 \\ 1 & 0 \end{bmatrix}\begin{bmatrix} 1 & 4 & 0 \\ 2 & -1 & 2 \end{bmatrix}$

19. $\begin{bmatrix} 5 & 1 \\ 2 & 3 \end{bmatrix}\begin{bmatrix} 3 & -1 & 0 \\ 1 & 0 & 2 \end{bmatrix}$

20. $\begin{bmatrix} 6 & 0 & -4 \\ 1 & 2 & 5 \\ 10 & -1 & 3 \end{bmatrix}\begin{bmatrix} 1 \\ 2 \\ 0 \end{bmatrix}$

21. $\begin{bmatrix} 2 & 2 & -1 \\ 3 & 0 & 1 \end{bmatrix} \begin{bmatrix} 0 & 2 \\ -1 & 4 \\ 0 & 2 \end{bmatrix}$

22. $\begin{bmatrix} -9 & 2 & 1 \\ 3 & 0 & 0 \end{bmatrix} \begin{bmatrix} 2 \\ -1 \\ 4 \end{bmatrix}$

23. $\begin{bmatrix} 1 & 2 \\ 3 & 4 \end{bmatrix} \begin{bmatrix} -1 & 5 \\ 7 & 0 \end{bmatrix}$

24. $\begin{bmatrix} -1 & 5 \\ 7 & 0 \end{bmatrix} \begin{bmatrix} 1 & 2 \\ 3 & 4 \end{bmatrix}$

25. $\begin{bmatrix} -2 & -3 & 7 \\ 1 & 5 & 6 \end{bmatrix} \begin{bmatrix} 1 \\ 2 \\ 3 \end{bmatrix}$

26. $\begin{bmatrix} 6 \\ 5 \\ 4 \end{bmatrix} [-1 \quad 1 \quad 1]$

27. $\left(\begin{bmatrix} 4 & 3 \\ 1 & 2 \\ 0 & -5 \end{bmatrix} \begin{bmatrix} 2 & -2 \\ 1 & -1 \end{bmatrix} \right) \begin{bmatrix} 10 \\ 0 \end{bmatrix}$

28. $\begin{bmatrix} 4 & 3 \\ 1 & 2 \\ 0 & -5 \end{bmatrix} \left(\begin{bmatrix} 2 & -2 \\ 1 & -1 \end{bmatrix} \begin{bmatrix} 10 \\ 0 \end{bmatrix} \right)$

29. $\begin{bmatrix} 2 & -2 \\ 1 & -1 \end{bmatrix} \left(\begin{bmatrix} 4 & 3 \\ 1 & 2 \end{bmatrix} + \begin{bmatrix} 7 & 0 \\ -1 & 5 \end{bmatrix} \right)$

30. $\begin{bmatrix} 2 & -2 \\ 1 & -1 \end{bmatrix} \begin{bmatrix} 4 & 3 \\ 1 & 2 \end{bmatrix} + \begin{bmatrix} 2 & -2 \\ 1 & -1 \end{bmatrix} \begin{bmatrix} 7 & 0 \\ -1 & 5 \end{bmatrix}$

31. Let $A = \begin{bmatrix} -2 & 4 \\ 1 & 3 \end{bmatrix}$ and $B = \begin{bmatrix} -2 & 1 \\ 3 & 6 \end{bmatrix}$.

(a) Find AB.

(b) Find BA.

(c) Did you get the same answer in parts (a) and (b)?

(d) In general, for matrices A and B such that AB and BA both exist, does AB always equal BA?

Given matrices $\quad P = \begin{bmatrix} m & n \\ p & q \end{bmatrix}, \quad X = \begin{bmatrix} x & y \\ z & w \end{bmatrix}, \quad and \quad T = \begin{bmatrix} r & s \\ t & u \end{bmatrix},$

verify that the statements in Exercises 32–36 are true. The statements are valid for any matrices whenever matrix multiplication and addition can be carried out. This, of course, depends on the order *of the matrices.*

32. $(PX)T = P(XT)$ (associative property: see Exercises 27 and 28.)

33. $P(X + T) = PX + PT$ (distributive property: see Exercises 29 and 30.)

34. PX is a 2×2 matrix   (closure property)

35. $k(X + T) = kX + kT$ for any real number k.

36. $(k + h)P = kP + hP$ for any real numbers k and h.

37. Let I be the matrix $I = \begin{bmatrix} 1 & 0 \\ 0 & 1 \end{bmatrix}$, and let matrices P, X, and T be defined as above.

(a) Find IP, PI, and IX.

(b) Without calculating, guess what the matrix IT might be.

(c) Suggest a reason for naming a matrix such as I an *identity matrix*.

38. Show that the system of linear equations

$$2x_1 + 3x_2 + x_3 = 5$$
$$x_1 - 4x_2 + 5x_3 = 8$$

can be written as the matrix equation

$$\begin{bmatrix} 2 & 3 & 1 \\ 1 & -4 & 5 \end{bmatrix} \begin{bmatrix} x_1 \\ x_2 \\ x_2 \end{bmatrix} = \begin{bmatrix} 5 \\ 8 \end{bmatrix}.$$

Solve the system and substitute into the matrix equation to check the results.

39. Let $A = \begin{bmatrix} 1 & 2 \\ -3 & 5 \end{bmatrix}$, $X = \begin{bmatrix} x_1 \\ x_2 \end{bmatrix}$, and $B = \begin{bmatrix} -4 \\ 12 \end{bmatrix}$.

Show that the equation $AX = B$ represents a linear system of two equations in two unknowns. Solve the system and substitute into the matrix equation to check your results.

▤ APPLICATIONS

BUSINESS AND ECONOMICS

Cost Analysis **40.** The Bread Box, a small neighborhood bakery, sells four main items: sweet rolls, bread, cake, and pie. The amount of eggs or of certain other main ingredients (in cups) required to make these items is given in matrix A.

	Eggs	Flour	Sugar	Shortening	Milk	
	1	4	$\frac{1}{4}$	$\frac{1}{4}$	1	Sweet Rolls (dozen)
$A =$	0	3	0	$\frac{1}{4}$	0	Bread (loaves)
	4	3	2	1	1	Cake (1)
	0	1	0	$\frac{1}{3}$	0	Pie (1)

The cost (in cents per egg or per cup) for each ingredient when purchased in large lots and in small lots is given by matrix B.

	Cost		
	Large Lot	Small Lot	
	5	5	Eggs
	8	10	Flour
$B =$	10	12	Sugar
	12	15	Shortening
	5	6	Milk

(a) Use matrix multiplication to find a matrix representing the comparative costs per item under the two purchase options.

Suppose a day's orders consist of 20 dozen sweet rolls, 200 loaves of bread, 50 cakes, and 60 pies.

(b) Represent these orders as a 1×4 matrix and use matrix multiplication to write as a matrix the amount of each ingredient required to fill the day's orders.

(c) Use matrix multiplication to find a matrix representing the costs under the two purchase options to fill the day's orders.

Cost Analysis **41.** The Mundo Candy Company makes three types of chocolate candy: Cheery Cherry, Mucho Mocha and Almond Delight. The company produces its products in San Diego, Mexico City, and Managua using two main ingredients: chocolate and sugar.

(a) Each kilogram of Cheery Cherry requires .5 kg of sugar and .2 kg of chocolate; each kilogram of Mucho Mocha requires .4 kg of sugar and .3 kg of chocolate; and each kilogram of Almond Delight requires .3 kg of sugar and .3 kg of chocolate. Put this information into a 2 x 3 matrix, labeling the rows and columns.

(b) The cost of 1 kg of sugar is $3 in San Diego, $2 in Mexico City, and $1 in Managua. The cost of 1 kg of chocolate is $3 in San Diego, $3 in Mexico City, and $4 in Managua. Put this information into a matrix in such a way that when you multiply it with your matrix from part (a), you get a matrix representing the ingredient cost of producing each type of candy in each city.

(c) Multiply the matrices in parts (a) and (b), labeling the product matrix.

(d) From part (c) what is the combined sugar-and-chocolate cost to product 1 kg of Mucho Mocha in Managua?

(e) Mundo Candy needs to quickly produce a special shipment of 100 kg of Cheery Cherry, 200 kg of Mucho Mocha, and 500 kg of Almond Delight, and it decides to select one factory to fill the entire order. Use matrix multiplication to determine in which city the total sugar-and-chocolate cost to produce the order is the smallest.

LIFE SCIENCES

Dietetics **42.** In Exercise 41 from Section 2.3, label the matrices found in parts (a), (b), and (c) respectively X, Y, and Z.

(a) Find the product matrix XY. What do the entries of this matrix represent?

(b) Find the product matrix YZ. What do the entries represent?

FOR THE COMPUTER

Use the following matrices to find the matrix products in Exercises 43–45.

$$A = \begin{bmatrix} 2 & 3 & -1 & 5 & 10 \\ 2 & 8 & 7 & 4 & 3 \\ -1 & -4 & -12 & 6 & 8 \\ 2 & 5 & 7 & 1 & 4 \end{bmatrix} \qquad B = \begin{bmatrix} 9 & 3 & 7 & -6 \\ -1 & 0 & 4 & 2 \\ -10 & -7 & 6 & 9 \\ 8 & 4 & 2 & -1 \\ 2 & -5 & 3 & 7 \end{bmatrix}$$

$$C = \begin{bmatrix} -6 & 8 & 2 & 4 & -3 \\ 1 & 9 & 7 & -12 & 5 \\ 15 & 2 & -8 & 10 & 11 \\ 4 & 7 & 9 & 6 & -2 \\ 1 & 3 & 8 & 23 & 4 \end{bmatrix} \qquad D = \begin{bmatrix} 5 & -3 & 7 & 9 & 2 \\ 6 & 8 & -5 & 2 & 1 \\ 3 & 7 & -4 & 2 & 11 \\ 5 & -3 & 9 & 4 & -1 \\ 0 & 3 & 2 & 5 & 1 \end{bmatrix}$$

43. **(a)** Find AC. **(b)** Find CA. **(c)** Does $AC = CA$?

44. **(a)** Find CD. **(b)** Find DC. **(c)** Does $CD = DC$?

45. **(a)** Find $C + D$. **(b)** Find $(C + D)B$. **(c)** Find CB. **(d)** Find DB.
  (e) Find $CB + DB$. **(f)** Does $(C + D)B = CB + DB$?

2.5 MATRIX INVERSES

In Section 2.3, we defined a zero matrix as an additive identity matrix with properties similar to those of the real number 0, the additive identity for real numbers. The real number 1 is the multiplicative identity for real numbers: for any real number a, $a \cdot 1 = 1 \cdot a$. In this section, we define a *multiplicative identity matrix I* that has properties similar to those of the number 1. We then use the definition of matrix I to find the *multiplicative inverse* of any square matrix that has an inverse.

If I is to be the identity matrix, both of the products AI and IA must equal A. This means that an identity matrix exists only for square matrices. Otherwise, IA and AI could not both be found. The **2 × 2 identity matrix** that satisfies these conditions is

$$I = \begin{bmatrix} 1 & 0 \\ 0 & 1 \end{bmatrix}.$$

To check that I, as defined above, is really the 2 × 2 identity matrix, let

$$A = \begin{bmatrix} a & b \\ c & d \end{bmatrix}.$$

Then AI and IA should both equal A.

$$AI = \begin{bmatrix} a & b \\ c & d \end{bmatrix}\begin{bmatrix} 1 & 0 \\ 0 & 1 \end{bmatrix} = \begin{bmatrix} a(1) + b(0) & a(0) + b(1) \\ c(1) + d(0) & c(0) + d(1) \end{bmatrix} = \begin{bmatrix} a & b \\ c & d \end{bmatrix} = A$$

$$IA = \begin{bmatrix} 1 & 0 \\ 0 & 1 \end{bmatrix}\begin{bmatrix} a & b \\ c & d \end{bmatrix} = \begin{bmatrix} 1(a) + 0(c) & 1(b) + 0(d) \\ 0(a) + 1(c) & 0(b) + 1(d) \end{bmatrix} = \begin{bmatrix} a & b \\ c & d \end{bmatrix} = A$$

This verifies that I has been defined correctly. (It also can be shown that I is the only 2 × 2 identity matrix.)

The identity matrices for 3 × 3 matrices and 4 × 4 matrices, respectively, are

$$I = \begin{bmatrix} 1 & 0 & 0 \\ 0 & 1 & 0 \\ 0 & 0 & 1 \end{bmatrix} \quad \text{and} \quad I = \begin{bmatrix} 1 & 0 & 0 & 0 \\ 0 & 1 & 0 & 0 \\ 0 & 0 & 1 & 0 \\ 0 & 0 & 0 & 1 \end{bmatrix}.$$

By generalizing, we can find an $n \times n$ identity matrix for any value of n.

Recall that the multiplicative inverse of the nonzero real number a is $1/a$. The product of a and its multiplicative inverse $1/a$ is 1. Given a matrix A, can a **multiplicative inverse matrix A^{-1}** (read "A-inverse") be found satisfying both

$$AA^{-1} = I \quad \text{and} \quad A^{-1}A = I?$$

For a given matrix, we often can find an inverse matrix by using the row operations of Section 2.2. Before we do this, note that A^{-1} does not mean $1/A$; here, A^{-1} is just the notation for the inverse of matrix A. Also, only square matrices can have inverses because both $A^{-1}A$ and AA^{-1} must exist and be equal to I. If an inverse exists, it is unique. That is, any given square matrix has no more than one inverse. The proof of this is left to Exercise 49 in this section.

As an example, let us find the inverse of

$$A = \begin{bmatrix} 2 & 4 \\ 1 & -1 \end{bmatrix}.$$

Let the unknown inverse matrix be

$$A^{-1} = \begin{bmatrix} x & y \\ z & w \end{bmatrix}.$$

By the definition of matrix inverse, $AA^{-1} = I$, or

$$AA^{-1} = \begin{bmatrix} 2 & 4 \\ 1 & -1 \end{bmatrix} \begin{bmatrix} x & y \\ z & w \end{bmatrix} = \begin{bmatrix} 1 & 0 \\ 0 & 1 \end{bmatrix}.$$

By matrix multiplication,

$$\begin{bmatrix} 2x + 4z & 2y + 4w \\ x - z & y - w \end{bmatrix} = \begin{bmatrix} 1 & 0 \\ 0 & 1 \end{bmatrix}.$$

Setting corresponding elements equal gives the system of equations

$$2x + 4z = 1 \qquad \textbf{(1)}$$
$$2y + 4w = 0 \qquad \textbf{(2)}$$
$$x - z = 0 \qquad \textbf{(3)}$$
$$y - w = 1. \qquad \textbf{(4)}$$

Since equations (1) and (3) involve only x and z, while equations (2) and (4) involve only y and w, these four equations lead to two systems of equations,

$$\begin{matrix} 2x + 4z = 1 \\ x - z = 0 \end{matrix} \quad \text{and} \quad \begin{matrix} 2y + 4w = 0 \\ y - w = 1. \end{matrix}$$

Writing the two systems as augmented matrices gives

$$\begin{bmatrix} 2 & 4 & | & 1 \\ 1 & -1 & | & 0 \end{bmatrix} \quad \text{and} \quad \begin{bmatrix} 2 & 4 & | & 0 \\ 1 & -1 & | & 1 \end{bmatrix}.$$

Each of these systems can be solved by the Gauss-Jordan method. Notice, however, that the elements to the left of the vertical bar are identical. The two systems can be combined into the single matrix

$$\begin{bmatrix} 2 & 4 & | & 1 & 0 \\ 1 & -1 & | & 0 & 1 \end{bmatrix}$$

and solved simultaneously as follows. Exchange the two rows to get a 1 in the upper left corner.

$$\begin{bmatrix} 1 & -1 & | & 0 & 1 \\ 2 & 4 & | & 1 & 0 \end{bmatrix}$$

Multiply row 1 by -2 and add the results to row 2 to get

$$\begin{bmatrix} 1 & -1 & 0 & 1 \\ 0 & 6 & 1 & -2 \end{bmatrix}. \qquad -2R_1 + R_2$$

Now, to get a 1 in the second-row, second-column position, multiply row 2 by 1/6.

$$\begin{bmatrix} 1 & -1 & 0 & 1 \\ 0 & 1 & \frac{1}{6} & -\frac{1}{3} \end{bmatrix} \qquad \frac{1}{6}R_2$$

Finally, add row 2 to row 1 to get a 0 in the second column above the 1.

$$\begin{bmatrix} 1 & 0 & \frac{1}{6} & \frac{2}{3} \\ 0 & 1 & \frac{1}{6} & -\frac{1}{3} \end{bmatrix}. \qquad R_2 + R_1$$

The numbers in the first column to the right of the vertical bar give the values of x and z. The second column gives the values of y and w. That is,

$$\begin{bmatrix} 1 & 0 & x & y \\ 0 & 1 & z & w \end{bmatrix} = \begin{bmatrix} 1 & 0 & \frac{1}{6} & \frac{2}{3} \\ 0 & 1 & \frac{1}{6} & -\frac{1}{3} \end{bmatrix}.$$

so that

$$A^{-1} = \begin{bmatrix} x & y \\ z & w \end{bmatrix} = \begin{bmatrix} \frac{1}{6} & \frac{2}{3} \\ \frac{1}{6} & -\frac{1}{3} \end{bmatrix}.$$

To check, multiply A by A^{-1}. The result should be I.

$$AA^{-1} = \begin{bmatrix} 2 & 4 \\ 1 & -1 \end{bmatrix} \begin{bmatrix} \frac{1}{6} & \frac{2}{3} \\ \frac{1}{6} & -\frac{1}{3} \end{bmatrix} = \begin{bmatrix} \frac{1}{3} + \frac{2}{3} & \frac{4}{3} - \frac{4}{3} \\ \frac{1}{6} - \frac{1}{6} & \frac{2}{3} + \frac{1}{3} \end{bmatrix} = \begin{bmatrix} 1 & 0 \\ 0 & 1 \end{bmatrix} = I$$

Verify that $A^{-1}A = I$, also. Finally,

$$A^{-1} = \begin{bmatrix} \frac{1}{6} & \frac{2}{3} \\ \frac{1}{6} & -\frac{1}{3} \end{bmatrix}.$$

FINDING A MULTIPLICATIVE INVERSE MATRIX

To obtain A^{-1} for any $n \times n$ matrix A for which A^{-1} exists, follow these steps.

1. Form the augmented matrix $[A|I]$, where I is the $n \times n$ identity matrix.
2. Perform row operations on $[A|I]$ to get a matrix of the form $[I|B]$.
3. Matrix B is A^{-1}.

EXAMPLE 1

Find A^{-1} if $A = \begin{bmatrix} 1 & 0 & 1 \\ 2 & -2 & -1 \\ 3 & 0 & 0 \end{bmatrix}.$

Write the augmented matrix $[A\,|\,I\,]$.

$$[A\,|\,I\,] = \begin{bmatrix} 1 & 0 & 1 & | & 1 & 0 & 0 \\ 2 & -2 & -1 & | & 0 & 1 & 0 \\ 3 & 0 & 0 & | & 0 & 0 & 1 \end{bmatrix}$$

Since 1 is already in the upper left-hand corner as desired, begin by selecting the row operation that will result in a 0 for the first element in row 2. Multiply row 1 by -2 and add the result to row 2. This gives

$$\begin{bmatrix} 1 & 0 & 1 & | & 1 & 0 & 0 \\ 0 & -2 & -3 & | & -2 & 1 & 0 \\ 3 & 0 & 0 & | & 0 & 0 & 1 \end{bmatrix}. \qquad -2R_1 + R_2$$

To get 0 for the first element in row 3, multiply row 1 by -3 and add to row 3. The new matrix is

$$\begin{bmatrix} 1 & 0 & 1 & | & 1 & 0 & 0 \\ 0 & -2 & -3 & | & -2 & 1 & 0 \\ 0 & 0 & -3 & | & -3 & 0 & 1 \end{bmatrix}. \qquad -3R_1 + R_3$$

To get 1 for the second element in row 2, multiply row 2 by $-1/2$, obtaining the new matrix

$$\begin{bmatrix} 1 & 0 & 1 & | & 1 & 0 & 0 \\ 0 & 1 & \frac{3}{2} & | & 1 & -\frac{1}{2} & 0 \\ 0 & 0 & -3 & | & -3 & 0 & 1 \end{bmatrix}. \qquad -\frac{1}{2}R_2$$

To get 1 for the third element in row 3, multiply row 3 by $-1/3$, with the result

$$\begin{bmatrix} 1 & 0 & 1 & | & 1 & 0 & 0 \\ 0 & 1 & \frac{3}{2} & | & 1 & -\frac{1}{2} & 0 \\ 0 & 0 & 1 & | & 1 & 0 & -\frac{1}{3} \end{bmatrix}. \qquad -\frac{1}{3}R_3$$

To get 0 for the third element in row 1, multiply row 3 by -1 and add to row 1 which gives

$$\begin{bmatrix} 1 & 0 & 0 & | & 0 & 0 & \frac{1}{3} \\ 0 & 1 & \frac{3}{2} & | & 1 & -\frac{1}{2} & 0 \\ 0 & 0 & 1 & | & 1 & 0 & -\frac{1}{3} \end{bmatrix}. \qquad -1R_3 + R_1$$

To get 0 for the third element in row 2, multiply row 3 by $-3/2$ and add to row 2.

$$\begin{bmatrix} 1 & 0 & 0 & | & 0 & 0 & \frac{1}{3} \\ 0 & 1 & 0 & | & -\frac{1}{2} & -\frac{1}{2} & \frac{1}{2} \\ 0 & 0 & 1 & | & 1 & 0 & -\frac{1}{3} \end{bmatrix} \qquad -\frac{3}{2}R_3 + R_2$$

From the last transformation, the desired inverse is

$$A^{-1} = \begin{bmatrix} 0 & 0 & \frac{1}{3} \\ -\frac{1}{2} & -\frac{1}{2} & \frac{1}{2} \\ 1 & 0 & -\frac{1}{3} \end{bmatrix}.$$

Confirm this by forming the products $A^{-1}A$ and AA^{-1}, both of which should equal I. ■

■ EXAMPLE 2

Find A^{-1} if $A = \begin{bmatrix} 2 & -4 \\ 1 & -2 \end{bmatrix}$.

Using row operations to transform the first column of the augmented matrix

$$\left[\begin{array}{cc|cc} 2 & -4 & 1 & 0 \\ 1 & -2 & 0 & 1 \end{array}\right]$$

gives the following results.

$$\left[\begin{array}{cc|cc} 1 & -2 & \frac{1}{2} & 0 \\ 1 & -2 & 0 & 1 \end{array}\right] \qquad \frac{1}{2}\,R_1$$

$$\left[\begin{array}{cc|cc} 1 & -2 & \frac{1}{2} & 0 \\ 0 & 0 & -\frac{1}{2} & 1 \end{array}\right] \qquad -R_1 + R_2$$

At this point, the matrix should be transformed so that the second element of row 2 will be 1. Since that element is now 0, there is no way to complete the desired transformation.

What is wrong? Just as the real number 0 has no multiplicative inverse, some matrices do not have inverses. Matrix A is an example of a matrix that has no inverse: there is no matrix A^{-1} such that $AA^{-1} = A^{-1}A = I$. ■

Solving Systems of Equations with Inverses   We used matrices to solve systems of linear equations by the Gauss-Jordan method in Section 2.2. Another way to use matrices to solve linear systems is to write the system as a matrix equation $AX = B$, where A is the matrix of the coefficients of the variables of the system, X is the matrix of the variables, and B is the matrix of the constants. Matrix A is called the **coefficient matrix.**

To solve the matrix equation $AX = B$, first see if A^{-1} exists. Assuming A^{-1} exists and using the facts that $A^{-1}A = I$ and $IX = X$ gives

$$
\begin{aligned}
AX &= B \\
A^{-1}(AX) &= A^{-1}B \qquad & \text{Multiply both sides by } A^{-1} \\
(A^{-1}A)X &= A^{-1}B \qquad & \text{Associative property} \\
IX &= A^{-1}B \qquad & \text{Multiplicative inverse property} \\
X &= A^{-1}B. \qquad & \text{Identity property}
\end{aligned}
$$

When multiplying by matrices on both sides of a matrix equation, be careful to multiply in the same order on both sides of the equation, since multiplication of matrices is not commutative (unlike multiplication of real numbers).

SOLVING A SYSTEM
$AX = B$ **USING**
MATRIX INVERSES

■ To solve a system of equations $AX = B$ where A is the matrix of coefficients, X is the matrix of variables, and B is the matrix of constants, first find A^{-1}. Then $X = A^{-1}B$.

This method is most practical in solving several systems that have the same coefficient matrix but different constants, as in Example 4 in this section. Then just one inverse matrix must be found.

■ EXAMPLE 3

Use the inverse of the coefficient matrix to solve the linear system

$$2x - 3y = 4$$
$$x + 5y = 2.$$

To represent the system as a matrix equation, use the coefficient matrix of the system together with the matrix of variables and the matrix of constants.

$$A = \begin{bmatrix} 2 & -3 \\ 1 & 5 \end{bmatrix}, \quad X = \begin{bmatrix} x \\ y \end{bmatrix}, \quad \text{and} \quad B = \begin{bmatrix} 4 \\ 2 \end{bmatrix}.$$

The system can now be written in matrix form as the equation $AX = B$ since

$$AX = \begin{bmatrix} 2 & -3 \\ 1 & 5 \end{bmatrix} \begin{bmatrix} x \\ y \end{bmatrix} = \begin{bmatrix} 2x - 3y \\ x + 5y \end{bmatrix} = \begin{bmatrix} 4 \\ 2 \end{bmatrix} = B.$$

To solve the system, first find A^{-1}. Do this by using row operations on matrix $[A \mid I]$ to get

$$\begin{bmatrix} 1 & 0 & \frac{5}{13} & \frac{3}{13} \\ 0 & 1 & -\frac{1}{13} & \frac{2}{13} \end{bmatrix}.$$

From this result,

$$A^{-1} = \begin{bmatrix} \frac{5}{13} & \frac{3}{13} \\ -\frac{1}{13} & \frac{2}{13} \end{bmatrix}.$$

Next, find the product $A^{-1} B$.

$$A^{-1} B = \begin{bmatrix} \frac{5}{13} & \frac{3}{13} \\ -\frac{1}{13} & \frac{2}{13} \end{bmatrix} \begin{bmatrix} 4 \\ 2 \end{bmatrix} = \begin{bmatrix} 2 \\ 0 \end{bmatrix}$$

Since $X = A^{-1} B$,

$$X = \begin{bmatrix} x \\ y \end{bmatrix} = \begin{bmatrix} 2 \\ 0 \end{bmatrix}.$$

The solution of the system is (2, 0). ■

■ EXAMPLE 4

Use the inverse of the coefficient matrix to solve the following systems.

(a) $-x - 2y + 2z = 9$
$2x + y - z = -3$
$3x - 2y + z = -6$

(b) $-x - 2y + 2z = 3$
$2x + y - z = 3$
$3x - 2y + z = 7$

(c) $\begin{aligned} -x - 2y + 2z &= 12 \\ 2x + y - z &= 0 \\ 3x - 2y + z &= 18 \end{aligned}$ (d) $\begin{aligned} -x - 2y + 2z &= 1 \\ 2x + y - z &= 7 \\ 3x - 2y + z &= 17 \end{aligned}$

Notice that the four systems all have the same matrix of coefficients and the same matrix of variables:

$$A = \begin{bmatrix} -1 & -2 & 2 \\ 2 & 1 & -1 \\ 3 & -2 & 1 \end{bmatrix} \quad \text{and} \quad X = \begin{bmatrix} x \\ y \\ z \end{bmatrix}.$$

Find A^{-1} first, then use it to solve all four systems.

To find A^{-1}, we start with matrix

$$[A|I] = \begin{bmatrix} -1 & -2 & 2 & | & 1 & 0 & 0 \\ 2 & 1 & -1 & | & 0 & 1 & 0 \\ 3 & -2 & 1 & | & 0 & 0 & 1 \end{bmatrix}$$

and use row operations to get $[I|A^{-1}]$, from which

$$A^{-1} = \begin{bmatrix} \frac{1}{3} & \frac{2}{3} & 0 \\ \frac{5}{3} & \frac{7}{3} & -1 \\ \frac{7}{3} & \frac{8}{3} & -1 \end{bmatrix}.$$

Now we can solve each of the four systems by using $X = A^{-1}B$.

(a) Here the matrix of coefficients is

$$B = \begin{bmatrix} 9 \\ -3 \\ -6 \end{bmatrix}.$$

Since $X = A^{-1}B$,

$$X = \begin{bmatrix} \frac{1}{3} & \frac{2}{3} & 0 \\ \frac{5}{3} & \frac{7}{3} & -1 \\ \frac{7}{3} & \frac{8}{3} & -1 \end{bmatrix} \begin{bmatrix} 9 \\ -3 \\ -6 \end{bmatrix} = \begin{bmatrix} 1 \\ 14 \\ 19 \end{bmatrix}.$$

From this result, $x = 1$, $y = 14$, and $z = 19$, and the solution is $(1, 14, 19.)$

(b) The matrix of coefficients is $B = \begin{bmatrix} 3 \\ 3 \\ 7 \end{bmatrix}$.

$$X = A^{-1}B = \begin{bmatrix} \frac{1}{3} & \frac{2}{3} & 0 \\ \frac{5}{3} & \frac{7}{3} & -1 \\ \frac{7}{3} & \frac{8}{3} & -1 \end{bmatrix} \begin{bmatrix} 3 \\ 3 \\ 7 \end{bmatrix} = \begin{bmatrix} 3 \\ 5 \\ 8 \end{bmatrix}.$$

The solution is $(3, 5, 8)$.

(c) This time $B = \begin{bmatrix} 12 \\ 0 \\ 18 \end{bmatrix}$.

$$X = A^{-1}B = \begin{bmatrix} \frac{1}{3} & \frac{2}{3} & 0 \\ \frac{5}{3} & \frac{7}{3} & -1 \\ \frac{7}{3} & \frac{8}{3} & -1 \end{bmatrix} \begin{bmatrix} 12 \\ 0 \\ 18 \end{bmatrix} = \begin{bmatrix} 4 \\ 2 \\ 10 \end{bmatrix}$$

The solution is (4, 2, 10).

(d) Since $B = \begin{bmatrix} 1 \\ 7 \\ 17 \end{bmatrix}$,

$$X = A^{-1}B = \begin{bmatrix} \frac{1}{3} & \frac{2}{3} & 0 \\ \frac{5}{3} & \frac{7}{3} & -1 \\ \frac{7}{3} & \frac{8}{3} & -1 \end{bmatrix} \begin{bmatrix} 1 \\ 7 \\ 17 \end{bmatrix} = \begin{bmatrix} 5 \\ 1 \\ 4 \end{bmatrix}.$$

The solution is (5, 1, 4). ■

In Example 4, using the matrix inverse method of solving the systems involved considerably less work than using row operations for each of the four systems.

■ EXAMPLE 5

Use the inverse of the coefficient matrix to solve the system

$$2x + 5y = 20$$
$$6x + 15y = 30.$$

The matrices needed are

$$A = \begin{bmatrix} 2 & 5 \\ 6 & 15 \end{bmatrix}, \qquad X = \begin{bmatrix} x \\ y \end{bmatrix}, \qquad \text{and} \qquad B = \begin{bmatrix} 20 \\ 30 \end{bmatrix}.$$

Find A^{-1}.

$$\begin{bmatrix} 2 & 5 & | & 1 & 0 \\ 6 & 15 & | & 0 & 1 \end{bmatrix}$$

$$\begin{bmatrix} 1 & \frac{5}{2} & | & \frac{1}{2} & 0 \\ 6 & 15 & | & 0 & 1 \end{bmatrix} \qquad \frac{1}{2}R_1$$

$$\begin{bmatrix} 1 & \frac{5}{2} & | & \frac{1}{2} & 0 \\ 0 & 0 & | & -3 & 1 \end{bmatrix} \qquad -6R_1 + R_2$$

As mentioned above, the zeros in the second row indicate that matrix A does not have an inverse. This means that the given system either has no solution or has dependent equations. Another method must be used to determine which is the case. Here, the system is inconsistent and there is no solution. ■

2.5 EXERCISES

In Exercises 1–8, decide whether the given matrices are inverse of each other. (Check to see if their product is the identity matrix I.)

1. $\begin{bmatrix} 2 & 3 \\ 1 & 1 \end{bmatrix}$ and $\begin{bmatrix} -1 & 3 \\ 1 & -2 \end{bmatrix}$

2. $\begin{bmatrix} 5 & 7 \\ 2 & 3 \end{bmatrix}$ and $\begin{bmatrix} 3 & -7 \\ -2 & 5 \end{bmatrix}$

3. $\begin{bmatrix} 2 & 1 \\ 3 & 2 \end{bmatrix}$ and $\begin{bmatrix} 2 & 1 \\ -3 & 2 \end{bmatrix}$

4. $\begin{bmatrix} -1 & 2 \\ 3 & -5 \end{bmatrix}$ and $\begin{bmatrix} -5 & -2 \\ -3 & -1 \end{bmatrix}$

5. $\begin{bmatrix} 1 & 2 & 0 \\ 0 & 1 & 0 \\ 0 & 1 & 0 \end{bmatrix}$ and $\begin{bmatrix} 1 & -2 & 0 \\ 0 & 1 & 0 \\ 0 & -1 & 1 \end{bmatrix}$

6. $\begin{bmatrix} 0 & 1 & 0 \\ 0 & 0 & -2 \\ 1 & -1 & 0 \end{bmatrix}$ and $\begin{bmatrix} 1 & 0 & 1 \\ 1 & 0 & 0 \\ 0 & -1 & 0 \end{bmatrix}$

7. $\begin{bmatrix} 1 & 3 & 3 \\ 1 & 4 & 3 \\ 1 & 3 & 4 \end{bmatrix}$ and $\begin{bmatrix} 7 & -3 & -3 \\ -1 & 1 & 0 \\ -1 & 0 & 1 \end{bmatrix}$

8. $\begin{bmatrix} -1 & 0 & 2 \\ 3 & 1 & 0 \\ 0 & 2 & -3 \end{bmatrix}$ and $\begin{bmatrix} -\frac{1}{5} & \frac{4}{15} & -\frac{2}{15} \\ \frac{3}{5} & \frac{1}{5} & \frac{2}{5} \\ \frac{2}{5} & \frac{2}{15} & -\frac{1}{15} \end{bmatrix}$

Find the inverse, if it exists, for each matrix.

9. $\begin{bmatrix} 1 & -1 \\ 2 & 0 \end{bmatrix}$

10. $\begin{bmatrix} -1 & 2 \\ -2 & -1 \end{bmatrix}$

11. $\begin{bmatrix} 3 & -1 \\ -5 & 2 \end{bmatrix}$

12. $\begin{bmatrix} -1 & -2 \\ 3 & 4 \end{bmatrix}$

13. $\begin{bmatrix} -6 & 4 \\ -3 & 2 \end{bmatrix}$

14. $\begin{bmatrix} 5 & 10 \\ -3 & -6 \end{bmatrix}$

15. $\begin{bmatrix} 1 & 0 & 0 \\ 0 & -1 & 0 \\ 1 & 0 & 1 \end{bmatrix}$

16. $\begin{bmatrix} 1 & 0 & 1 \\ 0 & -1 & 0 \\ 2 & 1 & 1 \end{bmatrix}$

17. $\begin{bmatrix} -1 & -1 & -1 \\ 4 & 5 & 0 \\ 0 & 1 & -3 \end{bmatrix}$

18. $\begin{bmatrix} 2 & 0 & 4 \\ 3 & 1 & 5 \\ -1 & 1 & -2 \end{bmatrix}$

19. $\begin{bmatrix} 1 & 2 & 3 \\ -3 & -2 & -1 \\ -1 & 0 & 1 \end{bmatrix}$

20. $\begin{bmatrix} 2 & 0 & 4 \\ 1 & 0 & -1 \\ 3 & 0 & -2 \end{bmatrix}$

21. $\begin{bmatrix} 2 & 4 & 6 \\ -1 & -4 & -3 \\ 0 & 1 & -1 \end{bmatrix}$

22. $\begin{bmatrix} 2 & 2 & -4 \\ 2 & 6 & 0 \\ -3 & -3 & 5 \end{bmatrix}$

23. $\begin{bmatrix} 1 & -2 & 3 & 0 \\ 0 & 1 & -1 & 1 \\ -2 & 2 & -2 & 4 \\ 0 & 2 & -3 & 1 \end{bmatrix}$

24. $\begin{bmatrix} 1 & 1 & 0 & 2 \\ 2 & -1 & 1 & -1 \\ 3 & 3 & 2 & -2 \\ 1 & 2 & 1 & 0 \end{bmatrix}$

Solve each system of equations by using the inverse of the coefficient matrix.

25. $\begin{aligned} 2x + 3y &= 10 \\ x - y &= -5 \end{aligned}$

26. $\begin{aligned} -x + 2y &= 15 \\ -2x - y &= 20 \end{aligned}$

27. $\begin{aligned} 2x + y &= 5 \\ 5x + 3y &= 13 \end{aligned}$

28. $\begin{aligned} -x - 2y &= 8 \\ 3x + 4y &= 24 \end{aligned}$

29. $\begin{aligned} -x + y &= 1 \\ 2x - y &= 1 \end{aligned}$

30. $\begin{aligned} 3x - 6y &= 1 \\ -5x + 9y &= -1 \end{aligned}$

31. $\begin{aligned} -x - 8y &= 12 \\ 3x + 24y &= -36 \end{aligned}$

32. $\begin{aligned} x + 3y &= -14 \\ 2x - y &= 7 \end{aligned}$

Solve each system of equations by using the inverse of the coefficient matrix. (The inverses for the first four problems were found in Exercises 17, 18, 21, and 22 above.)

33. $\begin{aligned} -x - y - z &= 1 \\ 4x + 5y &= -2 \\ y - 3z &= 3 \end{aligned}$

34. $\begin{aligned} 2x + 4z &= -8 \\ 3x + y + 5z &= 2 \\ -x + y - 2z &= 4 \end{aligned}$

35. $\begin{aligned} 2x + 4y + 6z &= 4 \\ -x - 4y - 3z &= 8 \\ y - z &= -4 \end{aligned}$

36. $\begin{aligned} 2x + 2y - 4z &= 12 \\ 2x + 6y &= 16 \\ -3x - 3y + 5z &= -20 \end{aligned}$

37. $\begin{aligned} x + 2y + 3z &= 5 \\ 2x + 3y + 2z &= 2 \\ -x - 2y - 4z &= -1 \end{aligned}$

38. $\begin{aligned} x + y - 3z &= 4 \\ 2x + 4y - 4z &= 8 \\ -x + y + 4z &= -3 \end{aligned}$

39. $\begin{aligned} 2x - 2y \quad\;\; &= 5 \\ 4y + 8z &= 7 \\ x \qquad\; + 2z &= 1 \end{aligned}$

40. $\begin{aligned} x \qquad\; + z &= 3 \\ y + 2z &= 8 \\ -x + y \qquad &= 4 \end{aligned}$

Solve the systems of equations in Exercises 41–42 by using the inverse of the coefficient matrix. (The inverses were found in Exercises 23 and 24.)

41. $\begin{aligned} x - 2y + 3z \qquad\quad &= 4 \\ y - z + w &= -8 \\ -2x + 2y - 2z + 4w &= 12 \\ 2y - 3z + w &= -4 \end{aligned}$

42. $\begin{aligned} x + y \qquad\;\; + 2w &= 3 \\ 2x - y + z - w &= 3 \\ 3x + 3y + 2z - 2w &= 5 \\ x + 2y + z \qquad &= 3 \end{aligned}$

Let $A = \begin{bmatrix} a & b \\ c & d \end{bmatrix}$ *in Exercises 43–47.*

43. Show that $IA = A$.

44. Show that $AI = A$.

45. Show that $A \cdot 0 = 0$.

46. Find A^{-1}. (Assume $ad - bc \neq 0$.)

47. Show that $A^{-1}A = I$.

48. Show that $AA^{-1} = I$.

49. Using the definitions and properties listed in this section, show that for square matrices A and B of the same order, if $AB = 0$ and if A^{-1} exists, then $B = 0$.

50. Prove that, if it exists, the inverse of a matrix is unique. (*Hint:* Assume there are two inverses B and C for some matrix A, so that $AB = BA = I$ and $AC = CA = I$. Multiply the first equation by C and the second by B.)

APPLICATIONS

BUSINESS AND ECONOMICS

Solve the following problems by using the inverse of the coefficient matrix to solve a system of equations.

Analysis of Orders   **51.** The Bread Box Bakery sells three types of cake, each requiring the amounts of the basic ingredients shown in the following matrix.

		I	II	III
	Flour (in cups)	2	4	2
Ingredient	Sugar (in cups)	2	1	2
	Eggs	2	1	3

Type of Cake

To fill its daily orders for these three kinds of cake, the bakery uses 72 cups of flour, 48 cups of sugar, and 60 eggs.

(a) Write a 3×1 matrix for the amounts used daily.

(b) Let the number of daily orders for cakes be a 3 × 1 matrix X with entries x_1, x_2, and x_3. Write a matrix equation that can be solved for X, using the given matrix and the matrix from part (a).

(c) Solve the equation from part (b) to find the number of daily orders for each type of cake.

Production Requirements   **52.** An electronics company produces transistors, resistors, and computer chips. Each transistor requires 3 units of copper, 1 unit of zinc, and 2 units of glass. Each resistor requires 3, 2, and 1 units of the three materials, and each computer chip requires 2, 1, and 2 units of these materials, respectively. How much of each material is needed to fill the following orders?

(a) 100 transistors, 110 resistors, 90 computer chips

(b) 95 transistors, 100 resistors, 90 computer chips

(c) 140 transistors, 130 resistors, 100 computer chips

Investments   **53.** An investment firm recommends that a client invest in AAA, A, and B rated bonds. The average yield on AAA bonds is 6%, on A bonds 7%, and on B bonds 10%. The client wants to invest twice as much in AAA bonds as in B bonds. How much should be invested in each type of bond under the following conditions?

(a) The total investment is $25,000, and the investor wants an annual return of $1810 on the three investments.

(b) The values in part (a) are changed to $30,000 and $2150 respectively.

(c) The values in part (a) are changed to $40,000 and $2900 respectively.

FOR THE COMPUTER

Use matrices C and D in Exercises 54–57.

$$C = \begin{bmatrix} -6 & 8 & 2 & 4 & -3 \\ 1 & 9 & 7 & -12 & 5 \\ 15 & 2 & -8 & 10 & 11 \\ 4 & 7 & 9 & 6 & -2 \\ 1 & 3 & 8 & 23 & 4 \end{bmatrix} \quad D = \begin{bmatrix} 5 & -3 & 7 & 9 & 2 \\ 6 & 8 & -5 & 2 & 1 \\ 3 & 7 & -4 & 2 & 11 \\ 5 & -3 & 9 & 4 & -1 \\ 0 & 3 & 2 & 5 & 1 \end{bmatrix}$$

54. Find C^{-1}.   **55.** Find $(CD)^{-1}$.   **56.** Find D^{-1}.   **57.** Is $C^{-1}D^{-1} = (CD)^{-1}$?

Solve the matrix equation $AX = B$ for X, given A and B as follows.

58. $A = \begin{bmatrix} 2 & 3 & 5 \\ 1 & 7 & 9 \\ -3 & 2 & 10 \end{bmatrix}$, $B = \begin{bmatrix} 3 \\ 4 \\ 1 \end{bmatrix}$

59. $A = \begin{bmatrix} 2 & 5 & 7 & 9 \\ 1 & 3 & -4 & 6 \\ -1 & 0 & 5 & 8 \\ 2 & -2 & 4 & 10 \end{bmatrix}$, $B = \begin{bmatrix} 3 \\ 7 \\ -1 \\ 5 \end{bmatrix}$

60. $A = \begin{bmatrix} 3 & 2 & -1 & -2 & 6 \\ -5 & 17 & 4 & 3 & 15 \\ 7 & 9 & -3 & -7 & 12 \\ 9 & -2 & 1 & 4 & 8 \\ 1 & 21 & 9 & -7 & 25 \end{bmatrix}$, $B = \begin{bmatrix} -2 \\ 5 \\ 3 \\ -8 \\ 25 \end{bmatrix}$

▤ 2.6   INPUT-OUTPUT MODELS

Nobel prize winner Wassily Leontief developed an interesting application of matrix theory to economics. His matrix models for studying the interdependencies in an economy are called *input-output* models. In practice these models are very complicated, with many variables. Only simple examples with few variables are discussed here.

Input-output models are concerned with the production and flow of goods (and perhaps services). In an economy with n basic commodities, or sectors, the production of each commodity uses some (perhaps all) of the commodities in the economy as inputs. The amounts of each commodity used in the production of one unit of each commodity can be written as an $n \times n$ matrix A, called the **technological matrix** or **input-output matrix** of the economy.

▬ EXAMPLE 1

Suppose a simplified economy involves just three commodity categories: agriculture, manufacturing, and transportation, all in appropriate units. Production of 1 unit of agriculture requires 1/2 of manufacturing and 1/4 unit of transportation; production of 1 unit of manufacturing requires 1/4 unit of agriculture and 1/4 unit of transportation; and production of 1 unit of transportation requires 1/3 unit of agriculture and 1/4 unit of manufacturing. Give the input-output matrix for this economy.

The matrix is shown below.

$$
\begin{array}{c}
 \\
\text{Agriculture} \\
\text{Manufacturing} \\
\text{Transportation}
\end{array}
\begin{array}{ccc}
\text{Agriculture} & \text{Manufacturing} & \text{Transportation} \\
\left[\begin{array}{ccc}
0 & \frac{1}{4} & \frac{1}{3} \\
\frac{1}{2} & 0 & \frac{1}{4} \\
\frac{1}{4} & \frac{1}{4} & 0
\end{array} \right] & & = A
\end{array}
$$

The first column of the input-output matrix represents the amount of each of the three commodities consumed in the production of 1 unit of agriculture. The second column gives the amounts required to produce 1 unit of manufacturing, and the last column gives the amounts required to produce 1 unit of transportation. (Although it is perhaps unrealistic that production of a unit of each commodity requires none of that commodity, the simpler matrix involved is useful for our purposes.) ▬

Another matrix used with the input-output matrix is the matrix giving the amount of each commodity produced, called the **production matrix,** or the matrix of gross output. In an economy producing n commodities, the production matrix can be represented by a column matrix X with entries $x_1, x_2, x_3, \ldots, x_n$.

■ EXAMPLE 2

In Example 1, suppose the production matrix is

$$X = \begin{bmatrix} 60 \\ 52 \\ 48 \end{bmatrix}.$$

Then 60 units of agriculture, 52 units of manufacturing, and 48 units of transportation are produced. Because 1/4 unit of agriculture is used for each unit of manufacturing produced, $1/4 \times 52 = 13$ units of agriculture must be used in the "production" of manufacturing. Similarly, $1/3 \times 48 = 16$ units of agriculture will be used in the "production" of transportation. Thus, $13 + 16 = 29$ units of agriculture are used for production in the economy. Look again at the matrices A and X. Since X gives the number of units of each commodity produced and A gives the amount (in units) of each commodity used to produce 1 unit of each of the various commodities, the matrix product AX gives the amount of each commodity used in the production process.

$$AX = \begin{bmatrix} 0 & \frac{1}{4} & \frac{1}{3} \\ \frac{1}{2} & 0 & \frac{1}{4} \\ \frac{1}{4} & \frac{1}{4} & 0 \end{bmatrix} \begin{bmatrix} 60 \\ 52 \\ 48 \end{bmatrix} = \begin{bmatrix} 29 \\ 42 \\ 28 \end{bmatrix}$$

From this result, 29 units of agriculture, 42 units of manufacturing, and 28 units of transportation are used to produce 60 units of agriculture, 52 units of manufacturing, and 48 units of transportation. ■

The matrix product AX represents the amount of each commodity used in the production process. The remainder (if any) must be enough to satisfy the demand for the various commodities from outside the production system. In an n-commodity economy, this demand can be represented by a **demand matrix** D with entries $d_1, d_2, \ldots, d_n$. The difference between the production matrix X and the amount AX used in the production process must equal the demand D, or

$$D = X - AX.$$

In Example 2,

$$D = \begin{bmatrix} 60 \\ 52 \\ 48 \end{bmatrix} - \begin{bmatrix} 29 \\ 42 \\ 28 \end{bmatrix} = \begin{bmatrix} 31 \\ 10 \\ 20 \end{bmatrix},$$

so production of 60 units of agriculture, 52 units of manufacturing, and 48 units of transportation would satisfy a demand of 31, 10, and 20 units of each commodity, respectively.

Another way to state the relationship between production X and demand D is to express X as $X = D + AX$. In practice, A and D usually are known and

X must be found. That is, we need to decide what amounts of production are needed to satisfy the required demands. Matrix algebra can be used to solve the equation $D = X - AX$ for *X*.

$$D = X - AX$$
$$D = IX - AX$$
$$D = (I - A)X$$

If the matrix $I - A$ has an inverse, then

$$X = (I - A)^{-1}D.$$

■ EXAMPLE 3

Suppose, in the 3-commodity economy from Examples 1 and 2, there is a demand for 516 units of agriculture, 258 units of manufacturing, and 129 units of transportation. What should production of each commodity be?

The demand matrix is

$$D = \begin{bmatrix} 516 \\ 258 \\ 129 \end{bmatrix}.$$

To find the production matrix *X*, first calculate $I - A$.

$$I - A = \begin{bmatrix} 1 & 0 & 0 \\ 0 & 1 & 0 \\ 0 & 0 & 1 \end{bmatrix} - \begin{bmatrix} 0 & \frac{1}{4} & \frac{1}{3} \\ \frac{1}{2} & 0 & \frac{1}{4} \\ \frac{1}{4} & \frac{1}{4} & 0 \end{bmatrix} = \begin{bmatrix} 1 & -\frac{1}{4} & -\frac{1}{3} \\ -\frac{1}{2} & 1 & -\frac{1}{4} \\ -\frac{1}{4} & -\frac{1}{4} & 1 \end{bmatrix}$$

Use row operations to find the inverse of $I - A$ (the entries are rounded to two decimal places).

$$(I - A)^{-1} = \begin{bmatrix} 1.40 & .50 & .59 \\ .84 & 1.36 & .62 \\ .56 & .47 & 1.30 \end{bmatrix}$$

Since $X = (I - A)^{-1}D$,

$$X = \begin{bmatrix} 1.40 & .50 & .59 \\ .84 & 1.36 & .62 \\ .56 & .47 & 1.30 \end{bmatrix} \begin{bmatrix} 516 \\ 258 \\ 129 \end{bmatrix} = \begin{bmatrix} 928 \\ 864 \\ 578 \end{bmatrix}.$$

(Each entry has been rounded to the nearest whole number).

The last result shows that production of 928 units of agriculture, 864 units of manufacturing, and 578 units of transportation is required to satisfy demands of 516, 258, and 129 units respectively.   ■

▬ EXAMPLE 4

An economy depends on two basic products, wheat and oil. To produce 1 metric ton of wheat requires .25 metric tons of wheat and .33 metric tons of oil. Production of 1 metric ton of oil consumes .08 metric tons of wheat and .11 metric tons of oil. Find the production that will satisfy a demand for 500 metric tons of wheat and 1000 metric tons of oil.

The input-outut matrix is

$$A = \begin{bmatrix} .25 & .08 \\ .33 & .11 \end{bmatrix}.$$

Also,

$$I - A = \begin{bmatrix} .75 & -.08 \\ -.33 & .89 \end{bmatrix}.$$

Next, calculate $(I - A)^{-1}$.

$$(I - A)^{-1} = \begin{bmatrix} 1.39 & .13 \\ .51 & 1.17 \end{bmatrix} \quad \text{(rounded)}$$

To find the production matrix X, use the equation $X = (I - A)^{-1}D$, with

$$D = \begin{bmatrix} 500 \\ 1000 \end{bmatrix}.$$

The production matrix is

$$X = \begin{bmatrix} 1.39 & .13 \\ .51 & 1.17 \end{bmatrix} \begin{bmatrix} 500 \\ 1000 \end{bmatrix} = \begin{bmatrix} 815 \\ 1425 \end{bmatrix}.$$

Production of 815 metric tons of wheat and 1425 metric tons of oil is required to satisfy the indicated demand. ▬

The input-output model discussed above is referred to as an **open model,** since it allows for a surplus from the production equal to D. In the **closed model,** all the production is consumed internally in the production process, so that $X = AX$. There is nothing left over to satisfy any outside demands from other parts of the economy or from other economies. In this case, the sum of each column in the input-output matrix equals 1.

To solve the equation $X = AX$ for X, first let O represent a column matrix with n rows, in which each element is equal to 0. Write $X = AX$ or $X - AX = O$; then

$$IX - AX = O$$
$$(I - A)X = O.$$

The system of equations that corresponds to $(I - A)X = O$ does not have a single unique solution, but it can be solved in terms of a parameter. As mentioned earlier, this means there are infinitely many solutions.

■ EXAMPLE 5

Use matrix A below to find the production of each commodity in a closed model.

$$A = \begin{bmatrix} \frac{1}{2} & \frac{1}{4} & \frac{1}{3} \\ 0 & \frac{1}{4} & \frac{1}{3} \\ \frac{1}{2} & \frac{1}{2} & \frac{1}{3} \end{bmatrix}$$

Find the value of $I - A$, then set $(I - A)X = O$ to find X.

$$I - A = \begin{bmatrix} \frac{1}{2} & -\frac{1}{4} & -\frac{1}{3} \\ 0 & \frac{3}{4} & -\frac{1}{3} \\ -\frac{1}{2} & -\frac{1}{2} & \frac{2}{3} \end{bmatrix}$$

$$(I - A)X = \begin{bmatrix} \frac{1}{2} & -\frac{1}{4} & -\frac{1}{3} \\ 0 & \frac{3}{4} & -\frac{1}{3} \\ -\frac{1}{2} & -\frac{1}{2} & \frac{2}{3} \end{bmatrix} \begin{bmatrix} x_1 \\ x_2 \\ x_3 \end{bmatrix} = \begin{bmatrix} 0 \\ 0 \\ 0 \end{bmatrix}$$

Multiply to get

$$\begin{bmatrix} \frac{1}{2} x_1 - \frac{1}{4} x_2 - \frac{1}{3} x_3 \\ 0 x_1 + \frac{3}{4} x_2 - \frac{1}{3} x_3 \\ -\frac{1}{2} x_1 - \frac{1}{2} x_2 + \frac{2}{3} x_3 \end{bmatrix} = \begin{bmatrix} 0 \\ 0 \\ 0 \end{bmatrix}.$$

The last matrix equation corresponds to the following system.

$$\frac{1}{2} x_1 - \frac{1}{4} x_2 - \frac{1}{3} x_3 = 0$$

$$\frac{3}{4} x_2 - \frac{1}{3} x_3 = 0$$

$$-\frac{1}{2} x_1 - \frac{1}{2} x_2 + \frac{2}{3} x_3 = 0$$

Solving the system with x_3 as the parameter gives a solution of the system that can be written as

$$x_1 = \frac{8}{9} x_3$$

$$x_2 = \frac{4}{9} x_3$$

$$x_3 \text{ arbitrary.}$$

If $x_3 = 9$, then $x_1 = 8$ and $x_2 = 4$, with the production of the three commodities in the ratio $8:4:9$. ■

═ 2.6 EXERCISES

Find the production matrix for the following input-output and demand matrices using the open model.

1. $A = \begin{bmatrix} .5 & .4 \\ .25 & .2 \end{bmatrix}$, $D = \begin{bmatrix} 2 \\ 4 \end{bmatrix}$

2. $A = \begin{bmatrix} .2 & .04 \\ .6 & .05 \end{bmatrix}$, $D = \begin{bmatrix} 3 \\ 10 \end{bmatrix}$

3. $A = \begin{bmatrix} .1 & .03 \\ .07 & .6 \end{bmatrix}$, $D = \begin{bmatrix} 5 \\ 10 \end{bmatrix}$

4. $A = \begin{bmatrix} .01 & .03 \\ .05 & .05 \end{bmatrix}$, $D = \begin{bmatrix} 100 \\ 200 \end{bmatrix}$

5. $A = \begin{bmatrix} .4 & 0 & .3 \\ 0 & .8 & .1 \\ 0 & .2 & .4 \end{bmatrix}$, $D = \begin{bmatrix} 1 \\ 3 \\ 2 \end{bmatrix}$

6. $A = \begin{bmatrix} .1 & .5 & 0 \\ 0 & .3 & .4 \\ .1 & .2 & .1 \end{bmatrix}$, $D = \begin{bmatrix} 10 \\ 4 \\ 2 \end{bmatrix}$

Find the ratios of products A, B, *and* C *using a closed model.*

7. $\begin{array}{c} \\ A \\ B \\ C \end{array} \begin{array}{ccc} A & B & C \\ \begin{bmatrix} .3 & .1 & .8 \\ .5 & .6 & .1 \\ .2 & .3 & .1 \end{bmatrix} \end{array}$

8. $\begin{array}{c} \\ A \\ B \\ C \end{array} \begin{array}{ccc} A & B & C \\ \begin{bmatrix} .2 & .1 & .5 \\ .4 & .3 & .4 \\ .4 & .6 & .1 \end{bmatrix} \end{array}$

▬ APPLICATIONS

BUSINESS AND ECONOMICS

Input-Ouput Open Model   *In Exercises 9 and 10, refer to Example 4.*

9. If the demand is changed to 690 metric tons of wheat and 920 metric tons of oil, how many units of each commodity should be produced?

10. Change the technological matrix so that production of 1 ton of wheat requires 1/5 metric ton of oil (and no wheat), and production of 1 metric ton of oil requires 1/3 metric tons of wheat (and no oil). To satisfy the same demand matrix, how many units of each commodity should be produced?

Input-Output Open Model   *In Exercises 11–14, refer to Example 3.*

11. If the demand is changed to 516 units of each commodity, how many units of each commodity should be produced?

12. Suppose 1/3 unit of manufacturing (no agriculture or transportation) is required to produce 1 unit of agriculture, 1/4 unit of transportation is required to produce 1 unit of manufacturing, and 1/2 unit of agriculture is required to produce 1 unit of transportation. How many units of each commodity should be produced to satisfy a demand of 1000 units for each commodity?

13. Suppose 1/4 unit of manufacturing and 1/2 unit of transportation are required to produce 1 unit of agriculture, 1/2 unit of agriculture and 1/4 unit of transportation to produce 1 unit of manufacturing, and 1/4 unit of agriculture and 1/4 unit of manufacturing to produce 1 unit of transportation. How many units of each commodity should be produced to satisfy a demand of 1000 units for each commodity?

14. If the input-output matrix is changed so that 1/4 unit of manufacturing and 1/2 unit of transportation are required to produce 1 unit of agriculture, 1/2 unit of agriculture and 1/4 unit of transportation are required to produce 1 unit of manufacturing, and 1/4 unit each of agriculture and manufacturing are required to produce 1 unit of transportation, find the number of units of each commodity that should be produced to satisfy a demand for 500 units of each commodity.

15. A primitive economy depends on two basic goods, yams and pork. Production of 1 bushel of yams requires 1/4 bushel of yams and 1/2 of a pig. To produce 1 pig requires 1/6 bushel of yams. Find the amount of each commodity that should be produced to get the following.

(a) 1 bushel of yams and 1 pig

(b) 100 bushels of yams and 70 pigs

Input-Output Closed Model   **16.** Use the input-output matrix

$$
\begin{array}{cc}
& \text{Yams} \quad \text{Pigs} \\
\begin{array}{c} \text{Yams} \\ \text{Pigs} \end{array} &
\left[\begin{array}{cc} \frac{1}{4} & \frac{1}{2} \\ \frac{3}{4} & \frac{1}{2} \end{array} \right]
\end{array}
$$

and the closed model to find the ratios of yams and pigs produced.

FOR THE COMPUTER

Find the production matrix X, given the following input-output and demand matrices.

17. $A = \begin{bmatrix} .25 & .25 & .25 & .05 \\ .01 & .02 & .01 & .1 \\ .3 & .3 & .01 & .1 \\ .2 & .01 & .3 & .01 \end{bmatrix}$ $D = \begin{bmatrix} 2930 \\ 3570 \\ 2300 \\ 580 \end{bmatrix}$

18. $A = \begin{bmatrix} .2 & .1 & .2 & .2 \\ .3 & .05 & .07 & .02 \\ .1 & .03 & .02 & .01 \\ .05 & .3 & .05 & .03 \end{bmatrix}$ $D = \begin{bmatrix} 5000 \\ 1000 \\ 8000 \\ 500 \end{bmatrix}$

19. A simple economy depends on three commodities: oil, corn, and coffee. Production of 1 unit of oil requires .1 unit of oil, .2 unit of corn, and no units of coffee. To produce 1 unit of corn requires .2 unit of oil, .1 unit of corn, and .05 unit of coffee. To produce 1 unit of coffee requires .1 unit of oil, .05 unit of corn, and .1 unit of coffee. Find the gross production required to give a net production of 1000 units each of oil, corn, coffee.

KEY WORDS

To understand the concepts presented in this chapter, you should know the meaning and use of the following words. For easy reference, the section in the chapter where a word (or expression) was first used is given with each item.

	system of equations	**2.3**	**order (dimension)**
2.1	**first-degree equation in n**		**square matrix**
	unknowns		**row matrix (row vector)**
	inconsistent system		**column matrix (column vector)**
	dependent equations		**additive inverse of a matrix**
	equivalent system		**zero matrix**
	echelon method		**additive identity**
	parameter	**2.4**	**scalar**
2.2	**matrix**	**2.5**	**identity matrix**
	element (entry)		**multiplicative inverse matrix**
	augmented matrix	**2.6**	**input-output (technological)**
	row operations		**matrix**
	Gauss-Jordan method		**production matrix**
			demand matrix

≡ CHAPTER 2   REVIEW EXERCISES

Solve each system by the echelon method.

1. $\begin{aligned} 2x + 3y &= 10 \\ -3x + y &= 18 \end{aligned}$

2. $\begin{aligned} \frac{x}{2} + \frac{y}{4} &= 3 \\ \frac{x}{4} - \frac{y}{2} &= 4 \end{aligned}$

3. $\begin{aligned} 2x - 3y + z &= -5 \\ x + 4y + 2z &= 13 \\ 5x + 5y + 3z &= 14 \end{aligned}$

4. $\begin{aligned} x - y &= 3 \\ 2x + 3y + z &= 13 \\ 3x - 2z &= 21 \end{aligned}$

Solve each system by the Gauss-Jordan method.

5. $\begin{aligned} 2x + 4y &= -6 \\ -3x - 5y &= 12 \end{aligned}$

6. $\begin{aligned} x + 2y &= -9 \\ 4x + 9y &= 41 \end{aligned}$

7. $\begin{aligned} x - y + 3z &= 13 \\ 4x + y + 2z &= 17 \\ 3x + 2y + 2z &= 1 \end{aligned}$

8. $\begin{aligned} x - 2z &= 5 \\ 3x + 2y &= 8 \\ -x + 2z &= 10 \end{aligned}$

9. $\begin{aligned} 3x - 6y + 9z &= 12 \\ -x + 2y - 3z &= -4 \\ x + y + 2z &= 7 \end{aligned}$

Find the order of each matrix, find the values of any variables, and identify any square, row, or column matrices.

10. $\begin{bmatrix} 2 & 3 \\ 5 & q \end{bmatrix} = \begin{bmatrix} a & b \\ c & 9 \end{bmatrix}$

11. $\begin{bmatrix} 2 & x \\ y & 6 \\ 5 & z \end{bmatrix} = \begin{bmatrix} a & -1 \\ 4 & 6 \\ p & 7 \end{bmatrix}$

12. $[m \quad 4 \quad z \quad -1] = [12 \quad k \quad -8 \quad r]$

13. $\begin{bmatrix} a+5 & 3b & 6 \\ 4c & 2+d & -3 \\ -1 & 4p & q-1 \end{bmatrix} \begin{bmatrix} -7 & b+2 & 2k-3 \\ 3 & 2d-1 & 4l \\ m & 12 & 8 \end{bmatrix}$

Given the matrices

$$A = \begin{bmatrix} 4 & 10 \\ -2 & -3 \\ 6 & 9 \end{bmatrix}, \quad B = \begin{bmatrix} 2 & 3 & -2 \\ 2 & 4 & 0 \\ 0 & 1 & 2 \end{bmatrix}, \quad C = \begin{bmatrix} 5 & 0 \\ -1 & 3 \\ 4 & 7 \end{bmatrix},$$

$$D = \begin{bmatrix} 6 \\ 1 \\ 0 \end{bmatrix}, \quad E = [1 \quad 3 \quad -4], \quad F = \begin{bmatrix} -1 & 4 \\ 3 & 7 \end{bmatrix}, \quad G = \begin{bmatrix} 2 & 5 \\ 1 & 6 \end{bmatrix},$$

find each of the following, if it exists.

14. $A + C$ **15.** $2G - 4F$ **16.** $3C + 2A$ **17.** $B - A$ **18.** $2A - 5C$

19. AF **20.** AC **21.** DE **22.** ED **23.** BD

24. EA **25.** F^{-1} **26.** B^{-1} **27.** $(A + C)^{-1}$

Find the inverse of each of the following matrices that has an inverse.

28. $\begin{bmatrix} 2 & 1 \\ 5 & 3 \end{bmatrix}$

29. $\begin{bmatrix} -4 & 2 \\ 0 & 3 \end{bmatrix}$

30. $\begin{bmatrix} 2 & 0 \\ -1 & 5 \end{bmatrix}$

31. $\begin{bmatrix} 6 & 4 \\ 3 & 2 \end{bmatrix}$

32. $\begin{bmatrix} 2 & -1 & 0 \\ 1 & 0 & 1 \\ 1 & -2 & 0 \end{bmatrix}$ **33.** $\begin{bmatrix} 2 & 0 & 4 \\ 1 & -1 & 0 \\ 0 & 1 & -2 \end{bmatrix}$ **34.** $\begin{bmatrix} 1 & 3 & 6 \\ 4 & 0 & 9 \\ 5 & 15 & 30 \end{bmatrix}$ **35.** $\begin{bmatrix} 2 & 3 & 5 \\ -2 & -3 & -5 \\ 1 & 4 & 2 \end{bmatrix}$

Solve the matrix equation AX = B for X using the given matrices.

36. $A = \begin{bmatrix} 2 & 4 \\ -1 & -3 \end{bmatrix}$, $B = \begin{bmatrix} 8 \\ 3 \end{bmatrix}$

37. $A = \begin{bmatrix} 1 & 3 \\ -2 & 4 \end{bmatrix}$, $B = \begin{bmatrix} 15 \\ 10 \end{bmatrix}$

38. $A = \begin{bmatrix} 1 & 0 & 2 \\ -1 & 1 & 0 \\ 3 & 0 & 4 \end{bmatrix}$, $B = \begin{bmatrix} 8 \\ 4 \\ -6 \end{bmatrix}$

39. $A = \begin{bmatrix} 2 & 4 & 0 \\ 1 & -2 & 0 \\ 0 & 0 & 3 \end{bmatrix}$, $B = \begin{bmatrix} 72 \\ -24 \\ 48 \end{bmatrix}$

Solve each of the following systems of equations by inverses.

40. $\begin{aligned} 2x + y &= 5 \\ 3x - 2y &= 4 \end{aligned}$

41. $\begin{aligned} 5x + 10y &= 80 \\ 3x - 2y &= 120 \end{aligned}$

42. $\begin{aligned} x + y + z &= 1 \\ 2x + y &= -2 \\ 3y + z &= 2 \end{aligned}$

Find each production matrix, given the following input-output and demand matrices.

43. $A = \begin{bmatrix} .01 & .05 \\ .04 & .03 \end{bmatrix}$, $D = \begin{bmatrix} 200 \\ 300 \end{bmatrix}$

44. $A = \begin{bmatrix} .2 & .1 & .3 \\ .1 & 0 & .2 \\ 0 & 0 & .4 \end{bmatrix}$, $D = \begin{bmatrix} 500 \\ 200 \\ 100 \end{bmatrix}$

■■ APPLICATIONS

BUSINESS AND ECONOMICS

Write each of Exercises 45–48 as a system of equations and solve.

Scheduling Production **45.** An office supply manufacturer makes two kinds of paper clips, standard and extra large. To make 1000 standard paper clips requires 1/4 hr on a cutting machine and 1/2 hr on a machine that shapes the clips. One thousand extra large paper clips require 1/3 hr on each machine. The manager of paper clip production has 4 hr per day available on the cutting machine and 6 hr per day on the shaping machine. How many of each kind of clip can he make?

Investment **46.** Gretchen Schmidt plans to buy shares of two stocks. One costs $32 per share and pays dividends of $1.20 per share. The other costs $23 per share and pays dividends of $1.40 per share. She has $10,100 to spend and wants to earn dividends of $540. How many shares of each stock should she buy?

Production Requirements **47.** The Waputi Indians make woven blankets, rugs, and skirts. Each blanket requires 24 hr for spinning the yarn, 4 hr for dying the yarn, and 15 hr for weaving. Rugs require 30, 5, and 18 hr and skirts 12, 3, and 9 hr, respectively. If there are 306, 59, and 201 hr available for spinning, dying, and weaving, respectively, how many of each item can be made? (*Hint:* Simplify the equations you write, if possible, before solving the system.)

Distribution   **48.** An oil refinery in Tulsa sells 50% of its production to a Chicago distributor, 20% to a Dallas distributor, and 30% to an Atlanta distributor. Another refinery in New Orleans sells 40% of its production to the Chicago distributor, 40% to the Dallas distributor, and 20% to the Atlanta distributor. A third refinery in Ardmore sells the same distributors 30%, 40%, and 30% of its production. The three distributors received 219,000, 192,000, and 144,000 gallons of oil respectively. How many gallons of oil were produced at each of the three plants?

Stock Reports   **49.** The New York Stock Exchange reports in daily newspapers give the dividend, price-to-earnings ratio, sale (in hundreds of shares), last price, and change in price for each company. Write the following stock reports as a 4×5 matrix: American Telephone & Telegraph: 5, 7, 2532, $52\frac{3}{8}$, $-1/4$; General Electric: 3, 9, 1464, 56, $+\frac{1}{8}$; Mobil Oil: 2.50, 5, 4974, 41, $-1\frac{1}{2}$; Sears: 1.36, 10, 1754, $18\frac{7}{8}$, $+\frac{1}{2}$.

Filling Orders   **50.** A printer has three orders for pamphlets that require three kinds of paper, as shown in the following matrix.

$$
\begin{array}{cc}
 & \begin{array}{ccc} & Order & \\ \text{I} & \text{II} & \text{III} \end{array} \\
Paper\ \begin{array}{r} \text{High-grade} \\ \text{Medium-grade} \\ \text{Coated} \end{array} & \begin{bmatrix} 10 & 5 & 8 \\ 12 & 0 & 4 \\ 0 & 10 & 5 \end{bmatrix}
\end{array}
$$

The printer has on hand 3170 sheets of high-grade paper, 2360 sheets of medium-grade paper, and 1800 sheets of coated paper. All the paper must be used in preparing the order.

(a) Write a 3×1 matrix for the amounts of paper on hand.

(b) Write a matrix of variables to represent the number of pamphlets that must be printed in each of the three orders.

(c) Write a matrix equation using the given matrix and your matrices from parts (a) and (b).

(d) Solve the equation from part (c).

Input-Output   **51.** An economy depends on two commodities, goats and cheese. It takes 2/3 of a unit of goats to produce 1 unit of cheese and 1/2 unit of cheese to produce 1 unit of goats.

(a) Write the input-output matrix for this economy.

(b) Find the production required to satisfy a demand of 400 units of cheese and 800 units of goats.

LIFE SCIENCES

Animal Activity   **52.** The activities of a grazing animal can be classified roughly into three categories: grazing, moving, and resting. Suppose horses spend 8 hr grazing, 8 moving, and 8 resting; cattle spend 10 grazing, 5 moving and 9 resting; sheep spend 7 grazing, 10 moving, and 7 resting; and goats spend 8 grazing, 9 moving, and 7 resting. Write this information as a 4×3 matrix.

EXTENDED
APPLICATION

ROUTING

The diagram in Figure 5 shows the roads connecting four citie . Another way of representing this information is shown in matrix A, where the entries represent the number of roads connecting two cities without passing through another city.* For example, from the diagram we see that there are two roads connecting city 1 to city 4 without passing through either city 2 or 3. This information is entered in row 1, column 4 and again in row 4, colum 1 of matrix A.

$$A = \begin{bmatrix} 0 & 1 & 2 & 2 \\ 1 & 0 & 1 & 0 \\ 2 & 1 & 0 & 1 \\ 2 & 0 & 1 & 0 \end{bmatrix}$$

Note that there are 0 roads connecting each city to itself. Also, there is one road connecting cities 3 and 2.

How many ways are there to go from city 1 to city 2, for example, by going through exactly one other city? Since we must go through one other city, we must go through either city 3 or city 4. On the diagram in Figure 5, we see that we can go from city 1 to city 2 through city 3 in 2 ways. We can go from city 1 to city 3 in 2 ways and then from city 3 to city 2 in one way, giving the $2 \cdot 1 = 2$ ways to get from city 1 to city 2 through city 3. It is not possible to go from city 1 to city 2 through city 4, because there is no direct route between cities 4 and 2.

Now multiply matrix A by itself, to get A^2. Let the first-row, second-column entry of A^2 be b_{12}. (We use a_{ij} to denote the entry in the ith row and jth column of matrix A.) The entry b_{12} is found as follows.

$$b_{12} = a_{11}a_{12} + a_{12}a_{22} + a_{13}a_{32} + a_{14}a_{42}$$
$$= 0 \cdot 1 + 1 \cdot 0 + 2 \cdot 1 + 2 \cdot 0$$
$$= 2$$

The matrix A^2 gives the number of ways to travel between any two cities by passing through exactly one other city. The first product $0 \cdot 1$ in the calculations above represents the number of ways to go from city 1 to city 1 (0) and then from city 1 to city 2 (1). The 0 result indicates that such a trip does not involve a third city. The only nonzero product $(2 \cdot 1)$ represents the two routes from city 1 to city 3 and the one route from city 3 to city 2 that result in the $2 \cdot 1$ or 2 routes from city 1 to city 2 by going through city 3.

Similarly, A^3 gives the number of ways to travel between any two cities by passing through exactly two cities. Also, $A + A^2$ represents the total number of ways to travel between two cities with at most one intermediate city.

The diagram can be given many other interpretations. For example, the lines could represent lines of mutual influence between people or nations, or they could represent communication lines such as telephone lines.

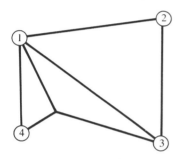

FIGURE 5

---

*Taken from Hugh G. Campbell, *Matrices With Applications*, © 1968, p. 50–51. Reprinted by permission of Prentice-Hall, Inc., Englewood Cliffs, N.J.

EXERCISES

1. Use matrix A from the text to find A^2. Then answer the following questions.
   (a) How many ways are there to travel from city 1 to city 3 by passing through exactly one city?
   (b) How many ways are there to travel from city 2 to city 4 by passing through exactly one city?
   (c) How many ways are there to travel from city 1 to city 3 by passing through at most one city?
   (d) How many ways are there to travel from city 2 to city 4 by passing through at most one city?

2. Find A^3. Then answer the following questions.
   (a) How many ways are there to travel between cities 1 and 4 by passing through exactly two cities?
   (b) How many ways are there to travel between cities 1 and 4 by passing through at most two cities?

3. A small telephone system connects three cities. There are four lines between cities 3 and 2, three lines connecting city 3 with city 1, and two lines between cities 1 and 2.
   (a) Write a matrix B to represent this information.
   (b) Find B^2.
   (c) How many lines that connect cities 1 and 2 go through exactly one other city (city 3)?
   (d) How many lines that connect cities 1 and 2 go through at most one other city?

4. The figure shows four southern cities served by Delta Airlines.

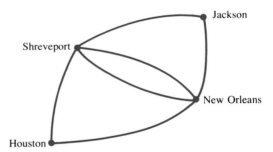

   (a) Write a matrix to represent the number of nonstop routes between cities.
   (b) Find the number of one-stop flights between Houston and Jackson.
   (c) Find the number of flights between Houston and Jackson that require at most one stop.
   (d) Find the number of one-stop flights between New Orleans and Houston.

5. The figure shows a food web. The arrows indicate the food sources of each population. For example, cats feed on rats and on mice.

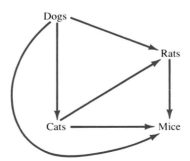

(a) Write a matrix C in which each row and corresponding column represents a population in the food chain. Enter a 1 when the population in a given row feeds on the population in the given column and a 0 otherwise.

(b) Calculate and interpret C^2.

6. Use a computer to find A^2, B^2, and C^2 for the matrices used in these problems.

EXTENDED APPLICATION

CONTAGION

Suppose that three people have contracted a contagious disease.[*] A second group of five people may have been in contact with the three infected persons. A third group of six people may have been in contact with the second group. We can form a 3 × 5 matrix P with rows representing the first group of three and columns representing the second group of five. We enter a 1 in the corresponding position if a person in the first group has contact with a person in the second group. These direct contacts are called *first-order contacts*. Similarly, we form a 5 × 6 matrix Q representing the first-order contacts between the second and third groups. For example, suppose.

$$P = \begin{bmatrix} 1 & 0 & 0 & 1 & 0 \\ 0 & 0 & 1 & 1 & 0 \\ 1 & 1 & 0 & 0 & 0 \end{bmatrix} \quad \text{and} \quad Q = \begin{bmatrix} 1 & 1 & 0 & 1 & 1 & 1 \\ 0 & 0 & 0 & 0 & 1 & 0 \\ 0 & 0 & 0 & 0 & 0 & 0 \\ 0 & 1 & 0 & 1 & 0 & 0 \\ 1 & 0 & 0 & 0 & 1 & 0 \end{bmatrix}.$$

From matrix P we see the first person in the first group had contact with the first and fourth persons in the second group. Also, no one in the first group had contact with the last person in the second group.

A *second-order contact* is an indirect contact between persons in the first and third groups through some person in the second group. The product matrix PQ indicates these contacts. Verify that the second-row, fourth-column entry of PQ is 1. That is, there is one second-order contact between the second person in group 1 and the fourth person in group 3. Let a_{ij} denote the element in the ith row and jth column of the matrix PQ. By looking at the products that form a_{24} below, we see that the common contact was with the fourth individual in group 2. (The p_{ij} are entries in P, and the q_{ij} are entries in Q.)

$$\begin{aligned} a_{24} &= p_{21}q_{14} + p_{22}q_{24} + p_{23}q_{34} + p_{24}q_{44} + p_{25}q_{54} \\ &= 0 \cdot 1 \quad + 0 \cdot 0 \quad + 1 \cdot 0 \quad + 1 \cdot 1 \quad + 0 \cdot 1 \\ &= 1 \end{aligned}$$

The second person in group 1 and the fourth person in group 3 both had contact with the fourth person in group 2.

This idea could be extended to third-, fourth-, and larger order contacts. It indicates a way to use matrices to trace the spread of a contagious disease. It could also pertain to the dispersal of ideas or anything that might pass from one individual to another.

[*]Reprinted by permission of Stanley I. Grossman and James E. Turner from *Mathematics for the Biological Sciences* (New York: Macmillan Publishing Company, Inc., 1974). See also Stanley I. Grossman, *Finite Mathematics with Applications to Business, Life Sciences, and Social Sciences* (Belmont, CA: Wadsworth Publishing Company, 1983), p. 103.

EXERCISES

1. Find the second-order contact matrix *PQ* mentioned in the text.
2. How many second-order contacts were there between the second contagious person and the third person in the third group?
3. Is there anyone in the third group who has had no contacts at all with the first group?
4. The totals of the columns in *PQ* give the total number of second-order contacts per person, while the column totals in *P* and *Q* give the total number of first-order contacts per person. Which person has the most contacts, counting both first- and second-order contacts?
5. Use a computer to find the matrix product in Exercise 1.

EXTENDED APPLICATION

LEONTIEF'S MODEL OF THE AMERICAN ECONOMY

In the April 1965 issue of *Scientific American,* Leontief explained his input-output system, using the 1958 American economy as an example.* He divided the economy into 81 sectors, grouped into six families of related sectors. In order to keep the discussion reasonably simple, we will treat each family of sectors as a single sector and so, in effect, work with a six-sector model. The sectors are listed in Table 1.

The workings of the American economy in 1958 are described in the input-output table (Table 2) based on Leontief's figures. We will demonstrate the meaning of Table 2 by considering the first left-hand column of numbers. The numbers in this column mean that 1 unit of final nonmetal production requires the consumption of .170 unit of (other) final nonmetal production, .003 unit of final metal output, .025 unit of basic metal products, and so on down the column. Since the unit of measurement that Leontief used for this table is millions of dollars, we conclude that $1 million worth of final nonmetal production consumes $.170 million, or $170,000, worth of other final nonmetal products, $3000 of final metal production, $25,000 of basic metal products, and so on. Similarly, the entry in the column headed FM and opposite S of .074 means that $74,000 worth of output from the service industries goes into the production of $1 million worth of final metal products, and the number .358 in the column headed E and opposite E means that $358,000 worth of energy must be consumed to produce $1 million worth of energy.

Table 1

Sector	Examples
Final nonmetal (FN)	Furniture, processed food
Final metal (FM)	Household appliances, motor vehicles
Basic metal (BM)	Machine-shop products, mining
Basic nonmetal (BN)	Agriculture, printing
Energy (E)	Petroleum, coal
Services (S)	Amusements, real estate

*From *Applied Finite Mathematics* by Robert F. Brown and Brenda W. Brown. (Belmont, California: Wadsworth Publishing Company, Inc., 1977). Reprinted by permission of the authors.

Table 2

	FN	FM	BM	BN	E	S
FN	.170	.004	0	.029	0	.008
FM	.003	.295	.018	.002	.004	.016
BM	.025	.173	.460	.007	.011	.007
BN	.348	.037	.021	.403	.011	.048
E	.007	.001	.039	.025	.358	.025
S	.120	.074	.104	.123	.173	.234

By the underlying assumption of Leontief's model, the production of *n* units (*n* = any number) of final nonmetal production consumes .170*n* unit of final nonmetal output, .003*n* unit of final metal output, .025*n* unit of basic metal production, and so on. Thus, production of $50 million worth of products from the final nonmetal section of the 1958 American economy required (.170)(50) = 8.5 units ($8.5 million) worth of final nonmetal output, (.003)(50) = .15 unit of final metal output, (.025)(50) = 1.25 units of basic metal production, and so on.

EXAMPLE 1

According to the simplified input-output table for the 1958 American economy, how many dollars' worth of final metal products, basic nonmetal products, and services are required to produce $120 million worth of basic metal products?

Each unit ($1 million worth) of basic metal products requires .018 unit of final metal products because the number in the BM column opposite FM is .018. Thus, $120 million, or 120 units, requires (.018)(120) = 2.16 units, or $2.16 million of final metal products. Similarly, 120 units of basic metal production uses (.021)(120) = 2.52 units of basic nonmetal production and (.104)(120) = 12.48 units of services, or $2.52 million and $12.48 million of basic nonmetal output and services, respectively.

The Leontief model also involves a *bill of demands,* that is, a list of requirements for units of output beyond that required for its inner workings as described in the input-output table. These demands represent exports, surpluses, government and individual consumption, and the like. The bill of demands for the simplified version of the 1958 American economy we have been using was (in millions)

FN	$99,640
FM	$75,548
BM	$14,444
BN	$33,501
E	$23,527
S	$263,985.

We can now use the methods developed above to answer the following question: How many units of output from each sector are needed in order to run the economy and fill the bill of demands? The units of output from each sector required to run the economy and fill the bill of demands is unknown, so we denote them by variables. In our example, there are six quantities that are, at the moment, unknown. The number of units of final nonmetal production required to solve the problem will be our first unknown, because this sector is represented by the first row of the input-output matrix. The unknown quantity of final nonmetal units will be represented by the symbol x_1. Following the same pattern, we represent the unknown quantities in the following manner:

$$x_1 = \text{units of final nonmetal production required,}$$
$$x_2 = \text{units of final metal production required,}$$
$$x_3 = \text{units of basic metal production required,}$$
$$x_4 = \text{units of basic nonmetal production required,}$$
$$x_5 = \text{units of energy required,}$$
$$x_6 = \text{units of services required.}$$

These six numbers are the quantities we are attempting to calculate.

To find these numbers, first let A be the 6×6 matrix corresponding to the input-output table.

$$A = \begin{bmatrix} .170 & .004 & 0 & .029 & 0 & .008 \\ .003 & .295 & .018 & .002 & .004 & .016 \\ .025 & .173 & .460 & .007 & .011 & .007 \\ .348 & .037 & .021 & .403 & .011 & .048 \\ .007 & .001 & .039 & .025 & .358 & .025 \\ .120 & .074 & .104 & .123 & .173 & .234 \end{bmatrix}$$

A is the input-output matrix. The bill of demands leads to a 6×1 demand matrix D, and X is the matrix of unknowns.

$$D = \begin{bmatrix} 99,640 \\ 75,548 \\ 14,444 \\ 33,501 \\ 23,527 \\ 263,985 \end{bmatrix} \quad \text{and} \quad X = \begin{bmatrix} x_1 \\ x_2 \\ x_3 \\ x_4 \\ x_5 \\ x_6 \end{bmatrix}$$

Now we need to find $I - A$.

$$I - A = \begin{bmatrix} 1 & 0 & 0 & 0 & 0 & 0 \\ 0 & 1 & 0 & 0 & 0 & 0 \\ 0 & 0 & 1 & 0 & 0 & 0 \\ 0 & 0 & 0 & 1 & 0 & 0 \\ 0 & 0 & 0 & 0 & 1 & 0 \\ 0 & 0 & 0 & 0 & 0 & 1 \end{bmatrix} - \begin{bmatrix} .170 & .004 & 0 & .029 & 0 & .008 \\ .003 & .295 & .018 & .002 & .004 & .016 \\ .025 & .173 & .460 & .007 & .011 & .007 \\ .348 & .037 & .021 & .403 & .011 & .048 \\ .007 & .001 & .039 & .025 & .358 & .025 \\ .120 & .074 & .104 & .123 & .173 & .234 \end{bmatrix}$$

$$= \begin{bmatrix} .830 & -.004 & 0 & -.029 & 0 & -.008 \\ -.003 & .705 & -.018 & -.002 & -.004 & -.016 \\ -.025 & -.173 & .540 & -.007 & -.011 & -.007 \\ -.348 & -.037 & -.021 & .597 & -.011 & -.048 \\ -.007 & -.001 & -.039 & -.025 & .642 & -.025 \\ -.120 & -.074 & -.104 & -.123 & -.173 & .766 \end{bmatrix}$$

Find the inverse (actually an approximation) by the methods of this chapter.

$$(I - A)^{-1} = \begin{bmatrix} 1.234 & 0.014 & .006 & .064 & .007 & .018 \\ .017 & 1.436 & .057 & .012 & .020 & .032 \\ .071 & .465 & 1.877 & .019 & .045 & .031 \\ .751 & .134 & .100 & 1.740 & .066 & .124 \\ .060 & .045 & .130 & .082 & 1.578 & .059 \\ .339 & .236 & .307 & .312 & .376 & 1.349 \end{bmatrix}$$

Therefore,

$$X = (I - A)^{-1}D = \begin{bmatrix} 1.234 & .014 & .006 & .064 & .007 & .018 \\ .017 & 1.436 & .057 & .012 & .020 & .032 \\ .071 & .465 & 1.877 & .019 & .045 & .031 \\ .751 & .134 & .100 & 1.740 & .066 & .124 \\ .060 & .045 & .130 & .082 & 1.578 & .059 \\ .339 & .236 & .307 & .312 & .376 & 1.349 \end{bmatrix} \begin{bmatrix} 99,640 \\ 75,548 \\ 14,444 \\ 33,501 \\ 23,527 \\ 263,985 \end{bmatrix} = \begin{bmatrix} 131,161 \\ 120,324 \\ 79,194 \\ 178,936 \\ 66,703 \\ 426,542 \end{bmatrix}$$

From this result,

$$x_1 = 131,161$$
$$x_2 = 120,324$$
$$x_3 = 79,194$$
$$x_4 = 178,936$$
$$x_5 = 66,703$$
$$x_6 = 426,542.$$

In other words, by this model 131,161 units ($131,161 million worth) of final nonmetal production, 120,324 units of final metal output, 79,194 units of basic metal products, and so on are required to run the 1958 American economy and completely fill the stated bill of demands. ▬

EXERCISES

1. A much simplified version of Leontief's 42-sector analysis of the 1947 American economy divides the economy into just three sectors: agriculture, manufacturing, and the household (i.e., the sector of the economy that produces labor). It consists of the following input-output table:

	Agriculture	Manufacturing	Household
Agriculture	.245	.102	.051
Manufacturing	.099	.291	.279
Household	.433	.372	.011

The bill of demands (in billions of dollars) is

Agriculture	2.88
Manufacturing	31.45
Household	30.91.

(a) Write the input-output matrix A, the demand matrix D, and the matrix X.

(b) Compute $I - A$.

(c) Check that

$$(I - A)^{-1} = \begin{bmatrix} 1.454 & .291 & .157 \\ .533 & 1.763 & .525 \\ .837 & .791 & 1.278 \end{bmatrix}$$

is an approximation to the inverse of $I - A$ by calculating $(I - A)^{-1}(I - A)$.

(d) Use the matrix of part (c) to compute X.

(e) Explain the meaning of the numbers in X in dollars.

2. An analysis of the 1958 Israeli economy* is here simplified by grouping the economy into three sectors: agriculture, manufacturing, and energy. The input-output table is the following.

	Agriculture	Manufacturing	Energy
Agriculture	.293	0	0
Manufacturing	.014	.207	.017
Energy	.044	.010	.216

Exports (in thousands of Israeli pounds) were

Agriculture	138,213
Manufacturing	17,597
Energy	1,786

(a) Write the input-output matrix A and the demand (export) matrix D.

(b) Compute $I - A$.

(c) Check that

$$(I - A)^{-1} = \begin{bmatrix} 1.414 & 0 & 0 \\ .027 & 1.261 & .027 \\ .080 & .016 & 1.276 \end{bmatrix}$$

is an approximation to the inverse of $I - A$ by calculating $(I - A)^{-1}(I - A)$.

(d) Use the matrix of part (c) to determine the number of Israeli pounds' worth of agricultural products, manufactured goods, and energy required to run this model of the Israeli economy and export the stated value of products.

---

*Wassily Leontief, *Input-Output Economics* (New York: Oxford University Press, 1966), pp. 54–57.

3 Linear Programming: The Graphical Method

3.1 Graphing Linear Inequalities

3.2 Mathematical Models for Linear Programming

3.3 Solving Linear Programming Problems Graphically

Review Exercises

Many realistic problems involve inequalities—a factory can manufacture *no more* than 12 items on a shift, or a medical researcher must interview *at least* a hundred patients to be sure that a new treatment for a disease is better than the old treatment. *Linear inequalities* of the form $ax + by \leq c$ (or with $\geq$, $<$, or $>$ instead of $\leq$) can be used in a process called *linear programming* to *optimize* (find the maximum or minimum value of a quantity for) a given situation.

In this chapter we introduce some *linear programming* problems that can be solved by graphical methods. Then, in Chapter 4, we discuss the simplex method, a general method for solving linear programming problems with many variables.

▬ 3.1 GRAPHING LINEAR INEQUALITIES

As mentioned above, a linear inequality is defined as follows.

LINEAR INEQUALITY

> ▬ A **linear inequality** in two variables has the form
> $$ax + by \leq c$$
> for real numbers a, b, and c, with a and b not both 0.

Although we will generally use the symbol $\leq$ in definitions and theorems involving inequalities, keep in mind that $\leq$ may be replaced with $\geq$, $<$, or $>$.

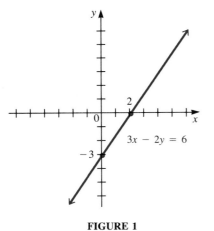

FIGURE 1

▬ EXAMPLE 1

Graph the linear inequality $3x - 2y \leq 6$.

Because of the "$=$" portion of $\leq$, the points of the line $3x - 2y = 6$ satisfy the linear inequality $3x - 2y \leq 6$ and are part of its graph. As in Chapter 1, find the intercepts by first letting $x = 0$ and then letting $y = 0$; use these points to get the graph of $3x - 2y = 6$ shown in Figure 1.

The points on the line satisfy "$3x - 2y$ *equals* 6." To locate the points satisfying "$3x - 2y$ *is less than* or equal to 6," first solve $3x - 2y \leq 6$ for y.

$$3x - 2y \leq 6$$
$$-2y \leq -3x + 6$$
$$y \geq \frac{3}{2}x - 3$$

(Recall that multiplying both sides of an inequality by a negative number reverses the direction of the inequality symbol.)

As shown in Figure 2, the points *above* the line $3x - 2y = 6$ satisfy

$$y > \frac{3}{2}x - 3,$$

while those below the line satisfy

$$y < \frac{3}{2}x - 3.$$

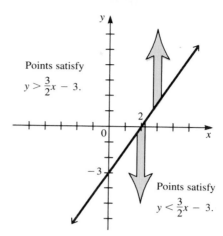

FIGURE 2

In summary, the inequality $3x - 2y \leq 6$ is satisfied by all points *on or above* the line $3x - 2y = 6$. Indicate the points above the line by shading, as in Figure 3. The line and shaded region in Figure 3 make up the graph of the linear inequality $3x - 2y \leq 6$. ▬▬

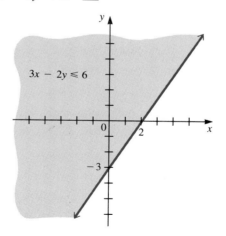

FIGURE 3

In Example 1, the line $3x - 2y = 6$, which separates the points in the solution from the points that are not in the solution, is called the **boundary.**

▰ EXAMPLE 2

Graph $x + 4y < 4$.

The boundary here is the line $x + 4y = 4$. Since the points on this line do not satisfy $x + 4y < 4$, the line is drawn dashed, as in Figure 4. To decide whether to shade the region above the line or the region below the line, solve for y.

$$x + 4y < 4$$
$$4y < -x + 4$$
$$y < -\frac{1}{4}x + 1$$

Since y is less than $(-1/4)x + 1$, the solution is the region below the boundary, as shown by the shaded region in Figure 4. ▰

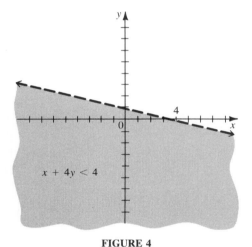

FIGURE 4

There is an alternative way to find the correct region to shade, or to check the method shown above. Choose as a test point any point not on the boundary line. For example, in Example 2 we could choose the point $(1, 0)$, which is not on the line $x + 4y = 4$. Substitute 1 for x and 0 for y in the given inequality.

$$x + 4y < 4$$
$$1 + 4(0) < 4 \qquad \text{Let } x = 1, y = 0$$
$$1 < 4 \qquad \text{True}$$

Since the result $1 < 4$ is true, the test point $(1, 0)$ belongs on the side of the boundary line where all the points satisfy $x + 4y < 4$. For this reason, we shade the side containing $(1, 0)$, as in Figure 4. Choosing a point on the other side of the line, such as $(1, 5)$, would produce a false result when the values $x = 1$ and $y = 5$ were substituted into the given inequality. In such a case, we would shade the side of the line *not including* the test point.

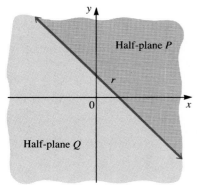

Half-plane *P*

r

0

Half-plane *Q*

FIGURE 5

As the examples above suggest, the graph of a linear inequality is represented as a shaded region in the plane, perhaps including the line that is the boundary of the region. Each of the shaded regions is an example of a **half-plane,** a region on one side of a line. For example, in Figure 5 line *r* divides the plane into half-planes *P* and *Q*. The points on *r* belong neither to *P* nor to *Q*. Line *r* is the boundary of each half-plane.

EXAMPLE 3

Graph $x \leq -1$.

Recall that the graph of $x = -1$ is the vertical line through $(-1, 0)$. To decide which half-plane belongs to the solution, choose a test point. Choosing $(2, 0)$ and replacing *x* with 2 gives a false statement:

$$x < -1$$
$$2 < -1. \quad \text{False}$$

The correct half-plane is the one that does *not* contain $(2, 0)$; it is shaded in Figure 6.

EXAMPLE 4

Graph $x \geq 3y$.

Start by graphing the boundary, $x = 3y$. If $x = 0$, then $y = 0$, giving the point $(0, 0)$. Setting $y = 0$ would produce $(0, 0)$ again. To get a second point on the line, choose another number for *x*. Choosing $x = 3$ gives $y = 1$, with $(3, 1)$ a point on the line $x = 3y$. The points $(0, 0)$ and $(3, 1)$ lead to the line graphed in Figure 7. To decide which half-plane to shade, let us choose $(6, 1)$ as a test point. (Any point that is not on the line $x = 3y$ may be used.) Replacing *x* with 6 and *y* with 1 in the original inequality gives a true statement, so the half-plane containing $(6, 1)$ is shaded, as shown in Figure 7.

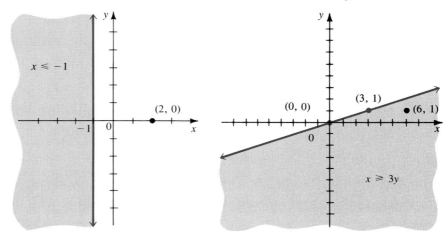

FIGURE 6   **FIGURE 7**

The steps in graphing a linear inequality are summarized below.

GRAPHING A LINEAR INEQUALITY

1. Draw the graph of the boundary line. Make the line solid if the inequality involves ≤ or ≥; make the line dashed if the inequality involves < or >.
2. Decide which half-plane to shade. Use either of the following methods.
   (a) Solve the inequality for y; shade the region above the line if the inequality uses > or ≥; shade the region below the line if the inequality uses < or ≤.
   (b) Choose any point not on the line as a test point. Shade the half-plane that includes the test point if the test point satisfies the original inequality; otherwise, shade the half-plane on the other side of the boundary line.

Systems of Inequalities   Realistic problems often involve many inequalities. For example, a manufacturing problem might produce inequalities resulting from production requirements, as well as inequalities about cost requirements. A collection of at least two inequalities is called a **system of inequalities.** The **solution** of a system of inequalities is made up of all those points that satisfy all the inequalities of the system at the same time. To graph the solution of a system of inequalities, graph all the inequalities on the same axes and identify, by heavy shading, the region common to all graphs. The next example shows how this is done.

■ EXAMPLE 5

Graph the system

$$y < -3x + 12$$
$$x < 2y.$$

The heavily shaded region in Figure 8 shows all the points that satisfy both inequalities of the system. Since the points on the boundary lines are not in the solution, the boundary lines are dashed.   ■

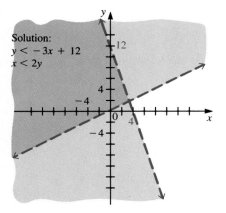

Solution:
$y < -3x + 12$
$x < 2y$

FIGURE 8

A region consisting of the overlapping parts of two or more graphs of inequalities in a system, such as the heavily shaded region in Figure 8, is sometimes called the **region of feasible solutions** or the **feasible region,** since it is made up of all the points that satisfy (are feasible for) each inequality of the system.

▬ EXAMPLE 6

Graph the feasible region for the system

$$2x - 5y \leq 10$$
$$x + 2y \leq 8$$
$$x \geq 0$$
$$y \geq 0.$$

On the same axes, graph each inequality by graphing the boundary and choosing the appropriate half-plane. Then find the feasible region by locating the overlap of all the half-planes. This feasible region is shaded in Figure 9. The inequalities $x \geq 0$ and $y \geq 0$ restrict the feasible region to the first quadrant. ▬

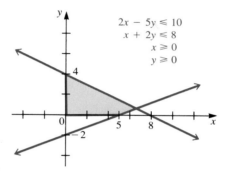

FIGURE 9

Applications   As shown in the rest of this chapter, many realistic problems lead to systems of linear inequalities. The next example is typical of such problems.

▬ EXAMPLE 7

Midtown Manufacturing Company makes plastic plates and cups, both of which require time on two machines. Manufacturing a unit of plates requires one hour on machine A and two on machine B, and producing a unit of cups requires three hours on machine A and one on machine B. Each machine is operated for at most 15 hours per day. Write a system of inequalities that expresses these restrictions, and sketch the feasible region.

Start by making a chart that summarizes the given information.

	Number Made	*Time on Machine* A	B
Plates	x	1	2
Cups	y	3	1
Maximum Time Available		15	15

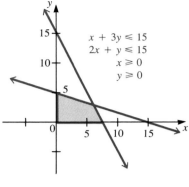

FIGURE 10

Here x represents the number of units of plates to be made, and y represents the number of units of cups.

On machine A, x units of plates require a total of $1 \cdot x = x$ hours, and y units of cups require $3 \cdot y = 3y$ hours. Since machine A is available no more than 15 hours per day,

$$x + 3y \leq 15.$$

The requirement that machine B be used no more than 15 hours per day gives

$$2x + y \leq 15.$$

Since it is not possible to produce a negative number of cups or plates,

$$x \geq 0 \quad \text{and} \quad y \geq 0.$$

The feasible region for this system of inequalities is shown in Figure 10. ▬

3.1 EXERCISES

Graph the following linear inequalities.

1. $x + y \leq 2$
2. $y \leq x + 1$
3. $x \geq 3 + y$
4. $y \geq x - 3$

5. $4x - y < 6$
6. $3y + x > 4$
7. $3x + y < 6$
8. $2x - y > 2$

9. $x + 3y \geq -2$
10. $2x + 3y \leq 6$
11. $4x + 3y > -3$
12. $5x + 3y > 15$

13. $2x - 4y < 3$
14. $4x - 3y < 12$
15. $x \leq 5y$
16. $2x \geq y$

17. $-3x < y$
18. $-x > 6y$
19. $x + y \leq 0$
20. $3x + 2y \geq 0$

21. $y < x$
22. $y > -2x$
23. $x < 4$
24. $y > 5$

25. $y \leq -2$
26. $x \geq 3$

Graph the feasible region for each system of inequalities.

27. $x + y \leq 1$
$\quad\;\; x - y \geq 2$

28. $3x - 4y < 6$
$\quad\;\; 2x + 5y > 15$

29. $2x - y < 1$
$\quad\;\; 3x + y < 6$

30. $\;\;\; x + 3y \leq 6$
$\quad\;\; 2x + 4y \geq 7$

31. $-x - y < 5$
$\quad\;\; 2x - y < 4$

32. $6x - 4y > 8$
$\quad\;\; 3x + 2y > 4$

33. $\;\; x + y \leq \;\;\; 4$
$\quad\;\; x - y \leq \;\;\; 5$
$\quad 4x + y \leq -4$

34. $3x - 2y \geq \;\;\; 6$
$\quad\;\; x + \;\; y \leq -5$
$\quad\quad\quad\quad\; y \leq \;\;\; 4$

35. $-2 < x < 3$
$\quad -1 \leq y \leq 5$
$\quad\; 2x + y < 6$

36. $-2 < x < 2$
$\quad\quad\quad\;\; y > 1$
$\quad\;\; x - y > 0$

37. $2y + x \geq -5$
$\quad\quad\;\; y \leq 3 + x$
$\quad\quad\quad\; x \geq 0$
$\quad\quad\quad\; y \geq 0$

38. $2x + 3y \leq \quad 12$
$\quad\; 2x + 3y > -6$
$\quad\; 3x + \;\; y < \quad 4$
$\quad\quad\quad\quad\; x \geq \quad 0$
$\quad\quad\quad\quad\; y \geq \quad 0$

39. $3x + 4y > 12$
$\quad\; 2x - 3y < \;\; 6$
$\quad\quad\; 0 \leq y \leq \;\; 2$
$\quad\quad\quad\quad x \geq \;\; 0$

40. $0 \leq \;\; x \leq \;\; 9$
$\quad\;\; x - 2y \geq \;\; 4$
$\quad 3x + 5y \leq 30$
$\quad\quad\quad\quad y \geq \;\; 0$

▤ APPLICATIONS

BUSINESS AND ECONOMICS

Production Scheduling **41.** A small pottery shop makes two kinds of planters, glazed and unglazed. The glazed type requires 1/2 hr to throw on the wheel and 1 hr in the kiln. The unglazed type takes 1 hr to throw on the wheel and 6 hr in the kiln. The wheel is available for at most 8 hr per day, and the kiln for at most 20 hr per day.

(a) Complete this chart.

	Number Made	Time on Wheel	Time in Kiln
Glazed	x		
Unglazed	y		

Maximum Time Available

(b) Set up a system of inequalities and graph the feasible region.

Time Management **42.** Carmella and Walt produce handmade shawls and afghans. They spin the yarn, dye it, and then weave it. A shawl requires 1 hr of spinning, 1 hr of dyeing, and 1 hr of weaving. An afghan needs 2 hr of spinning, 1 hr of dyeing, and 4 hr of weaving. Together, they spend at most 8 hr spinning, 6 hr dyeing, and 14 hr weaving.

(a) Complete this chart.

	Number Made	Spinning Time	Dyeing Time	Weaving Time
Shawls	x			
Afghans	y			
Maximum Time Available		8	6	14

(b) Set up a system of inequalities and graph the feasible region.

For Exercises 43–46, perform each of the following steps.

(a) *Write a system of inequalities to express the conditions of the problem.*

(b) *Graph the feasible region of the system.*

Transportation **43.** Southwestern Oil supplies two distributors located in the Northwest. One distributor needs at least 3000 barrels of oil, and the other needs at least 5000 barrels. Southwestern can send out at most 10,000 barrels. Let x = the number of barrels of oil sent to distributor 1 and y = the number sent to distributor 2.

Finance **44.** The loan department in a bank has set aside $25 million for commercial and home loans. The bank's policy is to allocate at least four times as much money to home loans as to commercial loans. The bank's return is 12% on a home loan and 10% on a commercial loan. The manager of the loan department wants to earn a return of at least $2.8 million on these loans. Let x = the amount for home loans and y = the amount for commercial loans.

Transportation 45. The California Almond Growers have 2400 boxes of almonds to be shipped from their plant in Sacramento to Des Moines and San Antonio. The Des Moines market needs at least 1000 boxes, while the San Antonio market must have at least 800 boxes. Let x = the number of boxes to be shipped to Des Moines and y = the number of boxes to be shipped to San Antonio.

Production Scheduling 46. A cement manufacturer produces at least 3.2 million barrels of cement annually. He is told by the Environmental Protection Agency that his operation emits 2.5 lb of dust for each barrel produced. The EPA has ruled that annual emissions must be reduced to 1.8 million pounds. To do this the manufacturer plans to replace the present dust collectors with two types of electronic precipitators. One type would reduce emissions to .5 lb per barrel and would cost 16¢ per barrel. The other would reduce the dust to .3 lb per barrel and would cost 20¢ per barrel. The manufacturer does not want to spend more than .8 million dollars on the precipitators. He needs to know how many barrels he should produce with each type. Let x = the number of barrels in millions produced with the first type and y = the number of barrels in millions produced with the second type.

3.2 MATHEMATICAL MODELS FOR LINEAR PROGRAMMING

Many mathematical models designed to solve problems in business, biology, and economics involve finding the optimum value (maximum or minimum) of a function, subject to certain restrictions. In a **linear programming** problem, we must find the maximum or minimum value of a function, called the **objective function,** and also satisfy a set of restrictions, or **constraints,** given by linear inequalities.

In this section we show how to set up linear programming problems in two variables. In the next section we will show how to solve such problems.

EXAMPLE 1

A farmer raises only geese and pigs. She wants to raise no more than 16 animals including no more than 10 geese. She spends $5 to raise a goose and $15 to raise a pig, and she has $180 available for this project. The farmer wishes to maximize her profits. Each goose produces $6 in profit and each pig $20 in profit. Set up the mathematical model and graph the feasible region.

First, set up a table that shows the information given in the problem.

	Number Raised	Cost to Raise	Profit Each
Geese	x	$5	$6
Pigs	y	$15	$20
Maximum Funds Available		$180	

Use the table to write the necessary constraints. Since the total number of animals cannot exceed 16, the first constraint is

$$x + y \leq 16.$$

"No more than 10 geese" means

$$x \leq 10.$$

The cost to raise x geese at \$5 per goose is $5x$ dollars, while the cost for y pigs at \$15 each is $15y$ dollars. Since only \$180 is available,

$$5x + 15y \leq 180.$$

The number of geese and pigs cannot be negative, so

$$x \geq 0 \quad \text{and} \quad y \geq 0.$$

The farmer wants to know how many geese and pigs to raise in order to produce maximum profit. Each goose yields \$6 profit, and each pig \$20. If z represents total profit, then

$$z = 6x + 20y.$$

In summary, the mathematical model for the given linear programming problem is as follows:

Maximize $z = 6x + 20y$ **(1)**
subject to: $x + y \leq 16$ **(2)**
                         $x \leq 10$ **(3)**
        $5x + 15y \leq 180$ **(4)**
                         $x \geq 0$ **(5)**
                         $y \geq 0.$ **(6)**

Here $z = 6x + 20y$ is the objective function, and inequalities (2)–(6) are the constraints.

Using the methods described in the previous section, graph the feasible region for the system of inequalities (2)–(6), as in Figure 11. As mentioned previously, in the next section we will see how to complete the solution and find the maximum profit. ▬

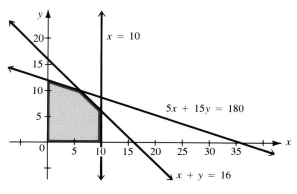

FIGURE 11

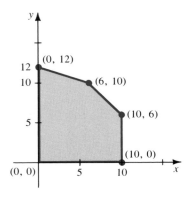

FIGURE 12

Any value in the feasible region of Figure 11 will satisfy all the constraints of the problem. In most problems, however, only one point in the feasible region will lead to maximum profit. In the next section we will see that maximum profit is always found at a **corner point,** a point in the feasible region where the boundary lines of two constraints cross. The feasible region of Figure 11 has been redrawn in Figure 12, with all the corner points identified.

Since corner points occur where two straight lines cross, the coordinates of a corner point are the solution of a system of two linear equations. For example, the corner point (6, 10) of Figure 12 is found by solving the system

$$x + y = 16$$
$$5x + 15y = 180.$$

By the methods of Chapter 2, the solution of this system is (6, 10).

■ EXAMPLE 2

An office manager needs to purchase new filing cabinets. He knows that Ace cabinets cost \$40 each, require 6 square feet of floor space, and hold 8 cubic feet of files. On the other hand, each Excello cabinet costs \$80, requires 8 square feet of floor space, and holds 12 cubic feet. His budget permits him to spend no more than \$560 on files, while the office has room for no more than 72 square feet of cabinets. The manager desires the greatest storage capacity within the limitations imposed by funds and space. Set up the mathematical model and sketch the feasible region, identifying all corner points.

Let x represent the number of Ace cabinets to be bought and let y represent the number of Excello cabinets. Summarize the given information in a table.

	Number of Cabinets	Cost of Each	Space Required	Storage Capacity
Ace	x	\$40	6 sq ft	8 cu ft
Excello	y	\$80	8 sq ft	12 cu ft
Maximum Available		\$560	72 sq ft	

The constraints imposed by cost and space are as follows.

$$40x + 80y \le 560 \qquad \text{Cost}$$
$$6x + 8y \le 72 \qquad \text{Floor space}$$

Since the number of cabinets cannot be negative, $x \ge 0$ and $y \ge 0$. The objective function to be maximized gives the amount of storage capacity provided by some combination of Ace and Excello cabinets. From the information in the table, the objective function is

$$\text{Storage Space} = z = 8x + 12y.$$

In summary, the given problem has produced the following mathematical model.

$$\text{Maximize} \qquad z = 8x + 12y$$
$$\text{subject to:} \qquad 40x + 80y \leq 560$$
$$6x + 8y \leq 72$$
$$x \geq 0$$
$$y \geq 0.$$

A graph of the feasible region is shown in Figure 13. Three of the corner points can be identified from the graph as (0, 0), (0, 7), and (12, 0). The fourth corner point, labeled Q in the figure, can be found by solving the system of equations

$$40x + 80y = 560$$
$$6x + 8y = 72.$$

Solve this system to find that Q is the point (8, 3). The maximum storage capacity under the given constraints will be found in the next section. ▬

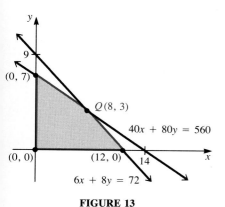

FIGURE 13

▬ **EXAMPLE 3**

Certain laboratory animals must have at least 30 grams of protein and at least 20 grams of fat per feeding period. These nutrients come from food A, which costs 18¢ per unit and supplies 2 grams of protein and 4 of fat, and food B, which costs 12¢ per unit and has 6 grams of protein and 2 of fat. Food B is bought under a long-term contract requiring that at least 2 units of B be used per serving. The laboratory wishes to minimize the cost per serving. Set up the mathematical model and graph the feasible region, identifying all corner points.

Let x represent the required amount of food A and y the amount of food B. Use the given information to prepare the following table.

Food	Number of Units	Grams of Protein	Grams of Fat	Cost
A	x	2	4	18¢
B	y	6	2	12¢
Minimum Required		30	20	

The mathematical model is as follows.

$$\text{Minimize} \qquad z = .18x + .12y$$
$$\text{subject to:} \qquad 2x + 6y \geq 30$$
$$4x + 2y \geq 20$$
$$y \geq 2$$
$$x \geq 0.$$

(The usual constraint $y \geq 0$ is redundant because of the constraint $y \geq 2$.) A graph of the feasible region is shown in Figure 14. The corner ponts are $(0, 10)$, $(3, 4)$, and $(9, 2)$. ■

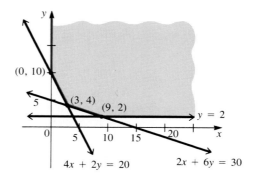

FIGURE 14

The feasible region in Figure 14 is an **unbounded** feasible region—the region extends indefinitely to the upper right. With this region it would not be possible to *maximize* the objective function, because the total cost of the food could always be increased by encouraging the animals to eat more. One of the corner points will produce a minimum daily cost, however.

═ 3.2 EXERCISES

Write Exercises 1–6 as linear inequalities. Identify all variables used. (Not all of the given information is used in Exercises 5 and 6.)

1. Product A requires 2 hr on machine I, while product B needs 3 hr on the same machine. The machine is available at most 45 hr per week.

2. A cow requires a third of an acre of pasture and a sheep needs a quarter acre. A rancher wants to use at least 120 acres of pasture.

3. Jesus Garcia needs at least 25 units of vitamin A per day. Green pills provide 4 units and red pills provide 1.

4. Pauline Wong spends 3 hr selling a small computer and 5 hr selling a larger model. She works no more than 45 hr per week.

5. Coffee costing $6 per pound is to be mixed with coffee costing $5 per pound to get at least 50 lb of a mixture.

6. A tank in an oil refinery holds 120 gal. The tank contains a mixture of light oil worth $1.25 per gallon and heavy oil worth $.80 per gallon.

■ APPLICATIONS

In Exercises 7–22, set up a mathematical model, graph the feasible region, and identify all corner points. Do not try to solve the problem.

BUSINESS AND ECONOMICS

Transportation   **7.** The Miers Company produces small engines for several manufacturers. The company receives orders from two assembly plants for their Topflight engine. Plant I needs at least 50 engines, and plant II needs at least 27 engines. The company can send at most 85 engines to these two assembly plants. It costs $20 per engine to ship to plant I and $35 per engine to ship to plant II. The company wishes to minimize shipping costs.

Transportation   **8.** A manufacturer of refrigerators must ship at least 100 refrigerators to its two West Coast warehouses. Each warehouse holds a maximum of 100 refrigerators. Warehouse A holds 25 refrigerators already, and warehouse B has 20 on hand. It costs $12 to ship a refrigerator to warehouse A and $10 to ship one to warehouse B. The manufacturer wants to minimize costs.

Insurance   **9.** A company is considering two insurance plans with the types of coverage (in thousands of dollars) and premiums per thousand dollars as shown below.

		Coverage		Premium
		Fire/Theft	Liability	
Policy	A	$10	$80	$50
	B	$15	$120	$40

The company wants at least $100,000 fire/theft insurance and at least $1,000,000 liability insurance from these plans at a minimum premium cost. The company must decide how many units of each policy to purchase.

Profit   **10.** Seall Manufacturing Company makes color television sets. It produces a bargain set that sells for $100 profit and a deluxe set that sells for $150 profit. On the assembly line the bargain set requires 3 hr, and the deluxe set takes 5 hr. The cabinet shop spends 1 hr on the cabinet for the bargain set and 3 hr on the cabinet for the deluxe set. Both sets require 2 hr of time for testing and packing. On a particular production run the Seall Company has available 3900 work hours on the assembly line, 2100 work hours in the cabinet shop, and 2200 work hours in the testing and packing department. The company needs to know how many sets of each type it should produce to maximize profit.

Revenue   **11.** A machine shop manufactures two types of bolts. Each can be made on any of three groups of machines, but the time required on each group differs, as shown in the table below.

		Machine Group		
		I	II	III
Bolts	Type 1	.1 min	.1 min	.1 min
	Type 2	.1 min	.4 min	.5 min

Production schedules are made up one day at a time. In a day, there are 240, 720, and 160 min available, respectively, on these machines. Type 1 bolts sell for 10¢

and type 2 bolts for 12¢. The shop foreman wants to determine how many of each type of bolt should be manufactured per day to maximize revenue.

Revenue 12. The manufacturing process requires that oil refineries must manufacture at least 2 gal of gasoline for every gallon of fuel oil. To meet the winter demand for fuel oil, at least 3 million gallons a day must be produced. The demand for gasoline is no more than 6.4 million gallons per day. The refinery sells gasoline for $1.25 per gallon and fuel oil for $1 per gallon, and it wishes to maximize revenue.

Revenue 13. A candy company has 100 kg of chocolate-covered nuts and 125 kg of chocolate-covered raisins to be sold as two different mixes. One mix will contain half nuts and half raisins and will sell for $6 per kilogram. The other mix will contain 1/3 nuts and 2/3 raisins and will sell for $4.80 per kilogram. The company wants to decide how many kilograms of each mix should be prepared to maximize revenue.

Profit 14. A small country can grow only two crops for export, coffee and cocoa. The country has 500,000 hectares of land available for the crops. Long term contracts require that at least 100,000 hectares be devoted to coffee and at least 200,000 hectares to cocoa. Cocoa must be processed locally, and production bottlenecks limit cocoa to 270,000 hectares. Coffee requires two workers per hectare, with cocoa requiring five. No more than 1,750,000 people are available for working with these crops. Coffee produces a profit of $220 per hectare and cocoa a profit of $310 per hectare. The country needs to determine how many hectares should be devoted to each crop in order to maximize profit.

Blending 15. The Mostpure Milk Company gets milk from two dairies and then blends the milk to get the desired amount of butterfat for the company's premier product. Dairy I can supply at most 50 gal of milk averaging 3.7% butterfat, and Dairy II can supply at most 80 gal of milk averaging 3.2% butterfat. Mostpure wants to find out how much milk from each supplier should be used to get at most 100 gal of milk with maximum butterfat.

Transportation 16. A greeting card manufacturer has 400 copies of a particular card in warehouse I and 500 copies of the same card in warehouse II. A greeting card shop in San Jose orders 350 copies of the card, and another shop in Memphis orders 300 copies. The shipping costs per copy to these shops from the two warehouses are shown in the following table.

		Destination	
		San Jose	Memphis
Warehouse	I	.25	.15
	II	.20	.30

The manufacturer wants to know how many cards should be shipped to each city from each warehouse to minimize shipping costs.

Finance 17. A pension fund manager decides to invest at most $40 million in U.S. Treasury Bonds paying 12% annual interest and in mutual funds paying 8% annual interest. He plans to invest at least $20 million in bonds and at least $15 million in mutual funds. He wants to know how much to invest in each to maximize annual interest.

LIFE SCIENCES

Health Care **18.** Mark, who is ill, takes vitamin pills. Each day he must have at least 16 units of vitamin A, 5 units of vitamin B_1, and 20 units of vitamin C. He can choose between pill #1, which contains 8 units of A, 1 of B_1, and 2 of C, and pill #2, which contains 2 units of A, 1 of B_1, and 7 of C. Pill #1 costs 15¢, and pill #2 costs 30¢. Mark wants to determine how many of each pill he should buy in order to minimize his cost.

Predator Food Requirements **19.** A certain predator requires at least 10 units of protein and 8 units of fat per day. One prey of species I provides 5 units of protein and 2 units of fat; one prey of species II provides 3 units of protein and 4 units of fat. Capturing and digesting each species II prey requires 3 units of energy, and capturing and digesting each species I prey requires 2 units of energy. The predator must meet its daily food requirements with the least expenditure of energy.

Health Care **20.** Ms. Oliveras was given the following advice. She should supplement her daily diet with at least 6000 USP units of vitamin A, at least 195 milligrams of vitamin C, and at least 600 USP units of vitamin D. Ms. Oliveras finds that Mason's Pharmacy carries Brand X vitamin pills at 5¢ each and Brand Y vitamins at 4¢ each. Each Brand X pill contains 3000 USP units of A, 45 milligrams of C, and 75 USP units of D, while the Brand Y pills contain 1000 USP units of A, 50 milligrams of C, and 200 USP units of D. Ms. Oliveras must decide what combination of vitamin pills she should buy to obtain the least possible cost.

Dietetics **21.** Sing, who is dieting, requires two food supplements, I and II. He can get these supplements from two different products, A and B, as shown in the following table.

		Grams of Supplement per Serving	
		I	II
Product	A	3	2
	B	2	4

Sing's physician has recommended that he include at least 15 g of each supplement in his daily diet. Product A costs 25¢ per serving and product B costs 40¢ per serving. Sing wants to know how he can satisfy his dietetic requirements most economically.

GENERAL

Construction **22.** In a small town in South Carolina, zoning rules require that the window space (in square feet) in a house be at least one-sixth of the space used up by solid walls. The monthly cost to heat the house is 20¢ for each square foot of solid walls and 80¢ for each square foot of windows. A homeowner wants to determine the maximum total area (windows plus walls) if $160 is available to pay for heat.

3.3 SOLVING LINEAR PROGRAMMING PROBLEMS GRAPHICALLY

In the last section we showed how to set up a linear programming problem by writing an objective function and listing the necessary constraints. Next, we described how to graph the feasible region and find all corner points. In this section we complete the process by showing how to solve linear programming problems by finding the required maximum or minimum value. The method of solving these problems from the graph of the feasible region is explained in the next example.

EXAMPLE 1
Solve the following linear programming problem.

$$\text{Maximize} \quad z = 2x + 5y$$
$$\text{subject to:} \quad 3x + 2y \leq 6$$
$$-2x + 4y \leq 8$$
$$x \geq 0$$
$$y \geq 0.$$

The feasible region is graphed in Figure 15. We can find the coordinates of point A, $(1/2, 9/4)$, by solving the system

$$3x + 2y = 6$$
$$-2x + 4y = 8$$

Every point in the feasible region satisfies all the constraints; however, we want to find those points that produce the maximum possible value of the objective function. To see how to find this maximum value, change the graph of Figure 15 by adding lines that represent the objective function $z = 2x + 5y$ for various sample values of z. By choosing the values 0, 5, 10, and 15 for z, the objective function becomes (in turn)

$$0 = 2x + 5y, \quad 5 = 2x + 5y, \quad 10 = 2x + 5y, \quad \text{and} \quad 15 = 2x + 5y.$$

These four lines are graphed in Figure 16. (Why are all of the lines parallel?) The figure shows that z cannot take on the value 15 because the graph for $z = 15$ is entirely outside the feasible region. The maximum possible value of z will be obtained from a line parallel to the others and between the lines representing the objective function when $z = 10$ and $z = 15$. The value of z will be as large as possible and all constraints will be satisfied if this line just touches the feasible region. This occurs at point A. We found above that A has coordinates (1/2, 9/4). The value of z at this point is

$$z = 2x + 5y = 2\left(\frac{1}{2}\right) + 5\left(\frac{9}{4}\right) = 12\frac{1}{4}.$$

The maximum possible value of z is 12¼. Of all the points in the feasible region, A leads to the largest possible value of z. ▬

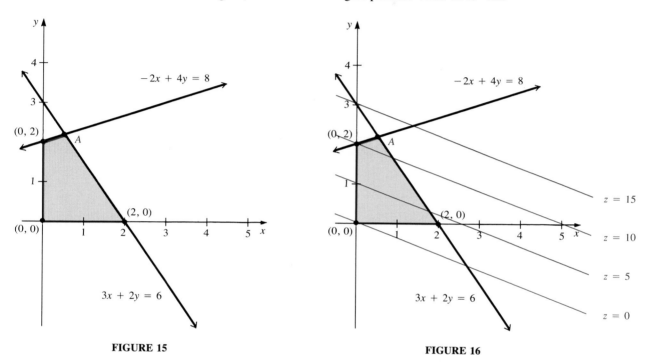

FIGURE 15   FIGURE 16

▬▬ EXAMPLE 2

Solve the following linear programming problem.

$$\begin{aligned}
\text{Minimize} \quad & z = 2x + 4y \\
\text{subject to:} \quad & x + 2y \geq 10 \\
& 3x + y \geq 10 \\
& x \geq 0 \\
& y \geq 0.
\end{aligned}$$

Figure 17 shows the feasible region and the lines that result when z in the objective function is replaced by 0, 10, 20, 40, and 50. The line representing the objective function touches the region of feasible solutions when $z = 20$.

Two corner points, (2, 4) and (10, 0), lie on this line. In this case, both (2, 4) and (10, 0), as well as all the points on the boundary line between them, give the same optimum value of z. There are infinitely many equally "good" values of x and y that will give the same minimum value of the objective function $z = 2x + 4y$. This minimum value is 20. ▄▄

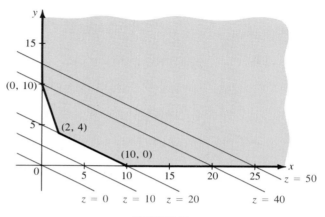

FIGURE 17

The feasible region in Example 1 is **bounded,** since the region is enclosed by boundary lines on all sides. Linear programming problems with bounded regions always have solutions. On the other hand, the feasible region in Example 2 is *unbounded,* and no solution will *maximize* the value of the objective function.

Some general conclusions can be drawn from the method of solution used in Examples 1 and 2. Figure 18 shows various feasible regions and the lines that result from various values of z. (We assume the lines are in order from left to right as z increases.) In part (a) of the figure, the objective function takes on its minimum value at corner point Q and its maximum value at P. The minimum is again at Q in part (b), but the maximum occurs at P_1 or P_2, or any point on the line segment connecting them. Finally, in part (c), the minimum value occurs at Q, but the objective function has no maximum value because the feasible region is unbounded.

The preceding discussion suggests the truth of the **corner point theorem.**

CORNER POINT THEOREM

▄▄
▄▄
▄▄ If an optimum value (either a maximum or a minimum) of the objective function exists, it will occur at one or more of the corner points of the feasible region.

This theorem simplifies the job of finding an optimum value. First, we graph the feasible region and find all corner points. Then we test each corner

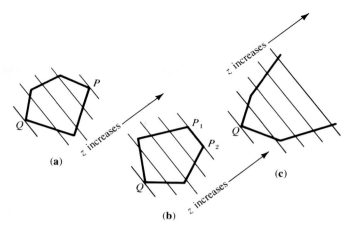

FIGURE 18

point in the objective function. Finally, we identify the corner point producing the optimum solution. For unbounded regions, we must decide whether the required optimum can be found; see Example 2.

With the theorem, we can solve the problem in Example 1 by first identifying the four corner points in Figure 15: (0, 0), (0, 2), (1/2, 9/4), and (2, 0). Then we substitute each of the four points into the objective function $z = 2x + 5y$ to identify the corner point that produces the maximum value of z.

Corner Point	Value of $z = 2x + 5y$
(0, 0)	$2(0) + 5(0) = 0$
(0, 2)	$2(0) + 5(2) = 10$
$(\frac{1}{2}, \frac{9}{4})$	$2(\frac{1}{2}) + 5(\frac{9}{4}) = 12\frac{1}{4}$ Maximum
(2, 0)	$2(2) + 5(0) = 4$

From these results, the corner point (1/2, 9/4) yields the maximum value of 12¼. This is the same result found earlier.

▬▬ EXAMPLE 3

Sketch the feasible region for the following set of constraints, and then find the maximum and minimum values of the objective function $z = 5x + 2y$.

$$3y - 2x \geq 0$$
$$y + 8x \leq 52$$
$$y - 2x \leq 2$$
$$x \geq 3$$

The graph in Figure 19 shows that the feasible region is bounded. Use the corner points from the graph to find the maximum and minimum values of the objective function.

Corner Point	Value of $z = 5x + 2y$	
(3, 2)	$5(3) + 2(2) = 19$	Minimum
(6, 4)	$5(6) + 2(4) = 38$	
(5, 12)	$5(5) + 2(12) = 49$	Maximum
(3, 8)	$5(3) + 2(8) = 31$	

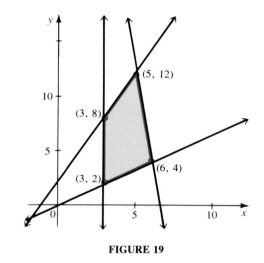

FIGURE 19

The minimum value of $z = 5x + 2y$ is 19 at the corner point (3, 2). The maximum value is 49 at (5, 12). ▬

The next example completes the solution of Example 2 from the previous section.

▬ EXAMPLE 4

An office manager needs to purchase new filing cabinets. He knows that Ace cabinets cost $40 each, require 6 square feet of floor space, and hold 8 cubic feet of files. On the other hand, each Excello cabinet costs $80, requires 8 square feet of floor space, and holds 12 cubic feet. His budget permits him to spend no more than $560 on files, while the office has room for no more than 72 square feet of cabinets. The manager desires the greatest storage capacity within the limitations imposed by funds and space. How many of each type of cabinet should he buy?

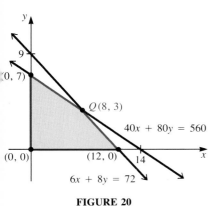

$40x + 80y = 560$

$6x + 8y = 72$

FIGURE 20

Figure 20 repeats the feasible region and corner points found in Example 2 of the previous section. Now test these four points in the objective function to determine the maximum value of z. The results are shown in the table.

Corner Point	Value of $z = 8x + 12y$
(0, 0)	0
(0, 7)	84
(12, 0)	96
(8, 3)	**100** Maximum

The objective function, which represents storage space, is maximized when $x = 8$ and $y = 3$. The manager should buy 8 Ace cabinets and 3 Excello cabinets.

The following summary gives the steps to use in solving a linear programming problem by the graphical method.

SOLVING A LINEAR PROGRAMMING PROBLEM

1. Write the objective function and all necessary constraints.
2. Graph the feasible region.
3. Identify all corner points.
4. Find the value of the objective function at each corner point.
5. For a bounded region, the solution is given by the corner point producing the optimum value of the objective function.
6. For an unbounded region, check that a solution actually exists. If it does, it will occur at a corner point.

3.3 EXERCISES

The graphs in Exercises 1–6 show regions of feasible solutions. Use these regions to find maximum and minimum values of the given objective functions.

1. $z = 3x + 5y$

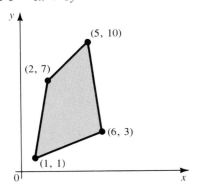

2. $z = 6x - y$

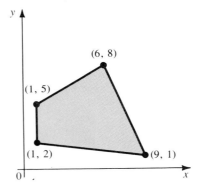

3. $z = .40x + .75y$

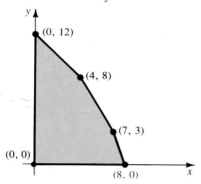

4. $z = .35x + 1.25y$

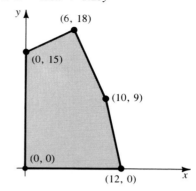

5. $z = 2x + 3y$

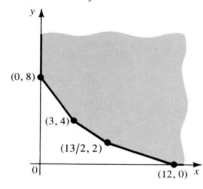

6. $z = 5x + 6y$

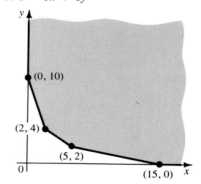

Use graphical methods to solve the linear programming problems in Exercises 7–14.

7. Maximize $z = 5x + 2y$
   subject to: $2x + 3y \le 6$
         $4x + y \le 6$
         $x \ge 0$
         $y \ge 0.$

8. Minimize $z = x + 3y$
   subject to: $x + y \le 10$
         $5x + 2y \ge 20$
         $-x + 2y \ge 0$
         $x \ge 0$
         $y \ge 0.$

9. Maximize $z = 2x - y$
   subject to: $3x - y \ge 12$
         $x + y \le 15$
         $x \ge 2$
         $y \ge 5.$

10. Maximize $z = x + 3y$
   subject to: $2x + 3y \le 100$
         $5x + 4y \le 200$
         $x \ge 10$
         $y \ge 20.$

11. Maximize $z = 4x + 2y$
   subject to: $x - y \le 10$
         $5x + 3y \le 75$
         $x \ge 0$
         $y \ge 0.$

12. Maximize $z = 4x + 5y$
   subject to: $10x - 5y \le 100$
         $20x + 10y \ge 150$
         $x \ge 0$
         $y \ge 0.$

13. Find values of $x \ge 0$ and $y \ge 0$ that maximize $z = 10x + 12y$ subject to each of the following sets of constraints.

(a) $x + y \le 20$
   $x + 3y \le 24$

(b) $3x + y \le 15$
   $x + 2y \le 18$

(c) $2x + 5y \ge 22$
   $4x + 3y \le 28$
   $2x + 2y \le 17$

14. Find values of $x \geq 0$ and $y \geq 0$ that minimize $z = 3x + 2y$ subject to each of the following sets of constraints.

(a) $10x + 7y \leq 42$
$4x + 10y \geq 35$

(b) $6x + 5y \geq 25$
$2x + 6y \geq 15$

(c) $x + 2y \geq 10$
$2x + y \geq 12$
$x - y \leq 8$

▄ APPLICATIONS

In Exercises 15–30, complete the solutions of the problems that were set up in Exercises 7–22 of the previous section.

BUSINESS AND ECONOMICS

Transportation **15.** The Miers Company produces small engines for several manufacturers. The company receives orders from two assembly plants for their Topflight engine. Plant I needs at least 50 engines, and plant II needs at least 27 engines. The company can send at most 85 engines to these two assembly plants. It costs $20 per engine to ship to plant I and $35 per engine to ship to plant II. How many engines should be shipped to each plant to minimize shipping costs? What is the minimum cost?

Transportation **16.** A manufacturer of refrigerators must ship at least 100 refrigerators to its two West Coast warehouses. Each warehouse holds a maximum of 100 refrigerators. Warehouse A holds 25 refrigerators already, and warehouse B has 20 on hand. It costs $12 to ship a refrigerator to warehouse A and $10 to ship one to warehouse B. How many refrigerators should be shipped to each warehouse to minimize costs? What is the minimum cost?

Insurance **17.** A company is considering two insurance plans with the types of coverage (in thousands of dollars) and premiums per thousand dollars as shown below.

		Coverage		
		Fire/Theft	Liability	Premium
Policy	A	$10	$80	$50
	B	$15	$120	$40

The company wants at least $100,000 fire/theft insurance and at least $1,000,000 liability insurance from these plans. How many units should be purchased from each plan to minimize the cost of the premiums? What is the minimum premium?

Profit **18.** Seall Manufacturing Company makes color television sets. It produces a bargain set that sells for $100 profit and a deluxe set that sells for $150 profit. On the assembly line the bargain set requires 3 hr, and the deluxe set takes 5 hr. The cabinet shop spends 1 hr on the cabinet for the bargain set and 3 hr on the cabinet for the deluxe set. Both sets require 2 hr of time for testing and packing. On a particular production run the Seall Company has available 3900 work hours on the assembly line, 2100 work hours in the cabinet shop, and 2200 work hours in the testing and packing department. How many sets of each type should it produce to make maximum profit? What is the maximum profit?

Revenue   **19.** A machine shop manufactures two types of bolts. Each can be made on any of three groups of machines, but the time required on each group differs, as shown in the table below.

		Machine Group		
		I	II	III
Bolts	Type 1	.1 min	.1 min	.1 min
	Type 2	.1 min	.4 min	.5 min

Production schedules are made up one day at a time. In a day there are 240, 720, and 160 min available, respectively, on these machines. Type 1 bolts sell for 10¢ and type 2 bolts for 12¢. How many of each type of bolt should be manufactured per day to maximize revenue? What is the maximum revenue?

Revenue   **20.** The manufacturing process requires that oil refineries must manufacture at least 2 gal of gasoline for every gallon of fuel oil. To meet the winter demand for fuel oil, at least 3 million gallons a day must be produced. The demand for gasoline is no more than 6.4 million gallons per day. If the refinery sells gasoline for $1.25 per gallon and fuel oil for $1 per gallon, how much of each should be produced to maximize revenue? Find the maximum revenue.

Revenue   **21.** A candy company has 100 kg of chocolate-covered nuts and 125 kg of chocolate-covered raisins to be sold as two different mixes. One mix will contain half nuts and half raisins and will sell for $6 per kilogram. The other mix will contain 1/3 nuts and 2/3 raisins and will sell for $4.80 per kilogram. How many kilograms of each mix should the company prepare for maximum revenue? Find the maximum revenue.

Profit   **22.** A small country can grow only two crops for export, coffee and cocoa. The country has 500,000 hectares of land available for the crops. Long-term contracts require that at least 100,000 hectares be devoted to coffee and at least 200,000 hectares to cocoa. Cocoa must be processed locally, and production bottlenecks limit cocoa to 270,000 hectares. Coffee requires two workers per hectare, with cocoa requiring five. No more than 1,750,000 people are available for working with these crops. Coffee produces a profit of $220 per hectare and cocoa a profit of $310 per hectare. How many hectares should the country devote to each crop in order to maximize profit? Find the maximum profit.

Blending   **23.** The Mostpure Milk Company gets milk from two dairies and then blends the milk to get the desired amount of butterfat for the company's premier product. If Dairy I can supply at most 50 gal of milk averaging 3.7% butterfat, and Dairy II can supply at most 80 gal of milk averaging 3.2% butterfat, how much milk from each supplier should Mostpure use to get at most 100 gal of milk with maximum butterfat? What is the maximum amount of butterfat?

Transportation   **24.** A greeting card manufacturer has 400 copies of a particular card in warehouse I and 500 copies of the same card in warehouse II. A greeting card shop in San Jose orders 350 copies of the card, and another shop in Memphis orders 300 copies.

The shipping costs per copy to these shops from the two warehouses are shown in the following table.

		Destination	
		San Jose	Memphis
Warehouse	I	.25	.15
	II	.20	.30

How many copies should be shipped to each city from each warehouse to minimize shipping costs? What is the minimum cost? (*Hint:* Use x, $350 - x$, y, and $300 - y$ as the variables.)

Finance   **25.** A pension fund manager decides to invest at most \$40 million in U.S. Treasury Bonds paying 12% annual interest and in mutual funds paying 8% annual interest. He plans to invest at least \$20 million in bonds and at least \$15 million in mutual funds. How much should be invested in each to maximize annual interest? What is the maximum annual interest?

LIFE SCIENCE

Health Care   **26.** Mark, who is ill, takes vitamin pills. Each day he must have at least 16 units of vitamin A, 5 units of vitamin B_1, and 20 units of vitamin C. He can choose between pill #1, which contains 8 units of A, 1 of B_1, and 2 of C, and pill #2, which contains 2 units of A, 1 of B_1, and 7 of C. Pill #1 costs 15¢, and pill #2 costs 30¢. How many of each pill should he buy in order to minimize his cost? What is the minimum cost?

Predator Food Requirements   **27.** A certain predator requires at least 10 units of protein and 8 units of fat per day. One prey of species I provides 5 units of protein and 2 units of fat; one prey of species II provides 3 units of protein and 4 units of fat. Capturing and digesting each species II prey requires 3 units of energy, and capturing and digesting each species I prey requires 2 units of energy. How many of each prey would meet the predator's daily food requirements with the least expenditure of energy? Are the answers reasonable? How could they be interpreted?

Health Care   **28.** Ms. Oliveras was given the following advice. She should supplement her daily diet with at least 6000 USP units of vitamin A, at least 195 milligrams of vitamin C, and at least 600 USP units of vitamin D. Ms. Oliveras finds that Mason's Pharmacy carries Brand X vitamin pills at 5¢ each and Brand Y vitamins at 4¢ each. Each Brand X pill contains 3000 USP units of A, 45 milligrams of C, and 75 USP units of D, while Brand Y pills contain 1000 USP units of A, 50 milligrams of C, and 200 USP units of D. What combination of vitamin pills should she buy to obtain the least possible cost? What is the least possible cost per day?

Dietetics   **29.** Sing, who is dieting, requires two food supplements, I and II. He can get these supplements from two different products, A and B, as shown in the following table.

		Grams of Supplement per Serving	
		I	II
Product	A	3	2
	B	2	4

Sing's physician has recommended that he include at least 15 g of each supplement in his daily diet. If product A costs 25¢ per serving and product B costs 40¢ per serving, how can he satisfy his dietetic requirements most economically? Find the minimum cost.

GENERAL

Construction   **30.** In a small town in South Carolina, zoning rules require that the window space (in square feet) in a house be at least one-sixth of the space used up by solid walls. The monthly cost to heat the house is 20¢ for each square foot of solid walls and 80¢ for each square foot of windows. Find the maximum total area (windows plus walls) if $160 is available to pay for heat.

*The importance of linear programming is shown by the inclusion of linear programming problems on most qualification examinations for Certified Public Accountants. Exercises 31–33 are reprinted from one such examination.**

The Random Company manufactures two products, Zeta and Beta. Each product must pass through two processing operations. All materials are introduced at the start of Process No. 1. There are no work in process inventories. Random may produce either one product exclusively or various combinations of both products subject to the following constraints:

	Process No. 1	Process No. 2	Contribution Margin Per Unit
Hours required to produce one unit of:			
Zeta	1 hour	1 hour	$4.00
Beta	2 hours	3 hours	5.25
Total capacity in hours per day	1,000 hours	1,275 hours	

A shortage of technical labor has limited Beta production to 400 units per day. There are no constraints on the production of Zeta other than the hour constraints in the above schedule. Assume that all relationships between capacity and production are linear.

31. Given the objective to maximize total contribution margin, what is the production constraint for Process No. 1?

   a.   Zeta + Beta ≤ 1,000.

   b.   Zeta + 2 Beta ≤ 1,000.

   c.   Zeta + Beta ≥ 1,000.

   d.   Zeta + 2 Beta ≥ 1,000.

32. Given the objective to maximize total contribution margin, what is the labor constraint for production of Beta?

   a. Beta ≤ 400.   **c.** Beta ≤ 425.

   b. Beta ≥ 400.   **d.** Beta ≥ 425.

---

*Material from *Uniform CPA Examinations and Unofficial Answers,* copyright © 1973, 1974, 1975 by the American Institute of Certified Public Accountants, Inc., is reprinted with permission.

33. What is the objective function of the data presented?
  a. Zeta + 2 Beta = $9.25.
  b. $4.00 Zeta + 3($5.25) Beta = Total Contribution Margin.
  c. $4.00 Zeta + $5.25 Beta = Total Contribution Margin.
  d. 2($4.00) Zeta + 3($5.25) Beta = Total Contribution Margin.

═══ **KEY WORDS** ═══

3.1 **linear inequality**
     boundary
     half-plane
     system of inequalities
     region of feasible solutions

3.2 **linear programming**
     objective function
     constraints
     corner point
     unbounded

3.3 **bounded**

═══ **CHAPTER 3**   **REVIEW EXERCISES** ═══

Graph each linear inequality.

1. $y \geq 2x + 3$

2. $3x - y \leq 5$

3. $3x + 4y \leq 12$

4. $2x - 6y \geq 18$

5. $y \geq x$

6. $y \leq 3$

Graph the solution of each system of inequalities. Find all corner points.

7. $x + y \leq 6$
  $2x - y \geq 3$

8. $4x + y \geq 8$
  $2x - 3y \leq 6$

9. $-4 \leq x \leq 2$
  $-1 \leq y \leq 3$
  $x + y \leq 4$

10. $2 \leq x \leq 5$
  $1 \leq y \leq 7$
  $x - y \leq 3$

11. $x + 3y \geq 6$
  $4x - 3y \leq 12$
  $x \geq 0$
  $y \geq 0$

12. $x + 2y \leq 4$
  $2x - 3y \leq 6$
  $x \geq 0$
  $y \geq 0$

Use the given regions to find the maximum and minimum values of the objective function $z = 2x + 4y$.

13.

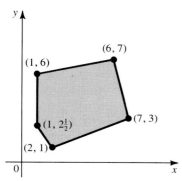

14.

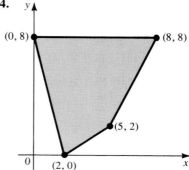

Use the graphical method to solve each linear programming problem.

15. Maximize $z = 2x + 4y$
subject to: $3x + 2y \leq 12$
$5x + y \geq 5$
$x \geq 0$
$y \geq 0.$

16. Minimize $z = 3x + 2y$
subject to: $8x + 9y \geq 72$
$6x + 8y \geq 72$
$x \geq 0$
$y \geq 0.$

17. Minimize $z = 4x + 2y$
subject to: $x + y \leq 50$
$2x + y \geq 20$
$x + 2y \geq 30$
$x \geq 0$
$y \geq 0.$

18. Maximize $z = 4x + 3y$
subject to: $2x + 7y \leq 14$
$2x + 3y \leq 10$
$x \geq 0$
$y \geq 0.$

▬ APPLICATIONS

BUSINESS AND ECONOMICS

Time Management **19.** A bakery makes both cakes and cookies. Each batch of cakes requires 2 hr in the oven and 3 hr in the decorating room. Each batch of cookies needs 1½ hr in the oven and ⅔ hr in the decorating room. The oven is available no more than 15 hr per day, and the decorating room can be used no more than 13 hr per day. Set up a system of inequalities, and then graph the solution of the system.

Cost Analysis **20.** A company makes two kinds of pizza, basic and plain. Basic contains cheese and beef, and plain contains onions and beef. The company sells at least 3 units a day of basic and at least 2 units of plain. The beef costs $5 per unit for basic and $4 per unit for plain. No more than $50 per day can be spent on beef. Dough for basic is $2 per unit, and dough for plain is $1 per unit. The company can spend no more than $16 per day on dough. Set up a system of inequalities, and then graph the solution of the system.

Profit **21.** How many batches of cakes and cookies should the bakery in Exercise 19 make in order to maximize profits if cookies produce a profit of $20 per batch and cakes produce a profit of $30 per batch?

Profit **22.** How many units of each kind of pizza should the company in Exercise 20 make in order to maximize profits if basic sells for $20 per unit and plain for $15 per unit?

4 Linear Programming: The Simplex Method

4.1 Slack Variables and the Pivot

4.2 Solving Maximization Problems

4.3 Mixed Constraints

4.4 Duality (Optional)

Review Exercises

Extended Application
Making Ice Cream

Extended Application
Merit Pay—The Upjohn Company

In the previous chapter, we discussed solving linear programming problems by the graphical method. This method illustrates the basic ideas of linear programming, but it is practical only for problems with two variables. For problems with more than two variables, or problems with two variables and many constraints, the *simplex method* is used. This method grew out of a practical problem faced by George B. Dantzig in 1947. Dantzig was concerned with finding the least expensive way to allocate supplies for the United States Air Force.

The **simplex method** starts with the selection of one corner point (often the origin) from the feasible region. Then, in a systematic way, another corner point is found that improves the value of the objective function. Finally, an optimum solution is reached, or it can be seen that no such solution exists.

The simplex method requires a number of steps. In this chapter we divide the presentation of these steps into two parts. First, a problem is set up in Section 4.1 and the method started, and then, in Section 4.2, the method is completed. Special situations are discussed in the remainder of the chapter.

▰ 4.1 SLACK VARIABLES AND THE PIVOT

Because the simplex method is used for problems with many variables, it usually is not convenient to use letters such as *x, y, z,* or *w* as variable names. Instead, the symbols x_1 (read "*x*-sub-one"), x_2, x_3, and so on, are used. These variable names lend themselves easily to use on the computer. In the simplex method, all constraints must be expressed in the linear form

$$a_1x_1 + a_2x_2 + a_3x_3 + \ldots \leq b$$

where $x_1, x_2, x_3, \ldots$ are variables and $a_1, a_2, \ldots$, and b are constants.

In this section we will use the simplex method only for problems such as the following:

$$\begin{aligned}
\text{Maximize} \quad & z = 2x_1 - 3x_2 \\
\text{subject to:} \quad & 2x_1 + x_2 \leq 10 \\
& x_1 - 3x_2 \leq 5 \\
\text{with} \quad & x_1 \geq 0, \quad x_2 \geq 0.
\end{aligned}$$

This type of problem is said to be in *standard maximum form*.

STANDARD MAXIMUM FORM

> ▰ A linear programming problem is in **standard maximum form** if the following conditions are satisfied.
>
> **1.** The objective function is to be maximized.
> **2.** All variables are nonnegative ($x_i \geq 0$).
> **3.** All constraints involve $\leq$.
> **4.** All of the constants in the constraints are nonnegative ($b \geq 0$).

(Problems that do not meet all of these conditions are discussed in Sections 4.3 and 4.4.)

To use the simplex method, we start by converting the constraints, which are linear inequalities, into linear equations. We do this by adding a nonnegative variable, called a **slack variable,** to each constraint. For example, the inequality $x_1 + x_2 \leq 10$ is converted into an equation by adding the slack variable x_3 to get

$$x_1 + x_2 + x_3 = 10, \quad \text{where } x_3 \geq 0.$$

The inequality $x_1 + x_2 \leq 10$ says that the sum $x_1 + x_2$ is less than or perhaps equal to 10. The variable x_3 "takes up any slack" and represents the amount by which $x_1 + x_2$ fails to equal 10. For example, if $x_1 + x_2$ equals 8, then x_3 is 2. If $x_1 + x_2 = 10$, the value of x_3 is 0.

■ EXAMPLE 1

Restate the following linear programming problem by introducing slack variables.

$$
\begin{array}{ll}
\text{Maximize} & z = 3x_1 + 2x_2 + x_3 \\
\text{subject to:} & 2x_1 + x_2 + x_3 \leq 150 \\
& 2x_1 + 2x_2 + 8x_3 \leq 200 \\
& 2x_1 + 3x_2 + x_3 \leq 320 \\
\text{with} & x_1 \geq 0, x_2 \geq 0, x_3 \geq 0.
\end{array}
$$

Rewrite the three constraints as equations by adding slack variables x_4, x_5, and x_6, one for each constraint. Then the problem can be restated as follows:

$$
\begin{array}{ll}
\text{Maximize} & z = 3x_1 + 2x_2 + x_3 \\
\text{subject to:} & 2x_1 + x_2 + x_3 + x_4 = 150 \\
& 2x_1 + 2x_2 + 8x_3 + x_5 = 200 \\
& 2x_1 + 3x_2 + x_3 + x_6 = 320 \\
\text{with} & x_1 \geq 0, x_2 \geq 0, x_3 \geq 0, x_4 \geq 0, x_5 \geq 0, x_6 \geq 0. \quad ■
\end{array}
$$

Adding slack variables to the constraints converts a linear programming problem into a system of linear equations. In each of these equations, all variables should be on the left side of the equals sign and all constants on the right. All the equations in Example 1 satisfy this condition except for the objective function, $z = 3x_1 + 2x_2 + x_3$, which may be written with all variables on the left as

$$-3x_1 - 2x_2 - x_3 + z = 0.$$

Now the equations in Example 1 can be written as the following augmented matrix.

$$
\begin{array}{ccccccc}
x_1 & x_2 & x_3 & x_4 & x_5 & x_6 & z
\end{array}
$$
$$
\left[
\begin{array}{ccccccc|c}
2 & 1 & 1 & 1 & 0 & 0 & 0 & 150 \\
2 & 2 & 8 & 0 & 1 & 0 & 0 & 200 \\
2 & 3 & 1 & 0 & 0 & 1 & 0 & 320 \\
\hline
-3 & -2 & -1 & 0 & 0 & 0 & 1 & 0
\end{array}
\right]
$$
$$
\text{Indicators}
$$

This matrix is called the **initial simplex tableau.** The numbers in the bottom row, which are from the objective function, are called **indicators** (except for the 0 at the far right).

■■ EXAMPLE 2

Set up the initial simplex tableau for the following problem:

A farmer has 100 acres of available land that he wishes to plant with a mixture of potatoes, corn, and cabbage. It costs him $400 to produce an acre of potatoes, $160 to produce an acre of corn, and $280 to produce an acre of cabbage. He has a maximum of $20,000 to spend. He makes a profit of $120 per acre of potatoes, $40 per acre of corn, and $60 per acre of cabbage. How many acres of each crop should he plant to maximize his profit?

Begin by summarizing the given information as follows.

Crop	Number of Acres	Cost per Acre	Profit per Acre
Potatoes	x_1	$400	$120
Corn	x_3	$160	$ 40
Cabbage	x_3	$280	$ 60
Maximum Available	100	$20,000	

If the number of acres allotted to each of the three crops is represented by x_1, x_2, and x_3, respectively, then the constraints of the example can be expressed as

$$x_1 + x_2 + x_3 \leq 100 \qquad \text{Number of acres}$$
$$400x_1 + 160x_2 + 280x_3 \leq 20{,}000 \qquad \text{Production costs}$$

where x_1, x_2, and x_3 are all nonnegative. The first of these constraints says that $x_1 + x_2 + x_3$ is less than or perhaps equal to 100. Use x_4 as the slack variable, giving the equation

$$x_1 + x_2 + x_3 + x_4 = 100.$$

Here x_4 represents the amount of the farmer's 100 acres that will not be used. (x_4 may be 0 or any value up to 100.)

In the same way, the constraint $400x_1 + 160x_2 + 280x_3 \leq 20{,}000$ can be converted into an equation by adding a slack variable, x_5:

$$400x_1 + 160x_2 + 280x_3 + x_5 = 20{,}000.$$

The slack variable x_5 represents any unused portion of the farmer's $20,000 capital. (Again, x_5 may be any value from 0 to 20,000.)

The objective function represents the profit. The farmer wants to maximize

$$z = 120x_1 + 40x_2 + 60x_3.$$

The linear programming problem can now be stated as follows:

Maximize $z = 120x_1 + 40x_2 + 60x_3$

subject to: $x_1 + x_2 + x_3 + x_4 = 100$

$400x_1 + 160x_2 + 280x_3 + x_5 = 20{,}000$

with $x_1 \geq 0, x_2 \geq 0, x_3 \geq 0, x_4 \geq 0, x_5 \geq 0.$

Rewrite the objective function as $-120x_1 - 40x_2 - 60x_3 + z = 0$, and complete the initial simplex tableau as follows.

$$
\begin{array}{cccccc}
x_1 & x_2 & x_3 & x_4 & x_5 & z \\
\end{array}
$$

$$
\left[
\begin{array}{cccccc|c}
1 & 1 & 1 & 1 & 0 & 0 & 100 \\
400 & 160 & 280 & 0 & 1 & 0 & 20{,}000 \\
\hline
-120 & -40 & -60 & 0 & 0 & 1 & 0
\end{array}
\right]
$$

The maximization problem in Example 2 consists of a system of two equations (describing the constraints) in five variables, together with the objective function. As with the graphical method, it is necessary to solve this system to find corner points of the region of feasible solutions. To produce a single, distinct solution, the number of variables must equal the number of equations in the system.

For example, the system of equations in Example 2 has an infinite number of solutions since there are more variables than equations. To see this, solve the system for x_4 and x_5.

$$x_4 = 100 - x_1 - x_2 - x_3$$
$$x_5 = 20{,}000 - 400 x_1 - 160x_2 - 280x_3$$

Each choice of values for x_1, x_2, and x_3 gives corresponding values for x_4 and x_5 that produce a solution of the system. But only some of these solutions are feasible. In a feasible solution all variables must be nonnegative. To get a unique feasible solution, we set three of the five variables equal to 0. In general, if there are m equations, then m variables can be nonzero. These are called **basic variables,** and the corresponding solutions are called **basic feasible solutions.** Each basic feasible solution corresponds to a corner point. In particular, if we choose the solution with $x_1 = 0$, $x_2 = 0$, and $x_3 = 0$, then $x_4 = 100$ and $x_5 = 20{,}000$. This solution, which corresponds to the corner point at the origin, is hardly optimal. It produces a profit of $0 for the farmer, since the objective function becomes

$$-120(0) - 40(0) - 60(0) + 0x_4 + 0x_5 + z = 0.$$

In the next section we will use the simplex method to start with this solution and improve it to find the maximum possible profit.

Each step of the simplex method produces a solution that corresponds to a corner point of the region of feasible solutions. These solutions can be read directly from the matrix, as shown in the next example.

■■ EXAMPLE 3

Read a solution from the matrix below.

$$
\begin{array}{cccccc}
x_1 & x_2 & x_3 & x_4 & x_5 & z \\
\end{array}
$$
$$
\left[
\begin{array}{ccccc|c}
2 & 1 & 8 & 5 & 0 & 0 & 27 \\
9 & 0 & 3 & 12 & 1 & 0 & 45 \\
\hline
-2 & 0 & -4 & 0 & 0 & 1 & 0
\end{array}
\right]
$$

The variables x_2 and x_5 are basic variables. They can quickly be identified because the columns for these variables (above the line) form an identity matrix. The 1 in the x_2 column is in the first row. This means that if x_1, x_3, and x_4 are zero, $x_2 = 27$ (the last number in the first row). Also, $x_5 = 45$ from the last number in the second row. The solution is thus $x_1 = 0$, $x_2 = 27$, $x_3 = 0$, $x_4 = 0$, and $x_5 = 45$, with $z = 0$ from the information in the last row. ■■

Pivots Solutions read directly from the initial simplex tableau are seldom, if ever, optimal. It is necessary to proceed to other solutions (corresponding to other corner points of the feasible region) until an optimum solution is found. To get these other solutions, use the row operations from Chapter 2 to change the tableau by using one of the nonzero entries of the tableau as a **pivot.** Pivoting, explained in the next example, produces a new tableau leading to another solution of the system of equations obtained from the original problem.

■■ EXAMPLE 4

Pivot about the indicated 2 of the initial simplex tableau given below.

$$
\begin{array}{ccccccc}
x_1 & x_2 & x_3 & x_4 & x_5 & x_6 & z \\
\end{array}
$$
$$
\left[
\begin{array}{ccccccc|c}
2 & 1 & 1 & 1 & 0 & 0 & 0 & 150 \\
2 & 2 & 8 & 0 & 1 & 0 & 0 & 200 \\
2 & 3 & 1 & 0 & 0 & 1 & 0 & 320 \\
\hline
-3 & -2 & -1 & 0 & 0 & 0 & 1 & 0
\end{array}
\right]
$$

To pivot about the indicated 2, change x_1 into a basic variable by getting a 1 where the 2 is now and changing all other entries in the x_1 column to 0. Start by multiplying each entry in row 1 by 1/2.

$$
\begin{array}{ccccccc}
x_1 & x_2 & x_3 & x_4 & x_5 & x_6 & z \\
\end{array}
$$
$$
\left[
\begin{array}{ccccccc|c}
1 & \frac{1}{2} & \frac{1}{2} & \frac{1}{2} & 0 & 0 & 0 & 75 \\
2 & 2 & 8 & 0 & 1 & 0 & 0 & 200 \\
2 & 3 & 1 & 0 & 0 & 1 & 0 & 320 \\
\hline
-3 & -2 & -1 & 0 & 0 & 0 & 1 & 0
\end{array}
\right]
\quad \frac{1}{2}R_1
$$

Now get 0 in row 2, column 1 by multiplying each entry in row 1 by -2 and adding the result to the corresponding entry in row 2.

$$
\begin{array}{ccccccc}
x_1 & x_2 & x_3 & x_4 & x_5 & x_6 & z
\end{array}
$$

$$
\left[
\begin{array}{ccccccc|c}
1 & \frac{1}{2} & \frac{1}{2} & \frac{1}{2} & 0 & 0 & 0 & 75 \\
0 & 1 & 7 & -1 & 1 & 0 & 0 & 50 \\
2 & 3 & 1 & 0 & 0 & 1 & 0 & 320 \\
-3 & -2 & -1 & 0 & 0 & 0 & 1 & 0
\end{array}
\right] \qquad -2R_1 + R_2
$$

Change the 2 in row 3, column 1 to a 0 by a similar process.

$$
\begin{array}{ccccccc}
x_1 & x_2 & x_3 & x_4 & x_5 & x_6 & z
\end{array}
$$

$$
\left[
\begin{array}{ccccccc|c}
1 & \frac{1}{2} & \frac{1}{2} & \frac{1}{2} & 0 & 0 & 0 & 75 \\
0 & 1 & 7 & -1 & 1 & 0 & 0 & 50 \\
0 & 2 & 0 & -1 & 0 & 1 & 0 & 170 \\
-3 & -2 & -1 & 0 & 0 & 0 & 1 & 0
\end{array}
\right] \qquad -2R_1 + R_3
$$

Finally, change the indicator -3 to 0 by multiplying each entry in row 1 by 3 and adding the result to the corresponding entry in row 4.

$$
\begin{array}{ccccccc}
x_1 & x_2 & x_3 & x_4 & x_5 & x_6 & z
\end{array}
$$

$$
\left[
\begin{array}{ccccccc|c}
1 & \frac{1}{2} & \frac{1}{2} & \frac{1}{2} & 0 & 0 & 0 & 75 \\
0 & 1 & 7 & -1 & 1 & 0 & 0 & 50 \\
0 & 2 & 0 & -1 & 0 & 1 & 0 & 170 \\
0 & -\frac{1}{2} & \frac{1}{2} & \frac{3}{2} & 0 & 0 & 1 & 225
\end{array}
\right] \qquad 3R_1 + R_4
$$

This simplex tableau gives the solution $x_1 = 75$, $x_2 = 0$, $x_3 = 0$, $x_4 = 0$, $x_5 = 50$, and $x_6 = 170$. Substituting these results into the objective function gives

$$
0(75) - \frac{1}{2}(0) + \frac{1}{2}(0) + \frac{3}{2}(0) + 0(50) + 0(170) + z = 225,
$$

or $z = 225$. (This shows that the value of z is always the number in the lower right-hand corner.) ■

In the simplex method, this process is repeated until an optimum solution is found, if one exists. In the next section we will see how to decide where to pivot to improve the value of the objective function and how to tell when an optimum solution either has been reached or does not exist.

4.1 EXERCISES

Convert each of the following inequalities into an equation by adding a slack variable.

1. $x_1 + 2x_2 \leq 6$

2. $3x_1 + 5x_2 \leq 100$

3. $2x_1 + 4x_2 + 3x_3 \leq 100$

4. $8x_1 + 6x_2 + 5x_3 \leq 250$

For Exercises 5–8, (a) determine the number of slack variables needed, (b) name them, and (c) use slack variables to convert each constraint into a linear equation.

5. Maximize $\quad z = 10x_1 + 12x_2$

subject to: $\quad 4x_1 + 2x_2 \le 20$
$\qquad\qquad 5x_1 + x_2 \le 50$
$\qquad\qquad 2x_1 + 3x_2 \le 25$

with $\qquad\quad x_1 \ge 0, x_2 \ge 0.$

6. Maximize $\quad z = 1.2x_1 + 3.5x_2$

subject to: $\quad 2.4x_1 + 1.5x_2 \le 10$
$\qquad\qquad 1.7x_1 + 1.9x_2 \le 15$

with $\qquad\quad x_1 \ge 0, x_2 \ge 0.$

7. Maximize $\quad z = 8x_1 + 3x_2 + x_3$

subject to: $\quad 7x_1 + 6x_2 + 8x_3 \le 118$
$\qquad\qquad 4x_1 + 5x_2 + 10x_3 \le 220$

with $\qquad\quad x_1 \ge 0, x_2 \ge 0, x_3 \ge 0.$

8. Maximize $\quad z = 12x_1 + 15x_2 + 10x_3$

subject to: $\quad 2x_1 + 2x_2 + x_3 \le 8$
$\qquad\qquad x_1 + 4x_2 + 3x_3 \le 12$

with $\qquad\quad x_1 \ge 0, x_2 \ge 0, x_3 \ge 0.$

Write the solutions that can be read from Exercises 9–12.

9.

$$
\begin{array}{cccccc}
x_1 & x_2 & x_3 & x_4 & x_5 & z \\
\end{array}
$$
$$
\left[\begin{array}{ccccc|c}
2 & 2 & 0 & 3 & 1 & 0 & 15 \\
3 & 4 & 1 & 6 & 0 & 0 & 20 \\
-2 & -1 & 0 & 1 & 0 & 1 & 10
\end{array}\right]
$$

10.

$$
\begin{array}{cccccc}
x_1 & x_2 & x_3 & x_4 & x_5 & z \\
\end{array}
$$
$$
\left[\begin{array}{ccccc|c}
0 & 2 & 1 & 1 & 3 & 0 & 5 \\
1 & 5 & 0 & 1 & 2 & 0 & 8 \\
0 & -2 & 0 & 1 & 1 & 1 & 10
\end{array}\right]
$$

11.

$$
\begin{array}{ccccccc}
x_1 & x_2 & x_3 & x_4 & x_5 & x_6 & z \\
\end{array}
$$
$$
\left[\begin{array}{cccccc|c}
6 & 2 & 1 & 3 & 0 & 0 & 0 & 8 \\
2 & 2 & 0 & 1 & 0 & 1 & 0 & 7 \\
2 & 1 & 0 & 3 & 1 & 0 & 0 & 6 \\
-3 & -2 & 0 & 2 & 0 & 0 & 1 & 12
\end{array}\right]
$$

12.

$$
\begin{array}{ccccccc}
x_1 & x_2 & x_3 & x_4 & x_5 & x_6 & z \\
\end{array}
$$
$$
\left[\begin{array}{cccccc|c}
0 & 2 & 0 & 1 & 2 & 2 & 0 & 3 \\
0 & 3 & 1 & 0 & 1 & 2 & 0 & 2 \\
1 & 4 & 0 & 0 & 3 & 5 & 0 & 5 \\
0 & -4 & 0 & 0 & 4 & 3 & 1 & 20
\end{array}\right]
$$

Pivot as indicated in each simplex tableau. Read the solution from the final result.

13.

$$
\begin{array}{cccccc}
x_1 & x_2 & x_3 & x_4 & x_5 & z \\
\end{array}
$$
$$
\left[\begin{array}{ccccc|c}
1 & 2 & 4 & 1 & 0 & 0 & 56 \\
2 & 2 & 1 & 0 & 1 & 0 & 40 \\
-1 & -3 & -2 & 0 & 0 & 1 & 0
\end{array}\right]
$$

14.

$$
\begin{array}{cccccc}
x_1 & x_2 & x_3 & x_4 & x_5 & z \\
\end{array}
$$
$$
\left[\begin{array}{ccccc|c}
5 & 4 & 1 & 1 & 0 & 0 & 50 \\
3 & 3 & 2 & 0 & 1 & 0 & 40 \\
-1 & -2 & -4 & 0 & 0 & 1 & 0
\end{array}\right]
$$

15.

$$
\begin{array}{ccccccc}
x_1 & x_2 & x_3 & x_4 & x_5 & x_6 & z \\
\end{array}
$$
$$
\left[\begin{array}{cccccc|c}
2 & 2 & 1 & 1 & 0 & 0 & 0 & 12 \\
1 & 2 & 3 & 0 & 1 & 0 & 0 & 45 \\
3 & 1 & 1 & 0 & 0 & 1 & 0 & 20 \\
-2 & -1 & -3 & 0 & 0 & 0 & 1 & 0
\end{array}\right]
$$

16.

$$
\begin{array}{ccccccc}
x_1 & x_2 & x_3 & x_4 & x_5 & x_6 & z \\
\end{array}
$$
$$
\left[\begin{array}{cccccc|c}
4 & 2 & 3 & 1 & 0 & 0 & 0 & 22 \\
2 & 2 & 5 & 0 & 1 & 0 & 0 & 28 \\
1 & 3 & 2 & 0 & 0 & 1 & 0 & 45 \\
-3 & -2 & -4 & 0 & 0 & 0 & 1 & 0
\end{array}\right]
$$

17.

$$
\begin{array}{ccccccc}
x_1 & x_2 & x_3 & x_4 & x_5 & x_6 & z \\
\end{array}
$$
$$
\left[\begin{array}{cccccc|c}
1 & 1 & 1 & 1 & 0 & 0 & 0 & 60 \\
3 & 1 & 2 & 0 & 1 & 0 & 0 & 100 \\
1 & 2 & 3 & 0 & 0 & 1 & 0 & 200 \\
-1 & -1 & -2 & 0 & 0 & 0 & 1 & 0
\end{array}\right]
$$

18.

$$
\begin{array}{cccccccc}
x_1 & x_2 & x_3 & x_4 & x_5 & x_6 & x_7 & z \\
\end{array}
$$
$$
\left[\begin{array}{ccccccc|c}
1 & 2 & 3 & 1 & 1 & 0 & 0 & 0 & 115 \\
2 & 1 & 8 & 5 & 0 & 1 & 0 & 0 & 200 \\
1 & 0 & 1 & 0 & 0 & 0 & 1 & 0 & 50 \\
-2 & -1 & -1 & -1 & 0 & 0 & 0 & 1 & 0
\end{array}\right]
$$

Introduce slack variables as necessary, and then write the initial simplex tableau for each of the following linear programming problems.

19. Find $x_1 \geq 0$ and $x_2 \geq 0$ such that

$$2x_1 + 3x_2 \leq 6$$
$$4x_1 + x_2 \leq 6$$

and $z = 5x_1 + x_2$ is maximized.

21. Find $x_1 \geq 0$ and $x_2 \geq 0$ such that

$$x_1 + x_2 \leq 10$$
$$5x_1 + 2x_2 \leq 20$$
$$x_1 + 2x_2 \leq 36$$

and $z = x_1 + 3x_2$ is maximized.

23. Find $x_1 \geq 0$ and $x_2 \geq 0$ such that

$$3x_1 + x_2 \leq 12$$
$$x_1 + x_2 \leq 15$$

and $z = 2x_1 + x_2$ is maximized.

20. Find $x_1 \geq 0$ and $x_2 \geq 0$ such that

$$2x_1 + 3x_2 \leq 100$$
$$5x_1 + 4x_2 \leq 200$$

and $z = x_1 + 3x_2$ is maximized.

22. Find $x_1 \geq 0$ and $x_2 \geq 0$ such that

$$x_1 + x_2 \leq 10$$
$$5x_1 + 3x_2 \leq 75$$

and $z = 4x_1 + 2x_2$ is maximized.

24. Find $x_1 \geq 0$ and $x_2 \geq 0$ such that

$$10x_1 + 4x_2 \leq 100$$
$$20x_1 + 10x_2 \leq 150$$

and $z = 4x_1 + 5x_2$ is maximized.

▀▀ APPLICATIONS

Set up Exercises 25–30 for solution by the simplex method. First express the linear constraints and objective function, then add slack variables, and then set up the initial simplex tableau. The solutions of some of these problems will be completed in the exercises for the next section.

BUSINESS AND ECONOMICS

Revenue   **25.** A candy company has 100 kg of chocolate-covered nuts and 125 kg of chocolate-covered raisins to be sold as two different mixtures. One mix will contain half nuts and half raisins and will sell for \$6 per kilogram. The other mix will contain 1/3 nuts and 2/3 raisins and will sell for \$4.80 per kilogram. How many kilograms of each mix should the company prepare for maximum revenue? (This is Exercise 13, Section 3.2.)

Profit   **26.** Seall Manufacturing Company makes color television sets. It produces a bargain set that sells for \$100 profit and a deluxe set that sells for \$150 profit. On the assembly line the bargain set requires 3 hr, and the deluxe set takes 5 hr. The cabinet shop spends 1 hr on the cabinet for the bargain set and 3 hr on the cabinet for the deluxe set. Both sets require 2 hr of time for testing and packing. On a particular production run the Seall Company has available 3900 work hours on the assembly line, 2100 work hours in the cabinet shop, and 2200 work hours in the testing and packing department. How many sets of each type should it produce to make maximum profit? What is the maximum profit? (See Exercise 10, Section 3.2.)

Profit   **27.** A small boat manufacturer builds three types of fiberglass boats: prams, runabouts, and trimarans. The pram sells at a profit of \$75, the runabout at a profit of \$90, and the trimaran at a profit of \$100. The factory is divided into two sections.

Section A does the molding and construction work, while section B does the painting, finishing and equipping. The pram takes 1 hr in section A and 2 hr in section B. The runabout takes 2 hr in A and 5 hr in B. The trimaran takes 3 hr in A and 4 hr in B. This year, section A has a total of 6240 hr available, and section B has 10,800 hr available. The manufacturer has ordered a supply of fiberglass that will build at most 3000 boats, figuring the average amount used per boat. How many of each type of boat should be made to produce maximum profit? What is the maximum profit?

Profit **28.** Caroline's Quality Candy Confectionery is famous for fudge, chocolate cremes, and pralines. Its candy-making equipment is set up to make 100-lb batches at a time. Currently there is a chocolate shortage and the company can get only 120 lb of chocolate in the next shipment. On a week's run, the confectionery's cooking and processing equipment is available for a total of 42 machine hours. During the same period the employees have a total of 56 work hours available for packaging. A batch of fudge requires 20 lb of chocolate, and a batch of cremes uses 25 lb of chocolate. The cooking and processing take 120 min for fudge, 150 min for chocolate cremes, and 200 min for pralines. The packaging times measured in minutes per one-pound box are 1, 2, and 3, respectively, for fudge, cremes, and pralines. Determine how many batches of each type of candy the confectionery should make, assuming that the profit per pound box is 50¢ on fudge, 40¢ on chocolate cremes, and 45¢ on pralines. What is the maximum profit?

Income **29.** A cat breeder has the following amounts of cat food: 90 units of tuna, 80 units of liver, and 50 units of chicken. To raise a Siamese cat, the breeder must use 2 units of tuna, 1 of liver, and 1 of chicken per day, while raising a Persian cat requires 1, 2, and 1 units, respectively, per day. If a Siamese cat sells for $12 and a Persian cat for $10, how many of each should be raised in order to obtain maximum gross income? What is the maximum gross income?

Profit **30.** Banal, Inc., produces art for motel rooms. Its painters can turn out mountain scenes, seascapes, and pictures of clowns. Each painting is worked on by three different artists, T, D, and H. Artist T works only 25 hr per week, while D and H work 45 hr and 40 hr per week, respectively. Artist T spends 1 hr on a mountain scene, 2 hr on a seascape, and 1 hr on a clown. Corresponding times for D and H are 3, 2, and 2 hr, and 2, 1, and 4 hr, respectively. Banal makes $20 on a mountain scene, $18 on a seascape, and $22 on a clown. The head painting packer can't stand clowns, so no more than 4 clown paintings may be done in a week. Find the number of each type of painting that should be made weekly in order to maximize profit. Find the maximum possible profit.

4.2 SOLVING MAXIMIZATION PROBLEMS

In the previous section we showed how to prepare a linear programming problem for solution. First, we converted the constraints to linear equations with slack variables; then we used the coefficients of the variables from the linear equations to write an augmented matrix. Finally, we used the pivot to go from one vertex to another vertex in the region of feasible solutions.

Now we are ready to put all this together and produce an optimum value for the objective function. To see how this is done, let us complete Example 2 from Section 4.1, the example about the farmer.

In the previous section we set up the following simplex tableau.

$$
\begin{array}{cccccc}
x_1 & x_2 & x_3 & x_4 & x_5 & z \\
\end{array}
$$

$$
\left[
\begin{array}{cccccc|c}
1 & 1 & 1 & 1 & 0 & 0 & 100 \\
400 & 160 & 280 & 0 & 1 & 0 & 20{,}000 \\
\hline
-120 & -40 & -60 & 0 & 0 & 1 & 0 \\
\end{array}
\right]
$$

This tableau leads to the solution $x_1 = 0$, $x_2 = 0$, $x_3 = 0$, $x_4 = 100$, and $x_5 = 20{,}000$, with x_4 and x_5 as the basic variables. These values produce a value of 0 for z. Since a value of 0 for the farmer's profit is not an optimum, we will try to improve this value.

The coefficients of x_1, x_2, and x_3 in the objective function are nonzero, so the profit could be improved by making any one of these variables take on a nonzero value in a solution. To decide which variable to use, look at the indicators in the initial simplex tableau above. The coefficient of x_1, -120, is the "most negative" of the indicators. This means that x_1 has the largest coefficient in the objective function, so that profit is increased the most by increasing x_1.

As we saw earlier, because there are two equations in the system, only two of the five variables can be basic variables (and be nonzero). If x_1 is nonzero in the solution, then x_1 will be a basic variable. This means that either x_4 or x_5 no longer will be a basic variable. To decide which variable will no longer be basic, start with the equations of the system,

$$
\begin{aligned}
x_1 + x_2 + x_3 + x_4 \qquad\qquad &= 100 \\
400x_1 + 160x_2 + 280x_3 \qquad + x_5 &= 20{,}000,
\end{aligned}
$$

and solve for x_4 and x_5 respectively.

$$
\begin{aligned}
x_4 &= 100 - x_1 - x_2 - x_3 \\
x_5 &= 20{,}000 - 400x_1 - 160x_2 - 280x_3
\end{aligned}
$$

Only x_1 is being changed to a nonzero value; both x_2 and x_3 keep the value 0. Replacing x_2 and x_3 with 0 gives

$$
\begin{aligned}
x_4 &= 100 - x_1 \\
x_5 &= 20{,}000 - 400x_1.
\end{aligned}
$$

Since both x_4 and x_5 must remain nonnegative, there is a limit to how much the value of x_1 can be increased. The equation $x_4 = 100 - x_1$ (or $x_4 = 100 - 1x_1$) shows that x_1 cannot exceed $100/1$, or 100. The second equation, $x_5 = 20{,}000 - 400x_1$, shows that x_1 cannot exceed $20{,}000/400$, or 50. To satisfy both these conditions, x_1 cannot exceed 50, the smaller of 50 and 100. If we let x_1 take the value 50, then $x_1 = 50$, $x_2 = 0$, $x_3 = 0$, and $x_5 = 0$. Since $x_4 = 100 - x_1$, then

$$
x_4 = 100 - 50 = 50.
$$

This solution gives a profit of

$$z = 120x_1 + 40x_2 + 60x_3 + 0x_4 + 0x_5$$
$$= 120(50) + 40(0) + 60(0) + 0(50) + 0(0) = 6000,$$

or $6000.

The same result could have been found from the initial simplex tableau given above. To use the tableau, select the most negative indicator. (If no indicator is negative, then the value of the objective function cannot be improved.)

$$\begin{array}{cccccc} x_1 & x_2 & x_3 & x_4 & x_5 & z \end{array}$$

$$\begin{bmatrix} 1 & 1 & 1 & 1 & 0 & 0 & | & 100 \\ 400 & 160 & 280 & 0 & 1 & 0 & | & 20{,}000 \\ -120 & -40 & -60 & 0 & 0 & 1 & | & 0 \end{bmatrix}$$

↑—Most negative indicator

The most negative indicator identifies the variable whose value is to be made nonzero. To find the variable that is now basic and will become nonbasic, calculate the quotients that were found above. Do this by dividing each number from the right side of the tableau by the corresponding number from the column with the most negative indicator.

Quotients

$$100/1 = 100$$

Smaller → $20{,}000/400 = 50$

$$\begin{array}{cccccc} x_1 & x_2 & x_3 & x_4 & x_5 & z \end{array}$$

$$\begin{bmatrix} 1 & 1 & 1 & 1 & 0 & 0 & | & 100 \\ 400 & 160 & 280 & 0 & 1 & 0 & | & 20{,}000 \\ -120 & -40 & -60 & 0 & 0 & 1 & | & 0 \end{bmatrix}$$

The smaller quotient is 50, from the second row. This identifies 400 as the pivot. Using 400 as the pivot, perform the appropriate row operations to get the second simplex tableau. First, get 1 in the pivot position by multiplying each element of the second row by 1/400. To make the other entries in the pivot column zeros, you must add multiples of the *pivot row* to the other rows. Multiply each of the entries in the second row by −1 and add the results to the corresponding entries in the first row, to get a 0 above the pivot. Get a 0 below the pivot, as the first indicator, in a similar way. The new tableau is

$$\begin{array}{cccccc} x_1 & x_2 & x_3 & x_4 & x_5 & z \end{array}$$

$$\begin{bmatrix} 0 & .6 & .3 & 1 & -.0025 & 0 & | & 50 \\ 1 & .4 & .7 & 0 & .0025 & 0 & | & 50 \\ 0 & 8 & 24 & 0 & .3 & 1 & | & 6000 \end{bmatrix}$$

and the solution read from this tableau is

$$x_1 = 50, \quad x_2 = 0, \quad x_3 = 0, \quad x_4 = 50, \quad x_5 = 0,$$

the same result found above. The entry 6000 (in color) in the lower right corner of the tableau gives the value of the objective function for this solution:

$$z = \$6000.$$

None of the indicators in the final simplex tableau is negative, which means that the value of z cannot be improved beyond $6000. To see why, recall that the last row gives the coefficients of the objective function so that

$$0x_1 + 8x_2 + 24x_3 + 0x_4 + .3x_5 + z = 6000,$$

or
$$z = 6000 - 0x_1 - 8x_2 - 24x_3 - 0x_4 - .3x_5.$$

Since x_2, x_3, and x_5 are zero, $z = 6000$, but if any of these three variables were to increase, z would decrease.

This result suggests that the optimal solution has been found as soon as no indicators are negative. As long as an indicator is negative, the value of the objective function can be improved. If any indicators are negative, we just find a new pivot and use row operations repeating the process until no negative indicators remain.

We can finally state the solution to the problem about the farmer: the optimum value of z is 6000, where $x_1 = 50$, $x_2 = 0$, $x_3 = 0$, $x_4 = 50$, and $x_5 = 0$. That is, the farmer will make a maximum profit of $6000 by planting 50 acres of potatoes. Another 50 acres should be left unplanted. It may seem strange that leaving assets unused can produce a maximum profit, but such results actually occur often.

▬▬ EXAMPLE 1

To compare the simplex method with the graphical method, use the simplex method to solve the problem in Example 2, Section 3.2. The graph is shown again in Figure 1. The objective function to be maximized was

$$z = 8x_1 + 12x_2. \qquad \text{Storage space}$$

(Since we are using the simplex method, we use x_1 and x_2 as variables instead of x and y.) The constraints were as follows:

$$40x_1 + 80x_2 \leq 560 \qquad \text{Cost}$$

$$6x_1 + 8x_2 \leq 72 \qquad \text{Floor space}$$

with
$$x_1 \geq 0, x_2 \geq 0.$$

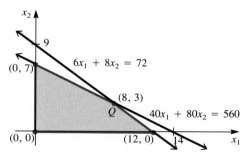

FIGURE 1

Add a slack variable to each constraint.

$$40x_1 + 80x_2 + x_3 \qquad\qquad = 560$$

$$6x_1 + \quad 8x_2 \qquad\quad + x_4 = 72$$

Write the initial simplex tableau.

$$
\begin{array}{ccccc}
x_1 & x_2 & x_3 & x_4 & z
\end{array}
$$

$$
\left[
\begin{array}{ccccc|c}
40 & 80 & 1 & 0 & 0 & 560 \\
6 & 8 & 0 & 1 & 0 & 72 \\
\hline
-8 & -12 & 0 & 0 & 1 & 0
\end{array}
\right]
$$

This tableau leads to the solution $x_1 = 0$, $x_2 = 0$, $x_3 = 560$, and $x_4 = 72$, with $z = 0$, which corresponds to the origin in Figure 1. The most negative indicator is -12. The necessary quotients are

$$\frac{560}{80} = 7 \qquad \text{and} \qquad \frac{72}{8} = 9.$$

The smaller quotient is 7, giving 80 as the pivot. Use row operations to get the new tableau.

$$
\left[
\begin{array}{ccccc|c}
\frac{1}{2} & 1 & \frac{1}{80} & 0 & 0 & 7 \\
2 & 0 & -\frac{1}{10} & 1 & 0 & 16 \\
-2 & 0 & \frac{3}{20} & 0 & 1 & 84
\end{array}
\right]
\quad
\begin{array}{l}
\frac{1}{80} R_1 \\
-8R_1 + R_2 \\
12R_1 + R_2
\end{array}
$$

The solution from this tableau is $x_1 = 0$, $x_2 = 7$, $x_3 = 0$, and $x_4 = 16$, with $z = 84$, which corresponds to the corner point $(0, 7)$ in Figure 1. Because of the indicator -2, the value of z can be improved. Use the 2 in row 2, column 1 as pivot to get the final tableau.

$$
\left[
\begin{array}{ccccc|c}
0 & 1 & \frac{3}{80} & -\frac{1}{4} & 0 & 3 \\
1 & 0 & -\frac{1}{20} & \frac{1}{2} & 0 & 8 \\
0 & 0 & \frac{1}{20} & 1 & 1 & 100
\end{array}
\right]
\quad
\begin{array}{l}
-\frac{1}{2}R_2 + R_1 \\
\frac{1}{2}R_2 \\
2R_2 + R_3
\end{array}
$$

Here the solution is $x_1 = 8$, $x_2 = 3$, $x_3 = 0$, and $x_4 = 0$, with $z = 100$. This solution, which corresponds to the corner point $(8, 3)$ in Figure 1, is the same as the solution found earlier.

Each simplex tableau above gave a solution corresponding to one of the corner points of the feasible region. As shown in Figure 2, the first solution corresponded to the origin, with $z = 0$. By choosing the appropriate pivot, we moved systematically to a new corner point, $(0, 7)$, which improved the value of z to 84. The next tableau took us to $(8, 3)$, producing the optimum value of $z = 100$. There was no reason to test the last corner point, $(12, 0)$, since the optimum value of z was found before that point was reached. ■

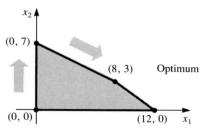

FIGURE 2

The next example shows how to handle zero or a negative number in the column containing the pivot.

■ EXAMPLE 2

Find the pivot for the following initial simplex tableau.

$$
\begin{array}{cccccc}
x_1 & x_2 & x_3 & x_4 & x_5 & z \\
\end{array}
$$

$$
\left[
\begin{array}{cccccc|c}
1 & -2 & 1 & 0 & 0 & 0 & 100 \\
3 & 4 & 0 & 1 & 0 & 0 & 200 \\
5 & 0 & 0 & 0 & 1 & 0 & 150 \\
\hline
-10 & -25 & 0 & 0 & 0 & 1 & 0 \\
\end{array}
\right]
$$

The most negative indicator is -25. To find the pivot, find the quotients formed by the entries in the rightmost column and in the x_2 column: $100/(-2)$, $200/4$, and $150/0$. Since division by 0 is meaningless, disregard the quotient $150/0$, which comes from the equation

$$5x_1 + 0x_2 + 0x_3 + 0x_4 + x_5 = 150.$$

If $x_1 = 0$, the equation reduces to

$$0x_2 + x_5 = 150 \qquad \text{or} \qquad x_5 = 150 - 0x_2.$$

Since the pivot is in the x_2 column, x_5 cannot be 0. In general, disregard quotients with 0 denominators.

The quotients predict the value of a variable in the solution, and thus cannot be negative. In this example, the quotient $100/(-2) = -50$ comes from the equation

$$x_1 - 2x_2 + x_3 + 0x_4 + 0x_5 = 100.$$

For $x_1 = 0$, this becomes

$$x_3 = 100 + 2x_2.$$

If x_3 is to be nonnegative, then

$$100 + 2x_2 \geq 0$$

$$x_2 \geq -50.$$

Since x_2 must be nonnegative anyway, this equation tells us nothing new. For this reason, disregard negative quotients also.

The only usable quotient is $200/4 = 50$, making 4 the pivot. If all the quotients either are negative or have zero denominators, no unique optimum solution will be found. Such a situation indicates an unbounded feasible region. The quotients, then, determine whether or not an optimum solution exists. ■

Let us now summarize the steps involved in solving a standard maximum linear programming problem by the simplex method.

SIMPLEX METHOD

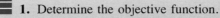

1. Determine the objective function.
2. Write all necessary constraints.
3. Convert each constraint into an equation by adding slack variables.
4. Set up the initial simplex tableau.
5. Locate the most negative indicator. If there are two such indicators, choose one.
6. Form the necessary quotients to find the pivot. Disregard any negative quotients or quotients with a 0 denominator. The smallest nonnegative quotient gives the location of the pivot. If all quotients must be disregarded, no maximum solution exists.* If two quotients are equally the smallest, let either determine the pivot.†
7. Use row operations to change the pivot to 1 and all other numbers in that column to 0.
8. If the indicators are all positive or 0, this is the final tableau. If not, go back to Step 5 above and repeat the process until a tableau with no negative indicators is obtained.
9. Read the solution from this final tableau. The maximum value of the objective function is the number in the lower right corner of the final tableau.

Although linear programming problems with more than a few variables would seem to be very complex, in practical applications many entries in the simplex tableau are zeros.

<hr>

4.2 EXERCISES

In Exercises 1–6, the initial tableau of a linear programming problem is given. Use the simplex method to solve each problem.

1.

x_1	x_2	x_3	x_4	x_5	z	
1	2	4	1	0	0	8
2	2	1	0	1	0	10
−2	−5	−1	0	0	1	0

2.

x_1	x_2	x_3	x_4	x_5	z	
2	2	1	1	0	0	10
1	2	3	0	1	0	15
−3	−2	−1	0	0	1	0

3.

x_1	x_2	x_3	x_4	x_5	z	
1	3	1	0	0	0	12
2	1	0	1	0	0	10
1	1	0	0	1	0	4
−2	−1	0	0	0	1	0

4.

x_1	x_2	x_3	x_4	x_5	x_6	z	
2	2	1	1	0	0	0	50
1	1	3	0	1	0	0	40
4	2	5	0	0	1	0	80
−2	−3	−5	0	0	0	1	0

<hr>

*Some special circumstances are noted at the end of Section 4.3

†If, on occasion, the first choice of a pivot does not produce a solution, try the other choice.

5.

$$\begin{array}{ccccccc} x_1 & x_2 & x_3 & x_4 & x_5 & x_6 & z \end{array}$$

$$\begin{bmatrix} 2 & 2 & 8 & 1 & 0 & 0 & 0 & 40 \\ 4 & -5 & 6 & 0 & 1 & 0 & 0 & 60 \\ 2 & -2 & 6 & 0 & 0 & 1 & 0 & 24 \\ \hline -14 & -10 & -12 & 0 & 0 & 0 & 1 & 0 \end{bmatrix}$$

6.

$$\begin{array}{cccccc} x_1 & x_2 & x_3 & x_4 & x_5 & z \end{array}$$

$$\begin{bmatrix} 3 & 2 & 4 & 1 & 0 & 0 & 18 \\ 2 & 1 & 5 & 0 & 1 & 0 & 8 \\ \hline -1 & -4 & -2 & 0 & 0 & 1 & 0 \end{bmatrix}$$

Use the simplex method to solve each linear programming problem.

7. Maximize $\quad z = 4x_1 + 3x_2$

subject to: $\quad 2x_1 + 3x_2 \le 11$
$\qquad\qquad\quad x_1 + 2x_2 \le 6,$

with $\qquad\quad x_1 \ge 0, x_2 \ge 0$

8. Maximize $\quad z = 2x_1 + 3x_2$

subject to: $\quad 3x_1 + 5x_2 \le 29$
$\qquad\qquad\quad 2x_1 + x_2 \le 10,$

with $\qquad\quad x_1 \ge 0, x_2 \ge 0$

9. Maximize $\quad z = 10x_1 + 12x_2$

subject to: $\quad 4x_1 + 2x_2 \le 20$
$\qquad\qquad\quad 5x_1 + x_2 \le 50$
$\qquad\qquad\quad 2x_1 + 2x_2 \le 24$

with $\qquad\quad x_1 \ge 0, x_2 \ge 0.$

10. Maximize $\quad z = 1.2x_1 + 3.5x_2$

subject to: $\quad 2.4x_1 + 1.5x_2 \le 10$
$\qquad\qquad\quad 1.7x_1 + 1.9x_2 \le 15$

with $\qquad\quad x_1 \ge 0, x_2 \ge 0.$

11. Maximize $\quad z = 8x_1 + 3x_2 + x_3$

subject to: $\quad x_1 + 6x_2 + 8x_3 \le 118$
$\qquad\qquad\quad x_1 + 5x_2 + 10x_3 \le 220$

with $\qquad\quad x_1 \ge 0, x_2 \ge 0, x_3 \ge 0.$

12. Maximize $\quad z = 12x_1 + 15x_2 + 5x_3$

subject to: $\quad 2x_1 + 2x_2 + x_3 \le 8$
$\qquad\qquad\quad x_1 + 4x_2 + 3x_3 \le 12$

with $\qquad\quad x_1 \ge 0, x_2 \ge 0, x_3 \ge 0.$

13. Maximize $\quad z = x_1 + 2x_2 + x_3 + 5x_4$

subject to: $\quad x_1 + 2x_2 + x_3 + x_4 \le 50$
$\qquad\qquad\quad 3x_1 + x_2 + 2x_3 + x_4 \le 100$

with $\qquad\quad x_1 \ge 0, x_2 \ge 0, x_3 \ge 0, x_4 \ge 0.$

14. Maximize $\quad z = x_1 + x_2 + 4x_3 + 5x_4$

subject to: $\quad x_1 + 2x_2 + 3x_3 + x_4 \le 115$
$\qquad\qquad\quad 2x_1 + x_2 + 8x_3 + 5x_4 \le 200$
$\qquad\qquad\quad x_1 \qquad\quad + x_3 \qquad\quad \le 50$

with $\qquad\quad x_1 \ge 0, x_2 \ge 0, x_3 \ge 0, x_4 \ge 0.$

▤ APPLICATIONS

Set up and solve Exercises 15–24 by the simplex method.

BUSINESS AND ECONOMICS

Charitable Contributions **15.** Jayanta is working to raise money for the homeless by sending information letters and making follow-up calls to local labor organizations and church groups. She discovered that each church group requires 2 hr of letter writing and 1 hr of follow-up, while for each labor union she needs 2 hr of letter writing and 3 hr of follow-up. Jayanta can raise $100 from each church group and $200 from each union local, and she has a maximum of 16 hr of letter-writing time and a maximum of 12 hr of follow-up time available per month. Determine the most profitable mixture of groups she should contact and the most money she can raise in a month.

Revenue **16.** A candy company has 100 kg of chocolate-covered nuts and 125 kg of chocolate-covered raisins to be sold as two different mixtures. One mix will contain half nuts and half raisins and will sell for $6 per kilogram. The other mix will contain 1/3 nuts and 2/3 raisins and will sell for $4.80 per kilogram. How many kilograms of each mix should the company prepare for maximum revenue? (See Exercise 25, Section 4.1.) Find the maximum revenue.

Profit   17. The Soul Sounds Recording company produces three main types of musical records: jazz, blues, and reggae. Each jazz album requires 4 hr of recording, 2 hr of mixing, and 6 hr of editing; each blues record requires 4 hr of recording, 8 hr of mixing, and 2 hr of editing; and each reggae album requires 10 hr of recording, 4 hr of mixing, and 6 hr of editing. The recording studio is available 80 hr per week, staff to operate the mixing board are available 52 hr a week, and the editing crew has at most 54 hr per week for work. Soul Sounds makes 80¢ per jazz record, 60¢ per blues album, and $1.20 for each reggae record. Use the simplex method to determine how many of each type of record the company should produce to maximize weekly profit. Find the maximum profit.

Income   18. A baker has 150 units of flour, 90 of sugar, and 150 of raisins. A loaf of raisin bread requires 1 unit of flour, 1 of sugar, and 2 of raisins, while a raisin cake needs 5, 2, and 1 units, respectively. If raisin bread sells for $1.75 a loaf and raisin cake for $4.00 each, how many of each should be baked so that gross income is maximized? What is the maximum gross income?

Profit   19. A manufacturer of bicycles builds one-, three-, and ten-speed models. The bicycles are made of both aluminum and steel. The company has available 91,800 units of steel and 42,000 units of aluminum. The one-, three-, and ten-speed models need 17, 27, and 34 units of steel, and 12, 21, and 15 units of aluminum, respectively. How many of each type of bicycle should be made in order to maximize profit if the company makes $8 per one-speed bike, $12 per three-speed, and $22 per ten-speed? What is the maximum possible profit?

Profit   20. Caroline's Quality Candy Confectionery is famous for fudge, chocolate cremes, and pralines. Its candy-making equipment is set up to make 100-lb batches at a time. Currently there is a chocolate shortage and the company can get only 120 lb of chocolate in the next shipment. On a week's run, the confectionery's cooking and processing equipment is available for a total of 42 machine hours. During the same period the employees have a total of 56 work hours available for packaging. A batch of fudge requires 20 lb of chocolate, and a batch of cremes uses 25 lb chocolate. The cooking and processing take 120 min for fudge, 150 min for chocolate cremes, and 200 min for pralines. The packaging times measured in minutes per one-pound box are 1, 2, and 3, respectively, for fudge, cremes and pralines. Determine how many batches of each type of candy the confectionery should make, assuming that the profit per pound box is 50¢ on fudge, 40¢ on chocolate cremes, and 45¢ on pralines. Also, find the maximum profit for the week. (See Exercise 28, Section 4.1.)

The next two problems come from past CPA examinations.* Select the appropriate answer for each question.

Profit   21. The Ball Company manufactures three types of lamps, labeled A, B, and C. Each lamp is processed in two departments, I and II. Total available man-hours per day for departments I and II are 400 and 600, respectively. No additional labor is

---

*Material from *Uniform CPA Examination Questions and Unofficial Answers,* copyright © 1973, 1974, 1975 by the American Institute of Certified Public Accountants, Inc., is reprinted with permission.

available. Time requirements and profit per unit for each lamp type is as follows:

	A	B	C
Man-hours in I	2	3	1
Man-hours in II	4	2	3
Profit per unit	$5	$4	$3

The company has assigned you as the accounting member of its profit planning committee to determine the numbers of types of A, B, and C lamps that it should produce in order to maximize its total profit from the sale of lamps. The following questions relate to a linear programming model that your group has developed.

(a) The coefficients of the objective function would be
  (1) 4, 2, 3. (2) 2, 3, 1.
  (3) 5, 4, 3. (4) 400, 600.

(b) The constraints in the model would be
  (1) 2, 3, 1. (2) 5, 4, 3.
  (3) 4, 2, 3. (4) 400, 600.

(c) The constraint imposed by the available man-hours in department I could be expressed as
  (1) $4X_1 + 2X_2 + 3X_3 \leq 400$. (2) $4X_1 + 2X_2 + 3X_3 \geq 400$.
  (3) $2X_1 + 3X_2 + 1X_3 \leq 400$. (4) $2X_1 + 3X_2 + 1X_3 \geq 400$.

Profit 22. The Golden Hawk Manufacturing Company wants to maximize the profits on products A, B, and C. The contribution margin for each product follows:

Product	Contribution margin
A	$2
B	$5
C	$4

The production requirements and departmental capacities, by departments, are as follows:

Department	Production requirements by product (hours)			Departmental capacity (total hours)
	A	B	C	
Assembling	2	3	2	30,000
Painting	1	2	2	38,000
Finishing	2	3	1	28,000

(a) What is the profit-maximization formula for the Golden Hawk Company?
  (1) $2A + $5B + $4C = X (where X = profit)
  (2) 5A + 8B + 5C ≤ 96,000
  (3) $2A + $5B + $4C ≤ X
  (4) $2A + $5B + $4C = 96,000

(b) What is the constraint for the Painting Department of the Golden Hawk Company?
  (1) 1A + 2B + 2C ≥ 38,000 (2) $2A + $5B + $4C ≥ 38,000
  (3) 1A + 2B + 2C ≤ 38,000 (4) 2A + 3B + 2C ≤ 30,000

LIFE SCIENCES

Blending Nutrients   **23.** A biologist has 500 kg of nutrient A, 600 kg of nutrient B, and 300 kg of nutrient C. These nutrients will be used to make 4 types of food, whose contents (in percent of nutrient per kilogram of food) and whose "growth values" are as shown below.

Food	Nutrient (%) A	B	C	Growth Value
P	0	0	100	90
Q	0	75	25	70
R	37.5	50	12.5	60
S	62.5	37.5	0	50

How many kilograms of each food should be produced in order to maximize total growth value? Find the maximum growth value.

SOCIAL SCIENCES

Politics   **24.** A political party is planning a half-hour television show. The show will have 3 min of direct requests for money from viewers. Three of the party's politicians will be on the show—a senator, a congresswoman, and a governor. The senator, a party "elder statesman," demands that he be on screen at least twice as long as the governor. The total time taken by the senator and the governor must be at least twice the time taken by the congresswoman. Based on a pre-show survey, it is believed that 40, 60, and 50 (in thousands) viewers will watch the program for each minute the senator, congresswoman, and governor, respectively are on the air. Find the time that should be allotted to each politician in order to get the maximum number of viewers. Find the maximum number of viewers.

FOR THE COMPUTER

Determine the constraints and the objective functions for Exercise 25 and 26, and then solve each problem.

Profit   **25.** A manufacturer makes two products, toy trucks and toy fire engines. Both are processed in four different departments, each of which has a limited capacity. The sheet metal department can handle at least 1½ times as many trucks as fire engines; the truck assembly department can handle at most 6700 trucks per week; and the fire engine assembly department assembles at most 5500 fire engines weekly. The painting department, which finishes both toys, has a maximum capacity of 12,000 per week. If the profit is $8.50 for a toy truck and $12.10 for a toy fire engine, how many of each item should the company produce to maximize profit?

Resource Management   **26.** The average weights of the three species stocked in the lake referred to in Section 2.2, Exercise 57 are 1.62, 2.14, and 3.01 kg for species A, B, and C, respectively. If the largest amounts of food that can be supplied each day are given as in Exercise 57, how should the lake be stocked to maximize the weight of the fish supported by the lake?

4.3   MIXED CONSTRAINTS

So far we have used the simplex method to solve linear programming problems in standard maximum form only. In this section, this work is extended to include linear programming problems with mixed $\le$ and $\ge$ constraints. The solution of these problems then gives a method of solving minimization problems. (An alternative approach for solving minimization problems is given in the next section.)

Problems with $\le$ and $\ge$ Constraints   Suppose a new constraint is added to the farmer problem from Example 2 of Section 4.1: to satisfy orders from regular buyers, the farmer must plant a total of at least 60 acres of the three crops. This constraint introduces the new inequality

$$x_1 + x_2 + x_3 \ge 60.$$

As before, this inequality must be rewritten as an equation in which the variables all represent nonnegative numbers. The inequality $x_1 + x_2 + x_3 \ge 60$ means that

$$x_1 + x_2 + x_3 - x_6 = 60$$

for some nonnegative variable x_6. (Remember that x_4 and x_5 are the slack variables in the problem.)

The new variable, x_6, is called a **surplus variable.** The value of this variable represents the excess number of acres (over 60) that may be planted. Since the total number of acres planted is to be no more than 100 but at least 60, the value of x_6 can vary from 0 to 40.

We must now solve the system of equations

$$
\begin{aligned}
x_1 + x_2 + x_3 + x_4 &= 100 \\
400x_1 + 160x_2 + 280x_3 + x_5 &= 20{,}000 \\
x_1 + x_2 + x_3 - x_6 &= 60 \\
-120x_1 - 40x_2 - 60x_3 + z &= 0,
\end{aligned}
$$

with x_1, x_2, x_3, x_4, x_5, and x_6 all nonnegative.

Set up the initial simplex tableau.

x_1	x_2	x_3	x_4	x_5	x_6	z	
1	1	1	1	0	0	0	100
400	160	280	0	1	0	0	20,000
1	1	1	0	0	−1	0	60
−120	−40	−60	0	0	0	1	0

This tableau gives the solution

$$x_1 = 0, \quad x_2 = 0, \quad x_3 = 0, \quad x_4 = 100, \quad x_5 = 20{,}000, \quad x_6 = -60.$$

But this is not a feasible solution, since x_6 is negative. All the variables in any feasible solution must be nonnegative if the solution is to correspond to a corner point of the region of feasible solutions.

When a negative value of a variable appears in the solution, we use row operations to transform the matrix until a solution is found in which all variables are nonnegative. Here the difficulty is caused by the -1 in row 3 of the matrix. We do not have the third column of the usual 3×3 identity matrix. To get around this, use row operations to change any column that has nonzero entries (such as the x_1, x_2, or x_3 columns) to one in which the third row entry is 1 and the other entries are 0. The choice of a column is arbitrary. We will choose the x_2 column, and if this choice does not lead to a feasible solution, we will try one of the other columns.

The third-row entry in the x_2 column is already 1. Using row operations to get 0's in the rest of the column gives the following tableau.

$$
\begin{array}{ccccccc}
x_1 & x_2 & x_3 & x_4 & x_5 & x_6 & z \\
\end{array}
$$

$$
\left[\begin{array}{ccccccc|c}
0 & 0 & 0 & 1 & 0 & 1 & 0 & 40 \\
240 & 0 & 120 & 0 & 1 & 160 & 0 & 10{,}400 \\
1 & 1 & 1 & 0 & 0 & -1 & 0 & 60 \\
-80 & 0 & -20 & 0 & 0 & -40 & 1 & 2400
\end{array}\right]
\begin{array}{l}
-R_3 + R_1 \\
-160R_3 + R_2 \\
\\
40R_3 + R_4
\end{array}
$$

This tableau gives the solution

$$x_1 = 0, \quad x_2 = 60, \quad x_3 = 0, \quad x_4 = 40, \quad x_5 = 10{,}400, \quad \text{and} \quad x_6 = 0,$$

which is feasible. The process of applying row operations to get a feasible solution is called **phase I** of the solution. When a feasible solution is reached, phase I is ended and **phase II** can begin. In phase II, the simplex method is applied as usual. Here, for the first step of phase II, the pivot is 240:

$$
\begin{array}{ccccccc}
x_1 & x_2 & x_3 & x_4 & x_5 & x_6 & z \\
\end{array}
$$

$$
\left[\begin{array}{ccccccc|c}
0 & 0 & 0 & 1 & 0 & 1 & 0 & 40 \\
240 & 0 & 120 & 0 & 1 & 160 & 0 & 10{,}400 \\
1 & 1 & 1 & 0 & 0 & -1 & 0 & 60 \\
-80 & 0 & -20 & 0 & 0 & -40 & 1 & 2400
\end{array}\right]
$$

$$
\begin{array}{ccccccc}
x_1 & x_2 & x_3 & x_4 & x_5 & x_6 & z \\
\end{array}
$$

$$
\left[\begin{array}{ccccccc|c}
0 & 0 & 0 & 1 & 0 & 1 & 0 & 40 \\
1 & 0 & .5 & 0 & .004 & .667 & 0 & 43.3 \\
0 & 1 & .5 & 0 & -.004 & -1.667 & 0 & 16.7 \\
0 & 0 & 20 & 0 & .32 & 13.4 & 1 & 5864
\end{array}\right]
\begin{array}{l}
\\
\frac{1}{240} R_2 \\
-1R_2 + R_3 \\
80R_2 + R_4
\end{array}
$$

The second matrix above has been obtained from the first by standard row operations; some numbers in it have been rounded. This final tableau gives

$$x_1 = 43.3, \quad x_2 = 16.7, \quad x_3 = 0, \quad x_4 = 40, \quad x_5 = 0, \quad \text{and} \quad x_6 = 0.$$

For maximum profit with this new constraint, the farmer should plant 43.3 acres of potatoes, 16.7 acres of corn, and no cabbage. Forty acres of the 100 available should not be planted. The profit will be $5864, less than the $6000 profit if the farmer were to plant only 50 acres of potatoes. Because of the additional constraint that at least 60 acres must be planted, the profit is reduced.

■ EXAMPLE 1

Maximize $\quad z = 10x_1 + 8x_2$

subject to: $\quad 4x_1 + 4x_2 \geq 60$

$\qquad\qquad 2x_1 + 5x_2 \leq 120$

with $\qquad x_1 \geq 0, x_2 \geq 0.$

Add slack or surplus variables to the constraints as needed to get the following system.

$$4x_1 + 4x_2 - x_3 \qquad\quad = 60$$
$$2x_1 + 5x_2 \qquad + x_4 \qquad = 120$$
$$-10x_1 - 8x_2 \qquad\qquad + z = \quad 0$$

Now write the first simplex tableau.

$$
\begin{array}{ccccc}
x_1 & x_2 & x_3 & x_4 & z \\
\end{array}
$$
$$
\left[
\begin{array}{ccccc|c}
4 & 4 & -1 & 0 & 0 & 60 \\
2 & 5 & 0 & 1 & 0 & 120 \\
\hline
-10 & -8 & 0 & 0 & 1 & 0 \\
\end{array}
\right]
$$

The solution here,

$$x_1 = 0, \quad x_2 = 0, \quad x_3 = -60, \quad \text{and} \quad x_4 = 120,$$

is not feasible, because of the -60. In phase I, use row operations to modify the tableau to produce a feasible solution. A column is needed with 1 in the first row and 0's in the rest of the rows. Choose the x_1 column. Multiply the entries in the first row by 1/4 to get 1 in the top row of the column. Then use row operations to get 0's in the other rows of that column.

$$
\begin{array}{ccccc}
x_1 & x_2 & x_3 & x_4 & z \\
\end{array}
$$
$$
\left[
\begin{array}{ccccc|c}
1 & 1 & -\frac{1}{4} & 0 & 0 & 15 \\
0 & 3 & \frac{1}{2} & 1 & 0 & 90 \\
\hline
0 & 2 & -\frac{5}{2} & 0 & 1 & 150 \\
\end{array}
\right]
\begin{array}{l}
\frac{1}{4}R_1 \\
-2R_1 + R_2 \\
10R_1 + R_3
\end{array}
$$

This solution,

$$x_1 = 15, \quad x_2 = 0, \quad x_3 = 0, \quad \text{and} \quad x_4 = 90,$$

is feasible, so phase I is completed. Perform phase II by completing the solution in the usual way. The pivot is 1/2. The next tableau is as follows.

$$
\begin{array}{ccccc}
x_1 & x_2 & x_3 & x_4 & z \\
\end{array}
$$
$$
\left[
\begin{array}{ccccc|c}
1 & \frac{5}{2} & 0 & \frac{1}{2} & 0 & 60 \\
0 & 6 & 1 & 2 & 0 & 180 \\
\hline
0 & 17 & 0 & 5 & 1 & 600 \\
\end{array}
\right]
\begin{array}{l}
\frac{1}{4}R_2 + R_1 \\
2R_2 \\
\frac{5}{2}R_2 + R_3
\end{array}
$$

No indicators are negative, so the optimum value of the objective function is

$$z = 600 \quad \text{when } x_1 = 60 \quad \text{and} \quad x_2 = 0. \quad \blacksquare$$

The approach discussed above is used to solve problems where the constraints are mixed $\leq$ and $\geq$ inequalities. The method also can be used to solve a problem where the constant term is negative, as in Example 3 of this section.

Minimization Problems   The definition of a problem in standard maximum form was given earlier in this chapter. Now we can define a linear programming problem in *standard minimum form,* as follows.

STANDARD MINIMUM FORM

> A linear programming problem is in **standard minimum form** if the following conditions are satisfied.
>
> **1.** The objective function is to be minimized.
> **2.** All variables are nonnegative.
> **3.** All constraints involve $\geq$.
> **4.** All of the constants in the constraints are nonnegative.

The difference between maximization and minimization problems is in conditions 1 and 3: in problems stated in standard minimum form the objective function is to be *minimized,* rather than maximized, and all constraints must have $\geq$ instead of $\leq$.

Problems in standard minimum form can be solved with the method of surplus variables presented above. To solve a problem in standard minimum form, first observe that the minimum of an objective function is the same number as the *maximum* of the *negative* of the function. For example, if $w = 3y_1 + 2y_2$, then $z = -w = -3y_1 - 2y_2$. For corner points $(0, 3)$, $(6, 0)$, and $(2, 1)$, the same ordered pair that minimizes w maximizes z, as shown below.

	$w = 3y_1 + 2y_2$		$z = -3y_1 - 2y_2$	
$(0, 3)$	6	Minimum	-6	Maximum
$(6, 0)$	18		-18	
$(2, 1)$	8		-8	

See Figure 3.

FIGURE 3

The next example illustrates the procedure described above. We use y_1 and y_2 as variables and w for the objective function as a reminder that this is a minimizing problem.

▰ EXAMPLE 2

Minimize $\quad w = 3y_1 + 2y_2$

subject to: $\quad y_1 + 3y_2 \geq 6$

$\qquad\qquad 2y_1 + \ \ y_2 \geq 3$

with $\qquad\quad y_1 \geq 0, y_2 \geq 0.$

Change this to a maximization problem by letting z equal the *negative* of the objective function: $z = -w$. Then find the *maximum* value of z.

$$z = -w = -3y_1 - 2y_2$$

The problem can now be stated as follows.

Maximize $\quad z = -3y_1 - 2y_2$

subject to: $\quad y_1 + 3y_2 \geq 6$

$\qquad\qquad 2y_1 + \ \ y_2 \geq 3$

with $\qquad\quad y_1 \geq 0, y_2 \geq 0.$

For phase I of the solution, add surplus variables and set up the first tableau.

$$
\begin{array}{ccccc}
y_1 & y_2 & y_3 & y_4 & z
\end{array}
$$

$$
\left[
\begin{array}{ccccc|c}
1 & 3 & -1 & 0 & 0 & 6 \\
2 & 1 & 0 & -1 & 0 & 3 \\
3 & 2 & 0 & 0 & 1 & 0
\end{array}
\right]
$$

The solution, $y_1 = 0$, $y_2 = 0$, $y_3 = -6$, and $y_4 = -3$, contains negative numbers. Row operations must be used to get a tableau with a feasible solution. Let's use the y_1 column since it already has a 1 in the first row. First get a 0 in the second row, and then get a 0 in the third row.

$$
\begin{array}{ccccc}
y_1 & y_2 & y_3 & y_4 & z
\end{array}
$$

$$
\left[
\begin{array}{ccccc|c}
1 & 3 & -1 & 0 & 0 & 6 \\
0 & -5 & 2 & -1 & 0 & -9 \\
0 & -7 & 3 & 0 & 1 & -18
\end{array}
\right]
\begin{array}{l}
\\ -2R_1 + R_2 \\ -3R_1 + R_3
\end{array}
$$

This tableau indicates a feasible solution. Now, in phase II, complete the solution as usual by the simplex method. The pivot is -5.

$$
\begin{array}{ccccc}
y_1 & y_2 & y_3 & y_4 & z
\end{array}
$$

$$
\left[
\begin{array}{ccccc|c}
1 & 0 & \frac{1}{5} & -\frac{3}{5} & 0 & \frac{3}{5} \\
0 & 1 & -\frac{2}{5} & \frac{1}{5} & 0 & \frac{9}{5} \\
0 & 0 & \frac{1}{5} & \frac{7}{5} & 1 & -\frac{27}{5}
\end{array}
\right]
\begin{array}{l}
-3R_2 + R_1 \\ -\frac{1}{5}R_2 \\ 7R_2 + R_3
\end{array}
$$

The solution is

$$y_1 = \frac{3}{5}, \quad y_2 = \frac{9}{5}, \quad y_3 = 0, \quad \text{and} \quad y_4 = 0.$$

This solution is feasible, and the tableau has no negative indicators. Since $z = -27/5$ and $z = -w$, the minimum value is $w = 27/5$, which is obtained when $y_1 = 3/5$ and $y_2 = 9/5$. ▬

In summary, the following steps are involved in the phase I and phase II method used to solve nonstandard problems in this section.

SOLVING NONSTANDARD PROBLEMS

1. If necessary, convert the problem to a maximization problem.
2. Add slack variables and subtract surplus variables as needed.
3. Write the initial simplex tableau.
4. If the solution from this tableau is not feasible, use row operations to get a feasible solution (phase I).
5. After a feasible solution is reached, solve by the simplex method (phase II).

An important application of linear programming is the problem of minimizing the cost of transporting goods. This type of problem is often referred to as a *transportation problem* or *warehouse problem*. Some problems of this type were included in the exercise sets in previous chapters. The next example illustrates the use of the simplex method in solving a transportation problem.

▬ **EXAMPLE 3**

A college textbook publisher has received orders from 2 colleges, C_1 and C_2. C_1 needs at least 500 books, and C_2 needs at least 1000. The publisher can supply the books from either of two warehouses. Warehouse W_1 has 900 books available and warehouse W_2 has 700. The costs to ship a book from each warehouse to each college are given below.

		To:	
		C_1	C_2
From:	W_1	\$1.20	1.80
	W_2	\$2.10	1.50

How many books should be sent from each warehouse to each college to minimize the shipping costs?

To begin, let

$$y_1 = \text{the number of books shipped from } W_1 \text{ to } C_1;$$

$$y_2 = \text{the number of books shipped from } W_2 \text{ to } C_1;$$

$$y_3 = \text{the number of books shipped from } W_1 \text{ to } C_2;$$

and $\quad y_4 = \text{the number of books shipped from } W_2 \text{ to } C_2.$

C_1 needs at least 500 books, so

$$y_1 + y_2 \geq 500.$$

Similarly,

$$y_3 + y_4 \geq 1000.$$

Since W_1 has 900 books available and W_2 has 700 available,

$$y_1 + y_3 \leq 900 \quad \text{and} \quad y_2 + y_4 \leq 700.$$

The company wants to minimize shipping costs, so the objective function is

$$w = 1.20y_1 + 2.10y_2 + 1.80y_3 + 1.50y_4.$$

Now write the problem as a system of linear equations, adding slack or surplus variables as needed, and let $z = -w$.

$$
\begin{aligned}
y_1 \quad + y_2 \qquad\qquad\qquad - y_5 \qquad\qquad\qquad &= 500 \\
y_3 + \quad y_4 \quad - y_6 \qquad\qquad &= 1000 \\
y_1 \qquad\qquad + \quad y_3 \qquad\qquad + y_7 \qquad &= 900 \\
y_2 \qquad + \quad y_4 \qquad\qquad + y_8 &= 700 \\
1.20y_1 + 2.10y_2 + 1.80y_3 + 1.50y_4 \qquad\qquad + z &= 0
\end{aligned}
$$

Set up the first simplex tableau.

y_1	y_2	y_3	y_4	y_5	y_6	y_7	y_8	z	
1	1	0	0	-1	0	0	0	0	500
0	0	1	1	0	-1	0	0	0	1000
1	0	1	0	0	0	1	0	0	900
0	1	0	1	0	0	0	1	0	700
1.20	2.10	1.80	1.50	0	0	0	0	1	0

The indicated solution is

$$y_5 = -500, \quad y_6 = -1000, \quad y_7 = 900, \quad y_8 = 700,$$

which is not feasible since y_5 and y_6 are negative.

Complete phase I and phase II to see that the solution is to ship 500 books from W_1 to C_1, 300 books from W_1 to C_2, and 700 books from W_2 to C_2 for a minimum shipping cost of \$2190. ■

When one or more of the constraints in a linear programming problem is an equation, rather than an inequality, there is no need for a slack or surplus variable. The simplex method requires an additional variable, however, for *each* constraint. To meet this condition, an **artificial variable** is added to each equation. These variables are called artificial variables because they have no meaning in the context of the original problem. The first goal of the simplex method is to eliminate any artificial variables as basic variables, since they must have a value of 0 in the solution.

■ EXAMPLE 4

In the transportation problem discussed in Example 3, it would be more realistic for the colleges to order exactly 500 and 1000 books respectively. Solve the problem with these two equality constraints.

Using the same variables, we can state the problem as follows.

$$\text{Minimize} \qquad w = 1.20y_1 + 2.10y_2 + 1.80y_3 + 1.50y_4$$

$$\text{subject to:} \qquad y_1 + y_2 = 500$$
$$y_3 + y_4 = 1000$$
$$y_1 + y_3 \leq 900$$
$$y_2 + y_4 \leq 700$$

with all variables nonnegative.

The corresponding system of equations requires slack variables y_5 and y_6 and two artificial variables that we shall call a_1 and a_2, to remind us that they require special handling. The system

$$y_1 + y_3 + y_5 = 900$$
$$y_2 + y_4 + y_6 = 700$$
$$y_1 + y_2 + a_1 = 500$$
$$y_3 + y_4 + a_2 = 1000$$
$$1.20y_1 + 2.10y_2 + 1.80y_3 + 1.50y_4 + z = 0$$

produces the following tableau.

y_1	y_2	y_3	y_4	y_5	y_6	a_1	a_2	z	
1	0	1	0	1	0	0	0	0	900
0	1	0	1	0	1	0	0	0	700
1	1	0	0	0	0	1	0	0	500
0	0	1	1	0	0	0	1	0	1000
1.20	2.10	1.80	1.50	0	0	0	0	1	0

The artificial variables, which are now basic variables, must be eliminated from the solution. To eliminate a_1, let us choose y_1 as the entering variable because it contributes less to the cost than y_2, the other possibility. The pivot is shown in the tableau above. The resulting tableau follows.

y_1	y_2	y_3	y_4	y_5	y_6	a_1	a_2	z	
0	−1	1	0	1	0	−1	0	0	400
0	1	0	1	0	1	0	0	0	700
1	1	0	0	0	0	1	0	0	500
0	0	1	1	0	0	0	1	0	1000
0	.90	1.80	1.50	0	0	−1.2	0	1	−600

As soon as an artificial variable is no longer basic, it can be dropped from the matrix, so the a_1 column can now be dropped. To eliminate a_2 from the solution, either y_3 or y_4 can enter. We choose y_4 as the entering variable because it contributes less to the cost than y_3 does.

y_1	y_2	y_3	y_4	y_5	y_6	a_2	z	
0	-1	1	0	1	0	-1	0	400
0	1	-1	0	0	1	0	0	-300
1	1	0	0	0	0	0	0	500
0	0	1	1	0	0	1	0	1000
0	0	.3	0	0	0	-1.5	1	-2100

Since a_2 is no longer a basic variable, it can be dropped from the matrix. At this point, we proceed with the solution as usual. The indicated solution, $y_1 = 500$, $y_4 = 1000$, $y_5 = 400$, and $y_6 = -300$, is not feasible. Complete phase I and phase II to get the feasible solution $y_1 = 500$, $y_3 = 300$, $y_4 = 700$, and $y_5 = 100$ with a minimum cost of $2190. This is the same result found in Example 3. ■

All of the possible complications involved in using the simplex method could not have been covered in this brief presentation. These potential difficulties include the following:

1. Occasionally, a transformation will cycle—that is, produce a "new" solution that was an earlier solution in the process. These situations are known as *degeneracies,* and special methods are available for handling them.
2. It may not be possible to convert a nonfeasible basic solution to a feasible basic solution. In that case, no solution can satisfy all the constraints. Graphically, this means there is no region of feasible solutions.

(These difficulties are covered in more detail in advanced texts. One example of a text that you might find helpful is *An Introduction to Management Science,* Fifth Edition, by David R. Anderson, Dennis J. Sweeney, and Thomas A. Williams, copyright 1988, West Publishing Company.)

Two linear programming models in actual use, one on making ice cream, the other on merit pay, are presented at the end of this chapter. These models illustrate the usefulness of linear programming. In most real applications, the number of variables is so large that these problems could not be solved without using methods (like the simplex method) that can be adapted to computers.

4.3 EXERCISES

Rewrite each system of inequalities, adding slack variables or subtracting surplus variables as necessary.

1. $2x_1 + 3x_2 \le 8$
   $x_1 + 4x_2 \ge 7$

2. $5x_1 + 8x_2 \le 10$
   $6x_1 + 2x_2 \ge 7$

3. $x_1 + x_2 + x_3 \le 100$
   $x_1 + x_2 + x_3 \ge 75$
   $x_1 + x_2 \ge 27$

4. $2x_1 \quad\;\; + x_3 \le 40$
   $x_1 + x_2 \quad\;\; \ge 18$
   $x_1 \quad\;\; + x_3 \ge 20$

Convert the following problems into maximization problems.

5. Minimize $\quad w = 4y_1 + 3y_2 + 2y_3$

subject to: $\quad y_1 + y_2 + y_3 \geq 5$
$ y_1 + y_2 \geq 4$
$ 2y_1 + y_2 + 3y_3 \geq 15$

with $\quad y_1 \geq 0, y_2 \geq 0, y_3 \geq 0.$

6. Minimize $\quad w = 8y_1 + 3y_2 + y_3$

subject to: $\quad 7y_1 + 6y_2 + 8y_3 \geq 18$
$ 4y_1 + 5y_2 + 10y_3 \geq 20$

with $\quad y_1 \geq 0, y_2 \geq 0, y_3 \geq 0.$

7. Minimize $\quad w = y_1 + 2y_2 + y_3 + 5y_4$

subject to: $\quad y_1 + y_2 + y_3 + y_4 \geq 50$
$ 3y_1 + y_2 + 2y_3 + y_4 \geq 100$

with $\quad y_1 \geq 0, y_2 \geq 0, y_3 \geq 0, y_4 \geq 0.$

8. Minimize $\quad w = y_1 + y_2 + 4y_3$

subject to: $\quad y_1 + 2y_2 + 3y_3 \geq 115$
$ 2y_1 + y_2 + y_3 \leq 200$
$ y_1 + y_3 \geq 50$

with $\quad y_1 \geq 0, y_2 \geq 0, y_3 \geq 0.$

Use the simplex method to solve each of the following.

9. Find $x_1 \geq 0$ and $x_2 \geq 0$ such that

$$x_1 + 2x_2 \geq 24$$
$$x_1 + x_2 \leq 40$$

and $z = 12x_1 + 10x_2$ is maximized.

10. Find $x_1 \geq 0$ and $x_2 \geq 0$ such that

$$3x_1 + 4x_2 \geq 48$$
$$2x_1 + 4x_2 \leq 60$$

and $z = 6x_1 + 8x_2$ is maximized.

11. Find $x_1 \geq 0$, $x_2 \geq 0$, and $x_3 \geq 0$ such that

$$x_1 + x_2 + x_3 \leq 150$$
$$x_1 + x_2 + x_3 \geq 100$$

and $z = 2x_1 + 5x_2 + 3x_3$ is maximized.

12. Find $x_1 \geq 0$, $x_2 \geq 0$, and $x_3 = 0$ such that

$$x_1 + x_2 + 2x_3 \leq 38$$
$$2x_1 + x_2 + x_3 \geq 24$$

and $z = 3x_1 + 2x_2 + 2x_3$ is maximized.

13. Find $x_1 \geq 0$ and $x_2 \geq 0$ such that

$$x_1 + x_2 \leq 100$$
$$x_1 + x_2 \geq 50$$
$$2x_1 + x_2 \leq 110$$

and $z = -2x_1 + 3x_2$ is maximized.

14. Find $x_1 \geq 0$ and $x_2 \geq 0$ such that

$$x_1 + 2x_2 \leq 18$$
$$x_1 + 3x_2 \geq 12$$
$$2x_1 + 2x_2 \leq 24$$

and $z = 5x_1 - 10x_2$ is maximized.

Solve each of the following by the two-phase method.

15. Find $y_1 \geq 0$, $y_2 \geq 0$ such that

$$10y_1 + 5y_2 \geq 100$$
$$20y_1 + 10y_2 \geq 150$$

and $w = 4y_1 + 5y_2$ is minimized.

16. Minimize $\quad w = 3y_1 + 2y_2$

subject to: $\quad 2y_1 + 3y_2 \geq 60$
$ y_1 + 4y_2 \geq 40$

with $\quad y_1 \geq 0, y_2 \geq 0.$

17. Minimize $\quad w = 2y_1 + y_2 + 3y_3$

subject to: $\quad y_1 + y_2 + y_3 \geq 100$
$ 2y_1 + y_2 \geq 50$

with $\quad y_1 \geq 0, y_2 \geq 0, y_3 \geq 0.$

18. Minimize $\quad w = 3y_1 + 2y_2$

subject to: $\quad y_1 + 2y_2 \geq 10$
$ y_1 + y_2 \geq 8$
$ 2y_1 + y_2 \geq 12$

with $\quad y_1 \geq 0, y_2 \geq 0.$

Solve each of the following using artificial variables.

19. Maximize $\quad z = 3x_1 + 2x_2$

subject to: $\quad x_1 + x_2 = 50$
$ 4x_1 + 2x_2 \geq 120$
$ 5x_1 + 2x_2 \leq 200$

with $\quad x_1 \geq 0, x_2 \geq 0.$

20. Maximize $\quad z = 10x_1 + 9x_2$

subject to: $\quad x_1 + x_2 = 30$
$ x_1 + x_2 \geq 25$
$ 2x_1 + x_2 \leq 40$

with $\quad x_1 \geq 0, x_2 \geq 0.$

21. Minimize $w = 15y_1 + 12y_2$

subject to: $y_1 + 2y_2 \leq 12$
$3y_1 + y_2 \geq 18$
$y_1 + y_2 = 10$

with $y_1 \geq 0, y_2 \geq 0$.

22. Minimize $w = 32y_1 + 40y_2$

subject to: $20y_1 + 10y_2 = 200$
$25y_1 + 40y_2 \leq 500$
$18y_1 + 24y_2 \geq 300$

with $y_1 \geq 0, y_2 \geq 0$.

 APPLICATIONS

Use the simplex method to solve Exercises 23–33.

BUSINESS AND ECONOMICS

Transportation **23.** Southwestern Oil supplies two distributors in the Northwest from two outlets. Distributor D_1 needs at least 3000 barrels of oil, and distributor D_2 needs at least 5000 barrels. The two outlets can each furnish 5000 barrels of oil. The costs per barrel to send the oil are given below.

		To:	
		D_1	D_2
From:	S_1	$30	$20
	S_2	$25	$22

How should the oil be supplied to minimize shipping costs?

Finance **24.** A bank has set aside a maximum of $25 million for commercial and home loans. The bank's policy is to loan at least four times as much for home loans as for commercial loans. Because of prior commitments, at least $10 million will be used for these two types of loans. The bank earns 12% on home loans and 10% on commercial loans. What amount of money should be loaned out for each type of loan to maximize the interest income?

Production Costs **25.** Brand X Canners produces canned whole tomatoes and tomato sauce. This season, the company has available 3,000,000 kg of tomatoes for these two products. To meet the demands of regular customers, it must produce at least 80,000 kg of sauce and 800,000 kg of whole tomatoes. The cost per kilogram is $4 to produce canned whole tomatoes and $3.25 to produce tomato sauce. How many kilograms of tomatoes should Brand X use for each product to minimize cost?

Production Costs **26.** A brewery produces regular beer and a lower-carbohydrate "light" beer. Steady customers of the brewery buy 12 units of regular beer and 10 units of light beer monthly. While setting up the brewery to produce the beers, the management decides to produce extra beer, beyond that needed to satisfy the steady customers. The cost per unit of regular beer is $36,000 and the cost per unit of light beer is $48,000. The number of units of light beer should not exceed twice the number of units of regular beer. At least 20 additional units of beer can be sold. How much of each type of beer should be made so as to minimize total production costs?

Supply Costs **27.** The chemistry department at a local college decides to stock at least 800 small test tubes and 500 large test tubes. It wants to buy at least 1500 test tubes to take advantage of a special price. Since the small tubes are broken twice as often as the larger, the department will order at least twice as many small tubes as large. If the small test tubes cost 15¢ each and large ones, made of a cheaper glass, cost 12¢ each, how many of each size should be ordered to minimize cost?

Blending Seed   28. Topgrade Turf lawn seed mixture contains three types of seeds: bluegrass, rye, and bermuda. The costs per pound of the three types of seed are 20¢, 15¢, and 5¢. In each batch there must be at least 20% bluegrass seed, and the amount of bermuda must be no more than the amount of rye. To fill current orders, the company must make at least 5000 pounds of the mixture. How much of each kind of seed should be used to minimize cost?

Investments   29. Virginia Keleske has decided to invest a $100,000 inheritance in government securities that earn 7% per year, municipal bonds that earn 6% per year, and mutual funds that earn an average of 10% per year. She will spend at least $40,000 on government securities, and she wants at least half the inheritance to go to bonds and mutual funds. How much should be invested in each way to maximize the interest yet meet the constraints? What is the maximum interest she can earn?

Transportation   30. The manufacturer of a popular personal computer has orders from two dealers. Dealer D_1 wants at least 32 computers, and dealer D_2 wants at least 20 computers. The manufacturer can fill the orders from either of two warehouses, W_1 or W_2. W_1 has 25 of the computers on hand, and W_2 has 30. The costs (in dollars) to ship one computer to each dealer from each warehouse are given below.

		To:	
		D_1	D_2
From:	W_1	14	22
	W_2	12	10

How should the orders be filled to minimize shipping costs?

LIFE SCIENCES

Health Care   31. Mark, who is ill, takes vitamin pills. Each day he must have at least 16 units of vitamin A, 5 units of vitamin B_1, and 20 units of vitamin C. He can choose between pill #1, which costs 10 cents and contains 8 units of A, 1 of B_1, and 2 of C; and pill #2, which costs 20 cents and contains 2 units of A, 1 of B_1, and 7 of C. How many of each pill should he buy in order to minimize his cost? (See Exercise 18 in Section 3.2.)

Dietetics   32. Sing, who is dieting, requires two food supplements, I and II. He can get these supplements from two different products, A and B, as shown below.

$$\begin{array}{cc} & \begin{matrix} \textit{Grams of} \\ \textit{Supplement} \\ \textit{per Serving} \end{matrix} \\ & \begin{matrix} \text{I} & \text{II} \end{matrix} \\ \textit{Product} \begin{matrix} \text{A} \\ \text{B} \end{matrix} & \begin{bmatrix} 3 & 2 \\ 2 & 4 \end{bmatrix} \end{array}$$

Sing's physician has recommended that he include at least 15 g of each supplement in his daily diet. If product A costs 25¢ per serving and product B costs 40¢ per serving, how can he satisfy his requirements most economically? Find the minimum cost. (See Exercise 21 in Section 3.2.)

Blending Nutrients 33. A biologist must make a nutrient for her algae. The nutrient must contain the three basic elements D, E, and F, and must contain at least 10 kg of D, 12 kg of E, and 20 kg of F. The nutrient is made from three ingredients, I, II, and III. The quantity of D, E, and F in one unit of each of the ingredients is as given in the following chart.

Ingredient	Kilograms (per Unit of Ingredients) of Elements:			Cost per Unit
	D	E	F	
I	4	3	0	4
II	1	2	4	7
III	10	1	5	5

How many units of each ingredient are required to meet the biologist's needs at minimum cost?

FOR THE COMPUTER

Determine the constraints and the objective function for Exercises 34–36, and then solve each problem.

Blending Chemicals 34. Natural Brand plant food is made from three chemicals. (See Section 2.2, Exercise 55.) In a batch of the plant food there must be at least 81 kg of the first chemical, and the other two chemicals must be in the ratio of 4 to 3. If the three chemicals cost $1.09, $.87, and $.65 per kilogram, respectively, how much of each should be used to minimize the cost of producing at least 750 kg of the plant food?

Blending Gasoline 35. A company is developing a new additive for gasoline. The additive is a mixture of three liquid ingredients, I, II, and III. For proper performance, the total amount of additive must be at least 10 oz per gallon of gasoline. However, for safety reasons, the amount of additive should not exceed 15 oz per gallon of gasoline. At least 1/4 oz of ingredient I must be used for every ounce of ingredient II, and at least 1 oz of ingredient III must be used for every ounce of ingredient I. If the cost of I, II, and III is $.30, $.09, and $.27 per ounce, respectively, find the mixture of the three ingredients that produces the minimum cost of the additive. How much of the additive should be used per gallon of gasoline?

Solve the following linear programming problem, which has both "greater than" and "less than" constraints.

Blending a Soft Drink 36. A popular soft drink called Sugarlo, which is advertised as having a sugar content of no more than 10%, is blended from five ingredients, each of which has some sugar content. Water may also be added to dilute the mixture. The sugar content of the ingredients and their costs per gallon are given below.

	Ingredient					
	1	2	3	4	5	Water
Sugar content (%)	.28	.19	.43	.57	.22	0
Cost ($/gal.)	.48	.32	.53	.28	.43	.04

At least .01 of the content of Sugarlo must come from ingredients 3 or 4, .01 must come from ingredients 2 or 5, and .01 from ingredients 1 or 4. How much of each ingredient should be used in preparing 15,000 gal of Sugarlo to minimize the cost?

▰▰ 4.4 DUALITY (OPTIONAL)

In this section we discuss an alternative method of solving minimizing problems in standard form. An interesting connection exists between standard maximizing and standard minimizing problems: any solution of a standard maximizing problem produces the solution of an associated standard minimizing problem, and vice versa. Each of these associated problems is called the **dual** of the other. One advantage of duals is that standard minimizing problems can be solved by the simplex methods already discussed. Let us explain the idea of a dual with an example. (This is similar to Example 2 in the previous section; compare this method of solution with the one given there.)

▰▰ EXAMPLE 1

Minimize $w = 8y_1 + 16y_2$

subject to: $y_1 + 5y_2 \geq 9$

$2y_1 + 2y_2 \geq 10$

with $y_1 \geq 0,\ y_2 \geq 0.$

(As mentioned earlier, y_1 and y_2 are used as variables and w as the objective function as a reminder that this is a minimization problem.)

Without considering slack variables just yet, write the augmented matrix of the system of inequalities, and include the coefficients of the objective function (not their negatives) as the last row in the matrix.

$$\begin{bmatrix} 1 & 5 & | & 9 \\ 2 & 2 & | & 10 \\ \hline 8 & 16 & | & 0 \end{bmatrix}$$

Now look at the following matrix, which we obtain from the one above by interchanging rows and columns.

$$\begin{bmatrix} 1 & 2 & | & 8 \\ 5 & 2 & | & 16 \\ \hline 9 & 10 & | & 0 \end{bmatrix}$$

The *rows* of the first matrix (for the minimization problem) are the *columns* of the second matrix.

The entries in this second matrix could be used to write the following maximizing problem in standard form (again ignoring the fact that the numbers in the last row are not negative):

Maximize $z = 9x_1 + 10x_2$

subject to: $x_1 + 2x_2 \leq 8$

$5x_1 + 2x_2 \leq 16.$

with all variables nonnegative.

Figure 4(a) shows the region of feasible solutions for the minimization problem given above, while Figure 4(b) shows the region of feasible solutions for the maximization problem produced by exchanging rows and columns. The solutions of the two problems are given below.

Corner Point	$w = 8y_1 + 16y_2$	
(0, 5)	80	
(4, 1)	**48**	Minimum
(9, 0)	72	

The minimum is 48 when $y_1 = 4$ and $y_2 = 1$.

Corner Point	$z = 9x_1 + 10x_2$	
(0, 0)	0	
(0, 4)	40	
(2, 3)	**48**	Maximum
$(\frac{16}{5}, 0)$	28.8	

The maximum is 48 when $x_1 = 2$ and $x_2 = 3$.

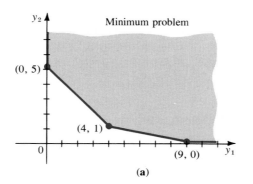

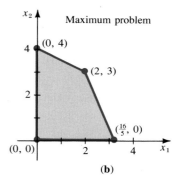

FIGURE 4

The two feasible regions in Figure 4 are different and the corner points are different, but the values of the objective functions are equal—both are 48. An even closer connection between the two problems is shown by using the simplex method to solve the maximization problem given above.

Maximization Problem

$$
\begin{array}{ccccc|c}
x_1 & x_2 & x_3 & x_4 & z & \\
\hline
1 & 2 & 1 & 0 & 0 & 8 \\
5 & 2 & 0 & 1 & 0 & 16 \\
\hline
-9 & -10 & 0 & 0 & 1 & 0
\end{array}
$$

$$
\begin{array}{ccccc|c}
x_1 & x_2 & x_3 & x_4 & z & \\
\hline
\frac{1}{2} & 1 & \frac{1}{2} & 0 & 0 & 4 \\
4 & 0 & -1 & 1 & 0 & 8 \\
\hline
-4 & 0 & 5 & 0 & 1 & 40
\end{array}
\quad
\begin{array}{l}
\frac{1}{2}R_1 \\
-2R_1 + R_2 \\
10R_1 + R_3
\end{array}
$$

$$\begin{array}{ccccc} x_1 & x_2 & x_3 & x_4 & z \end{array}$$

$$\left[\begin{array}{ccccc|c} 0 & 1 & \frac{5}{8} & -\frac{1}{8} & 0 & 3 \\ 1 & 0 & -\frac{1}{4} & \frac{1}{4} & 0 & 2 \\ 0 & 0 & 4 & 1 & 1 & 48 \end{array}\right] \quad \begin{array}{l} -\frac{1}{2}R_2 + R_1 \\ \frac{1}{4}R_2 \\ 4R_2 + R_3 \end{array}$$

The maximum is 48 when
$$x_1 = 2 \text{ and } x_2 = 3.$$

Notice that the solution to the *minimization problem* is found in the bottom row and slack variable columns of the final simplex tableau for the maximization problem. This result suggests that standard minimization problems can be solved by forming the dual standard maximization problem, solving it by the simplex method, and then reading the solution for the minimization problem from the bottom row of the final simplex tableau. ▬

Before using this method to actually solve a minimization problem, let us find the duals of some typical linear programming problems. The process of exchanging the rows and columns of a matrix, which is used to find the dual, is called **transposing** the matrix, and each of the two matrices is the **transpose** of the other.

▬ **EXAMPLE 2**

Find the transpose of each matrix.

(a) $A = \begin{bmatrix} 2 & -1 & 5 \\ 6 & 8 & 0 \\ -3 & 7 & -1 \end{bmatrix}$.

Write the rows of matrix A as the columns of the transpose.

$$\text{Transpose of } A = \begin{bmatrix} 2 & 6 & -3 \\ -1 & 8 & 7 \\ 5 & 0 & -1 \end{bmatrix}$$

(b) $B = \begin{bmatrix} 1 & 2 & 4 & 0 \\ 2 & 1 & 7 & 6 \end{bmatrix}$

$$\text{Transpose of } B = \begin{bmatrix} 1 & 2 \\ 2 & 1 \\ 4 & 7 \\ 0 & 6 \end{bmatrix} \quad ▬$$

▬ **EXAMPLE 3**

Write the dual of each of the following standard maximization linear programming problems.

(a) Maximize $\quad z = 2x_1 + 5x_2$

subject to: $\qquad x_1 + x_2 \leq 10$

$\qquad\qquad\quad 2x_1 + x_2 \leq 8$

with $\qquad\qquad x_1 \geq 0, x_2 \geq 0.$

Begin by writing the augmented matrix for the given problem.

$$\begin{bmatrix} 1 & 1 & | & 10 \\ 2 & 1 & | & 8 \\ 2 & 5 & | & 0 \end{bmatrix}$$

Form the transpose of this matrix:

$$\begin{bmatrix} 1 & 2 & | & 2 \\ 1 & 1 & | & 5 \\ 10 & 8 & | & 0 \end{bmatrix}.$$

The dual problem is stated from this second matrix as follows (using y instead of x):

$$\text{Minimize} \qquad w = 10y_1 + 8y_2$$

$$\text{subject to:} \qquad y_1 + 2y_2 \geq 2$$

$$y_1 + y_2 \geq 5$$

$$\text{with} \qquad\qquad y_1 \geq 0, y_2 \geq 0.$$

(b) Minimize $\quad w = 7y_1 + 5y_2 + 8y_3$

subject to: $\quad 3y_1 + 2y_2 + y_3 \geq 10$

$\qquad\qquad\quad y_1 + y_2 + y_3 \geq 8$

$\qquad\qquad\quad 4y_1 + 5y_2 \geq 25$

with $\qquad\qquad y_1 \geq 0, y_2 \geq 0, \quad y_3 \geq 0.$

The dual problem is stated as follows.

Maximize $\qquad z = 10x_1 + 8x_2 + 25x_3$

subject to: $\qquad 3x_1 + x_2 + 4x_3 \leq 7$

$\qquad\qquad\quad 2x_1 + x_2 + 5x_3 \leq 5$

$\qquad\qquad\quad x_1 + x_2 \leq 8$

with $\qquad\qquad x_1 \geq 0, x_2 \geq 0, x_3 \geq 0.$ ▬

In Example 3, all the constraints of the given standard maximization problems were $\leq$ inequalities, while all those in the dual minimization problems were $\geq$ inequalities. This is generally the case; inequalities are reversed when the dual problem is stated.

The following table shows the close connection between a problem and its dual.

Given Problem	Dual Problem
m variables	*n* variables
n constraints	*m* constraints
Coefficients from objective function	Constants
Constants	Coefficients from objective function

The next theorem, whose proof requires advanced methods, guarantees that a standard minimization problem can be solved by forming a dual standard maximization problem.

THEOREM OF DUALITY

The objective function w of a minimizing linear programming problem takes on a minimum value if and only if the objective function z of the corresponding dual maximizing problem takes on a maximum value. The maximum value of z equals the minimum value of w.

This method is illustrated in the following example. (This is Example 2 of the previous section; compare this solution with the one given there.)

EXAMPLE 4

Minimize $\quad w = 3y_1 + 2y_2$

subject to: $\quad y_1 + 3y_2 \geq 6$

$\qquad\qquad 2y_1 + y_2 \geq 3$

with $\qquad y_1 \geq 0, y_2 \geq 0.$

Use the given information to write the matrix.

$$\begin{bmatrix} 1 & 3 & 6 \\ 2 & 1 & 3 \\ \hline 3 & 2 & 0 \end{bmatrix}$$

Transpose to get the following matrix for the dual problem.

$$\begin{bmatrix} 1 & 2 & 3 \\ 3 & 1 & 2 \\ \hline 6 & 3 & 0 \end{bmatrix}$$

Write the dual problem from this matrix, as follows:

$$\text{Maximize} \quad z = 6x_1 + 3x_2$$
$$\text{subject to:} \quad x_1 + 2x_2 \leq 3$$
$$3x_1 + x_2 \leq 2$$
$$\text{with} \quad x_1 \geq 0, x_2 \geq 0.$$

Solve this standard maximization problem using the simplex method. Start by introducing slack variables to give the system

$$x_1 + 2x_2 + x_3 \qquad\qquad = 3$$
$$3x_1 + x_2 \qquad + x_4 \qquad = 2$$
$$-6x_1 - 3x_2 - 0x_3 - 0x_4 + z = 0$$
$$\text{with } x_1 \geq 0, x_2 \geq 0, x_3 \geq 0, x_4 \geq 0.$$

The first tableau for this system is given below, with the pivot as indicated.

Quotients	x_1	x_2	x_3	x_4	z	
$3/1 = 3$	1	2	1	0	0	3
$2/3$	3	1	0	1	0	2
	-6	-3	0	0	1	0

The simplex method gives the following as the final tableau.

x_1	x_2	x_3	x_4	z	
0	1	$\frac{3}{5}$	$-\frac{1}{5}$	0	$\frac{7}{5}$
1	0	$-\frac{1}{5}$	$\frac{2}{5}$	0	$\frac{1}{5}$
0	0	$\frac{3}{5}$	$\frac{9}{5}$	1	$\frac{27}{5}$

The last row of this final tableau shows that the solution of the given *standard minimization problem* is as follows:

The minimum value of $w = 3y_1 + 2y_2$, subject to the given constraints, is 27/5 and occurs when $y_1 = 3/5$ and $y_2 = 9/5$.

The minimum value of w, 27/5, is the same as the maximum value of z. ▰

Let us summarize the steps in solving a standard minimization linear programming problem by the method of duals.

SOLVING MINIMUM PROBLEMS WITH DUALS

1. Find the dual standard maximization problem.
2. Solve the maximization problem using the simplex method.
3. The minimum value of the objective function w is the maximum value of the objective function z.
4. The optimum solution is given by the entries in the bottom row of the columns corresponding to the slack variables.

Further Uses of the Dual   The dual is useful not only in solving minimum problems, but also in seeing how small changes in one variable will affect the value of the objective function. For example, suppose an animal breeder needs at least 6 units per day of nutrient A and at least 3 units of nutrient B and that the breeder can choose between two different feeds, feed 1 and feed 2. Find the minimum cost for the breeder if each bag of feed 1 costs \$3 and provides 1 unit of nutrient A and 2 units of B, while each bag of feed 2 costs \$2 and provides 3 units of nutrient A and 1 of B.

If y_1 represents the number of bags of feed 1 and y_2 represents the number of bags of feed 2, the given information leads to the following problem.

$$\text{Minimize} \qquad w = 3y_1 + 2y_2$$
$$\text{subject to:} \qquad y_1 + 3y_2 \geq 6$$
$$2y_1 + y_2 \geq 3$$
$$\text{with} \qquad y_1 \geq 0, \ y_2 \geq 0.$$

This standard minimization linear programming problem is the one solved in Example 4 of this section. In that example, the dual was formed and the following tableau was found.

$$
\begin{array}{ccccc}
x_1 & x_2 & x_3 & x_4 & z \\
\end{array}
$$
$$
\left[
\begin{array}{ccccc|c}
0 & 1 & \frac{3}{5} & -\frac{1}{5} & 0 & \frac{7}{5} \\
1 & 0 & -\frac{1}{5} & \frac{2}{5} & 0 & \frac{1}{5} \\
0 & 0 & \frac{3}{5} & \frac{9}{5} & 1 & \frac{27}{5}
\end{array}
\right]
$$

This final tableau shows that the breeder will obtain minimum feed costs by using 3/5 bag of feed 1 and 9/5 bags of feed 2 per day, for a daily cost of $27/5 = 5.40$ dollars.

Now look at the data from the problem shown in the table below.

	Units of Nutrient (per Bag): A B	Cost per Bag
Feed 1	1 2	\$3
Feed 2	3 1	\$2
	6 3	

If x_1 and x_2 are the cost per unit of nutrients A and B, the constraints of the dual problem can be stated as follows.

$$\text{Cost of Feed 1:} \qquad x_1 + 2x_2 \leq 3$$
$$\text{Cost of Feed 2:} \qquad 3x_1 + x_2 \leq 2$$

The solution of the dual problem, which maximizes nutrients, can be read from the final tableau:

$$x_1 = \frac{1}{5} = .20 \quad \text{and} \quad x_2 = \frac{7}{5} = 1.40,$$

which means that a unit of nutrient A costs $1/5 = .20$ dollars, while a unit of nutrient B costs $7/5 = 1.40$ dollars. The minimum daily cost, $5.40, is found by the following procedure.

($.20 per unit of A) × (6 units of A) = $1.20
+ ($1.40 per unit of B) × (3 units of B) = $4.20

Minimum Daily Cost = $5.40

The numbers .20 and 1.40 are called the **shadow costs** of the nutrients. These two numbers from the dual, $.20 and $1.40, also allow the breeder to estimate feed costs for "small" changes in nutrient requirements. For example, an increase of one unit in the requirement for each nutrient would produce a total cost of

$5.40	6 units of A, 3 of B
.20	1 extra unit of A
1.40	1 extra unit of B
$7.00.	Total cost per day

4.4 EXERCISES

Find the transpose of each matrix.

1. $\begin{bmatrix} 1 & 2 & 3 \\ 3 & 2 & 1 \\ 1 & 10 & 0 \end{bmatrix}$

2. $\begin{bmatrix} 2 & 5 & 8 & 6 & 0 \\ 1 & -1 & 0 & 12 & 14 \end{bmatrix}$

3. $\begin{bmatrix} -1 & 4 & 6 & 12 \\ 13 & 25 & 0 & 4 \\ -2 & -1 & 11 & 3 \end{bmatrix}$

4. $\begin{bmatrix} 1 & 11 & 15 \\ 0 & 10 & -6 \\ 4 & 12 & -2 \\ 1 & -1 & 13 \\ 2 & 25 & -1 \end{bmatrix}$

State the dual problem for each of the following.

5. Maximize $z = 4x_1 + 3x_2 + 2x_3$
   subject to:
   $x_1 + x_2 + x_3 \le 5$
   $x_1 + x_2 \le 4$
   $2x_1 + x_2 + 3x_3 \le 15$
   with $x_1 \ge 0, x_2 \ge 0, x_3 \ge 0.$

6. Maximize $z = 8x_1 + 3x_2 + x_3$
   subject to:
   $7x_1 + 6x_2 + 8x_3 \le 18$
   $4x_1 + 5x_2 + 10x_3 \le 20$
   with $x_1 \ge 0, x_2 \ge 0, x_3 \ge 0.$

7. Minimize $w = y_1 + 2y_2 + y_3 + 5y_4$

subject to: $y_1 + y_2 + y_3 + y_4 \geq 50$
$3y_1 + y_2 + 2y_3 + y_4 \geq 100$

with $y_1 \geq 0, y_2 \geq 0, y_3 \geq 0, y_4 \geq 0.$

Use the simplex method to solve Exercises 9–14.

9. Find $y_1 \geq 0$ and $y_2 \geq 0$ such that

$2y_1 + 3y_2 \geq 6$
$2y_1 + y_2 \geq 7$

and $w = 5y_1 + 2y_2$ is minimized.

11. Find $y_1 \geq 0$ and $y_2 \geq 0$ such that

$10y_1 + 5y_2 \geq 100$
$20y_1 + 10y_2 \geq 150$

and $w = 4y_1 + 5y_2$ is minimized.

13. Minimize $w = 2y_1 + y_2 + 3y_3$

subject to: $y_1 + y_2 + y_3 \geq 100$
$2y_1 + y_2 \qquad \geq 50$

with $y_1 \geq 0, y_2 \geq 0, y_3 \geq 0.$

8. Minimize $w = y_1 + y_2 + 4y_3$

subject to: $y_1 + 2y_2 + 3y_3 \geq 115$
$2y_1 + y_2 + 8y_3 \geq 200$
$y_1 \qquad + y_3 \geq 50$

with $y_1 \geq 0, y_2 \geq 0, y_3 \geq 0.$

10. Find $y_1 \geq 0$ and $y_2 \geq 0$ such that

$3y_1 + y_2 \geq 12$
$y_1 + 4y_2 \geq 16$

and $w = 2y_1 + y_2$ is minimized.

12. Minimize $w = 3y_1 + 2y_2$

subject to: $2y_1 + 3y_2 \geq 60$
$y_1 + 4y_2 \geq 40$

with $y_1 \geq 0, y_2 \geq 0.$

14. Minimize $w = 3y_1 + 2y_2$

subject to: $y_1 + 2y_2 \geq 10$
$y_1 + y_2 \geq 8$
$2y_1 + y_2 \geq 12$

with $y_1 \geq 0, y_2 \geq 0.$

▤ APPLICATIONS

Solve each of the following by using the method of duals.

BUSINESS AND ECONOMICS

Production Costs **15.** Brand X Canners produces canned whole tomatoes and tomato sauce. This season, the company has available 3,000,000 kg of tomatoes for these two products. To meet the demands of regular customers, it must produce at least 80,000 kg of sauce and 800,000 kg of whole tomatoes. The cost per kilogram is $4 to produce canned whole tomatoes and $3.25 to produce tomato sauce. How many kilograms of tomatoes should Brand X use for each product to minimize cost? (See Section 4.3, Exercise 25.)

Profit **16.** Refer to the problem about the farmer solved at the beginning of Section 4.2.

(a) Give the dual problem.

(b) Use the shadow values to estimate the farmer's profit if land is cut to 90 acres but capital increases to $21,000.

(c) Suppose the farmer has 110 acres but only $19,000. Find the optimum profit and the planting strategy that will produce this profit.

Profit **17.** A small toy manufacturing firm has 200 squares of felt, 600 oz of stuffing, and 90 ft of trim available to make two types of toys, a small bear and a monkey. The bear requires 1 square of felt and 4 oz of stuffing. The monkey requires 2 squares of felt, 3 oz of stuffing, and 1 ft of trim. The firm makes $1 profit on each bear and $1.50 profit on each monkey. The linear program to maximize profit is

$$\text{Maximize} \qquad x_1 + 1.5x_2 = z$$
$$\text{subject to:} \qquad x_1 + 2x_2 \leq 200$$
$$4x_1 + 3x_2 \leq 600$$
$$x_2 \leq 90.$$

The final simplex tableau is

$$\begin{bmatrix} 0 & 1 & .8 & -.2 & 0 & 40 \\ 1 & 0 & -.6 & .4 & 0 & 120 \\ 0 & 0 & -.8 & .2 & 1 & 50 \\ \hline 0 & 0 & .6 & .1 & 0 & 180 \end{bmatrix}.$$

(a) What is the corresponding dual problem?

(b) What is the optimal solution to the dual problem?

(c) Use the shadow values to estimate the profit the firm will make if its supply of felt increases to 210 squares.

(d) How much profit will the firm make if its supply of stuffing is cut to 590 oz and its supply of trim is cut to 80 ft?

LIFE SCIENCES

Dietetics **18.** Sing, who is dieting, requires two food supplements, I and II. He can get these supplements from two different products, A and B, as shown below.

$$\begin{array}{c} \textit{Grams of} \\ \textit{Supplement} \\ \textit{per Serving} \end{array}$$

$$\begin{array}{cc} & \text{I} \quad \text{II} \\ \textit{Product} \begin{array}{c} \text{A} \\ \text{B} \end{array} & \begin{bmatrix} 3 & 2 \\ 2 & 4 \end{bmatrix} \end{array}$$

Sing's physician has recommended that he include at least 15 g of each supplement in his daily diet. If product A costs 25¢ per serving and product B costs 40¢ per serving, how can he satisfy his requirements most economically? (See Section 4.3, Exercise 32.)

Health Care **19.** Mark, who is ill, takes vitamin pills. Each day he must have at least 16 units of vitamin A, 5 units of vitamin B_1, and 20 units of vitamin C. He can choose between pill #1, which costs 10 cents and contains 8 units of A, 1 of B_1, and 2 of C; and pill #2, which costs 20 cents and contains 2 units of A, 1 of B_1, and 7 of C. How many of each pill should he buy in order to minimize his cost? (See Section 4.3, Exercise 31.)

Feed Costs **20.** Refer to the example at the end of this section on minimizing the daily cost of feeds.

(a) Find a combination of feeds that will cost $7.00 and give 7 units of A and 4 units of B.

(b) Use the dual variables to predict the daily cost of feed if the requirements change to 5 units of A and 4 units of B. Find a combination of feeds to meet these requirements at the predicted price.

FOR THE COMPUTER

Determine the constraints and the objective functions for Exercises 21 and 22, and then use the method of duals to solve each problem by the simplex method.

Blending Chemicals   **21.** Natural Brand plant food is made from three chemicals. In a batch of the plant food there must be at least 81 kg of the first chemical, and the other two chemicals must be in the ratio of 4 to 3. If the three chemicals cost $1.09, $.87, and $.65 per kilogram, respectively, how much of each should be used to minimize the cost of producing at least 750 kg of the plant food? (See Section 4.3, Exercise 34.)

Blending Gasoline   **22.** A company is developing a new additive for gasoline. The additive is a mixture of three liquid ingredients, I, II, III. For proper performance, the total amount of additive must be at least 10 oz per gallon of gasoline. For safety reasons, the amount of additive should not exceed 15 oz per gallon of gasoline. At least 1/4 oz of ingredient I must be used for every ounce of ingredient II, and at least 1 oz of ingredient III must be used for every ounce of ingredient I. If the cost of I, II, and III is $.30, $.09, and $.27 per ounce, respectively, find the mixture of the three ingredients that produces the minimum cost of the additive. How much of the additive should be used per gallon of gasoline? (See Section 4.3, Exercise 35.)

KEY WORDS

	simplex method	**4.3**	surplus variable
4.1	slack variable		phase I
	simplex tableau		phase II
	indicators	**4.4**	dual
	basic variables		transpose
	pivot		

CHAPTER 4   REVIEW EXERCISES

For Exercises 1–4, **(a)** *add slack variables or subtract surplus variables, and* **(b)** *set up the initial simplex tableau.*

1. Maximize $\quad z = 5x_1 + 3x_2$

subject to: $\quad 2x_1 + 5x_2 \le 50$
$x_1 + 3x_2 \le 25$
$4x_1 + x_2 \le 18$
$x_1 + x_2 \le 12$

with $\quad x_1 \ge 0, x_2 \ge 0.$

2. Maximize $\quad z = 25x_1 + 30x_2$

subject to: $\quad 3x_1 + 5x_2 \le 47$
$x_1 + x_2 \le 25$
$5x_1 + 2x_2 \le 35$
$2x_1 + x_2 \le 30$

with $\quad x_1 \ge 0, x_2 \ge 0.$

3. Maximize $\quad z = 5x_1 + 8x_2 + 6x_3$

subject to: $\quad x_1 + x_2 + x_3 \le 90$
$2x_1 + 5x_2 + x_3 \le 120$
$x_1 + 3x_2 \ge 80$

with $\quad x_1 \ge 0, x_2 \ge 0, x_3 \ge 0.$

4. Maximize $\quad z = 2x_1 + 3x_2 + 4x_3$

subject to: $\quad x_1 + x_2 + x_3 \ge 100$
$2x_1 + 3x_2 \le 500$
$x_1 + 2x_3 \le 350$

with $\quad x_1 \ge 0, x_2 \ge 0, x_3 \ge 0.$

Use the simplex method to solve the maximizing linear programming problems with initial tableaus as given in Exercises 5–8.

5.

x_1	x_2	x_3	x_4	x_5	z	
1	2	3	1	0	0	28
2	4	1	0	1	0	32
-5	-2	-3	0	0	1	0

6.

x_1	x_2	x_3	x_4	z	
2	1	1	0	0	10
1	3	0	1	0	16
-2	-3	0	0	1	0

7.

x_1	x_2	x_3	x_4	x_5	x_6	z	
1	2	2	1	0	0	0	50
3	1	0	0	1	0	0	20
1	0	2	0	0	-1	0	15
-5	-3	-2	0	0	0	1	0

8.

x_1	x_2	x_3	x_4	x_5	z	
3	6	-1	0	0	0	28
1	1	0	1	0	0	12
2	1	0	0	1	0	16
-1	-2	0	0	0	1	0

Convert the problems of Exercises 9–11 into maximization problems.

9. Minimize $w = 10y_1 + 15y_2$

    subject to: $y_1 + y_2 \geq 17$
                     $5y_1 + 8y_2 \geq 42$

    with $y_1 \geq 0, y_2 \geq 0.$

10. Minimize $w = 20y_1 + 15y_2 + 18y_3$

     subject to: $2y_1 + y_2 + y_3 \geq 112$
                      $y_1 + y_2 + y_3 \geq 80$
                      $y_1 + y_2 \quad\; \geq 45$

     with $y_1 \geq 0, y_2 \geq 0, y_3 \geq 0.$

11. Minimize $w = 7y_1 + 2y_2 + 3y_3$

    subject to: $y_1 + y_2 + 2y_3 \geq 48$
                   $y_1 + y_2 \quad\;\; \geq 12$
                         $y_3 \geq 10$
             $3y_1 \quad\; + \; y_3 \geq 30$

    with $y_1 \geq 0, y_2 \geq 0, y_3 \geq 0.$

The tableaus in Exercises 12–14 are the final tableaus of minimizing problems. State the solution and the minimum value of the objective function for each problem.

12.

x_1	x_2	x_3	x_4	x_5	x_6	z	
1	0	0	3	1	2	0	12
0	0	1	4	5	3	0	5
0	1	0	-2	7	-6	0	8
0	0	0	5	7	3	1	-172

13.

x_1	x_2	x_3	x_4	x_5	x_6	z	
0	0	3	0	1	1	0	2
1	0	-2	0	2	0	0	8
0	1	7	0	0	0	0	12
0	0	1	1	-4	0	0	1
0	0	5	0	8	0	1	-62

14.

x_1	x_2	x_3	x_4	x_5	z	
5	1	0	7	-1	0	100
-2	0	1	1	3	0	27
12	0	0	7	2	1	-640

≣ APPLICATIONS

For Exercises 15–18, (**a**) *select appropriate variables;* (**b**) *write the objective function;* (**c**) *write the constraints as inequalities.*

BUSINESS AND ECONOMICS

Profit   **15.** Roberta Hernandez sells three items, A, B, and C, in her gift shop. Each unit of A costs her $5 to buy, $1 to sell, and $2 to deliver. For each unit of B, the costs are $3, $2, and $1 respectively, and for each unit of C the costs are $6, $2, and $5 respectively. The profit on A is $4; on B, $3; and on C, $3. How many of each should she get to maximize her profit if she can spend $1200 on buying costs, $800 on selling costs, and $500 on delivery costs?

Investments   **16.** An investor is considering three types of investment: a high-risk venture into oil leases with a potential return of 15%, a medium-risk investment in bonds with a 9% return, and a relatively safe stock investment with a 5% return. He has $50,000 to invest. Because of the risk, he will limit his investment in oil leases and bonds to 30% and his investment in oil leases and stock to 50%. How much should he invest in each to maximize his return, assuming investment returns are as expected?

Profit   **17.** The Aged Wood Winery makes two white wines, Fruity and Crystal, from two kinds of grapes and sugar. The wines require the following amounts of each ingredient per gallon and produce a profit per gallon as shown below.

	Grape A (Bushels)	Grape B (Bushels)	Sugar (Pounds)	Profit (Dollars)
Fruity	2	2	2	12
Crystal	1	3	1	15

The winery has available 110 bushels of grape A, 125 bushels of grape B, and 90 lb of sugar. How much of each wine should be made to maximize profit?

Profit   **18.** A company makes three sizes of plastic bags: 5-gallon, 10-gallon and 20-gallon. The production time in hours for cutting, sealing, and packaging a unit of each size is shown below.

Size	Cutting Time	Sealing Time	Packaging Time
5 gallon	1	1	2
10 gallon	1.1	1.2	3
20 gallon	1.5	1.3	4

There are at most 8 hr available each day for each of the three operations. If the profit on a unit of 5-gallon bags is $1; 10-gallon bags, $.90; and 20-gallon bags, $.95, how many of each size should be made per day to maximize profit?

19. Solve Exercise 15.   **20.** Solve Exercise 16.

21. Solve Exercise 17.   **22.** Solve Exercise 18.

EXTENDED
APPLICATION

MAKING ICE CREAM*

The first step in the commercial manufacture of ice cream is to blend several ingredients (such as dairy products, eggs, and sugar) to obtain a mix that meets the necessary minimum quality restrictions regarding butterfat content, serum solids, and so on.

Usually many different combinations of ingredients may be blended to obtain a mix of the necessary quality. Within this range of possible substitutes, the firm desires the combination that produces minimum total cost. This problem is quite suitable for solution by linear programming methods.

A "mid-quality" line of ice cream requires the following minimum percentages of constituents by weight:

Fat	16
Serum Solids	8
Sugar Solids	16
Egg Solids	.35
Stabilizer	.25
Emulsifier	.15
Total	40.75%

The balance of the mix is water. A batch of mix is made by blending a number of ingredients, each of which contains one or more of the necessary constituents, and nothing else. The following chart shows the possible ingredients and their costs.

Ingredients	1	2	3	4	5	6	7	8	9
	40% Cream	23% Cream	Butter	Plastic Cream	Butter Oil	4% Milk	Skim Condensed Milk	Skim Milk Powder	Liquid Sugar
Cost ($/lb)	.298	.174	.580	.576	.718	.045	.052	.165	.061
Constituents									
(1) Fat	.400	.230	.805	.800	.998	.040			
(2) Serum Solids	.054	.069		.025		.078	.280	.970	
(3) Sugar Solids									.707
(4) Egg Solids									
(5) Stabilizer									
(6) Emulsifier									

Ingredients	10	11	12	13	14	Requirements
	Sugared Egg Yolk	Powdered Egg Yolk	Stabilizer	Emulsifier	Water	
Cost ($/lb)	.425	1.090	.600	.420	0	
Constituents						
(1) Fat	.500	.625				16
(2) Serum Solids						8
(3) Sugar Solids	.100					16
(4) Egg Solids	.350	.315				.35
(5) Stabilizer			1			.25
(6) Emulsifier				1		.15

In setting up the mathematical model, use $c_1, c_2, \ldots, c_{14}$ as the cost per unit of the ingredients, and $x_1, x_2, \ldots, x_{14}$ for the quantities of ingredients. The table shows the composition of each ingredient. For example, one pound of ingredient 1 (the 40% cream) contains .400 pounds of fat and .054 pounds of serum solids, with the balance being water.

The ice cream mix is made up in batches of 100 pounds at a time. For a batch to contain 16% fat, at least $100(.16) = 16$ pounds of fat is necessary. Fat is contained in ingredients 1, 2, 3, 4, 5, 6, 10, and 11. The requirement of at least 16 pounds of fat produces the constraint

$$.400x_1 + .230x_2 + .805x_3 + .800x_4 + .998x_5 + .040x_6 + .500x_{10} + .625x_{11} \geq 16.$$

Similar constraints can be obtained for the other constituents. Because of the minimum requirements, the total of the ingredients will be at least 40.75 pounds, or

$$x_1 + x_2 + \ldots + x_{13} \geq 40.75.$$

The balance of the 100 pounds is water, making $x_{14} \leq 59.25$. Also, for a 100-pound batch,

$$x_1 + x_2 + \ldots x_{13} \leq 100.$$

The table shows that ingredients 12 and 13 must be used—there is no alternate way of getting these constituents into the final mix. For a 100-pound batch of ice cream, $x_{12} = .25$ pound and $x_{13} = .15$ pound. Removing x_{12} and x_{13} as variables permits a substantial simplification of the model. The problem is now reduced to the following system of five constraints:

$$.400x_1 + .230x_2 + .805x_3 + .800x_4 + .998x_5 + .040x_6 + .500x_{10} + .625x_{11} \geq 16$$
$$.054x_1 + .069x_2 + .025x_4 + .078x_6 + .280x_7 + .970x_8 \geq 8$$
$$.707x_9 + .100x_{10} \geq 16$$
$$.350x_{10} + .315x_{11} \geq .35$$
$$x_1 + x_2 + x_3 + x_4 + x_5 + x_6 + x_7 + x_8 + x_9 + x_{10} + x_{11} \leq 99.6$$

The objective function, which is to be minimized, is

$$c = .298x_1 + .174x_2 + \ldots + 1.090x_{11}.$$

As usual, $x_1 \geq 0, x_2 \geq 0, \ldots, x_{14} \geq 0$.

Solving this linear programming problem with the simplex method gives

$$x_2 = 67.394, \quad x_8 = 3.459, \quad x_9 = 22.483, \quad x_{10} = 1.000.$$

The total cost of these ingredients is $14.094. To this, we must add the cost of ingredients 12 and 13, producing total cost of

$$14.094 + (.600)(.250) + (.420)(.150) = 14.307,$$

or $14.31.

EXERCISES

1. Find the mix of ingredients needed for a 100-pound batch of the following grades of ice cream.

   (a) "Generic," sold in a white box, with at least 14% fat and at least 6% serum solids (all other ingredients the same);

   (b) "Premium," sold with a funny foreign name, with at least 17% fat and at least 16.5% sugar solids (all other ingredients the same).

2. Solve the problem in the text by using a computer.

EXTENDED
APPLICATION

MERIT PAY—THE UPJOHN COMPANY

Individuals doing the same job within the management of a company often receive different salaries. These salaries may differ because of length of service, productivity of an individual worker, and so on. However, for each job there is usually an established minimum and maximum salary.

Many companies make annual reviews of the salary of each of their management employees. At these reviews, an employee may receive a general cost of living increase, an increase based on merit, both, or neither.

In this case, we look at a mathematical model for distributing merit increases in an optimum way.* An individual who is due for salary review may be described as shown in Figure 5, where i represents the number of the employee whose salary is being reviewed. Here the salary ceiling is the maximum salary for the job classification, x_1 is the merit increase to be awarded to the individual ($x_i \geq 0$), d_i is the present distance of the current salary from the salary ceiling, and the difference, $d_i - x_i$, is the remaining gap.

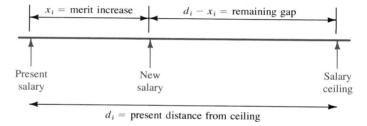

FIGURE 5

We let w_i be a measure of the relative worth of the individual to the company. This is the most difficult variable of the model to actually calculate. One way to evaluate w_i is to give a rating sheet to a number of co-workers and supervisors of employee i. An average rating can be obtained and then divided by the highest rating received by an employee with that same job. This number then gives the worth of employee i relative to all other employees with that job.

The best way to allocate money available for merit pay increases is to minimize $w_1(d_1 - x_1) + w_2(d_2 - x_2) + \ldots + w_n(d_n - x_n)$. This sum can be written in a shorthand notation called *summation notation* as

$$\sum_{i=1}^{n} w_i(d_i - x_i).$$

The sum is found by multiplying the relative worth of employee i and the distance of employee i from the salary gap, after any merit increase. Here n represents the total number of employees who have this job. The one constraint here is that the total of all merit increases cannot exceed P, the total amount available for merit increases. That is,

$$\sum_{i=1}^{n} x_i \leq P.$$

---

*Based on part of a paper by Jack Northam, Head, Mathematical Services Department, The Upjohn Company, Kalamazoo, Michigan.

Also, the increases for an employee must not put that employee over the maximum salary. That is, for employee i,

$$x_i \leq d_i.$$

We can simplify the objective function by using rules from algebra.

$$\sum_{i=1}^{n} w_i(d_i - x_i) = \sum_{i=1}^{n} (w_i d_i - w_i x_i)$$

$$= \sum_{i=1}^{n} w_i d_i - \sum_{i=1}^{n} w_i x_i$$

For a given individual, w_i and d_i are constant. Therefore, $\sum_{i=1}^{n} w_i d_i$ is some constant, say Z, and

$$\sum_{i=1}^{n} w_i(d_i - x_i) = Z - \sum_{i=1}^{n} w_i x_i.$$

We want to minimize the sum on the left; we can do so by *maximizing* the sum on the right. (Why?) Thus, the original model simplifies to maximizing

$$\sum_{i=1}^{n} w_i x_i,$$

subject to: $\sum_{i=1}^{n} x_i \leq P$ and $x_i \leq d_i$ for each i.

EXERCISES

Here are the current salary information and job evaluation averages for six employees who have the same job. The salary ceiling is $1700 per month.

Employee Number	Evaluation Average	Current Salary
1	570	$1600
2	500	$1550
3	450	$1500
4	600	$1610
5	520	$1530
6	565	$1420

1. Find w_i for each employee by dividing that employee's evaluation average by the highest evaluation average.

2. Use the simplex method with a computer to find the merit increase for each employee. Assume that p is 400.

5

Sets and Counting

5.1   Sets

5.2   Applications of Venn Diagrams

5.3   Permutations

5.4   Combinations

5.5   The Binomial Theorem   (Optional)

Review Exercises

Suppose that a personnel director has interviewed 20 job applicants, of which 9 have related work experience, 5 have MBA degrees and experience, and 12 have either MBA degrees or experience. The director may want to know how many applicants have MBA degrees only, or perhaps find the probability that any given applicant has an MBA degree. The terminology and concepts of sets have proved useful in solving problems of this type. Sets, which are collections of objects, help to clarify and classify many mathematical ideas, making them easier to understand. Sets are particularly useful in presenting the topics of probability. The principles of counting, discussed later in this chapter, also will be used to find probabilities.

5.1 SETS

Think of a **set** as a collection of objects. A set of coins might include one of each type of coin now put out by the United States government. Another set might be made up of all the students in your English class. In mathematics, sets are often made up of numbers. The set consisting of the numbers 3, 4, and 5 is written

$$\{3, 4, 5\},$$

with set braces, { }, enclosing the numbers belonging to the set. The numbers 3, 4, and 5 are called the **elements** or **members** of this set. To show that 4 is an element of the set $\{3, 4, 5\}$ use the symbol $\in$ and write

$$4 \in \{3, 4, 5\},$$

read "4 is an element of the set containing 3, 4, and 5." Also, $5 \in \{3, 4, 5\}$. To show that 8 is *not* an element of this set, place a slash through the symbol:

$$8 \notin \{3, 4, 5\}.$$

Sets often are named with capital letters, so that if

$$B = \{5, 6, 7\},$$

then, for example, $6 \in B$ and $10 \notin B$.

Two sets are **equal** if they contain the same elements. The sets $\{5, 6, 7\}$, $\{7, 6, 5\}$, and $\{6, 5, 7\}$ all contain exactly the same elements and are equal. In symbols,

$$\{5, 6, 7\} = \{7, 6, 5\} = \{6, 5, 7\}.$$

Sets that do not contain exactly the same elements are *not equal*. For example, the sets $\{5, 6, 7\}$ and $\{7, 8, 9\}$ do not contain exactly the same elements and thus are not equal. To show that these sets are not equal, we write

$$\{5, 6, 7\} \neq \{7, 8, 9\}.$$

Sometimes we are interested in a common property of the elements in a set, rather than a list of the elements. This common property can be expressed by using **set-builder notation;** for example,

$$\{x | x \text{ has property } P\}$$

(read "the set of all elements x such that x has property P") represents the set of all elements x having some stated property P.

■ EXAMPLE 1

Write the elements belonging to each of the following sets.

(a) $\{x | x$ is a natural number less than 5$\}$

The natural numbers less than 5 make up the set $\{1, 2, 3, 4\}$.

(b) $\{x | x$ is a state that borders Florida$\}$ = $\{$Alabama, Georgia$\}$ ■

In discussing a particular situation or problem, it may be necessary to identify a **universal set,** also called the **universe of discourse,** that contains all the elements appearing in any set used in that particular problem. The letter U is used to represent the universal set. For example, when discussing the set of company workers who favor a certain pension proposal, the universal set might be the set of all company workers. In discussing the species of animals found by Charles Darwin on the Galápagos Islands, the universal set might be the set of all species of animals on all Pacific islands.

Sometimes every element of one set also belongs to another set. For example, if

$$A = \{3, 4, 5, 6\}$$

and

$$B = \{2, 3, 4, 5, 6, 7, 8\},$$

then every element of A is also an element of B. This means that A is a *subset* of B, written $A \subset B$. For example, the set of all presidents of corporations is a subset of the set of all executives of corporations.

SUBSET

■ A set A is a **subset** of a set B if every element of A is also an element of B.

■ EXAMPLE 2

Decide whether the following statements are true or false.

(a) $\{3, 4, 5, 6\}$ = $\{4, 6, 3, 5\}$

Both sets contain exactly the same elements; the sets are equal. The given statement is true. (The fact that the elements are listed in a different order does not matter.)

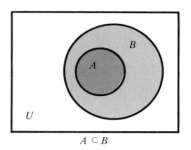

$A \subset B$

FIGURE 1

(b) $\{5, 6, 9, 10\} \subset \{5, 6, 7, 8, 9, 10, 11\}$

Every element of the first set is also an element of the second, making the given statement true. ▬

Figure 1 shows a set A that is a subset of a set B. The rectangle represents the universal set, U. Such diagrams, called **Venn diagrams,** are used to help clarify relationships among sets.

By the definition of subset, the **empty set** (which contains no elements) is a subset of every set. That is, if A is any set, and the symbol $\emptyset$ represents the empty set, then $\emptyset \subset A$. Also, the definition of subset can be used to show that every set is a subset of itself: that is, if A is any set, then $A \subset A$.

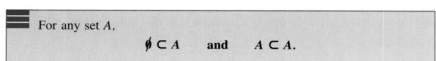

For any set A,

$$\emptyset \subset A \quad \text{and} \quad A \subset A.$$

Be careful to distinguish between the symbols 0, $\emptyset$, $\{0\}$, and $\{\emptyset\}$. The symbol 0 represents a *number;* $\emptyset$ represents a *set* with 0 elements; $\{0\}$ represents a set with one element, 0; and $\{\emptyset\}$ represents a set with one element, $\emptyset$.

▬ **EXAMPLE 3**

List all possible subsets for each of the following sets.

(a) $\{7, 8\}$

There are 4 subsets of $\{7, 8\}$:

$$\emptyset, \quad \{7\}, \quad \{8\}, \quad \text{and} \quad \{7, 8\}.$$

(b) $\{a, b, c\}$

There are 8 subsets of $\{a, b, c\}$:

$$\emptyset, \quad \{a\}, \quad \{b\}, \quad \{c\}, \quad \{a, b\}, \quad \{a, c\}, \quad \{b, c\}, \quad \text{and} \quad \{a, b, c\}. \quad ▬$$

The **cardinal number** of a set is the number of distinct elements in the set. The cardinal number of the set A is written $n(A)$. For example, the set

$$A = \{a, b, c, d, e\}$$

has cardinal number 5, written

$$n(A) = 5.$$

Since the empty set has no elements, its cardinal number is 0, by definition. Thus, $n(\emptyset) = 0$.

▬ EXAMPLE 4

Give the cardinal number of each of the following sets.

(a) $B = \{2, 5, 7, 9, 10, 12, 15\}$.

This set has 7 elements, so $n(B) = 7$.

(b) $C = \{x, y, z\}$.

Since set C has 3 elements, $n(C) = 3$. ▬

In Example 3, all the subsets of $\{7, 8\}$ and all the subsets of $\{a, b, c\}$ were found by trial and error. An alternate method uses a **tree diagram,** a systematic way of listing all the subsets of a given set. Figure 2 shows tree diagrams for finding the subsets of $\{7, 8\}$ and $\{a, b, c\}$.

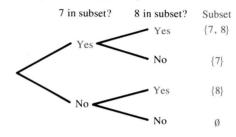

(a) Find the subsets of $\{7, 8\}$.

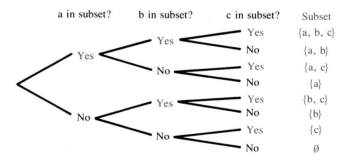

(b) Find the subsets of $\{a, b, c\}$.

FIGURE 2

Examples and tree diagrams similar to the ones above suggest the following rule, which depends on the fact that there are two ways of treating each element: either it is in a given subset or it is not.

> A set of n distinct elements has 2^n subsets.

A proof of this statement is given in Section 5.5.

▬ EXAMPLE 5

Find the number of subsets for each of the following sets.

(a) {3, 4, 5, 6, 7}

This set has 5 elements; thus, it has 2^5 or 32 subsets.

(b) {−1, 2, 3, 4, 5, 6, 12, 14}

This set has 8 elements and therefore has 2^8 or 256 subsets.

(c) ∅

Since the empty set has 0 elements, it has $2^0 = 1$ subset—itself. ▬

Given a set A and a universal set U, the set of all elements of U that do *not* belong to A is called the **complement** of set A. For example, if set A is the set of all the female students in a class, and U is the set of all students in the class, then the complement of A would be the set of all male students in the class. The complement of set A is written A', read "A-prime." The Venn diagram in Figure 3 shows a set B. Its complement, B', is shown in color.

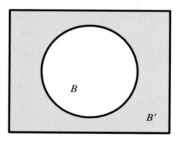

FIGURE 3

▬ EXAMPLE 6

Let $U = \{1, 2, 3, 4, 5, 6, 7\}$, $A = \{1, 3, 5, 7\}$, and $B = \{3, 4, 6\}$. Find each of the following sets.

(a) A'

Set A' contains the elements of U that are not in A.

$$A' = \{2, 4, 6\}$$

(b) $B' = \{1, 2, 5, 7\}$

(c) $∅' = U$ and $U' = ∅$

(d) $(A')' = A$ ▬

Given two sets A and B, the set of all elements belonging both to set A and to set B is called the **intersection** of the two sets, written $A \cap B$. For

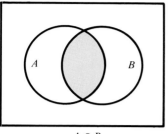

$A \cap B$

FIGURE 4

example, the elements that belong both to set $A = \{1, 2, 4, 5, 7\}$ and to set $B = \{2, 4, 5, 7, 9, 11\}$ are 2, 4, 5, and 7, so that

$$A \cap B = \{1, 2, 4, 5, 7\} \cap \{2, 4, 5, 7, 9, 11\} = \{2, 4, 5, 7\}.$$

The Venn diagram in Figure 4 shows two sets A and B, with their intersection, $A \cap B$, shown in color.

■ EXAMPLE 7

(a) Find the intersection of $\{9, 15, 25, 36\}$ and $\{15, 20, 25, 30, 35\}$.

$$\{9, 15, 25, 36\} \cap \{15, 20, 25, 30, 35\} = \{15, 25\}$$

The elements 15 and 25 are the only ones belonging to both sets.

(b) $\{-2, -3, -4, -5, -6\} \cap \{-4, -3, -2, -1, 0, 1, 2\}$
$= \{-4, -3, -2\}$ ■

Two sets that have no elements in common are called **disjoint sets.** For example, there are no elements common to both $\{50, 51, 54\}$ and $\{52, 53, 55, 56\}$, so these two sets are disjoint, and

$$\{50, 51, 54\} \cap \{52, 53, 55, 56\} = \emptyset.$$

This result can be generalized: for any sets A and B, if A and B are disjoint sets, then $A \cap B = \emptyset$.

The set of all elements belonging to set A, to set B, or to both sets is called the **union** of the two sets, written $A \cup B$. For example,

$$\{1, 3, 5\} \cup \{3, 5, 7, 9\} = \{1, 3, 5, 7, 9\}.$$

The Venn diagram in Figure 5 shows two sets A and B; their union, $A \cup B$, is shown in color.

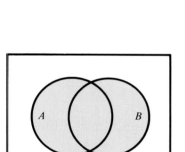

$A \cup B$

FIGURE 5

■ EXAMPLE 8

(a) Find the union of $\{1, 2, 5, 9, 14\}$ and $\{1, 3, 4, 8\}$.

Begin by listing the elements of the first set, $\{1, 2, 5, 9, 14\}$. Then include any elements from the second set that are not already listed. Doing this gives

$$\{1, 2, 5, 9, 14\} \cup \{1, 3, 4, 8\} = \{1, 2, 3, 4, 5, 8, 9, 14\}.$$

(b) $\{1, 3, 5, 7\} \cup \{2, 4, 6\} = \{1, 2, 3, 4, 5, 6, 7\}$ ■

Finding the complement of a set, the intersection of two sets, and the union of two sets are examples of **set operations.** These are similar to operations on numbers, such as addition, subtraction, multiplication, and division.

The various operations on sets are summarized below.

OPERATIONS ON SETS

> Let A and B be any sets, with U representing the universal set. Then the complement of A, written A', is
> $$A' = \{x \mid x \notin A \text{ and } x \in U\};$$
> the intersection of A and B is
> $$A \cap B = \{x \mid x \in A \text{ and } x \in B\};$$
> the union of A and B is
> $$A \cup B = \{x \mid x \in A \text{ or } x \in B\}.$$

As shown in the definitions given above, an element is in the *intersection* of sets A and B if it is in A *and* B. On the other hand, an element is in the *union* of sets A and B if it is in A *or* B (or both).

■ EXAMPLE 9

The table below gives the current dividend, price-to-earnings ratio (the quotient of the price per share and the annual earnings per share), and price change at the end of a day for six companies, as listed on the New York Stock Exchange.

Stock	Dividend	Price-to-Earnings Ratio	Price Change
ATT	5	6	$+\frac{1}{8}$
GE	3	9	0
Hershey	1.4	6	$+\frac{3}{8}$
IBM	3.44	12	$-\frac{3}{8}$
Mobil	3.40	6	$-1\frac{5}{8}$
RCA	1.80	7	$-\frac{1}{4}$

Let set A include all stocks with a dividend greater than \$3, B all stocks with a price-to-earnings ratio of at least 10, and C all stocks with a positive price change. Find the following.

(a) A'

Set A' contains all the listed stocks outside set A, those with a dividend less than or equal to \$3, so $A' = \{\text{GE, Hershey, RCA}\}$.

(b) $A \cap B$

The intersection of A and B will contain those stocks that offer a dividend greater than \$3 *and* have a price-to-earnings ratio of at least 10.

$$A \cap B = \{\text{IBM}\}$$

(c) $A \cup C$

The union of A and C is the set of all stocks with a dividend greater than $3 or a positive price change (or both).

$$A \cup C = \{\text{ATT, Hershey, IBM, Mobil}\} \quad \blacksquare$$

■ EXAMPLE 10

A department store classifies credit applicants by sex, marital status, and employment status. Let the universal set be the set of all applicants, M be the set of male applicants, S be the set of single applicants, and E be the set of employed applicants. Describe the following sets in words.

(a) $M \cap E$

The set $M \cap E$ includes all applicants who are both male *and* employed, that is, employed male applicants.

(b) $M' \cup S$

This set includes all applicants who are female (not male) *or* single. All female applicants and all single applicants are in this set.

(c) $M' \cap S'$

These applicants are female *and* married (not single); thus, $M' \cap S'$ is the set of all married female applicants.

(d) $M \cup E'$

$M \cup E'$ is the set of applicants that are male *or* unemployed. The set includes *all* male applicants and *all* unemployed applicants. ■

▭ 5.1 EXERCISES

Write true *or* false *for each statement.*

1. $3 \in \{2, 5, 7, 9, 10\}$ **2.** $6 \in \{-2, 6, 9, 5\}$ **3.** $9 \notin \{2, 1, 5, 8\}$

4. $3 \notin \{7, 6, 5, 4\}$ **5.** $\{2, 5, 8, 9\} = \{2, 5, 9, 8\}$ **6.** $\{3, 7, 12, 14\} = \{3, 7, 12, 14, 0\}$

7. {all whole numbers greater than 7 and less than 10} $= \{8, 9\}$

8. {all counting numbers not greater than 3} $= \{0, 1, 2\}$

9. $\{x \mid x \text{ is an odd integer}, 6 \le x \le 18\} = \{7, 9, 11, 15, 17\}$

10. $\{x \mid x \text{ is an even counting number}, x \le 9\} = \{0, 2, 4, 6, 8\}$

Let $A = \{2, 4, 6, 8, 10, 12\}$; $B = \{2, 4, 8, 10\}$; $C = \{4, 10, 12\}$; $D = \{2, 10\}$;
$U = \{2, 4, 6, 8, 10, 12, 14\}$.

Write true *or* false *for each statement.*

11. $A \subset U$ **12.** $D \subset A$ **13.** $A \subset B$ **14.** $B \subset C$

15. $\emptyset \subset A$ **16.** $\{0, 2\} \subset D$ **17.** $D \not\subset B$ **18.** $A \not\subset C$

19. There are exactly 32 subsets of A. **20.** There are exactly 4 subsets of D.

Find the number of subsets for each set.

21. $\{4, 5, 6\}$ **22.** $\{3, 7, 9, 10\}$

23. $\{x | x$ is a counting number between 6 and 12$\}$ **24.** $\{x | x$ is a whole number between 8 and 12$\}$

Give the cardinal number of each set.

25. $\{m, p, q, n\}$ **26.** $\{a, b, c, d, e, f\}$ **27.** $\emptyset$ **28.** $\{0, \emptyset\}$

Write true *or* false *for each statement.*

29. $\{5, 7, 9, 19\} \cap \{7, 9, 11, 15\} = \{7, 9\}$ **30.** $\{8, 11, 15\} \cap \{8, 11, 19, 20\} = \{8, 11\}$

31. $\{2, 1, 7\} \cup \{1, 5, 9\} = \{1\}$ **32.** $\{6, 12, 14, 16\} \cup \{6, 14, 19\} = \{6, 14\}$

33. $\{3, 5, 9, 10\} \cap \emptyset = \{3, 5, 9, 10\}$ **34.** $\{3, 5, 9, 10\} \cup \emptyset = \{3, 5, 9, 10\}$

35. $\{1, 2, 4\} \cup \{1, 2, 4\} = \{1, 2, 4\}$ **36.** $\{1, 2, 4\} \cap \{1, 2, 4\} = \emptyset$

Let $U = \{2, 3, 4, 5, 7, 9\}$, $X = \{2, 3, 4, 5\}$, $Y = \{3, 5, 7, 9\}$, *and* $Z = \{2, 4, 5, 7, 9\}$.
List the members of each of the following sets, using set braces.

37. $X \cap Y$ **38.** $X \cup Y$ **39.** X' **40.** Y'

41. $X' \cap Y'$ **42.** $X' \cap Z$ **43.** $X \cup (Y \cap Z)$ **44.** $Y \cap (X \cup Z)$

Let $U = \{$all students in this school$\}$;
      $M = \{$all students taking this course$\}$;
      $N = \{$all students taking accounting$\}$;
      $P = \{$all students taking zoology$\}$.

Describe each of the following sets in words.

45. M' **46.** $M \cup N$ **47.** $N \cap P$ **48.** $N' \cap P'$

Given $U = \{1, 2, 3, 4, 5, 6, 7, 8, 9, 10\}$, $P = \{2, 4, 6, 8, 10\}$, $Q = \{4, 5, 6\}$, *and*
$R = \{4\}$, *find the cardinal number of each of the following sets.*

49. $P \cup Q$ **50.** $P \cap Q$ **51.** P'

52. Q' **53.** $P' \cap Q$ **54.** $P \cup R'$

*Refer to Example 9 in the text. Describe each of the sets in Exercises 55–58 in words;
then list the elements of each set.*

55. B' **56.** $A \cup B$ **57.** $(A \cap B)'$ **58.** $(A \cup C)'$

Let A *and* B *be sets with cardinal numbers* $n(A) = a$ *and* $n(B) = b$, *respectively.*
Write true *or* false *for the following statements.*

59. $n(A \cup B) = n(A) + n(B)$ **60.** $n(A \cup B) = n(A) + n(B) - n(A \cap B)$

APPLICATIONS

LIFE SCIENCES

Health   **61.** The lists below show some symptoms of an overactive thyroid and an underactive thyroid.

Underactive Thyroid	Overactive Thyroid
Sleepiness, s	Insomnia, i
Dry hands, d	Moist hands, m
Intolerance of cold, c	Intolerance of heat, h
Goiter, g	Goiter, g

(a) Find the smallest possible universal set U that includes all of the symptoms listed. Let N be the set of symptoms for an underactive thyroid, and let O be the set of symptoms for an overactive thyroid. Find each of the following sets.

(b) O' (c) N' (d) $N \cap O$ (e) $N \cup O$ (f) $N \cap O'$

FOR THE COMPUTER

Hazardous Waste   **62.** The table below lists the industries that produce most (93%) of the hazardous wastes in the United States.

Industry	%
Inorganic Chemicals (1)	11
Organic Chemicals (2)	34
Electroplating (3)	12
Petroleum Refining (4)	5
Smelting and Refining (5)	26
Textile Dyeing and Finishing (6)	5

Let x be the number associated with an industry in the table. The universal set is $U = \{1, 2, 3, 4, 5, 6\}$. List the elements of the following subsets of U.

(a) $\{x|\text{the industry produces more than 15\% of the wastes}\}$

(b) $\{x|\text{the industry produces less than 5\% of the wastes}\}$

(c) Find all subsets of U containing industries that together produce at least 80% of the waste.

5.2   APPLICATIONS OF VENN DIAGRAMS

Venn diagrams were used in the last section to illustrate set union and intersection. The rectangular region of a Venn diagram represents the universal set U. Including only a single set A inside the universal set, as in Figure 6, divides U into two regions. Region 1 represents those elements of U outside set A, and region 2 represents those elements belonging to set A. (Our numbering of these regions is arbitrary.)

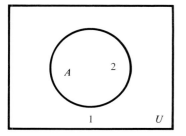

One set leads to 2 regions (numbering is arbitrary).

FIGURE 6

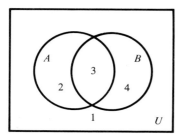

Two sets lead to 4 regions (numbering is arbitrary).

FIGURE 7

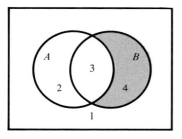

FIGURE 8

The Venn diagram in Figure 7 shows two sets inside U. These two sets divide the universal set into four regions. As labeled in Figure 7, region 1 represents the set whose elements are outside both set A and set B. Region 2 shows the set whose elements belong to A and not to B. Region 3 represents the set whose elements belong to both A and B. Which set is represented by region 4? (Again, the labeling is arbitrary.)

▆ EXAMPLE 1

Draw Venn diagrams similar to Figure 7 and shade the regions representing the following sets.

(a) $A' \cap B$

Set A' contains all the elements outside set A. As labeled in Figure 7, A' is represented by regions 1 and 4. Set B is represented by regions 3 and 4. The intersection of sets A' and B, the set $A' \cap B$, is given by the region common to regions 1 and 4 and regions 3 and 4. The result is the set represented by region 4, which is shaded in Figure 8.

(b) $A' \cup B'$

Again, set A' is represented by regions 1 and 4, and set B' by regions 1 and 2. To find $A' \cup B'$, identify the region that represents the set of all elements in A', B', or both. The result, which is shaded in Figure 9, includes regions 1, 2, and 4. ▆

Venn diagrams also can be drawn with three sets inside U. These three sets divide the universal set into eight regions, which can be numbered (arbitrarily) as in Figure 10.

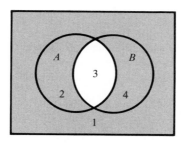

FIGURE 9

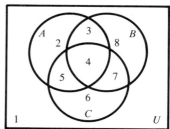

Three sets lead to 8 regions.

FIGURE 10

▆ EXAMPLE 2

On a Venn diagram, shade the region that represents $A' \cup (B \cap C')$.

First find $B \cap C'$. Set B is represented by regions 3, 4, 7, and 8, and set C' by regions 1, 2, 3, and 8. The overlap of these regions, regions 3 and 8,

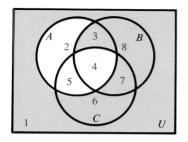

FIGURE 11

represents the set $B \cap C'$. Set A' is represented by regions 1, 6, 7, and 8. The union of the set represented by regions 3 and 8 and the set represented by regions 1, 6, 7, and 8 is the set represented by regions 1, 3, 6, 7, and 8, which are shaded in Figure 11. ■

Applications   Venn diagrams and the cardinal number of a set can be used to solve problems that result from surveying groups of people. As an example, suppose a survey of 60 freshman business students at a large university produced the following results.

> 19 of the students read *Business Week;*
> 18 read *The Wall Street Journal;*
> 50 read *Fortune;*
> 13 read *Business Week and The Wall Street Journal;*
> 11 read *The Wall Street Journal and Fortune;*
> 13 read *Business Week* and *Fortune;*
> 9 read all three.

Now suppose we want to use this information to answer the following questions.

> **(a)** How many students read none of the publications?
> **(b)** How many read only *Fortune?*
> **(c)** How many read *Business Week* and *The Wall Street Journal,* but not *Fortune?*

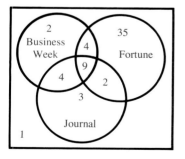

FIGURE 12

Many of the students are listed more than once in the given data. For example, some of the 50 students who read *Fortune* also read *Business Week.* The 9 students who read all three are counted in the 13 who read *Business Week* and *Fortune,* and so on.

A Venn diagram like the one in Figure 12 can be used to interpret the available data. Since 9 students read all three publications, begin by placing 9 in the area that belongs to all three regions, as shown in Figure 13. Of the 13 students who read *Business Week* and *Fortune,* 9 also read *The Wall Street Journal.* Therefore, only 13 − 9 = 4 read just *Business Week* and *Fortune.* Place the number 4 in the area of Figure 13 common to *Business Week* and *Fortune* readers. In the same way, place 4 in the region common only to *Business Week* and *The Wall Street Journal,* and 2 in the region common only to *Fortune* and *The Wall Street Journal.*

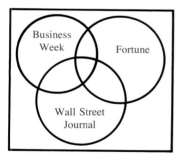

FIGURE 13

The data show that 19 students read *Business Week.* However, 4 + 9 + 4 = 17 readers already have been placed in the region representing *Business Week.* The balance of this region will contain only 19 − 17 = 2 students. These 2 students read *Business Week* only—not *Fortune* and not *The Wall Street Journal.* In the same way, 3 students read only *The Wall Street Journal* and 35 read only *Fortune.*

A total of $2 + 4 + 3 + 4 + 9 + 2 + 35 = 59$ students are placed in the three circles in Figure 13. Since 60 students were surveyed, $60 - 59 = 1$ student reads none of the three publications, and 1 is placed outside all the three regions.

Now Figure 13 can be used to answer the questions asked above.

(a) Only 1 student reads none of the three publications.

(b) From Figure 13, there are 35 students who read only *Fortune*.

(c) The overlap of the regions representing readers of *Business Week* and *The Wall Street Journal* shows that 4 students read *Business Week* and *The Wall Street Journal* but not *Fortune*.

▬ EXAMPLE 3

Jeff Friedman is a section chief for an electric utility company. The employees in his section cut down trees, climb poles, and splice wire. Friedman reported the following information to the management of the utility.

"Of the 100 employees in my section,

45 can cut trees;

50 can climb poles;

57 can splice wire;

28 can cut trees and climb poles;

20 can climb poles and splice wire;

25 can cut trees and splice wire;

11 can do all three;

9 can't do any of the three (management trainees)."

The data supplied by Friedman lead to the numbers shown in Figure 14. Add the numbers from all of the regions to get the total number of employees:

$$9 + 3 + 14 + 23 + 11 + 9 + 17 + 13 = 99.$$

Friedman claimed to have 100 employees, but his data indicate only 99. The management decided that Friedman didn't qualify as a section chief, and he was reassigned as a night-shift meter reader in Guam. (Moral: he should have taken this course.) ▬

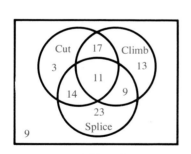

FIGURE 14

Note that in both examples above, we started in the innermost region, the intersection of the three categories. This is usually the best way to begin solving problems of this type.

Venn diagrams are useful for proving general statements about sets. For example, let us show that

$$n(A \cup B) = n(A) + n(B) - n(A \cap B). \qquad (*)$$

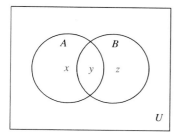

FIGURE 15

In Figure 15, let $x + y$ represent $n(A)$, y represent $n(A \cap B)$, and $y + z$ represent $n(B)$. Then

$$n(A \cup B) = x + y + z,$$
$$n(A) + n(B) - n(A \cap B) = (x + y) + (y + z) - y = x + y + z,$$

so the two sides of equation (*) are equal.

■ EXAMPLE 4

If $n(A) = 5$, $n(B)$ is 7, and $n(A \cup B)$ is 10, find $n(A \cap B)$.

We have

$$n(A \cup B) = n(A) + n(B) - n(A \cap B)$$
$$10 = 5 + 7 - n(A \cap B),$$

so

$$n(A \cap B) = 5 + 7 - 10 = 2. \quad ■$$

5.2 EXERCISES

Use a Venn diagram similar to Figure 7 to shade the regions representing the following sets.

1. $B \cap A'$ **2.** $A \cup B'$ **3.** $A' \cup B$ **4.** $A' \cap B'$

5. $B' \cup (A' \cap B')$ **6.** $(A \cap B) \cup B'$ **7.** U' **8.** $\emptyset'$

Use a Venn diagram similar to Figure 10 to shade the regions representing the following sets.

9. $(A \cap B) \cap C$ **10.** $(A \cap C') \cup B$ **11.** $A \cap (B \cup C')$ **12.** $A' \cap (B \cap C)$

13. $(A' \cap B') \cap C$ **14.** $(A \cap B') \cup C$ **15.** $(A \cap B') \cap C$ **16.** $A' \cap (B' \cup C)$

Find each of the following cardinal numbers. See Example 4.

17. If $n(A) = 5$, $n(B) = 8$, and $n(A \cap B) = 4$, what is $n(A \cup B)$?

18. If $n(A) = 12$, $n(B) = 27$, and $n(A \cup B) = 30$, what is $n(A \cap B)$?

19. Suppose $n(B) = 7$, $n(A \cap B) = 3$, and $n(A \cup B) = 20$. What is $n(A)$?

20. Suppose $n(A \cap B) = 5$, $n(A \cup B) = 35$, and $n(A) = 13$. What is $n(B)$?

Draw a Venn diagram and use the given information to fill in the number of elements for each region.

21. $n(U) = 38$, $n(A) = 16$, $n(A \cap B) = 12$, $n(B') = 20$

22. $n(A) = 26$, $n(B) = 10$, $n(A \cup B) = 30$, $n(A') = 17$

23. $n(A \cup B) = 17$, $n(A \cap B) = 3$, $n(A) = 8$, $n(A' \cup B') = 21$

24. $n(A') = 28$, $n(B) = 25$, $n(A' \cup B') = 45$, $n(A \cap B) = 12$

25. $n(A) = 28$, $n(B) = 34$, $n(C) = 25$, $n(A \cap B) = 14$, $n(B \cap C) = 15$, $n(A \cap C) = 11$, $n(A \cap B \cap C) = 9$, $n(U) = 59$

26. $n(A) = 54$, $n(A \cap B) = 22$, $n(A \cup B) = 85$, $n(A \cap B \cap C) = 4$, $n(A \cap C) = 15$, $n(B \cap C) = 16$, $n(C) = 44$, $n(B') = 63$

27. $n(A \cap B) = 6$, $n(A \cap B \cap C) = 4$, $n(A \cap C) = 7$, $n(B \cap C) = 4$, $n(A \cap C') = 11$, $n(B \cap C') = 8$, $n(C) = 15$, $n(A' \cap B' \cap C') = 5$

28. $n(A) = 13$, $n(A \cap B \cap C) = 4$, $n(A \cap C) = 6$, $n(A \cap B') = 6$, $n(B \cap C) = 6$, $n(B \cap C') = 11$, $n(B \cup C) = 22$, $n(A' \cap B' \cap C') = 5$

In Exercises 29–32, prove that the statements are true by drawing Venn diagrams and shading the regions representing the sets on each side of the equals signs. *

29. $(A \cup B)' = A' \cap B'$

30. $(A \cap B)' = A' \cup B'$

31. $A \cap (B \cup C) = (A \cap B) \cup (A \cap C)$

32. $A \cup (B \cap C) = (A \cup B) \cap (A \cup C)$

33. Verify that $n(A \cup B) = n(A) + n(B) - n(A \cap B)$ for sets $A = \{1, 2, 3, 4, 5\}$ and $B = \{3, 5, 7\}$.

34. Use Venn diagrams to show that
$$n(A \cup B \cup C) = n(A) + n(B) + n(C) - n(A \cap B) - n(A \cap C) - n(B \cap C) + n(A \cap B \cap C).$$

≡ APPLICATIONS

Use Venn diagrams to answer the following questions.

GENERAL

Native American Ceremonies

35. At a pow-wow in Arizona, Native Americans from all over the Southwest came to participate in the ceremonies. A coordinator of the pow-wow took a survey and found that

15 families brought food, costumes, and crafts;
25 families brought food and crafts;
42 families brought food;
20 families brought costumes and food;
6 families brought costumes and crafts, but not food;
4 families brought crafts, but neither food nor costumes;
10 families brought none of the three items;
18 families brought costumes but not crafts.

(a) How many families were surveyed?

(b) How many families brought costumes?

(c) How many families brought crafts, but not costumes?

(d) How many families did not bring crafts?

(e) How many families brought food or costumes?

---

*The statements in Exercises 29 and 30 are known as De Morgan's laws. They are named for the English mathematician Augustus De Morgan (1806—71).

Chinese New Year **36.** A survey of people attending a Lunar New Year celebration in Chinatown yielded the following results:

> 120 were women;
> 150 spoke Cantonese;
> 170 lit firecrackers;
> 108 of the men spoke Cantonese;
> 100 of the men did not light firecrackers;
> 18 of the non-Cantonese-speaking women lit firecrackers;
> 78 non-Cantonese-speaking men did not light firecrackers;
> 30 of the women who spoke Cantonese lit firecrackers.

(a) How many attended?

(b) How many of those who attended did not speak Cantonese?

(c) How many women did not light firecrackers?

(d) How many of those who lit firecrackers were Cantonese-speaking men?

Poultry Analysis **37.** A chicken farmer surveyed his flock with the following results. The farmer had

> 9 fat red roosters;
> 2 fat red hens;
> 37 fat chickens;
> 26 fat roosters;
> 7 thin brown hens;
> 18 thin brown roosters;
> 6 thin red roosters;
> 5 thin red hens.

Answer the following questions about the flock. [*Hint:* You need a Venn diagram with regions for fat, for male (a rooster is a male, a hen is a female), and for red (assume that brown and red are opposites in the chicken world).] How many chickens were

(a) fat?   **(b)** red?

(c) male?   **(d)** fat, but not male?

(e) brown, but not fat?   **(f)** red and fat?

Music **38.** Country-western songs seem to emphasize three basic themes: love, prison, and trucks. A survey of the local country-western radio station produced the following data.

> 12 songs were about a truck driver who was in love while in prison;
> 13 were about a prisoner in love;
> 28 were about a person in love;
> 18 were about a truck driver in love;
> 3 were about a truck driver in prison who was not in love;
> 2 were about a prisoner who was not in love and did not drive a truck;
> 8 were about a person who was not in prison, not in love, and did not drive a truck;
> 16 were about truck drivers who were not in prison.

(a) How many songs were surveyed?

Find the number of songs about

(b) truck drivers;   **(c)** prisoners;

(d) truck drivers in prison;   **(e)** people not in prison;

(f) people not in love.

LIFE SCIENCES

Genetics **39.** After a genetics experiment, the number of pea plants having certain characteristics was tallied, with the following results.

  22 were tall;
  25 had green peas;
  39 had smooth peas;
  9 were tall and had green peas;
  17 were tall and had smooth peas;
  20 had green peas and smooth peas;
  6 had all three characteristics;
  4 had none of the characteristics.

(a) Find the total number of plants counted.

(b) How many plants were tall and had peas that were neither smooth nor green?

(c) How many plants were not tall but had peas that were smooth and green?

Blood Antigens **40.** Human blood can contain either no antigens, the A antigen, the B antigen, or both the A and B antigens. A third antigen, called the Rh antigen, is important in human reproduction, and again may or may not be present in an individual. Blood is called type A-positive if the individual has the A and Rh, but not the B antigen. A person having only the A and B antigens is said to have type AB-negative blood. A person having only the Rh antigen has type O-positive blood. Other blood types are defined in a similar manner. Identify the blood types of the individuals in regions (a)–(g) below. (One region has no letter assigned to it.)

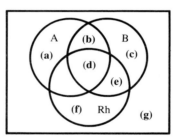

Blood Antigens **41.** (Use the diagram from Exercise 40.) In a certain hospital, the following data were recorded.

  25 patients had the A antigen;
  17 had the A and B antigens;
  27 had the B antigen;
  22 had the B and Rh antigens;
  30 had the Rh antigen;
  12 had none of the antigens;
  16 had the A and Rh antigens;
  15 had all three antigens.

How many patients

(a) were represented?   **(b)** had exactly one antigen?

(c) had exactly two antigens?   **(d)** had O-positive blood?

(e) had AB-positive blood?   **(f)** had B-negative blood?

(g) had O-negative blood?   **(h)** had A-positive blood?

SOCIAL SCIENCES

Education   **42.** A survey of 80 sophomores at a western college showed that

> 36 took English;
> 32 took history;
> 32 took political science;
> 16 took political science and history;
> 16 took history and English;
> 14 took political science and English;
> 6 took all three.

How many students

(a) took English and neither of the other two?

(b) took none of the three courses?

(c) took history, but neither of the other two?

(d) took political science and history, but not English?

(e) did not take political science?

BUSINESS AND ECONOMICS

Cola Consumption   **43.** Market research showed that the adult residents of a certain small town in Georgia fit into the following categories of cola consumption.

Age	Drink Regular Cola (R)	Drink Diet Cola (D)	Drink No Cola (N)	Totals
21–25 *(Y)*	40	15	15	70
26–35 *(M)*	30	30	20	80
Over 35 *(0)*	10	50	10	70
Totals	80	95	45	220

Using the letters given in the table, find the number of people in each of the following sets.

(a) $Y \cap R$ **(b)** $M \cap D$ **(c)** $M \cup (D \cap Y)$

(d) $Y' \cap (D \cup N)$ **(e)** $O' \cup N$ **(f)** $M' \cap (R' \cap N')$

Investment Habits   **44.** The following table shows the results of a survey taken by a bank in a medium-sized town in Tennessee. The survey asked questions about the investment habits of bank customers.

Age	Stocks (S)	Bonds (B)	Savings Accounts (A)	Totals
18–29 *(Y)*	6	2	15	23
30–49 *(M)*	14	5	14	33
50 or over *(O)*	32	20	12	64
Totals	52	27	41	120

Using the letters given in the table, find the number of people in each of the following sets.

(a) $Y \cap B$ **(b)** $M \cup A$ **(c)** $Y \cap (S \cup B)$

(d) $O' \cup (S \cup A)$ **(e)** $(M' \cup O') \cap B$

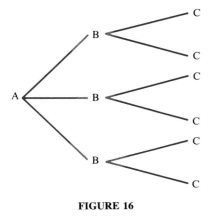

FIGURE 16

5.3 PERMUTATIONS

After making do with your old automobile for several years, you finally decide to replace it with a new super-economy model. You drive over to Ned's New Car Emporium to choose the car that's just right for you. Once there, you find that you can select from 5 models, each with 4 power options and a choice of 8 exterior color combinations and 3 interior colors. How many different new cars are available to you? Problems of this sort are best solved by the counting principles discussed in this section. These counting methods are very useful in probability.

Let us begin with a simpler example. If there are 3 roads from town A to town B and 2 roads from town B to town C, in how many ways can a person travel from A to C by way of B? For each of the 3 roads from A there are 2 different routes leading from B to C, or a total of $3 \cdot 2 = 6$ different ways for the trip, as shown in Figure 16. This example illustrates a general principle of counting, called the **multiplication principle.**

MULTIPLICATION PRINCIPLE

Suppose n independent choices must be made, with

m_1 ways to make choice 1,
m_2 ways to make choice 2,

and so on, with

m_n ways to make choice n.

Then there are

$$m_1 \cdot m_2 \cdots m_n$$

different ways to make the entire sequence of choices.

Using the multiplication principle, there are

$$5 \cdot 4 \cdot 8 \cdot 3 = 480$$

ways of selecting a new car at Ned's.

EXAMPLE 1

An ancient Chinese philosophical work known as the *I Ching* (*Book of Changes*) is often used as an oracle from which people can seek and obtain advice. The philosophy describes the duality of the universe in terms of two primary forces: *yin* (passive, dark, receptive) and *yang* (active, light, creative). See Figure 17. The yin energy is represented by a broken line (– –) and the yang by a solid line (—) These lines are written on top of one another in groups of three, known as *trigrams*. For example, the trigram ☱ is called *Tui,* the Joyous, and has the image of a lake.

yin   yang

FIGURE 17

(a) How many trigrams are there altogether?

Think of choosing between the 2 types of line for each of the 3 positions in the trigram. There will be 2 choices for each position, so there are $2 \cdot 2 \cdot 2 = 8$ different trigrams.

(b) The trigrams are grouped together, one on top of the other, in pairs known as *hexagrams*. Each hexagram represents one aspect of the *I Ching* philosophy. How many hexagrams are there?

For each position in the hexagram there are 8 possible trigrams, giving $8 \cdot 8 = 64$ hexagrams. ▬

▬ EXAMPLE 2

A certain combination lock can be set to open to any 3-letter sequence. How many such sequences are possible?

Since there are 26 letters in the alphabet, there are 26 choices for each of the 3 letters. By the multiplication principle, there are $26 \cdot 26 \cdot 26 = 17,576$ different sequences. ▬

▬ EXAMPLE 3

Morse code uses a sequence of dots and dashes to represent letters and words. How many sequences are possible with at most 3 symbols?

"At most 3" means "1 or 2 or 3." Each symbol may be either a dot or a dash. Thus the following number of sequences are possible in each case.

Number of Symbols	*Number of Sequences*
1	2
2	$2 \cdot 2 = 4$
3	$2 \cdot 2 \cdot 2 = 8$

Altogether, $2 + 4 + 8 = 14$ different sequences are possible. ▬

▬ EXAMPLE 4

A teacher has 5 different books that he wishes to arrange side by side. How many different arrangements are possible?

Five choices will be made, one for each space that will hold a book. Any of the 5 books could be chosen for the first space. There are 4 choices for the second space, since 1 book has already been placed in the first space; there are 3 choices for the third space, and so on. By the multiplication principle, the number of different possible arrangements is $5 \cdot 4 \cdot 3 \cdot 2 \cdot 1 = 120$. ▬

The use of the multiplication principle often leads to products such as $5 \cdot 4 \cdot 3 \cdot 2 \cdot 1$. For convenience, the symbol $n!$ (read "*n-factorial*") is used for such products.

FACTORIAL NOTATION

> For any natural number n,
> $$n! = n(n - 1)(n - 2) \cdots (3)(2)(1).$$
> Also,
> $$0! = 1.$$

With this symbol, the product $5 \cdot 4 \cdot 3 \cdot 2 \cdot 1$ can be written as $5!$. Also, $3! = 3 \cdot 2 \cdot 1 = 6$. The definition of $n!$ could be used to show that $n[(n - 1)!] = n!$ for all natural numbers $n \geq 2$. It is helpful if this result also holds for $n = 1$. This can happen only if $0!$ equals 1, as defined above.

■ EXAMPLE 5

Suppose the teacher in Example 4 wishes to place only 3 of the 5 books on his desk. How many arrangements of 3 books are possible?

The teacher again has 5 ways to fill the first space, 4 ways to fill the second space, and 3 ways to fill the third. Since he wants to use only 3 books, only 3 spaces can be filled (3 events) instead of 5, for $5 \cdot 4 \cdot 3 = 60$ arrangements. ■

Permutations The answer 60 in Example 5 is called the number of *permutations* of 5 things taken 3 at a time. A **permutation** of r (where $r \geq 1$) elements from a set of n elements is any arrangement, *without repetition,* of the r elements. The number of permutations of n things taken r at a time (with $r \leq n$) is written $P(n, r)$. Based on the work in Example 5,

$$P(5, 3) = 5 \cdot 4 \cdot 3 = 60.$$

Factorial notation can be used to express this product as follows.

$$5 \cdot 4 \cdot 3 = 5 \cdot 4 \cdot 3 \cdot \frac{2 \cdot 1}{2 \cdot 1} = \frac{5 \cdot 4 \cdot 3 \cdot 2 \cdot 1}{2 \cdot 1} = \frac{5!}{2!} = \frac{5!}{(5 - 3)!}$$

This example illustrates the general rule of permutations, which is stated below.

PERMUTATIONS

> If $P(n, r)$ (where $r \leq n$) is the number of permutations of n elements taken r at a time, then
> $$P(n, r) = \frac{n!}{(n - r)!}.$$

The proof of this rule follows the discussion in Example 5. There are n ways to choose the first of the r elements, $n - 1$ ways to choose the second, and $n - r + 1$ ways to choose the rth element, so that

$$P(n, r) = n(n - 1)(n - 2) \cdots (n - r + 1).$$

Now multiply on the right by $(n - r)!/(n - r)!$.

$$P(n, r) = n(n - 1)(n - 2) \cdots (n - r + 1) \cdot \frac{(n - r)!}{(n - r)!}$$

$$= \frac{n(n - 1)(n - 2) \cdots (n - r + 1)(n - r)!}{(n - r)!}$$

$$= \frac{n!}{(n - r)!}$$

To find $P(n, r)$, we can use either the permutations formula or direct application of the multiplication principle, as the following example shows.

■ EXAMPLE 6

Find the number of permutations of 8 elements taken 3 at a time.

Since there are 3 choices to be made, the multiplication principle gives $P(8, 3) = 8 \cdot 7 \cdot 6 = 336$. Alternatively, use the permutations formula to get

$$P(8, 3) = \frac{8!}{(8 - 3)} = \frac{8!}{5!} = \frac{8 \cdot 7 \cdot 6 \cdot 5!}{5!} = 8 \cdot 7 \cdot 6 = 336. \quad ■$$

■ EXAMPLE 7

Find each of the following.

(a) The number of permutations of the letters A, B, and C

By the formula for $P(n, r)$ with both n and r equal to 3,

$$P(3, 3) = \frac{3!}{(3 - 3)!} = \frac{3!}{0!} = \frac{3!}{1} = 3 \cdot 2 \cdot 1 = 6.$$

The 6 permutations (or arrangements) are

$$ABC, \quad ACB, \quad BAC, \quad BCA, \quad CAB, \quad CBA.$$

(b) The number of permutations if just 2 of the letters A, B, and C are to be used

Find $P(3, 2)$.

$$P(3, 2) = \frac{3!}{(3 - 2)!} = \frac{3!}{1!} = 3! = 6$$

This result is exactly the same answer as in part (a). This is because, in the case of $P(3, 3)$, after the first 2 choices are made, the third is already determined, as shown in the table below.

First Two Letters	AB	AC	BA	BC	CA	CB
Third Letter	C	B	C	A	B	A

▬ EXAMPLE 8

A televised talk show will include 4 women and 3 men as panelists.

(a) In how many ways can the panelists be seated in a row of 7 chairs?

Find $P(7, 7)$, the total number of ways to seat 7 panelists in 7 chairs.

$$P(7, 7) = \frac{7!}{(7 - 7)!} = \frac{7!}{0!} = \frac{7!}{1} = 7 \cdot 6 \cdot 5 \cdot 4 \cdot 3 \cdot 2 \cdot 1 = 5040$$

There are 5040 ways to seat the 7 panelists.

(b) In how many ways can the panelists be seated if the men and women are to be alternated?

In order to alternate men and women, a woman must be seated in the first chair (since there are 4 women and only 3 men), any of the men next, and so on. Thus there are 4 ways to fill the first seat, 3 ways to fill the second seat, 3 ways to fill the third seat (with any of the 3 remaining women), and so on. This gives

$$4 \cdot 3 \cdot 3 \cdot 2 \cdot 2 \cdot 1 \cdot 1 = 144$$

ways to seat the panelists. ▬

If the n objects in a permutation are not all distinguishable—that is, if there are n_1 of type 1, n_2 of type 2, and so on for r different types, then the number of **distinguishable permutations** is

$$\frac{n!}{n_1!n_2! \ldots n_r!}.$$

Since n_1 objects are indistinguishable and these n_1 objects could be arranged in $n_1!$ ways, the total number of permutations must be divided by $n_1!$. The same reasoning requires dividing by $n_2!$, and so on.

▬ EXAMPLE 9

In how many ways can the letters in the word *Mississippi* be arranged?

This word contains 1 m, 4 i's, 4 s's, and 2 p's. To use the formula, let $n = 11$, $n_1 = 1$, $n_2 = 4$, $n_3 = 4$, and $n_4 = 2$ to get

$$\frac{11!}{1!4!4!2!} = 34,650$$

arrangements. ▬

If Example 9 had asked for the number of ways that the letters in a word with 11 *different* letters could be arranged, the answer would be 11! = 39,916,800.

▆▆▆ 5.3 EXERCISES

Evaluate the following factorials and permutations.

1. $P(4, 2)$ **2.** 3!   **3.** 6!   **4.** $P(8, 0)$ **5.** 0!

6. $P(7, 1)$ **7.** 4!   **8.** $P(4, 4)$ **9.** $P(13, 2)$ **10.** $P(12, 3)$

Evaluate each of the following permutations. use a computer if one is available.

11. $P(25, 12)$ **12.** $P(38, 17)$ **13.** $P(14, 5)$ **14.** $P(17, 12)$

Use the multiplication principle to solve the following problems.

15. How many different types of homes are available if a builder offers a choice of 5 basic plans, 3 roof styles, and 2 exterior finishes?

16. An automobile manufacturer produces 7 models, each available in 6 different exterior colors, with 4 different upholstery fabrics and 5 interior colors. How many varieties of automobile are available?

17. How many different 4-letter radio station call letters can be made if
   (a) the first letter must be K or W and no letter may be repeated?
   (b) repeats are allowed (but the first letter is K or W)?
   (c) the first letter is K or W, there are no repeats, and the last letter is R?

18. A menu offers a choice of 3 salads, 8 main dishes, and 5 desserts. How many different meals consisting of one salad, one main dish, and one dessert are possible?

19. A couple has narrowed down the choice of a name for their new baby to 3 first names and 5 middle names. How many different first- and middle-name arrangements are possible?

20. A concert to raise money for an economics prize is to consist of 5 works: 2 overtures, 2 sonatas, and a piano concerto.
   (a) In how many ways can the program be arranged?
   (b) In how many ways can the program be arranged if an overture must come first?

21. A certain state has license plate numbers consisting of 3 letters followed by 3 digits.
   (a) How many different license plates are possible?
   (b) How many different license plate numbers can be formed if no repeats are allowed?
   (c) How many different license plate numbers are possible if there are no repeats and either numbers or letters can come first?

22. How many 7-digit telephone numbers are possible if the first digit cannot be zero and
    (a) only odd digits may be used?
    (b) the telephone number must be a multiple of 10 (that is, it must end in zero)?
    (c) the telephone number must be a multiple of 100?
    (d) the first 3 digits are 481?
    (e) no repetitions are allowed?

Use permutations to solve each of the following problems.

23. In an experiment on social interaction, 6 people will sit in 6 seats in a row. In how many ways can this be done?

24. In how many ways can 7 of 10 monkeys be arranged in a row for a genetics experiment?

25. A business school gives courses in typing, shorthand, transcription, business English, technical writing, and accounting. In how many ways can a student arrange a schedule if 3 courses are taken?

26. If your college offers 400 courses, 20 of which are in mathematics, and your counselor arranges your schedule of 4 courses by random selection, how many schedules are possible that do not include a math course?

27. In a club with 15 members, how many ways can a slate of 3 officers consisting of president, vice-president, and secretary/treasurer be chosen?

28. A baseball team has 20 players. How many 9-player batting orders are possible?

29. A chapter of union Local 715 has 35 members. In how many different ways can the chapter select a president, a vice-president, a treasurer, and a secretary?

30. A zydeco band from Louisiana will play 5 traditional and 3 original Cajun compositions at a concert. In how many ways can they arrange the program if
    (a) they begin with a traditional piece?
    (b) an original piece will be played last?

31. In an election with 3 candidates for one office and 5 candidates for another office, how many different ballots are possible?

32. The television schedule for a certain evening shows 8 choices from 8 to 9 P.M., 5 choices from 9 to 10 P.M., and 6 choices from 10 to 11 P.M.? In how many different ways could a person schedule that evening of television viewing from 8 to 11 P.M.? (Assume each program that is selected is watched for an entire hour.)

33. Find the number of distinguishable permutations of the letters in each of the following words.
    (a) initial   (b) little   (c) decreed

34. A printer has 5 A's, 4 B's, 2 C's, and 2 D's. How many different "words" are possible that use all these letters? (A "word" does not have to have any meaning here.)

35. Wing has 4 blue, 3 green, and 2 red books to arrange on a shelf.
    (a) In how many ways can this be done if the books can be arranged in any order?
    (b) How many arrangements are possible if books of the same color are identical and must be grouped together?
    (c) In how many distinguishable ways can the books be arranged if books of the same color are identical but need not be grouped together?

36. A child has a set of differently shaped plastic objects. There are 3 pyramids, 4 cubes, and 7 spheres.

   (a) In how many ways can she arrange the objects in a row if they are all different colors?

   (b) How many arrangements are possible if objects of the same shape must be grouped together?

   (c) In how many distinguishable ways can the objects be arranged in a row if objects of the same shape are also the same color, but need not be grouped together?

■ APPLICATIONS

FOR THE COMPUTER

Drug Sequencing **37.** Eleven drugs have been found to be effective in the treatment of a disease. It is believed that the sequence in which the drugs are administered is important in the effectiveness of the treatment. In how many different orderings can 5 of the 11 drugs be administered?

Insect Classification **38.** A biologist is attempting to classify 52,000 species of insects by assigning 3 initials to each species. Is it possible to classify all the species in this way? If not, how many initials should be used?

■ 5.4 COMBINATIONS

In the previous section we saw that there are 60 ways that a teacher can arrange 3 of 5 different books on his desk. That is, there are 60 permutations of 5 things taken 3 at a time. Suppose now that the teacher does not wish to arrange the books on his desk, but rather wishes to choose, without regard to order, any 3 of the 5 books for a book sale to raise money for his school. In how many ways can this be done?

At first glance, we might say 60 again, but this is incorrect. The number 60 counts all possible *arrangements* of 3 books chosen from 5. The following 6 arrangements, however, would all lead to the same set of 3 books being given to the book sale.

mystery-biography-textbook	biography-textbook-mystery
mystery-textbook-biography	textbook-biography-mystery
biography-mystery-textbook	textbook-mystery-biography

The list shows 6 different *arrangements* of 3 books but only one *set* of 3 books. A subset of items selected *without regard to order* is called a **combination.** The number of combinations of 5 things taken 3 at a time is written $\binom{5}{3}$, or $C(5, 3)$. In this book, we will use the more common notation $\binom{5}{3}$.

To evaluate $\binom{5}{3}$, start with the $5 \cdot 4 \cdot 3$ *permutations* of 5 things taken 3 at a time. Since order doesn't matter, and each subset of 3 items from the set of 5 items can have its elements rearranged in $3 \cdot 2 \cdot 1 = 3!$ ways, $\binom{5}{3}$ can be found by dividing the number of permutations by 3!, or

$$\binom{5}{3} = \frac{5 \cdot 4 \cdot 3}{3!} = \frac{5 \cdot 4 \cdot 3}{3 \cdot 2 \cdot 1} = 10.$$

There are 10 ways that the teacher can choose 3 books for the book sale.

Generalizing this discussion gives the following formula for the number of combinations of n elements taken r at a time:

$$\binom{n}{r} = \frac{P(n, r)}{r!}.$$

A more useful version of this formula is found as follows.

$$\binom{n}{r} = \frac{P(n, r)}{r!}$$

$$= \frac{n!}{(n - r)!} \cdot \frac{1}{r!}$$

$$= \frac{n!}{(n - r)!r!}$$

This last version is the most useful for calculation. The steps above lead to the following result.

COMBINATIONS

> If $\binom{n}{r}$ is the number of combinations of n elements taken r at a time, then
>
> $$\binom{n}{r} = \frac{n!}{(n - r)!r!}.$$

The table of combinations in the Appendix gives the value of $\binom{n}{r}$ for $n \leq 20$.

▬ EXAMPLE 1

How many committees of 3 people can be formed from a group of 8 people?

A committee is an unordered set, so use the combinations formula for $\binom{8}{3}$.

$$\binom{8}{3} = \frac{8!}{5!3!} = \frac{8 \cdot 7 \cdot 6 \cdot 5 \cdot 4 \cdot 3 \cdot 2 \cdot 1}{5 \cdot 4 \cdot 3 \cdot 2 \cdot 1 \cdot 3 \cdot 2 \cdot 1} = \frac{8 \cdot 7 \cdot 6}{3 \cdot 2 \cdot 1} = 56 \quad ▬$$

▬ EXAMPLE 2

Three lawyers are to be selected from a group of 30 to work on a special project.

(a) In how many different ways can the lawyers be selected?

Here we wish to know the number of 3-element combinations that can be formed from a set of 30 elements. (We want combinations, not permutations, since order within the group of 3 doesn't matter.)

$$\binom{30}{3} = \frac{30!}{27!3!} = \frac{30 \cdot 29 \cdot 28 \cdot 27!}{27! \cdot 3 \cdot 2 \cdot 1}$$

$$= \frac{30 \cdot 29 \cdot 28}{3 \cdot 2 \cdot 1}$$

$$= 4060$$

There are 4060 ways to select the project group.

(b) In how many ways can the group of 3 be selected if a certain lawyer must work on the project?

Since 1 lawyer already has been selected for the project, the problem is reduced to selecting 2 more from the remaining 29 lawyers.

$$\binom{29}{2} = \frac{29!}{27!2!} = \frac{29 \cdot 28 \cdot 27!}{27! \cdot 2 \cdot 1} = \frac{29 \cdot 28}{2 \cdot 1} = 29 \cdot 14 = 406$$

In this case, the project group can be selected in 406 ways.

(c) In how many ways can a group of at most 3 lawyers be selected from these 30 lawyers?

Here "at most 3" means "0 or 1 or 2 or 3." Find the number of ways for each case.

Case	Number of Ways
0	$\binom{30}{0} = \dfrac{30!}{30!0!} = 1$
1	$\binom{30}{1} = \dfrac{30!}{29!1!} = \dfrac{30 \cdot 29!}{29!(1)} = 30$
2	$\binom{30}{2} = \dfrac{30!}{28!2!} = \dfrac{30 \cdot 29 \cdot 28!}{28! \cdot 2 \cdot 1} = 435$
3	$\binom{30}{3} = \dfrac{30!}{27!3} = \dfrac{30 \cdot 29 \cdot 28 \cdot 27!}{27! \cdot 3 \cdot 2 \cdot 1} = 4060$

The total number of ways to select at most 3 lawyers will be the sum

$$1 + 30 + 435 + 4060 = 4526. \quad \blacksquare$$

■ EXAMPLE 3

A salesman has 10 accounts in a certain city.

(a) In how many ways can he select 3 accounts to call on?

Within a selection of 3 accounts, the arrangement of the calls is not important, so there are

$$\binom{10}{3} = \frac{10!}{3!7!} = \frac{10 \cdot 9 \cdot 8}{3 \cdot 2 \cdot 1} = 120$$

ways he can make a selection of 3 accounts.

(b) In how many ways can he select at least 8 of the 10 accounts to use in preparing a report?

"At least 8" means "8 or more than 8," which is "8 or 9 or 10." First find the number of ways to choose in each case.

Case	Number of Ways
8	$\binom{10}{8} = \dfrac{10!}{8!2!} = \dfrac{10 \cdot 9}{2 \cdot 1} = 45$
9	$\binom{10}{9} = \dfrac{10!}{9!1!} = \dfrac{10}{1} = 10$
10	$\binom{10}{10} = \dfrac{10!}{10!0!} = 1$

He can select at least 8 of the 10 accounts in $45 + 10 + 1 = 56$ ways. ■

The formulas for permutations and combinations given in this section and the previous section will be very useful in solving probability problems in the next chapter. Any difficulty in using these formulas usually comes from being unable to differentiate between them. Both permutations and combinations give the number of ways to choose r objects from a set of n objects. The differences between permutations and combinations are outlined below.

Permutations	*Combinations*
Different orderings or arrangements of the r objects are different permutations.	Each choice or subset of r objects gives one combination. Order within the group of r objects does not matter.
$$P(n, r) = \frac{n!}{(n - r)!}$$	$$\binom{n}{r} = \frac{n!}{(n - r)!r!}$$
Clue words: Arrangement, Schedule, Order	Clue words: Group, Committee, Sample

In the next examples, concentrate on recognizing which formula should be applied.

■ EXAMPLE 4

For each of the following problems, tell whether permutations or combinations should be used to solve the problem.

(a) How many 4-digit code numbers are possible if no digits are repeated?

Since changing the order of the 4 digits results in a different code, use permutations.

(b) A sample of 3 light bulbs is randomly selected from a batch of 15. How many different samples are possible?

The order in which the 3 light bulbs are selected is not important. The sample is unchanged if the items are rearranged, so combinations should be used.

(c) In a baseball conference with 8 teams, how many games must be played so that each team plays every other team exactly once?

Selection of two teams for a game is an unordered subset of 2 from the set of 8 teams. Use combinations again.

(d) In how many ways can 4 patients be assigned to 6 hospital rooms so that each patient has a private room?

The room assignments are an ordered selection of 4 rooms from the 6 rooms. Exchanging the rooms of any two patients within a selection of 4 rooms gives a different assignment, so permutations should be used.

(e) Solve the problems in parts (a)–(d) above. The answers are given in the footnote.* ■

■ EXAMPLE 5

A manager must select 4 employees for promotion; 12 employees are eligible.

(a) In how many ways can 4 be chosen?

Since there is no reason to differentiate among the 4 who are selected, use combinations.

$$\binom{12}{4} = \frac{12!}{4!8!} = 495$$

---

*(a) 5040   **(b)** 455   **(c)** 28   **(d)** 360

(b) In how many ways can 4 employees be chosen (from 12) to be placed in 4 different jobs?

In this case, once a group of 4 is selected, they can be assigned in many different ways (or arrangements) to the 4 jobs. Therefore, this problem requires permutations.

$$P(12, 4) = \frac{12!}{8!} = 11{,}880 \quad \blacksquare$$

The following problems involve a standard deck of 52 playing cards, shown in Figure 18.

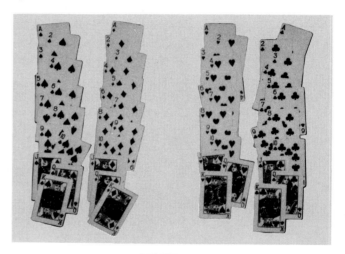

FIGURE 18

▬ EXAMPLE 6

In how many ways can a full house of aces and 8's (three aces and two 8's) occur in 5-card poker?

The arrangement of the three aces or the two 8's does not matter. Use combinations and the multiplication principle. There are $\binom{4}{3}$ ways to get 3 aces from the 4 aces in the deck, and $\binom{4}{2}$ ways to get 2 8's. The number of ways to get 3 aces and 2 8's is

$$\binom{4}{3} \cdot \binom{4}{2} = 4 \cdot 6 = 24. \quad \blacksquare$$

▬ EXAMPLE 7

Five cards are dealt from a standard 52-card deck.

(a) How many such hands have only face cards?

The face cards are the king, queen, and jack of each suit. Since there are 4 suits, there are 12 face cards. The arrangement of the 5 cards is not important, so use combinations to get

$$\binom{12}{5} = \frac{12!}{5!7!} = 792.$$

(b) How many such hands have cards of a single suit?

The total number of ways that 5 cards of a particular suit of 13 cards can occur is $\binom{13}{5}$. Since the arrangement of the five cards is not important, use combinations. There are four different suits, so the multiplication principle gives

$$4 \cdot \binom{13}{5} = 4 \cdot 1287 = 5148$$

ways to deal 5 cards of the same suit. As this example shows, many times both combinations and the multiplication principle must be used in the same problem. ▬

▬ **EXAMPLE 8**

To illustrate the differences between permutations and combinations in another way, suppose 2 cans of soup are to be selected from 4 cans on a shelf: noodle (N), bean (B), mushroom (M), and tomato (T). As shown in Figure 19(a), there are 12 ways to select 2 cans from the 4 cans if the order matters (if noodle first and bean second is considered different from bean, then noodle, for example). On the other hand, if order is unimportant, then there are 6 ways to choose 2 cans of soup from the 4, as illustrated in Figure 19(b). ▬

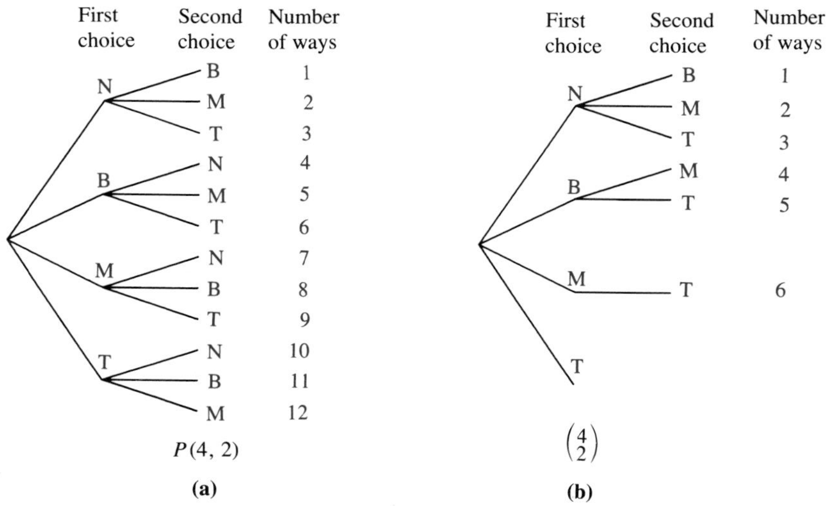

First choice	Second choice	Number of ways
N	B	1
	M	2
	T	3
B	N	4
	M	5
	T	6
M	N	7
	B	8
	T	9
T	N	10
	B	11
	M	12

$P(4, 2)$

(a)

First choice	Second choice	Number of ways
N	B	1
	M	2
	T	3
B	M	4
	T	5
M	T	6

$\binom{4}{2}$

(b)

FIGURE 19

It should be stressed that not all counting problems lend themselves to solution either by permutations or combinations. Whenever a tree diagram or the multiplication principle can be used directly, this should be done.

5.4 EXERCISES

Evaluate the following combinations.

1. $\begin{pmatrix} 8 \\ 3 \end{pmatrix}$ **2.** $\begin{pmatrix} 8 \\ 1 \end{pmatrix}$ **3.** $\begin{pmatrix} 12 \\ 5 \end{pmatrix}$ **4.** $\begin{pmatrix} 10 \\ 8 \end{pmatrix}$ **5.** $\begin{pmatrix} 6 \\ 0 \end{pmatrix}$ **6.** $\begin{pmatrix} 5 \\ 3 \end{pmatrix}$

Evaluate each combination. Use a computer if one is available.

7. $\begin{pmatrix} 21 \\ 10 \end{pmatrix}$ **8.** $\begin{pmatrix} 34 \\ 25 \end{pmatrix}$ **9.** $\begin{pmatrix} 25 \\ 16 \end{pmatrix}$ **10.** $\begin{pmatrix} 30 \\ 15 \end{pmatrix}$ **11.** $\begin{pmatrix} 17 \\ 10 \end{pmatrix}$ **12.** $\begin{pmatrix} 28 \\ 5 \end{pmatrix}$

Use combinations to solve each of the following problems.

13. How many different 2-card hands can be dealt from an ordinary deck (52 cards)?

14. How many different 13-card bridge hands can be selected from an ordinary deck?

15. In how many ways can a hand of 5 red cards be chosen from an ordinary deck?

16. In how many ways can a hand of 6 clubs be chosen from an ordinary deck?

17. Five cards are marked with the numbers 1, 2, 3, 4, and 5, then shuffled, and 2 cards are drawn.
   (a) How many different 2-card combinations are possible?
   (b) How many 2-card hands contain a number less than 3?

18. Five cards are chosen from an ordinary deck. In how many ways is it possible to get the following results?
   (a) All queens
   (b) All face cards (face cards are the jack, queen, and king)
   (c) No face card
   (d) Exactly 2 face cards
   (e) At least 2 face cards
   (f) 1 heart, 2 diamonds, and 2 clubs

19. If a baseball coach has 5 good hitters and 4 poor hitters on the bench and chooses 3 players at random, in how many ways can he choose at least 2 good hitters?

20. An economics club has 30 members.
   (a) If a committee of 4 is to be selected, in how many ways can the selection be made?
   (b) In how many ways can a committee of at least 3 be selected?

21. Use a tree diagram to find the number of ways 2 letters can be chosen from the set {L, M, N} if order is important and
   (a) if repetition is allowed;   **(b)** if no repeats are allowed.
   (c) Find the number of combinations of 3 elements taken 2 at a time. Does this answer differ from (a) or (b)?

22. Repeat Exercise 21 using the set {L, M, N, P}.

Decide whether each of the following problems involves permutations or combinations, and then solve the problem.

23. In how many ways can an employer select 2 new workers from a group of 4 applicants?

24. Hal's Hamburger Palace sells hamburgers with cheese, relish, lettuce, tomato, mustard, or catsup. How many different kinds of hamburgers can be made using any three of the extras?

25. In a club with 8 male and 11 female members, how many 5-member committees can be chosen that have

    (a) all men?   **(b)** all women?   **(c)** 3 men and 2 women?

26. In Exercise 25, how many committees can be selected that have

    (a) at least 4 women?   **(b)** no more than 2 men?

27. A group of 3 students is to be selected from a group of 12 students to take part in a class in cell biology.

    (a) In how many ways can this be done?

    (b) In how many ways can the group who will not take part be chosen?

28. The coach of the Morton Valley Softball Team has 6 good hitters and 8 poor hitters. He chooses 3 hitters at random.

    (a) In how many ways can he choose 2 good hitters and 1 poor hitter?

    (b) In how many ways can he choose 3 good hitters?

    (c) In how many ways can he choose at least 2 good hitters?

29. In a game of musical chairs, 12 children will sit in 11 chairs arranged in a row (one will be left out). In how many ways can the 11 children find seats?

30. Marbles are drawn without replacement from a bag containing 15 marbles.

    (a) How many samples of 2 marbles can be drawn?

    (b) How many samples of 4 marbles can be drawn?

    (c) If the bag contains 3 yellow, 4 white, and 8 blue marbles, how many samples of 2 marbles can be drawn in which both marbles are blue?

31. A city council is composed of 5 liberals and 4 conservatives. A delegation of 3 is to be selected to attend a convention.

    (a) How many delegations are possible?

    (b) How many delegations could have all liberals?

    (c) How many delegations could have 2 liberals and 1 conservative?

    (d) If one member of the council serves as mayor, how many delegations that include the mayor are possible?

32. In an experiment on plant hardiness, a researcher gathers 6 wheat plants, 3 barley plants, and 2 rye plants. She wishes to select 4 plants at random.

    (a) In how many ways can this be done?

    (b) In how many ways can this be done if 2 wheat plants must be included?

33. A group of 7 workers decides to send a delegation of 2 to their supervisor to discuss their grievances.

  (a) How many delegations are possible?

  (b) If it is decided that a particular worker must be in the delegation, how many different delegations are possible?

  (c) If there are 2 women and 5 men in the group, how many delegations would include at least 1 woman?

34. From a pool of 7 secretaries, 3 are selected to be assigned to 3 managers. In how many ways can they be selected?

35. From 10 names on a ballot, 4 will be elected to a political party committee. In how many ways can the committee of 4 be formed if each person will have a different responsibility?

36. There are 5 rotten apples in a crate of 25 apples.

  (a) How many samples of 3 apples can be drawn from the crate?

  (b) How many samples of 3 could be drawn in which all 3 are rotten?

  (c) How many samples of 3 could be drawn in which there are two good apples and one rotten one?

37. A bag contains 5 black, 1 red, and 3 yellow jelly beans; you take 3 at random. How many samples are possible in which the jelly beans are

  (a) all black?   **(e)** 2 black and 1 yellow?

  (b) all red?   **(f)** 2 yellow and 1 black?

  (c) all yellow?   **(g)** 2 red and 1 yellow?

  (d) 2 black and 1 red?

38. In how many ways can 5 out of 9 plants be arranged in a row on a windowsill?

39. A salesperson has the names of 6 prospects.

  (a) In how many ways can she arrange her schedule if she calls on all 6?

  (b) In how many ways can she arrange her schedule if she can call on only 4 of the 6?

40. Five orchids from a collection of 20 are to be selected for a flower show.

  (a) In how many ways can this be done?

  (b) In how many ways can the 5 be selected if 2 special plants must be included?

▤ APPLICATIONS

FOR THE COMPUTER

41. How many 5-card poker hands are possible with a regular deck of 52 cards?

42. How many 5-card poker hands with all cards from the same suit are possible?

43. How many 13-card bridge hands are possible with a regular deck of 52 cards?

44. How many 13-card bridge hands with 4 aces are possible?

▰ 5.5 THE BINOMIAL THEOREM (OPTIONAL)

Evaluating the expression $(x + y)^n$ for various values of n gives a family of expressions, called **binomial expansions,** that are important in mathematics, particularly in the study of probability (discussed in the next chapter). Some expansions of $(x + y)^n$ are listed below.

$$(x + y)^1 = x + y$$
$$(x + y)^2 = x^2 + 2xy + y^2$$
$$(x + y)^3 = x^3 + 3x^2y + 3xy^2 + y^3$$
$$(x + y)^4 = x^4 + 4x^3y + 6x^2y^2 + 4xy^3 + y^4$$
$$(x + y)^5 = x^5 + 5x^4y + 10x^3y^2 + 10x^2y^3 + 5xy^4 + y^5$$

Inspection of these expansions shows a pattern. Let us try to identify the pattern so that we can write a general expression for $(x + y)^n$.

First, each expansion begins with x raised to the same power as the binomial itself. That is, the expansion of $(x + y)^1$ has first term x^1, that of $(x + y)^2$ starts with x^2, the expansion of $(x + y)^3$ has first term x^3, and so on. The last term in each expansion is y raised to the same power as the binomial. Based on this, the expansion of $(x + y)^n$ should begin with x^n and end with y^n.

Also, the exponents on x decrease by 1 in each term after the first, while the exponents on y, beginning with y in the second term, increase by 1 in each succeeding term, with the *variables* in the expansion of $(x + y)^n$ having the following pattern:

$$x^n, \quad x^{n-1}y, \quad x^{n-2}y^2, \quad x^{n-3}y^3, \quad \ldots, \quad x^2y^{n-2}, \quad xy^{n-1}, \quad y^n.$$

This pattern shows that the sum of the exponents on x and y in each term is n. For example, in the third term above, the variable is $x^{n-2}y^2$, and the sum of the exponents, $n - 2 + 2$, is n.

Now let us try to find a pattern for the *coefficients* in the terms of the expansions shown above. In the product

$$(x + y)^5 = (x + y)(x + y)(x + y)(x + y)(x + y), \qquad (*)$$

the variable x occurs 5 times, once in each factor. To get the first term of the expansion, form the product of these 5 x's to get x^5. The product x^5 can occur in just one way, by taking an x from each factor in (*), so that the coefficient of x^5 is 1. The term with x^4y can occur in more than one way. For example, the x's could come from the first 4 factors and the y from the last factor, or the x's might be taken from the last 4 factors and the y from the first, and so on. Since there are 5 factors of $x + y$ in (*), from which exactly 4 x's must be selected for the term x^4y, there are $\binom{5}{4} = 5$ of the x^4y terms. Therefore, the term x^4y has coefficient 5. In this manner, combinations can be used to find the coefficients for each term of the expansion:

$$(x + y)^5 = x^5 + \binom{5}{4}x^4y + \binom{5}{3}x^3y^2 + \binom{5}{2}x^2y^3 + \binom{5}{1}xy^4 + y^5.$$

The coefficient 1 of the first and last terms could be written $\binom{5}{5}$ and $\binom{5}{0}$ to complete the pattern.

Generalizing from this special case, the coefficient for any term of $(x + y)^n$ in which the variable is $x^{n-r}y^r$ is $\binom{n}{n-r}$. The **binomial theorem** gives the general binomial expansion.

BINOMIAL THEOREM

For any positive integer n,

$$(x + y)^n = \binom{n}{n}x^n + \binom{n}{n-1}x^{n-1}y + \binom{n}{n-2}x^{n-2}y^2$$

$$+ \binom{n}{n-3}x^{n-3}y^3 + \cdots + \binom{n}{1}xy^{n-1} + \binom{n}{0}y^n.$$

A proof of the binomial theorem requires the method of mathematical induction. Details are given in most college algebra texts.

■ EXAMPLE 1

Write out the binomial expansion of $(a + b)^7$.

Use the binomial theorem.

$$(a + b)^7 = a^7 + \binom{7}{6}a^6b + \binom{7}{5}a^5b^2 + \binom{7}{4}a^4b^3 + \binom{7}{3}a^3b^4$$

$$+ \binom{7}{2}a^2b^5 + \binom{7}{1}ab^6 + b^7$$

$$= a^7 + 7a^6b + 21a^5b^2 + 35a^4b^3 + 35a^3b^4$$

$$+ 21a^2b^5 + 7ab^6 + b^7 \quad ■$$

■ EXAMPLE 2

Expand $\left(a - \dfrac{b}{2} \right)^4$.

Use the binomial theorem to write

$$\left(a - \frac{b}{2} \right)^4 = \left[a + \left(-\frac{b}{2} \right) \right]^4$$

$$= a^4 + \binom{4}{3}a^3\left(-\frac{b}{2} \right) + \binom{4}{2}a^2\left(-\frac{b}{2} \right)^2 + \binom{4}{1}a\left(-\frac{b}{2} \right)^3 + \left(-\right.$$

$$= a^4 + 4a^3\left(-\frac{b}{2} \right) + 6a^2\left(\frac{b^2}{4} \right) + 4a\left(-\frac{b^3}{8} \right) + \frac{b^4}{16}$$

$$= a^4 - 2a^3b + \frac{3}{2}a^2b^2 - \frac{1}{2}ab^3 + \frac{1}{16}b^4. \quad ■$$

Pascal's Triangle   Another method for finding the coefficients of the terms in a binomial expansion is by **Pascal's triangle,** shown in Figure 20. The nth row in the triangle gives the coefficients for the expansion of $(x + y)^n$. To see this, compare the numbers in the rows shown below with the coefficients of the expansions given at the beginning of this section. Each number in the triangle is found by adding the two numbers directly above it. Two illustrations of this are shown in color on the triangle below. A disadvantage of this method of finding coefficients is that the entire triangle must be produced down to the row that gives the desired coefficients.

$$
\begin{array}{ccccccccccc}
& & & & & 1 & & & & & \\
& & & & 1 & & 1 & & & & \\
& & & 1 & & 2 & & 1 & & & \\
& & 1 & & 3 & & 3 & & 1 & & \\
& 1 & & 4 & & 6 & & 4 & & 1 & \\
1 & & 5 & & 10 & & 10 & & 5 & & 1
\end{array}
$$

FIGURE 20

The binomial theorem can be used to prove the following result, used in Section 5.1.

A set of n distinct elements has 2^n subsets.

A proof is given for $n = 6$. Subsets of a set of 6 elements can be chosen as follows: there are $\binom{6}{6}$ subsets with 6 elements, $\binom{6}{5}$ subsets with 5 elements, $\binom{6}{4}$ subsets with 4 elements, and so on. Altogether, there are

$$
\binom{6}{6} + \binom{6}{5} + \binom{6}{4} + \binom{6}{3} + \binom{6}{2} + \binom{6}{1} + \binom{6}{0}
$$

subsets. By the binomial theorem,

$$
(x + y)^6 = \binom{6}{6}x^6 + \binom{6}{5}x^5y + \binom{6}{4}x^4y^2 + \binom{6}{3}x^3y^3 + \binom{6}{2}x^2y^4
$$
$$
+ \binom{6}{1}xy^5 + \binom{6}{0}y^6.
$$

If $x = 1$ and $y = 1$,

$$
(1 + 1)^6 = \binom{6}{6} \cdot 1^6 + \binom{6}{5} \cdot 1^5 \cdot 1 + \binom{6}{4} \cdot 1^4 \cdot 1^2 + \binom{6}{3} \cdot 1^3 \cdot 1^3
$$
$$
+ \binom{6}{2} \cdot 1^2 \cdot 1^4 + \binom{6}{1} \cdot 1 \cdot 1^5 + \binom{6}{0} \cdot 1^6
$$

or $$2^6 = \binom{6}{6} + \binom{6}{5} + \binom{6}{4} + \binom{6}{3} + \binom{6}{2} + \binom{6}{1} + \binom{6}{0}.$$

Thus, the total number of subsets of a set of 6 elements is $2^6 = 64$. In the general case, using the binomial theorem in the same way,

$$2^n = \binom{n}{n} + \binom{n}{n-1} + \binom{n}{n-2} + \cdots + \binom{n}{0},$$

so the total number of subsets is 2^n.

▬ EXAMPLE 3

The Yummy Yogurt Shoppe offers either chocolate or vanilla frozen yogurt with a choice of 3 fruit toppings, chocolate topping, and chopped nuts. How many different servings are possible with one flavor of yogurt and any combination of toppings?

Use the multiplication principle first. There are really 2 basic choices—a flavor of yogurt and a combination of toppings. A flavor can be selected in 2 ways. The 4 toppings plus nuts form a set of 5 elements. The number of different subsets that can be selected from a set of 5 elements is 2^5, making the number of different servings

$$2 \cdot 2^5 = 2^6 = 64. \quad \blacksquare$$

▬ 5.5 EXERCISES

Write out each binomial expansion and simplify its terms.

1. $(m + n)^4$ **2.** $(p - q)^5$ **3.** $(3x - 2y)^6$ **4.** $(2x + t^3)^4$ **5.** $\left(\dfrac{m}{2} - 3n\right)^5$ **6.** $\left(2p + \dfrac{q}{3}\right)^3$

In Exercises 7–10, write out the first four terms of the binomial expansion and simplify.

7. $(p + q)^{10}$ **8.** $(r + 5)^9$ **9.** $(a + 2b)^{15}$ **10.** $(3c + d)^{12}$

11. How many different subsets can be chosen from a set of ten elements?

12. How many different subsets can be chosen from a set of eight elements?

13. How many different pizzas can be chosen if you can have cheese, beef, sausage, pepper, tomatoes, and onion?

14. How many different vanilla sundaes can be chosen if the available trimmings are chocolate, strawberry, apricot, pineapple, peanuts, walnuts, and pecan crunch?

15. How many different school programs can be selected from 20 course offerings if at least two courses and no more than six courses can be selected? (Assume no courses overlap.)

16. How many different committees can be selected from a group of 16 people if the committee must have between 2 and 5 people (inclusive)?

17. A buffet offers 4 kinds of salad to any of which can be added sliced beets, bean sprouts, chopped egg, and sliced mushrooms. How many different salads are possible?

18. The buffet in Exercise 17 offers 3 meat and 2 fish entrees. How many different entree combinations are possible?

▆▆ KEY WORDS ▆▆	5.1			

5.1			
set		disjoint sets	
element (member)		union	
set-builder notation		set operations	
universal set	5.3	multiplication principle	
subset		factorial	
Venn diagram		permutations	
empty set		distinguishable permutations	
cardinal number	5.4	combinations	
tree diagram	5.5	binomial theorem	
complement		binomial expansion	
intersection		Pascal's triangle	

CHAPTER 5   REVIEW EXERCISES

Write true *or* false *for each statement.*

1. $9 \in \{8, 4, -3, -9, 6\}$

2. $4 \notin \{3, 9, 7\}$

3. $2 \notin \{0, 1, 2, 3, 4\}$

4. $0 \in \{0, 1, 2, 3, 4\}$

5. $\{3, 4, 5\} \subset \{2, 3, 4, 5, 6\}$

6. $\{1, 2, 5, 8\} \subset \{1, 2, 5, 10, 11\}$

7. $\{3, 6, 9, 10\} \subset \{3, 9, 11, 13\}$

8. $\emptyset \subset \{1\}$

9. $\{2, 8\} \not\subset \{2, 4, 6, 8\}$

10. $0 \subset \emptyset$

List the elements in each set, and give the cardinal number.

11. $\{x \mid x$ is a counting number more than 5 and less than 8$\}$

12. $\{x \mid x$ is an integer, $-3 \le x < 1\}$

13. $\{$all counting numbers less than 5$\}$

14. $\{$all whole numbers not greater than 2$\}$

Let $U = \{a, b, c, d, e, f, g\}$, $K = \{c, d, f, g\}$, *and* $R = \{a, c, d, e, g\}$. *Find the following.*

15. the number of subsets of K

16. the number of subsets of R

17. K'

18. R'

19. $K \cap R$

20. $K \cup R$

21. $(K \cap R)'$

22. $(K \cup R)'$

23. $\emptyset'$

24. U'

Let $U = \{$all employees of the K.O. Brown Company$\}$;
    $A = \{$employees in the accounting department$\}$;
    $B = \{$employees in the sales department$\}$;
    $C = \{$employees with at least 10 years in the company$\}$;
    $D = \{$employees with an MBA degree$\}$.

Describe the following sets in words.

25. $A \cap C$ **26.** $B \cap D$ **27.** $A \cup D$ **28.** $A' \cap D$ **29.** $B' \cap C'$ **30.** $(B \cup C)'$

Draw a Venn diagram and shade each set in Exercises 31–34.

31. $A \cup B'$ **32.** $A' \cap B$ **33.** $(A \cap B) \cup C$ **34.** $(A \cup B)' \cap C$

A telephone survey of television viewers revealed the following information. Use this information for Exercises 35–38.

    20 watch situation comedies.
    19 watch game shows.
    27 watch movies.
     5 watch both situation comedies and game shows.
     8 watch both game shows and movies.
    10 watch both situation comedies and movies.
     3 watch all three.
     6 watch none of these.

35. How many viewers were interviewed?

36. How many viewers watch comedies and movies but not game shows?

37. How many viewers watch only movies?

38. How many viewers watch comedies and game shows but not movies?

39. In how many ways can 6 business tycoons line up their golf carts at their country club?

40. In how many ways can a sample of 3 oranges be taken from a bag of a dozen oranges?

41. If 2 of the oranges in Exercise 40 are rotten, in how many ways can the sample of 3 include

   (a) 1 rotten orange?

   (b) 2 rotten oranges?

   (c) no rotten oranges?

   (d) at most 2 rotten oranges?

42. In how many ways can 2 pictures be selected from a group of 5 different pictures and then arranged side by side on a wall?

43. In how many ways can the 5 pictures in Exercise 42 be arranged side by side if a certain one must be first?

44. In how many ways can the 5 pictures in Exercise 42 be arranged if 2 are landscapes and 3 are puppies and if

   (a) like types must be kept together?

   (b) landscapes and puppies are alternated?

45. In a Chinese restaurant the menu lists 8 items in column A and 6 items in column B.

    (a) To order a dinner, the diner is told to select 3 items from column A and 2 from column B. How many dinners are possible?

    (b) How many dinners are possible if the diner can select up to 3 from column A and up to 2 from column B? Assume at least one item must be included from either A or B.

46. A representative is to be selected from each of 3 departments in a small college. There are 7 people in the first department, 5 in the second department, and 4 in the third department.

    (a) How many different groups of 3 representatives are possible?

    (b) How many groups are possible if any number (at least 1) up to 3 representatives can form a group?

47. Write out the binomial expansion of $(2m + n)^5$.

48. Write out the first 3 terms of the binomial expansion of $(a + b)^{16}$.

49. Write out the first 4 terms of the binomial expansion of $(x - y/2)^{20}$.

50. How many different sums can be formed from combinations of 2 or more of the numbers 2, 5, 10, and 13?

51. How many different collections can be formed from a set of 8 old coins, if each collection must contain at least 2 coins?

52. How many sets of 3 or more books can be formed from a collection of 6 books?

6 Probability

6.1 Basics of Probability

6.2 Further Probability Rules

6.3 Conditional Probability

6.4 Bayes' Formula

6.5 Applications of Counting

6.6 Bernoulli Trials

Review Exercises

Extended Application
Making a First Down

Extended Application
Medical Diagnosis

If you go to a supermarket and buy five pounds of peaches at 54¢ per pound, you can easily find the *exact* price of your purchase: $2.70. Such an activity is *deterministic;* the result can be predicted *exactly.*

On the other hand, the produce manager of the market is faced with the problem of ordering peaches. The manager may have a good estimate of the number of pounds of peaches that will be sold during the day, but it is impossible to predict the *exact* amount. The number of items that customers will purchase during a day is *random:* the quantity cannot be predicted exactly.

A great many problems that come up in applications of mathematics are random phenomena—those for which exact prediction is impossible. The best that we can do is construct a mathematical model that gives the *probability* of certain events. The basics of probability are discussed in this chapter, and applications of probability are explored in succeeding chapters.

6.1 BASICS OF PROBABILITY

In this section we introduce the terminology of probability theory and the use of sets to solve problems involving probability. In probability, an **experiment** is an activity or occurrence with an observable result. Each repetition of an experiment is called a **trial.** The possible results of each trial are **outcomes.** An example of a probability experiment is the tossing of a coin. Each trial of the experiment (each toss) has 2 possible outcomes, heads (*h*) and tails (*t*). If the outcomes *h* and *t* are equally likely to occur, then the coin is not "loaded" to favor one side over the other. Such a coin is called **fair.** For a coin that is not loaded, this "equally likely" assumption is made for each trial.

Since *two* equally likely outcomes are possible, *h* and *t,* and just *one* of them is heads, we would expect that a coin tossed many, many times would come up heads approximately 1/2 of the time. We also would expect that the more times the coin was tossed, the closer the occurrence of heads should be to 1/2.

Imagine an experiment that could be repeated again and again under unchanging conditions. Suppose the experiment is repeated *n* times and that a certain outcome happens *m* times. The ratio *m/n* is called the **relative frequency** of the outcome after *n* trials.

If this ratio *m/n* approaches closer and closer to some fixed number *p* as *n* gets larger and larger, then *p* is called the **probability** of the outcome. If *p* exists, then

$$p \approx \frac{m}{n}$$

as *n* gets larger and larger.

This approach to probability is consistent with most people's intuitive feeling that probability is a way to measure the likelihood of occurrence of a certain outcome. For example, since 1/4 of the cards in an ordinary deck are

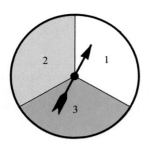

FIGURE 1

diamonds, we would assume that if we drew a card from a well-shuffled deck, kept track of whether it was a diamond or not, and then replaced the card in the deck, after a large number of trials about 1/4 of the cards drawn would have been diamonds.

This definition of probability, on the other hand, has the disadvantage of not being precise, since it uses phrases such as "approaches closer and closer" and "gets larger and larger." In the first few sections of this chapter a more precise meaning is given to the terms associated with probability.

▬ EXAMPLE 1

Suppose a spinner like the one shown in Figure 1 is spun.

(a) Find the probability that the spinner will point to 1.

It is natural to assume that if this spinner were spun many, many times, it would point to 1 about 1/3 of the time, so that

$$P(1) = \frac{1}{3}.$$

(b) Find the probability that the spinner will point to 2.

Since 2 is one of three possible outcomes, $P(2) = 1/3$. ▬

Sometimes we are interested in a result that is satisfied by more than one of the possible outcomes. To find the probability that the spinner in Example 1 will point to an odd number, notice that two of the three possible outcomes are odd numbers, 1 and 3, so that

$$P(\text{odd}) = \frac{2}{3}.$$

An ordinary die is a cube whose six different faces show the following numbers of dots: 1, 2, 3, 4, 5, and 6. If the die is not "loaded" to favor certain faces over others, then each of the faces is equally likely to come up when the die is rolled.

▬ EXAMPLE 2

(a) If a single fair die is rolled, find the probability of rolling a 4.

Since one out of six faces shows a 4, $P(4) = 1/6$.

(b) Using the same die, find the probability of rolling a 6.

Since one out of six faces shows a 6, $P(6) = 1/6$. ▬

Sample Spaces   The set of all possible outcomes for an experiment is the **sample space** for that experiment. A sample space for the experiment of tossing a coin is made up of the outcomes heads (h) and tails (t). If S represents this sample space, then

$$S = \{h, t\}.$$

In the same way, a sample space for tossing a single fair die as in Example 2 is

$$S = \{1, 2, 3, 4, 5, 6\}.$$

◼ EXAMPLE 3

Give the sample space for each experiment.

(a) A spinner like the one in Figure 1 is spun.

The three outcomes are 1, 2, and 3, so the sample space is

$$\{1, 2, 3\}.$$

(b) For the purposes of a public opinion poll, respondents are classified as young, middle-aged, or older, and as male or female.

A sample space for this poll could be written as a set of ordered pairs:

{(young, male), (young, female), (middle-aged, male),
(middle-aged, female), (older, male), (older, female)}.

(c) A manufacturer tests automobile tires by running a tire until it fails or until tread depth reaches a certain unsafe level. The number of miles that the tire lasts, m, is recorded.

At least in theory, m can be any nonnegative real number, so that the sample space is

$$\{m \mid m \geq 0\}.$$

For a particular type of tire, however, there would be practical limits on m, and the sample space might thus be

$$\{m \mid 0 \leq m \leq 50{,}000\}.$$

(d) An experiment consists of studying the numbers of boys and girls in families with exactly 3 children. Let b represent *boy* and g represent *girl*.

A three-child family can have 3 boys, written *bbb*, 3 girls, *ggg*, or various combinations such as *bgg*. A sample space with four outcomes (not equally likely) is

$$S_1 = \{3 \text{ boys}, 2 \text{ boys and 1 girl}, 1 \text{ boy and 2 girls}, 3 \text{ girls}\}.$$

Another sample space might take into consideration the ordering of the births of the children (so that *bgg* is different from *gbg* or *ggb*, for example):

$$S_2 = \{bbb, bbg, bgb, gbb, bgg, gbg, ggb, ggg\}.$$

The second sample space, S_2, has equally likely outcomes; S_1 does not, since there is more than one way to get a family with 2 boys and 1 girl or a family with 2 girls and 1 boy, but only one way to get 3 boys or 3 girls. For the purpose of assigning responsibilities, S_2 is preferable. ◼

An experiment may have more than one sample space, as shown in Example 3(d). The most useful sample spaces have equally likely outcomes, but it is not always possible to choose such a sample space.

Events   An **event** is a subset of a sample space. If the sample space for tossing a coin is $S = \{h, t\}$, then one event is $E = \{h\}$, which represents the outcome "heads." For the sample space of rolling a single fair die, $S = \{1, 2, 3, 4, 5, 6\}$, some possible events are listed below.

The die shows an even number: $E_1 = \{2, 4, 6\}$.

The die shows a 1: $E_2 = \{1\}$.

The die shows a number less than 5: $E_3 = \{1, 2, 3, 4\}$.

The die shows a multiple of 3: $E_4 = \{3, 6\}$.

■■ EXAMPLE 4

For the sample space S_2 in Example 3(d), write the following events.

(a) Event H: the three-child family has exactly 2 girls

Families with three children can have exactly 2 girls with either *bgg, gbg,* or *ggb,* so event H is

$$H = \{bgg, gbg, ggb\}.$$

(b) Event K: the three children are the same sex

Two outcomes satisfy this condition, all boys or all girls.

$$K = \{bbb, ggg\}$$

(c) Event J: the family has three girls

Only *ggg* satisfies this condition, so

$$J = \{ggg\}. \quad ■■$$

In Example 4(c), event J had only one possible outcome, *ggg*. Such an event, with only one possible outcome, is a **simple event.** If event E equals the sample space S, then E is called a **certain event.** If event $E = \emptyset$, then E is called an **impossible event.**

■■ EXAMPLE 5

Suppose a die is rolled. As shown above, the sample space is $\{1, 2, 3, 4, 5, 6\}$.

(a) The event "the die shows a 4" $\{4\}$, has only one possible outcome. It is a simple event.

(b) The event "the number showing is less than ten" equals the sample space. $S = \{1, 2, 3, 4, 5, 6\}$. This event is a certain event; if a die is rolled the number showing (either 1, 2, 3, 4, 5, or 6), must be less than ten.

(c) The event "the die shows a 7" is the empty set, $\emptyset$; this is an impossible event. ■■

Probability   We now need to assign to each event E in a sample space S a number called the *probability of the event,* which indicates the relative likelihood of the event. For sample spaces with equally likely outcomes, the probability of an event can be found with the following *basic probability principle*. (Recall from Chapter 5 that $n(E)$ represents the number of elements in a finite set E.)

BASIC PROBABILITY PRINCIPLE

Let S be a sample space with n equally likely outcomes. Let event E contain m of these outcomes. Then the probability that event E occurs, written $P(E)$, is

$$P(E) = \frac{m}{n} = \frac{n(E)}{n(S)}.$$

■ EXAMPLE 6

Suppose a single die is rolled. Use the sample space $S = \{1, 2, 3, 4, 5, 6\}$ and give the probability of each of the following events.

(a) E: the die shows an even number

Here, $E = \{2, 4, 6\}$, a set with three elements. Since S contains six elements,

$$P(E) = \frac{3}{6} = \frac{1}{2}.$$

(b) F: the die shows a number greater than 4

Since F contains two elements, 5 and 6,

$$P(F) = \frac{2}{6} = \frac{1}{3}.$$

(c) G: the die shows a number less than 10

Event G is a certain event, with

$$G = \{1, 2, 3, 4, 5, 6\},$$

so that

$$P(G) = \frac{6}{6} = 1.$$

(d) H: the die shows an 8

This event is impossible, so

$$P(H) = 0. \quad ■$$

The Addition Principle   Suppose event E is the union of several simple events, say

$$E = \{s_1, s_2, s_3\} = \{s_1\} \cup \{s_2\} \cup \{s_3\}.$$

Then $P(E)$, the probability of event E, is found by adding the probabilities for each of the simple events making up E. For $E = \{s_1, s_2, s_3\}$,

$$P(E) = P(\{s_1\}) + P(\{s_2\}) + P(\{s_3\}).$$

The generalization of this result is called the *addition principle*.

ADDITION PRINCIPLE FOR SIMPLE EVENTS

Suppose $E = \{s_1, s_2, s_3, \cdots, s_m\}$, where $\{s_1\}, \{s_2\}, \{s_3\}, \cdots, \{s_m\}$ are distinct simple events. Then

$$P(E) = P(\{s_1\}) + P(\{s_2\}) + P(\{s_3\}) + \cdots + P(\{s_m\}).$$

Be careful! The addition rule *does not necessarily apply* to the addition of probabilities of events *that are not simple,* that is, events that include more than one possible outcome.

◾ EXAMPLE 7

If a single playing card is drawn at random from an ordinary 52-card bridge deck, find the probability of each of the following events.

(a) Drawing an ace

There are 4 aces in the deck. The event "drawing an ace" is

{heart ace, diamond ace, club ace, spade ace},

where drawing each type of ace is a simple event. Each ace has a probability of 1/52 of being drawn, so by the addition principle,

$$P(\text{ace}) = \frac{1}{52} + \frac{1}{52} + \frac{1}{52} + \frac{1}{52} = \frac{4}{52} = \frac{1}{13}.$$

(b) Drawing a face card

Since there are 12 face cards, the addition principle gives

$$P(\text{face card}) = \frac{12}{52} = \frac{3}{13}.$$

(c) Drawing a spade

The deck contains 13 spades, so

$$P(\text{spade}) = \frac{13}{52} = \frac{1}{4}.$$

(d) Drawing a spade or a heart

Besides the 13 spades, the deck contains 13 hearts, so

$$P(\text{spade or heart}) = \frac{26}{52} = \frac{1}{2}. \quad ◾$$

The definition of probability given above leads to the following properties of probability. (Recall that a simple event contains only *one* of the *equally likely* outcomes in a sample space.)

PROPERTIES OF PROBABILITY

Let $S = \{s_1, s_2, s_3, \ldots, s_n\}$ be the sample space obtained from the union of the n distinct simple events $\{s_1\}, \{s_2\}, \{s_3\}, \ldots, \{s_n\}$ with associated probabilities $p_1, p_2, p_3, \ldots, p_n$. Then

1. $0 \le p_1 \le 1, \quad 0 \le p_2 \le 1, \cdots, \quad 0 \le p_n \le 1$
   (all probabilities are between 0 and 1, inclusive);

2. $p_1 + p_2 + p_3 + \cdots + p_n = 1$
   (the sum of all probabilities for a sample space is 1);

3. $P(S) = 1$;

4. $P(\emptyset) = 0$.

6.1 EXERCISES

Write sample spaces for the following experiments. (The sample spaces in Exercises 11–12 are infinite.)

1. A month of the year is chosen for a wedding.

2. A day in April is selected for a bicycle race.

3. A student is asked how many points she earned on a recent 80-point test.

4. A person is asked the number of hours (to the nearest hour) he watched television yesterday.

5. The management of an oil company must decide whether to go ahead with a new oil shale plant or to cancel it.

6. A record is kept for 3 days about whether a particular stock goes up or down.

7. The length of life of a light bulb is measured to the nearest hour; the bulbs never last more than 5000 hr.

8. A coin is tossed, and a die is rolled.

9. A coin is tossed 4 times.

10. A box contains 5 balls, numbered 1, 2, 3, 4, and 5. A ball is drawn at random, the number on it recorded, and the ball replaced. The box is shaken, a second ball is drawn, and its number is recorded.

11. A coin is tossed until a head appears.

12. A die is rolled, but only "odd" (1, 3, 5) or "even" (2, 4, 6) is recorded. The die is rolled until an odd number shows.

For the experiments in Exercises 13–16, write out the sample space, and then write the indicated events in set notation.

13. A die is tossed twice, with the tosses recorded as ordered pairs.
   (a) The first die shows a 3.
   (b) The sum of the numbers showing is 8.
   (c) The sum of the numbers showing is 13.

14. One urn contains four balls, labeled 1, 2, 3, and 4. A second urn contains five balls, labeled 1, 2, 3, 4, and 5. An experiment consists of taking one ball from the first urn, and then taking a ball from the second urn.
   (a) The number on the first ball is even.
   (b) The number on the second ball is even.
   (c) The sum of the numbers on the two balls is 5.
   (d) The sum of the numbers on the two balls is 1.

15. A coin is tossed until two heads appear, or until the coin is tossed five times, whichever comes first.
   (a) The coin is tossed exactly 2 times.
   (b) The coin is tossed exactly 3 times.
   (c) The coin is tossed exactly 5 times, without getting two heads.

16. A committee of 2 people is selected from 5 executives, Alam, Bartolini, Chinn, Dickson, and Ellsberg.
   (a) Chinn is on the committee.
   (b) Dickson and Ellsberg are not both on the committee.
   (c) Both Alam and Chinn are on the committee.

17. The management of a firm wishes to check on the opinions of its assembly line workers. Before the workers are interviewed, they are divided into various categories. Define events E, F, and G as follows.

   E: worker is female
   F: worker has worked less than 5 years
   G: worker contributes to a voluntary retirement plan

Describe each of the following events in words.
   (a) E' **(b)** F' **(c)** $E \cap F$ **(d)** $F \cup G$ **(e)** $E \cup G'$ **(f)** $F' \cap G'$

18. For a medical experiment, people are classified as to whether they smoke, have a family history of heart disease, or are overweight. Define events E, F, and G as follows.

   $E:$ person smokes
   $F:$ person has a family history of heart disease
   $G:$ person is overweight

Describe each of the following events in words.
   (a) G' **(b)** $E \cup F$ **(c)** $F \cap G$ **(d)** $E' \cap F$ **(e)** $E \cup G'$ **(f)** $F' \cup G'$

A single fair die is rolled. Find the probabilities of the following events.

19. Getting a 2

20. Getting an odd number

21. Getting a number less than 5

22. Getting a number greater than 2

A card is drawn from a well-shuffled deck of 52 cards. Find the probability of drawing each of the following.

23. A 9

24. A black card

25. A black 9

26. A heart

27. The 9 of hearts

28. A face card

List the simple events whose union forms each of the events in Exercise 29–36.

29. Getting an even number on a roll of a die

30. Getting a number greater than 4 on a roll of a die

31. Drawing an ace of spades from a standard deck

32. Drawing a card that is both a heart and a face card from a 52-card deck

33. Making a record of whether a company's annual profit goes up, goes down, or stays the same

34. Checking a batch of 7 items and recording the number of defectives

35. Getting a double (identical faces showing) when two dice are rolled

36. Getting a sum of 11 when two dice are rolled

An experiment is conducted for which the sample space is $S = \{s_1, s_2, s_3, s_4, s_5\}$. Which of the probability assignments in Exercises 37–42 is possible for this experiment? If an assignment is not possible, tell why.

37.

Outcomes	s_1	s_2	s_3	s_4	s_5
Probabilities	.09	.32	.21	.25	.13

38.

Outcomes	s_1	s_2	s_3	s_4	s_5
Probabilities	.92	.03	0	.02	.03

39.

Outcomes	s_1	s_2	s_3	s_4	s_5
Probabilities	$\frac{1}{3}$	$\frac{1}{4}$	$\frac{1}{6}$	$\frac{1}{8}$	$\frac{1}{10}$

40.

Outcomes	s_1	s_2	s_3	s_4	s_5
Probabilities	$\frac{1}{5}$	$\frac{1}{3}$	$\frac{1}{4}$	$\frac{1}{5}$	$\frac{1}{10}$

41.

Outcomes	s_1	s_2	s_3	s_4	s_5
Probabilities	.64	$-.08$	.30	.12	.02

42.

Outcomes	s_1	s_2	s_3	s_4	s_5
Probabilities	.05	.35	.5	.2	$-.3$

6.2 FURTHER PROBABILITY RULES

We extend the rules for probability to more complex events in this section. Since events are sets, we can use set operations to find unions, intersections, and complements of events.

EXAMPLE 1

Suppose a die is tossed. Let E be the event "the die shows a number greater than 3," and let F be the event "the die shows an even number." Then

$$E = \{4, 5, 6\} \quad \text{and} \quad F = \{2, 4, 6\}.$$

Find each of the following.

(a) $E \cap F$

The event $E \cap F$ is "the die shows a number that is greater than 3 *and* is even." Event $E \cap F$ includes the outcomes common to *both* E and F.

$$E \cap F = \{4, 6\}$$

(b) $E \cup F$

The event $E \cup F$ is "the die shows a number that is greater than 3 *or* is even." Event $E \cup F$ includes the outcomes of E *or* F, or both.

$$E \cup F = \{2, 4, 5, 6\}$$

(c) E'

Event E' is "the die shows a number that is *not* greater than 3." E' includes the elements of the sample space that are not in E. (Event E' is the complement of event E.)

$$E' = \{1, 2, 3\}$$

The Venn diagrams in Figure 2 show the events $E \cap F$, $E \cup F$, and E'. ▬

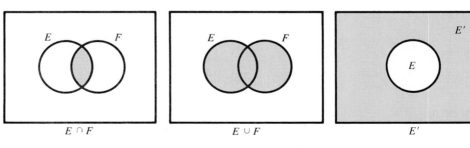

$E \cap F$ $E \cup F$ E'

FIGURE 2

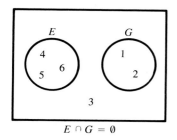

$E \cap G = \emptyset$

FIGURE 3

Two events that cannot both occur at the same time, such as getting both a head and a tail on the same toss of a coin, are called **mutually exclusive events.** Events E and F are mutually exclusive events if $E \cap F = \emptyset$. For any event E, E and E' are mutually exclusive.

▬ EXAMPLE 2

Let $S = \{1, 2, 3, 4, 5, 6\}$, the sample space for tossing a single die. Let $E = \{4, 5, 6\}$, and let $G = \{1, 2\}$. Then E and G are mutually exclusive events since they have no outcomes in common: $E \cap G = \emptyset$. See Figure 3. ▬

A summary of the set operations for events is given below.

EVENTS

> ▬ Let E and F be events for a sample space S. Then
> $E \cap F$ occurs when both E and F occur;
> $E \cup F$ occurs when E or F or both occur;
> E' occurs when E does not occur;
> E and F are mutually exclusive if $E \cap F = \emptyset$.

Remember that a simple event includes only *one* of the possible outcomes of an experiment.

The following formula for the number of elements in the union of two sets was given in Chapter 5.

$$n(A \cup B) = n(A) + n(B) - n(A \cap B).$$

This formula can now be used to prove the *union rule* for the probability of the union of two events.

UNION RULE

> ▬ For any two events E and F from a sample space S,
> $$P(E \cup F) = P(E) + P(F) - P(E \cap F).$$

Notice the similarity of this result to the formula from Chapter 5. The union rule follows from this formula and from the definition $P(E) = n(E)/n(S)$.

$$P(E \cup F) = \frac{n(E \cup F)}{n(S)}$$

$$= \frac{n(E) + n(F) - n(E \cap F)}{n(S)}$$

$$= \frac{n(E)}{n(S)} + \frac{n(F)}{n(S)} - \frac{n(E \cap F)}{n(S)}$$

$$= P(E) + P(F) - P(E \cap F)$$

Venn diagrams can be useful in finding probabilities, as the next example shows.

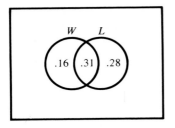

FIGURE 4

▬ EXAMPLE 3

Susan is a college student who receives heavy sweaters from her aunt at the first sign of cold weather. The probability that a sweater is the wrong size is .47, the probability that it is a loud color is .59, and the probability that it is both the wrong size and a loud color is .31. Let W represent the event "wrong size," and L represent "loud color." Place the given information on a Venn diagram by starting with .31 in the intersection of the regions for W and L. As stated earlier, event W has probability .47. Since .31 has already been placed inside the intersection of W and L in Figure 4,

$$.47 - .31 = .16$$

goes inside region W, but outside the intersection of W and L. In the same way,

$$.59 - .31 = .28$$

goes inside the region for L, and outside the overlap.

(a) Find the probability that the sweater is the correct size and not a loud color.

Using regions W and L, this event becomes $W' \cap L'$. From the Venn diagram in Figure 4, the labeled regions have probability

$$.16 + .31 + .28 = .75.$$

Since the entire region of the Venn diagram must have probability 1, the region outside W and L, or $W' \cap L'$, has probability

$$1 - .75 = .25.$$

The probability is .25 that the sweater is the correct size and not a loud color.

(b) Find the probability that the sweater is the correct size or is not loud.

The region $W' \cup L'$ has probability

$$.25 + .16 + .28 = .69. \quad ▬$$

▬ EXAMPLE 4

Find the probability that a family with three children has at least two girls.

Event E, "the family has *at least* two girls," is the union of two mutually exclusive events, F "the family has two girls," and G "the family has three girls." Events F and G are mutually exclusive, so $E \cap F = \emptyset$, and $P(E \cap F) = 0$. From the sample space,

$$\{ggg, \ ggb, \ gbg, \ bgg, \ gbb, \ bgb, \ bbg, \ bbb\},$$

$P(2 \text{ girls}) = 3/8$ and $P(3 \text{ girls}) = 1/8$. Using the union rule,

$$
\begin{aligned}
P(E) &= P(\text{at least 2 girls}) \\
&= P(2 \text{ girls}) + P(3 \text{ girls}) \\
&= \frac{3}{8} + \frac{1}{8} = \frac{1}{2}. \quad ▬
\end{aligned}
$$

■ EXAMPLE 5

If a single card is drawn from an ordinary 52-card deck, find the probability that it will be red or a face card.

Let R and F represent the events "red" and "face card," respectively. Then

$$P(R) = \frac{26}{52}, \qquad P(F) = \frac{12}{52}, \qquad \text{and} \qquad P(R \cap F) = \frac{6}{52}.$$

(There are 6 red face cards in a deck.) By the union rule,

$$P(R \cup F) = P(R) + P(F) - P(R \cap F)$$

$$= \frac{26}{52} + \frac{12}{52} - \frac{6}{52}$$

$$= \frac{32}{52} = \frac{8}{13}. \quad ■$$

■ EXAMPLE 6

Suppose two fair dice are rolled. Find each of the following probabilities.

(a) The first die shows a 2 or the sum of the results is 6 or 7.

The sample space for the throw of two dice is shown in Figure 5. The events are labeled A and B. From the diagram,

$$P(A) = \frac{6}{36}, \qquad P(B) = \frac{11}{36}, \qquad \text{and} \qquad P(A \cap B) = \frac{2}{36}.$$

By the extended addition principle,

$$P(A \cup B) = P(A) + P(B) - P(A \cap B),$$

$$P(A \cup B) = \frac{6}{36} + \frac{11}{36} - \frac{2}{36} = \frac{15}{36} = \frac{5}{12}.$$

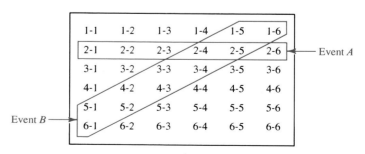

FIGURE 5

(b) The sum of the results is 11 or the second die shows a 5.

P(sum is 11) $= 2/36$, P(second die shows a 5) $= 6/36$, and P(sum is 11 and second die shows a 5) $= 1/36$, so

$$P(\text{sum is 11 or second die shows a 5}) = \frac{2}{36} + \frac{6}{36} - \frac{1}{36} = \frac{7}{36}.$$

Recall that the set of all outcomes in a sample space that do not belong to an event E is called the *complement* of E, written E'. For example, in the experiment of drawing a single card from a well-shuffled deck of 52 cards, let E be the event "the card is an ace." Then E' is the event "the card is not an ace." Using this definition of E', for any event E from a sample space S,

$$E \cup E' = S \qquad \text{and} \qquad E \cap E' = \emptyset.$$

Since $E \cap E' = \emptyset$, events E and E' are mutually exclusive, so that

$$P(E \cup E') = P(E) + P(E').$$

However, $E \cup E' = S$, the sample space, and $P(S) = 1$. Thus

$$P(E \cup E') = P(E) + P(E') = 1,$$

giving two alternate and useful results:

COMPLEMENTS

$$P(E) = 1 - P(E') \qquad \text{and} \qquad P(E') = 1 - P(E).$$

EXAMPLE 7

In a particular experiment, $P(E) = 2/7$. Find $P(E')$.

$$P(E') = 1 - P(E) = 1 - \frac{2}{7} = \frac{5}{7}$$

The next example shows that it is sometimes easier to find $P(E)$ by first finding $P(E')$ and then finding $P(E) = 1 - P(E')$.

EXAMPLE 8

In Example 6, find the probability that the sum of the numbers rolled is greater than 3.

To calculate this probability directly, we must find the probabilities that the sum is 4, 5, 6, 7, 8, 9, 10, 11, or 12 and then add them. It is much simpler to first find the probability of the complement, the event that the sum is less than or equal to 3.

$$P(\text{sum} \leq 3) = P(\text{sum is 2}) + P(\text{sum is 3})$$
$$= \frac{1}{36} + \frac{2}{36}$$
$$= \frac{3}{36} = \frac{1}{12}$$

Now use the fact that $P(E) = 1 - P(E')$ to get

$$P(\text{sum} > 3) = 1 - P(\text{sum} \le 3)$$

$$= 1 - \frac{1}{12} = \frac{11}{12}.$$ ▬

Odds Sometimes probability statements are given in terms of *odds,* a comparison of $P(E)$ with $P(E')$.

ODDS

> If $P(E') \ne 0$, the **odds in favor** of an event E are defined as the ratio of $P(E)$ to $P(E')$, or
> $$\frac{P(E)}{P(E')}.$$

▬ EXAMPLE 9

Suppose the weather forecaster says that the probability of rain tomorrow is 1/3. Find the odds in favor of rain tomorrow.

Let E be the event "rain tomorrow." Then E' is the event "no rain tomorrow." Since $P(E) = 1/3$, $P(E') = 2/3$. By the definition of odds, the *odds in favor of rain* are

$$\frac{1/3}{2/3} = \frac{1}{2}, \qquad \text{written} \qquad 1 \text{ to } 2, \text{ or } 1:2.$$

On the other hand, the odds that it will *not* rain, or the *odds against rain,* are

$$\frac{2/3}{1/3} = \frac{2}{1}, \qquad \text{written} \qquad 2 \text{ to } 1, \text{ or } 2:1. \ ▬$$

If the odds in favor of an event are, say, 3 to 5, then the probability of the event is 3/8, while the probability of the complement of the event is 5/8. (Odds of 3 to 5 indicate 3 outcomes in favor of the event out of a total of 8 possible outcomes.) This example suggests the following generalization.

> If the odds favoring event E are m to n, then
> $$P(E) = \frac{m}{m + n} \qquad \text{and} \qquad P(E') = \frac{n}{m + n}.$$

▬ EXAMPLE 10

The odds that a particular bid will be the low bid are 4 to 5.

(a) Find the probability that the bid will be the low bid.

Odds of 4 to 5 show 4 favorable chances out of $4 + 5 = 9$ chances altogether:

$$P(\text{bid will be low bid}) = \frac{4}{4 + 5} = \frac{4}{9}.$$

(b) Find the odds against that bid being the low bid.

There is a 5/9 chance that the bid will not be the low bid, so the odds against a low bid are

$$\frac{P(\text{bid will not be low})}{P(\text{bid will be low})} = \frac{5/9}{4/9} = \frac{5}{4}$$

or 5:4. ▬

▬ EXAMPLE 11

If the odds in favor of a particular horse's winning a race are 5 to 7, what is the probability that the horse will win the race?

The odds indicate chances of 5 out of 12 (5 + 7 = 12) that the horse will win, so

$$P(\text{winning}) = \frac{5}{12}. \quad ▬$$

Empirical Probability   In many realistic problems, it is not possible to establish exact probabilities for events. Useful approximations, however, often can be found by using past experience as a guide to the future. The next example shows one approach to such **empirical probabilities.**

▬ EXAMPLE 12

The manager of a store has decided to make a study of the amounts of money spent by people coming into the store. To begin, he chooses a day that seems fairly typical and gathers the following data. (Purchases have been rounded to the nearest dollar, with sales tax ignored.)

Amount Spent	Number of Customers
$0	158
$1–$5	94
$6–$9	203
$10–$19	126
$20–$49	47
$50–$99	38
$100 and over	53

First, the manager might add the numbers of customers to find that 719 people came into the store that day. Of these 719 people, $126/719 \approx .175$ made a purchase of at least $10 but no more than $19. Also, $53/719 \approx .074$ of the customers spent $100 or more. For this day, the probability that a customer entering the store will spend from $10 to $19 is .175, and the probability that the customer will spend $100 or more is .074. Probabilities for the other purchase amounts can be assigned in the same way, giving the results shown in the following table.

Amount Spent	Probability
$0	.220
$1–$5	.131
$6–$9	.282
$10–$19	.175
$20–$49	.065
$50–$99	.053
$100 and over	.074
	Total: 1.000

From the table, .282 of the customers spend from $6 to $9, inclusive. Since this price range attracts more than a quarter of the customers, perhaps the store's advertising should emphasize items in, or near, this price range.

The manager should use this table of probabilities to help in predicting the results on other days only if the manager is reasonably sure that the day when the measurements were made is fairly typical of the other days the store is open. For example, on the last few days before Christmas the probabilities might be quite different.

▆▆ EXAMPLE 13

Refer to Example 12 and find the probability that a customer spends at least $6 but less than $50.

This event is the union of three simple events, spending from $6 to $9, spending from $10 to $19, or spending from $20–$49. The probability of spending at least $6 but less than $50 can thus be found by the union rule.

P(spending at least $6 but less than $50)

$$= P(\text{spending } \$6–\$9) + P(\text{spending } \$10–\$19) + P(\text{spending } \$20–\$49)$$
$$= .282 + .175 + .065 = .522 \quad ▆▆$$

A table of probabilities, as in Example 12, sets up a **probability distribution:** that is, for each possible outcome of an experiment, we assign a number called the probability of that outcome. Probability distributions are discussed further in Section 7.1.

Another example of empirical probability occurs in weather forecasting. In one method, the forecaster compares atmospheric conditions with similar conditions in the past, then predicts the weather according to what happened previously under such conditions. For example, if certain kinds of atmospheric conditions have been followed by rain 80 times out of 100, the forecaster will announce an 80% chance of rain under those conditions.

One difficulty with empirical probability is that different people may assign different probabilities to the same event. Nevertheless, empirical probabilities can be assigned to many occurrences where the objective approach to probability cannot be used. In later chapters the use of empirical probability is shown in more detail.

6.2 EXERCISES

Decide whether the events in Exercises 1–6 are mutually exclusive.

1. Owning a car and owning a truck
2. Wearing glasses and wearing sandals
3. Being married and being over 30 years old
4. Being a teenager and being over 30 years old
5. Rolling a die once, getting a 4, and getting an odd number
6. Being a male and being a postal worker

Two dice are rolled. Find the probabilities of rolling the given sums.

7. **(a)** 2   **(b)** 4   **(c)** 5   **(d)** 6
8. **(a)** 8   **(b)** 9   **(c)** 10   **(d)** 13
9. **(a)** 9 or more   **(b)** Less than 7   **(c)** Between 5 and 8
10. **(a)** Not more than 5   **(b)** Not less than 8   **(c)** Between 3 and 7

One card is drawn from an ordinary deck of 52 cards. Find the probabilities of drawing the following cards.

11. **(a)** A 9 or 10
    (b) A red card or a 3
    (c) A 9 or a black 10
    (d) A heart or a black card

12. **(a)** Less than a 4 (count aces as ones)
    (b) A diamond or a 7
    (c) A black card or an ace
    (d) A heart or a jack

Ms. Elliott invites 10 relatives to a party: her mother, 2 aunts, 3 uncles, 2 brothers, 1 male cousin, and 1 female cousin. If the chances of any 1 guest arriving first are equally likely, find the probabilities that the first guest to arrive is as follows.

13. **(a)** A brother or uncle
    (b) A brother or a cousin
    (c) A brother or her mother

14. **(a)** An uncle or a cousin
    (b) A male or a cousin
    (c) A female or a cousin

The numbers 1, 2, 3, 4, and 5 are written on slips of paper, and 2 slips are drawn at random without replacement. Find each of the probabilities in Exercises 15–16.

15. **(a)** The sum of the numbers is 9.
    (b) The sum of the numbers is 5 or less.
    (c) The first number is 2 or the sum is 6.

16. **(a)** Both numbers are even.
    (b) One of the numbers is even or greater than 3.
    (c) The sum is 5 or the second number is 2.

17. At the first meeting of a committee to plan a local Lunar New Year celebration, the persons attending are 3 Chinese men, 4 Chinese women, 3 Vietnamese women, 2 Vietnamese men, 4 Korean women, and 2 Korean men. A chairperson is selected at random. Find the probabilities that the chairperson is

(a) Chinese;   (d) Chinese or Vietnamese;

(b) Korean or a woman;   (e) Korean and a woman.

(c) a man or Vietnamese;

18. In a refugee camp in southern Mexico, it is found that 90% of the refugees came to escape political oppression, 80% came to escape abject poverty, and 70% came to escape both. What is the probability that a refugee in the camp was not poor nor seeking political asylum?

Which of the following are examples of empirical probability?

19. The probability of heads on 5 consecutive tosses of a coin

20. The probability that a freshman entering college will graduate with a degree

21. The probability that a person is allergic to penicillin

22. The probability of drawing an ace from a standard deck of 52 cards

23. The probability that a person will get lung cancer from smoking cigarettes

24. A weather forecast that predicts a 70% chance of rain tomorrow

25. A gambler's claim that on a roll of a fair die, $P(\text{even}) = 1/2$

26. A surgeon's prediction that a patient has a 90% chance of a full recovery

27. A bridge player's 1/4 chance of being dealt a diamond

28. A forest ranger's statement that the probability of a short fire season this year is only 3 in 10

The table below gives a certain golfer's probabilities of scoring in various ranges on a par-70 course.

Range	Probability
Below 60	.01
60–64	.08
65–69	.15
70–74	.28
75–79	.22
80–84	.08
85–89	.06
90–94	.04
95–99	.02
100 or more	.06

In a given round, find the probability that the golfer's score will be as follows.

29. (a) 90 or higher

   (b) In the 90s

   (c) Below par of 70

30. (a) In the 70s

   (b) Not in the 60s

   (c) Not in the 60s or 70s

The table below shows the probability that a customer of a department store will make a purchase in the indicated range.

Amount Spent	Probability
Below $2	.07
$2–$4.99	.18
$5–$9.99	.21
$10–$19.99	.16
$20–$39.99	.11
$40–$69.99	.09
$70–$99.99	.07
$100–$149.99	.08
$150 or over	.03

Find the probabilities that a customer makes a purchase in the following ranges.

31. (a) Less than $5

(b) More than $4.99

(c) $10 to $69.99

32. (a) Less than $100

(b) $20 or more

(c) $100 or more

Suppose $P(E) = .26$, $P(F) = .41$, and $P(E \cap F) = .17$. Use a Venn diagram to help find each of the following.

33. (a) $P(E \cup F)$ (b) $P(E' \cap F)$ (c) $P(E \cap F')$ (d) $P(E' \cup F')$

Let $P(Z) = .42$, $P(Y) = .38$, and $P(Z \cup Y) = .61$. Find each of the following probabilities.

34. (a) $P(Z' \cap Y')$ (b) $P(Z' \cup Y')$ (c) $P(Z' \cup Y)$ (d) $P(Z \cap Y')$

Work the following exercises on odds.

A single fair die is rolled. Find the odds in favor of getting the following results.

35. 5   **36.** 3, 4, or 5   **37.** 1, 2, 3, or 4   **38.** Some number less than 2

39. Refer to Exercises 29 and 30. Find the following odds.

(a) The odds in favor of the golfer shooting below par

(b) The odds against the golfer shooting in the 70s

40. A marble is drawn from a box containing 3 yellow, 4 white, and 8 blue marbles. Find the odds in favor of drawing the following.

(a) A yellow marble

(b) A blue marble

(c) A white marble

41. Find the odds of *not* drawing a white marble in Exercise 40.

42. The probability that a company will make a profit this year is .74. Find the odds against the company making a profit.

43. If the odds that it will rain are 4 to 7, what is the probability of rain?

44. If the odds that a given candidate will win an election are 3 to 2, what is the probability that the candidate will lose?

45. On page 134 of Roger Staubach's autobiography, *First Down, Lifetime to Go,* Staubach makes the following statement regarding his experience in Vietnam:

"Odds against a direct hit are very low but when your life is in danger, you don't worry too much about the odds." Is this wording consistent with our definition of odds, for and against? How could it have been said so as to be technically correct?

46. Let E, F, and G be events from a sample space S. Show that

$$P(E \cup F \cup G) = P(E) + P(F) + P(G) - P(E \cap F) - P(E \cap G) - P(F \cap G) + P(E \cap F \cap G).$$

(*Hint:* Let $H = E \cup F$ and use the union rule.)

47. If E is an event, must E and E' be mutually exclusive?

48. Let E and F be mutually exclusive events. Let F and G be mutually exclusive events. Must E and G be mutually exclusive events?

49. Suppose E and F are mutually exclusive events. Must E' and F' be mutually exclusive events?

50. If E and F are mutually exclusive events, must E and $(E \cup F)'$ be mutually exclusive events?

 APPLICATIONS

LIFE SCIENCES

Color Blindness

51. Color blindness is an inherited characteristic that is more common in males than in females. If M represents male and C represents red-green color blindness, we use the relative frequencies of the incidence of males and of red-green color blindness as probabilities to get $P(C) = .049$, $P(M \cap C) = .042$, $P(M \cup C) = .534$. Find the following probabilities.

(a) $P(C')$ **(b)** $P(M)$ **(c)** $P(M')$

(d) $P(M' \cap C')$ **(e)** $P(C \cap M')$ **(f)** $P(C \cup M')$

Genetics

52. Gregor Mendel, an Austrian monk, was the first to use probability in the study of genetics. In an effort to understand the mechanism of character transmittal from one generation to the next in plants, he counted the number of occurrences of various characteristics. Mendel found that the flower color in certain pea plants obeyed this scheme:

Pure red crossed with pure white produces red.

The red offspring received from its parents genes for both red (R) and white (W) but in this case red is *dominant* and white *recessive,* so the offspring exhibits the color red. However, the offspring still carries both genes, and when two such offspring are crossed, several things can happen in the third generation. The table below, which is called a *Punnet square,* shows the possibilities.

		Second Parent	
		R	W
First Parent	R	RR	RW
	W	WR	WW

Use the fact that red is dominant over white to find each of the following.

(a) $P(\text{red})$ **(b)** $P(\text{white})$

Genetics   **53.** Mendel found no dominance in snapdragons, with one red gene and one white gene producing pink-flowered offspring. These second generation pinks, however, still carry one red and one white gene, and when they are crossed, the next generation still yields the Punnet square above. Find each of the following probabilities.

(a) P(red)   **(b)** P(pink)   **(c)** P(white)

(Mendel verified these probability ratios experimentally and did the same for many character units other than flower color. His work, published in 1866, was not recognized until 1890.)

Genetics   **54.** In most animals and plants, it is very unusual for the number of main parts of the organism (such as arms, legs, toes, or flower petals) to vary from generation to generation. Some species, however, have *meristic variability,* in which the number of certain body parts varies from generation to generation. One researcher studied the front feet of certain guinea pigs and produced the following probabilities.*

$$P(\text{only four toes, all perfect}) = .77$$
$$P(\text{one imperfect toe and four good ones}) = .13$$
$$P(\text{exactly five good toes}) = .10$$

Find the probability of each of the following events.

(a) No more than four good toes

(b) Five toes, whether perfect or not

FOR THE COMPUTER

One way to solve a probability problem is to repeat the experiment (or a simulation of the experiment) many times, keeping track of the results. Then the probability can be approximated using the basic definition of the probability of an event E: P(E) = m/n, where m favorable outcomes occur in n trials of an experiment. This is called the Monte Carlo method of finding probabilities.

55. Suppose a coin is tossed 5 times. Use the Monte Carlo method to approximate the following probabilities. Then calculate the theoretical probabilities using the methods of the text and compare the results.

(a) P(4 heads)   **(b)** P(2 heads, 1 tail, 2 heads) (in the order given)

56. Use the Monte Carlo method to approximate the following probabilities if 4 cards are drawn from 52.

(a) P(any 2 cards and then 2 kings)   **(b)** P(2 kings)

Use the Monte Carlo method to approximate the following probabilities.

57. A jeweler received 8 identical watches, boxed individually, with the serial number of the watch on each box. An assistant, who does not know that the boxes are marked, is told to polish the watches and then put them back in the boxes. She puts them in the boxes at random. What is the probability that she gets at least 1 watch in the right box?

58. A checkroom attendant has 10 hats but has lost the numbers identifying them. If he gives them back randomly, what is the probability that at least 1 hat is given back correctly?

---

*From "An Analysis of Variability in Guinea Pigs" by J. R. Wright in *Genetics* 19, pp. 506–536. Reprinted by permission.

6.3   CONDITIONAL PROBABILITY

The training manager for a large stockbrokerage firm has noticed that some of the firm's brokers use the firm's research advice, while other brokers tend to follow their own feelings of which stocks will go up. To see whether the research department performs better than the feelings of the brokers, the manager conducted a survey of 100 brokers, with results as shown in the following table.

	Picked Stocks That Went Up	*Didn't Pick Stocks That Went Up*	*Totals*
Used Research	30	15	45
Didn't Use Research	30	25	55
Totals	60	40	100

Letting *A* represent the event "picked stocks that went up," and letting *B* represent the event "used research," we can find the following probabilities.

$$P(A) = \frac{60}{100} = .6 \qquad P(A') = \frac{40}{100} = .4$$

$$P(B) = \frac{45}{100} = .45 \qquad P(B') = \frac{55}{100} = .55$$

Suppose we want to find the probability that a broker using research will pick stocks that go up. From the table above, of the 45 brokers who use research, 30 picked stocks that went up, with

$$P(\text{broker who uses research picks stocks that go up}) = \frac{30}{45} = .667.$$

This is a different number than the probability that a broker picks stocks that go up, .6, since we have additional information (the broker uses research) that has *reduced the sample space*. In other words, we found the probability that a broker picks stocks that go up, *A*, given the additional information that the broker uses research, *B*. This is called the *conditional probability* of event *A*, given that event *B* has occurred, written $P(A|B)$. In the example above,

$$P(A|B) = \frac{30}{45},$$

which can be written as

$$P(A|B) = \frac{30/100}{45/100} = \frac{P(A \cap B)}{P(B)},$$

where $P(A \cap B)$ represents, as usual, the probability that both *A* and *B* will occur.

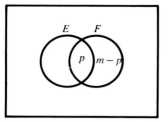

Event *F* has a total of *m* elements.

FIGURE 6

To generalize this result, assume that E and F are two events for a particular experiment. Assume that the sample space S for this experiment has n possible equally likely outcomes. Suppose event F has m elements, and $E \cap F$ has p elements ($p \le m$). Using the fundamental principle of probability,

$$P(F) = \frac{m}{n} \quad \text{and} \quad P(E \cap F) = \frac{p}{n}.$$

We now want $P(E|F)$, the probability that E occurs given that F has occurred. Since we assume F has occurred, reduce the sample space to F: look only at the m elements inside F. (See Figure 6). Of these m elements, there are p elements where E also occurs, since $E \cap F$ has p elements. This makes

$$P(E|F) = \frac{p}{m}.$$

Divide numerator and denominator by n to get

$$P(E|F) = \frac{\dfrac{p}{n}}{\dfrac{m}{n}} = \frac{P(E \cap F)}{P(F)}.$$

This last result gives the definition of conditional probability.

DEFINITION OF CONDITIONAL PROBABILITY

The **conditional probability** of event E given event F, written $P(E|F)$, is

$$P(E|F) = \frac{P(E \cap F)}{P(F)}, \quad \text{where } P(F) \ne 0.$$

EXAMPLE 1
Use the information given in the chart at the beginning of this section to find the following probabilities.

(a) $P(B|A)$

By the definition of conditional probability,

$$P(B|A) = \frac{P(B \cap A)}{P(A)}.$$

In the example, $P(B \cap A) = 30/100$, and $P(A) = 60/100$, with

$$P(B|A) = \frac{30/100}{60/100} = \frac{1}{2}.$$

If a broker picked stocks that went up, then the probability is 1/2 that the broker used research.

(b) $P(A'|B)$

$$P(A'|B) = \frac{P(A' \cap B)}{P(B)} = \frac{15/100}{45/100} = \frac{1}{3}$$

(c) $P(B'|A')$

$$P(B'|A') = \frac{P(B' \cap A')}{P(A')} = \frac{25/100}{40/100} = \frac{5}{8}$$

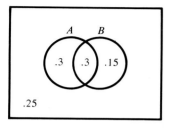

$$P(A) = .3 + .3 = .6$$

FIGURE 7

Venn diagrams are useful for illustrating problems in conditional probability. A Venn diagram for Example 1, in which the probabilities are used to indicate the number in the set defined by each region, is shown in Figure 7. In the diagram, $P(B|A)$ is found by reducing the sample space to just set A. Then $P(B|A)$ is the ratio of the number in that part of set B which is also in A to the number in set A, or $.3/.6 = .5$.

▬ EXAMPLE 2
Given $P(E) = .4$, $P(F) = .5$, and $P(E \cup F) = .7$, find $P(E|F)$.

Find $P(E \cap F)$ first. Then use a Venn diagram to find $P(E|F)$. By the extended addition principle,

$$P(E \cup F) = P(E) + P(F) - P(E \cap F)$$
or
$$.7 = .4 + .5 - P(E \cap F)$$
so that
$$P(E \cap F) = .2.$$

Now use the probabilities to indicate the number in each region of the Venn diagram in Figure 8. $P(E|F)$ is the ratio of the probability of that part of E which is in F to the probability of F, or

$$P(E|F) = \frac{P(E \cap F)}{P(F)} = \frac{.2}{.5} = \frac{2}{5}.$$

Notice that the reduced sample space has probability .5 and within that reduced sample space $P(E) = .2$, from which

$$P(E|F) = \frac{.2}{.5} = \frac{2}{5},$$

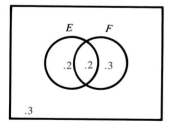

FIGURE 8

the same result. ▬

▬ EXAMPLE 3
Two fair coins were tossed, and it is known that at least one was a head. Find the probability that both were heads.

The sample space has four equally likely outcomes, $S = \{hh, ht, th, tt\}$. Define two events:

$$E_1 = \text{at least 1 head (or } E_1 = \{hh, ht, th\})$$
and
$$E_2 = 2 \text{ heads (or } E_2 = \{hh\}).$$

Since there are four equally likely outcomes, $P(E_1) = 3/4$. Also, $P(E_1 \cap E_2)$ $= 1/4$. We want the probability that both were heads, given that at least one was a head; that is, we want to find $P(E_2|E_1)$. Because of the condition that at least one coin was a head, the reduced sample space is

$$\{hh, \ ht, \ th\}.$$

Since only one outcome in this reduced sample space is 2 heads,

$$P(E_2|E_1) = \frac{1}{3}.$$

Alternatively, use the definition given above.

$$P(E_2|E_1) = \frac{P(E_2 \cap E_1)}{P(E_1)} = \frac{1/4}{3/4} = \frac{1}{3} \quad \blacksquare$$

Product Rule   In the definition of conditional probability given earlier, both sides of the equation for $P(E|F)$ can be multiplied by $P(F)$ to get the following *product rule* for probability.

PRODUCT RULE

> For any events E and F,
> $$P(E \cap F) = P(F) \cdot P(E|F).$$

The product rule gives a method for finding the probability that events E and F both occur, as illustrated by the next few examples.

■ EXAMPLE 4

A class is 2/5 women and 3/5 men. Of the women, 25% are business majors. Find the probability that a student chosen at random is a female business major.

Let B and W represent the events "business major" and "woman," respectively. We want to find $P(B \cap W)$. By the product rule,

$$P(B \cap W) = P(W) \cdot P(B|W).$$

Using the given information, $P(W) = 2/5 = .4$ and $P(B|W) = .25$. Thus,

$$P(B \cap W) = .4(.25) = .10. \quad \blacksquare$$

Tree diagrams are helpful in many conditional probability problems. The following examples illustrate how they are used.

■ EXAMPLE 5

A company needs to hire a new director of advertising. It has decided to try to hire either person A or person B, who are assistant advertising directors for its major competitor. In trying to decide between A and B, the company does research on the campaigns managed by either A or B (no campaign is managed by both), and finds that A is in charge of twice as many advertising campaigns as B. Also, A's campaigns have satisfactory results 3 out of 4 times, while B's campaigns have satisfactory results only 2 out of 5 times. Suppose one of the competitor's advertising campaigns (managed by A or B) is selected. Find the probabilities of the following events.

(a) Person A is in charge of an advertising campaign that produces satisfactory results.

First construct a tree diagram showing the various possible outcomes for this experiment, as in Figure 9. (Recall the discussion of tree diagrams in Section 5.1.) Let *A* represent the event "person A runs the campaign," *B* represent the event "person B runs the campaign," *S* represent "the campaign is satisfactory," and *U* represent "the campaign is unsatisfactory." Since A does twice as many jobs as B, the probabilities of A and B having done the job are 2/3 and 1/3 respectively, as shown on the first stage of the tree. The second stage shows four different conditional probabilities. For example, along the branch from *B* to *S*,

$$P(S|B) = \frac{2}{5}$$

since person B has satisfactory results in 2 out of 5 campaigns.

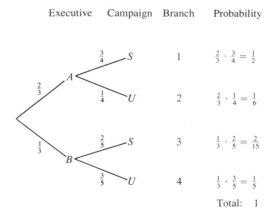

FIGURE 9

Each of the four composite branches in the tree is numbered, with its probability given on the right. At each point where the tree branches, the sum of the probabilities is 1. The event that A has a campaign with a satisfactory result (event $A \cap S$) is associated with branch 1, so

$$P(A \cap S) = \frac{1}{2}.$$

(b) B runs the campaign and produces satisfactory results.

This event, $B \cap S$, is shown on branch 3:

$$P(B \cap S) = \frac{2}{15}.$$

(c) The campaign is satisfactory.

The result S combines branches 1 and 3, so

$$P(S) = \frac{1}{2} + \frac{2}{15} = \frac{19}{30}.$$

(d) The campaign is unsatisfactory.

Event U combines branches 2 and 4, so

$$P(U) = \frac{1}{6} + \frac{1}{5} = \frac{11}{30}.$$

Alternatively, $P(U) = 1 - P(S) = 1 - 19/30 = 11/30$.

(e) Either A runs the campaign or the results are satisfactory (or both).

Event A combines branches 1 and 2, while event S combines branches 1 and 3. Thus, we use branches 1, 2, and 3.

$$P(A \cup S) = \frac{1}{2} + \frac{1}{6} + \frac{2}{15} = \frac{4}{5} \quad \blacksquare$$

▬ EXAMPLE 6

From a box containing 3 white, 2 green, and 1 red marble, 2 marbles are drawn one at a time without replacing the first before the second is drawn. Find the probability that 1 white and 1 green marble are drawn.

A tree diagram showing the various possible outcomes is given in Figure 10. In this diagram, W represents the event "drawing a white marble" and G represents "drawing a green marble." The probabilities for each draw are determined from the given information. For example, on the first draw, $P(W$ first$)$ $= 3/6 = 1/2$ because 3 of the 6 marbles in the box are white. On the second draw, $P(G$ second$|W$ first$) = 2/5$. One white marble has been removed, leaving 5, of which 2 are green.

Now we want to find the probability of drawing one white marble and one green marble. This event can occur in two ways: drawing a white marble first and then a green one (branch 2 of the tree diagram), or drawing a green marble first and then a white one (branch 4). For branch 2,

$$P(W \text{ first}) \cdot P(G \text{ second}|W \text{ first}) = \frac{1}{2} \cdot \frac{2}{5} = \frac{1}{5}.$$

For branch 4, where the green marble is drawn first,

$$P(G \text{ first}) \cdot P(W \text{ second}|G \text{ first}) = \frac{1}{3} \cdot \frac{3}{5} = \frac{1}{5}.$$

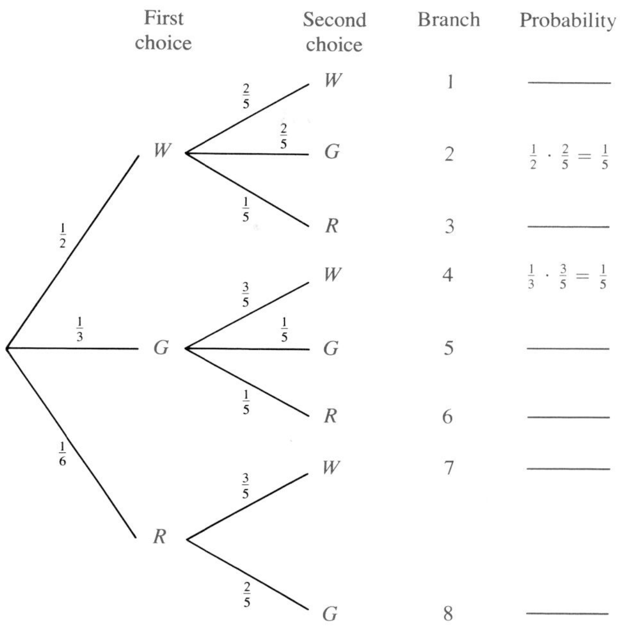

First choice   Second choice   Branch   Probability

$\frac{2}{5}$ W   1   ——————

W $\frac{2}{5}$ G   2 $\frac{1}{2} \cdot \frac{2}{5} = \frac{1}{5}$

$\frac{1}{5}$ R   3   ——————

$\frac{1}{2}$

$\frac{3}{5}$ W   4 $\frac{1}{3} \cdot \frac{3}{5} = \frac{1}{5}$

$\frac{1}{3}$ G $\frac{1}{5}$ G   5   ——————

$\frac{1}{5}$ R   6   ——————

$\frac{1}{6}$

$\frac{3}{5}$ W   7   ——————

R

$\frac{2}{5}$ G   8   ——————

FIGURE 10

Since the two events are mutually exclusive, the final probability is the sum of these two probabilities, or

$$P(1\ W,\ 1\ G) = P(W \text{ first}) \cdot P(G \text{ second}|W \text{ first})$$

$$+ P(G \text{ first}) \cdot P(W \text{ second}|G \text{ first}) = \frac{2}{5}. \quad \blacksquare\blacksquare$$

The product rule is often used with *stochastic processes,* where the outcome of an experiment depends on the outcomes of previous experiments. For example, the outcome of a draw of a card from a deck depends on any cards previously drawn. (Stochastic processes are studied in more detail in a later chapter.)

■■ EXAMPLE 7

Two cards are drawn, without replacement, from an ordinary deck. Find the probability that the first card is a heart and the second card is red.

Start with the tree diagram in Figure 11. On the first draw, since there are 13 hearts in the 52 cards, the probability of drawing a heart is $13/52 = 1/4$. On the second draw, since a heart has been drawn already, there are 25 red

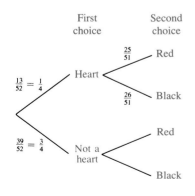

First choice   Second choice

$\frac{13}{52} = \frac{1}{4}$ Heart $\frac{25}{51}$ Red $\frac{26}{51}$ Black

$\frac{39}{52} = \frac{3}{4}$ Not a heart   Red   Black

FIGURE 11

cards in the remaining 51 cards. Thus, the probability of drawing a red card on the second draw, given that the first is a heart, is 25/51. Therefore,

$$P(\text{heart first and red second})$$
$$= P(\text{heart first}) \cdot P(\text{red second}|\text{heart first})$$
$$= \frac{1}{4} \cdot \frac{25}{51}$$
$$= \frac{25}{204} \approx .1225. \quad \blacksquare$$

■■ EXAMPLE 8

Three cards are drawn, without replacement, from an ordinary deck. Find the probability that exactly 2 of the cards are red.

Here we need a tree diagram with 3 stages, as shown in Figure 12. The three branches indicated with arrows produce exactly 2 red cards from the draws. Multiply the probabilities along each of these branches and then add.

$$P(\text{exactly 2 red cards}) = \frac{26}{52} \cdot \frac{25}{51} \cdot \frac{26}{50} + \frac{26}{52} \cdot \frac{26}{51} \cdot \frac{25}{50} + \frac{26}{52} \cdot \frac{26}{51} \cdot \frac{25}{50}$$
$$= \frac{50,700}{132,600}$$
$$= \frac{13}{34} \approx .382 \quad \blacksquare$$

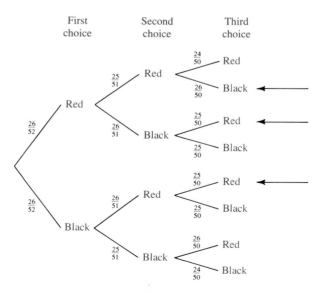

FIGURE 12

Independent Events   Suppose a fair coin is tossed and gives heads. The probability of heads on the next toss is still 1/2; the fact that heads was obtained on a given toss has no effect on the outcome of the next toss. Coin tosses are *independent events,* since knowledge of the outcome of one toss does not help decide on the outcome of the next toss. Rolls of a fair die are independent events; the fact that a two came up on one roll does not increase our knowledge of the outcome of the next roll. On the other hand, the events "today is cloudy" and "today is rainy" are *dependent events;* if we know that it is cloudy, we know that there is an increased chance of rain.

If events E and F are independent, then the knowledge that E has occurred gives no (probability) information about the occurrence or nonoccurrence of event F. That is, $P(F)$ is exactly the same as $P(F|E)$, or

$$P(F|E) = P(F).$$

This, in fact, is the formal definition of independent events.

DEFINITION OF INDEPENDENT EVENTS

> E and F are **independent events** if
> $$P(F|E) = P(F).$$

Using this definition, the product rule can be simplified for independent events.

PRODUCT RULE FOR INDEPENDENT EVENTS

> If E and F are independent events, then
> $$P(E \cap F) = P(E) \cdot P(F).$$

■ EXAMPLE 9

A calculator requires a keystroke assembly and a logic circuit. Assume that 99% of the keystroke assemblies are satisfactory and 97% of the logic circuits are satisfactory. Find the probability that a finished calculator will be satisfactory.

If the failure of a keystroke assembly and the failure of a logic circuit are independent events, then

P(satisfactory calculator)

   $= P$(satisfactory keystroke assembly) $\cdot P$(satisfactory logic circuit)

   $= (.99)(.97) \approx .96.$

The probability of a defective calculator is $1 - .96 = .04$.   ■

■ EXAMPLE 10

When black-coated mice are crossed with brown-coated mice, a pair of genes, one from each parent, determines the coat color of the offspring. Let b represent the gene for brown and B the gene for black. If a mouse carries either one B gene and one b gene (Bb or bB) or two B genes (BB), the coat will be black.

If the mouse carries two *b* genes (*bb*), the coat will be brown. Find the probability that a mouse born to a brown-coated female and a black-coated male who is known to carry the *Bb* combination will be brown.

To be brown-coated, the offspring must receive one *b* gene from each parent. The brown-coated parent carries two *b* genes, so that the probability of getting one *b* gene from the mother is 1. The probability of getting one *b* gene from the black-coated father is 1/2. Therefore, since these are independent events, the probability that a brown-coated offspring will be produced by these parents is $1 \cdot 1/2 = 1/2$. ▬

It is common for students to confuse the ideas of *mutually exclusive* events and *independent* events. Events *E* and *F* are mutually exclusive if $E \cap F = \emptyset$. For example, if a family has exactly one child, the only possible outcomes are $B = \{boy\}$ and $G = \{girl\}$. These two events are mutually exclusive. The events are *not* independent, however, since $P(G|B) = 0$ (if a family with only one child has a boy, the probability it has a girl is then 0). Since $P(G|B) \neq P(G)$, the events are not independent.

Of all the families with exactly two children, the events $G_1 = \{first\ child\ is\ a\ girl\}$ and $G_2 = \{second\ child\ is\ a\ girl\}$ are independent, since $P(G_2|G_1)$ equals $P(G_2)$. However, G_1 and G_2 are not mutually exclusive, since $G_1 \cap G_2 = \{both\ children\ are\ girls\} \neq \emptyset$.

The only way to show that two events *E* and *F* are independent is to show that $P(F|E) = P(F)$.

6.3 EXERCISES

If a single fair die is rolled, find the probabilities of the following results.

1. A 2, given that the number rolled was odd
2. A 4, given that the number rolled was even
3. An even number, given that the number rolled was 6

If two fair dice are rolled, find the probabilities of the following results.

4. A sum of 8, given that the sum is greater than 7
5. A sum of 6, given that the roll was a "double" (two identical numbers)
6. A double, given that the sum was 9

If two cards are drawn without replacement from an ordinary deck, find the probabilities of the following results.

7. The second card is a heart, given that the first is a heart.
8. Both cards are hearts.
9. The second card is black, given that the first is a spade.
10. The second is a face card, given that the first is a jack.

If five cards are drawn without replacement from an ordinary deck, find the probabilities of the following results.

11. All diamonds

12. All diamonds, given that the first and second were diamonds

13. All diamonds, given that the first four were diamonds

14. All clubs, given that the third was a spade

15. All the same suit

A smooth-talking young man has a 1/3 probability of talking a policeman out of giving him a speeding ticket. The probability that he is stopped for speeding during a given weekend is 1/2. Find the probabilities of the following events.

16. He will receive no speeding tickets on a given weekend.

17. He will receive no speeding tickets on 3 consecutive weekends.

Slips of paper marked with the digits 1, 2, 3, 4, and 5 are placed in a box and mixed well. If two slips are drawn (without replacement), find the probabilities of the following results.

18. The first number drawn is even and the second is odd.

19. The first number drawn is a 3 and the second a number greater than 3.

20. Both numbers drawn are even.

21. Both slips drawn are marked 3.

Two marbles are drawn without replacement from a jar with 4 black and 3 white marbles. Find the probabilities of the following results.

22. Both marbles drawn are white.

23. Both marbles drawn are black.

24. The second marble drawn is white, given that the first is black.

25. The first marble drawn is black and the second is white.

26. One black marble and 1 white marble are drawn.

The Motor Vehicle Department has found that the probability of a person passing the test for a driver's license on the first try is .75. The probability that an individual who fails on the first test will pass on the second try is .80, and the probability that an individual who fails the first and second tests will pass the third time is .70. Find the probabilities that an individual will do the following.

27. Fail both the first and second test

28. Fail three times in a row

29. Require at least two tries to pass the test

According to a booklet put out by Eastwest Airlines, 98% of all scheduled Eastwest flights actually take place. (The other flights are cancelled due to weather, equipment problems, and so on.) Assume that the event that a given flight takes place is independent of the event that another flight takes place.

30. Elisabeta Guervara plans to visit her company's branch offices; her journey requires 3 separate flights on Eastwest Airlines. What is the probability that all of these flights will take place?

31. Based on the reasons we gave for a flight to be cancelled, how realistic is the assumption of independence that we made?

32. In one area, 4% of the population drive luxury cars. However, 17% of the CPAs drive luxury cars. Are the events "person drives a luxury car" and "person is a CPA" independent?

33. Corporations where a computer is essential to day-to-day operations, such as banks, often have a second backup computer in case of failure by the main computer. Suppose there is a .003 chance that the main computer will fail in a given time period, and a .005 chance that the backup computer will fail while the main computer is being repaired. Assume these failures represent independent events, and find the fraction of the time that the corporation can assume it will have computer service. How realistic is our assumption of independence?

34. The probability that a key component of a space rocket will fail is .03. How many such components must be used as backups to ensure that the probability of at least one of the components' working is .999999?

Let E and F be events that are neither the empty set nor the sample space S. Identify the following as true *or* false.

35. $P(E|E) = 1$

36. $P(E|E') = 1$

37. $P(\emptyset|F) = 0$

38. $P(S|E) = P(E)$

39. $P(F|S) = P(F)$

40. $P(E|F) = P(F|E)$

41. $P(E|E \cap F) = 0$

42. If $P(E|F) = P(E \cap F)$, then $P(F) = 1$

43. Let E and F be mutually exclusive events such that $P(F) > 0$. Find $P(E|F)$.

44. If $E \subset F$, where $E \neq \emptyset$, find $P(F|E)$ and $P(E|F)$.

45. Let F_1, F_2, and F_3 be a set of pairwise mutually exclusive events (that is, $F_1 \cap F_2 = \emptyset$, $F_1 \cap F_3 = \emptyset$, and $F_2 \cap F_3 = \emptyset$), with sample space $S = F_1 \cup F_2 \cup F_3$. Let E be any event. Show that

$$P(E) = P(F_1) \cdot P(E|F_1) + P(F_2) \cdot P(E|F_2) + P(F_3) \cdot P(E|F_3).$$

46. Show that for 3 events, E, F, and G,

$$P(E \cap F \cap G) = P(E) \cdot P(F|E) \cdot P(G|E \cap F).$$

▰ APPLICATIONS

BUSINESS AND ECONOMICS

Banking *The Midtown Bank has found that most customers at the tellers' windows either cash a check or make a deposit. The chart below indicates the transactions for one teller for one day.*

	Cash Check	No Check	Totals
Make Deposit	50	20	70
No Deposit	30	10	40
Totals	80	30	110

Letting C represent "cashing a check" and D represent "making a deposit," express each of the following probabilities in words and find its value.

47. $P(C|D)$

48. $P(D'|C)$

49. $P(C'|D')$

50. $P(C'|D)$

51. $P[(C \cap D)']$

Pet Store Stock

A pet shop has 10 puppies, 6 of them males. There are 3 beagles (1 male), 1 cocker spaniel (male), and 6 poodles. Construct a table similar to the one above and find the probabilities that 1 of these puppies, chosen at random, is the following.

52. A beagle

53. A beagle, given that it is a male

54. A male, given that it is a beagle

55. A cocker spaniel, given that it is a female

56. A poodle, given that it is a male

57. A female, given that it is a beagle

Quality Control

A bicycle factory runs two assembly lines, A and B. If 95% of line A's products pass inspection, while only 90% of line B's products pass inspection, and 60% of the factory's bikes come off assembly line B (the rest off A), find the probabilities that one of the factory's bikes did not pass inspection and came off the following.

58. Assembly line A

59. Assembly line B

LIFE SCIENCES

Genetics

60. Both of a certain pea plant's parents had a gene for red and a gene for white flowers. (See the exercises for Section 6.2.) If the offspring has red flowers, find the probability that it combined a gene for red and a gene for white (rather than 2 for red).

Genetics

Assuming that boy and girl babies are equally likely, fill in the remaining probabilities on the tree diagram and use the following information to find the probability that a family with 3 children has all girls, given the following.

61. The first is a girl.

62. The third is a girl.

63. The second is a girl.

64. At least 2 are girls.

65. At least 1 is a girl.

Color Blindness *The following table shows frequencies for red-green color blindness, where M represents "person is male" and C represents "person is color-blind." Use this table to find the following probabilities.*

	M	M'	Totals
C	.042	.007	.049
C'	.485	.466	.951
Totals	.527	.473	1.000

66. $P(M)$ **67.** $P(C)$ **68.** $P(M \cap C)$ **69.** $P(M \cup C)$

70. $P(M|C)$ **71.** $P(C|M)$ **72.** $P(M'|C)$

73. Are the events C and M, described above, dependent?

Color Blindness **74.** A scientist wishes to determine whether there is a relationship between color blindness (C) and deafness (D). Given the probabilities listed in the table below, what should his findings be? (See Exercises 66–73.)

	D	D'	Totals
C	.0004	.0796	.800
C'	.0046	.9154	.9200
Totals	.0050	.9950	1.0000

Drug Screening *In searching for a new drug with commercial possibilities, drug company researchers use the ratio*

$$N_S : N_A : N_P : 1.$$

That is, if the company gives preliminary screening to N_S substances, it may find that N_A of them are worthy of further study, with N_P of these surviving into full-scale development. Finally, 1 of the substances will result in a marketable drug. Typical numbers used by Smith, Kline, and French Laboratories in planning research budgets might be 2000:30:8:1.* Use this ratio in the following exercises.

75. Suppose a compound has been chosen for preliminary screening. Find the probability that the compound will survive and become a marketable drug.

76. Find the probability that the compound will not lead to a marketable drug.

77. Suppose the number of such compounds receiving preliminary screening is a. Set up the probability that none of them produces a marketable drug. (Assume independence throughout these exercises.)

---

*Reprinted by permission of E. B. Pyle, III, B. Douglas, G. W. Ebright, W. J. Westlake, A. B. Bender, "Scientific Manpower Allocation to New Drug Screening Programs," *Management Science*, Vol. 19, No. 12, August 1973, copyright © 1973 The Institute of Management Sciences.

78. Use your results from Exercise 77 to set up the probability that at least one of the drugs will prove marketable.

79. Suppose now that N scientists are employed in the preliminary screening, and that each scientist can screen c compounds per year. Set up the probability that no marketable drugs will be discovered in a year.

80. Set up the probability that at least one marketable drug will be discovered.

FOR THE COMPUTER

For the following exercises, evaluate your answer in Exercise 80 for the following values of N and c. Use a calculator with a y^x key, or a computer.

81. $N = 100, c = 6$ **82.** $N = 25, c = 10$

6.4 BAYES' FORMULA

Suppose the probability that a person gets lung cancer, given that the person smokes a pack or more of cigarettes daily, is known. For a research project, it might be necessary to know the probability that a person smokes a pack or more of cigarettes daily, given that the person has lung cancer. More generally, if $P(E|F)$ is known for two events E and F, can $P(F|E)$ be found? The answer is yes, we can find $P(F|E)$ by using a formula that will be developed in this section. To find this formula, let us start with the product rule:

$$P(E \cap F) = P(E) \cdot P(F|E),$$

which can also be written as $P(F \cap E) = P(F) \cdot P(E|F)$. From the fact that $P(E \cap F) = P(F \cap E)$,

$$P(E) \cdot P(F|E) = P(F) \cdot P(E|F),$$

or
$$P(F|E) = \frac{P(F) \cdot P(E|F)}{P(E)}. \tag{1}$$

Given the two events E and F, if E occurs, then either F also occurs or F' also occurs. The probabilities of $E \cap F$ and $E \cap F'$ can be expressed as follows.

$$P(E \cap F) = P(F) \cdot P(E|F)$$
$$P(E \cap F') = P(F') \cdot P(E|F')$$

Since $(E \cap F) \cup (E \cap F') = E$ (because F and F' form the sample space),

$$P(E) = P(E \cap F) + P(E \cap F')$$
or
$$P(E) = P(F) \cdot P(E|F) + P(F') \cdot P(E|F').$$

From this, equation (1) produces the following result, a special case of Bayes' formula, which is generalized later in this section.

**BAYES' FORMULA
(SPECIAL CASE)**

$$P(F|E) = \frac{P(F) \cdot P(E|F)}{P(F) \cdot P(E|F) + P(F') \cdot P(E|F')} \tag{2}$$

EXAMPLE 1

For a fixed length of time, the probability of worker error on a production line is .1, the probability that an accident will occur when there is a worker error is .3, and the probability that an accident will occur when there is no worker error is .2. Find the probability of a worker error if there is an accident.

Let E represent the event of an accident, and let F represent the event of worker error. From the information above,

$$P(F) = .1, \qquad P(E|F) = .3, \qquad \text{and} \qquad P(E|F') = .2.$$

These probabilities are shown on the tree diagram in Figure 13.

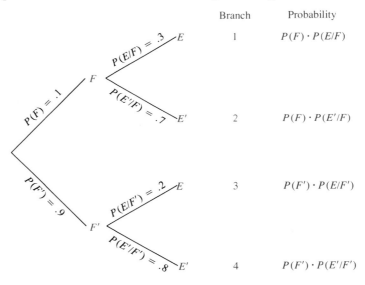

FIGURE 13

Find $P(F|E)$ using equation (2) above:

$$\begin{aligned}
P(F|E) &= \frac{P(F) \cdot P(E|F)}{P(F) \cdot P(E|F) + P(F') \cdot P(E|F')} \\
&= \frac{(.1)(.3)}{(.1)(.3) + (.9)(.2)} \\
&= \frac{1}{7}.
\end{aligned}$$

In a similar manner, the probability of no worker error if an accident occurs is $P(F'|E)$, or

$$P(F'|E) = \frac{P(F') \cdot F(E|F')}{P(F') \cdot P(E|F') + P(F) \cdot P(E|F)}$$

$$= \frac{(.9)(.2)}{(.9)(.2) + (.1)(.3)} = \frac{6}{7}. \quad \blacksquare$$

Equation (2) above can be generalized to more than two possibilities. To do so, use the tree diagram of Figure 14. This diagram shows the paths that can produce some event E. We assume that the events $F_1, F_2, \cdots, F_n$ are pairwise mutually exclusive events (that is, events which, taken two at a time, are disjoint) whose union is the sample space, and that E is an event that has occurred. See Figure 15.

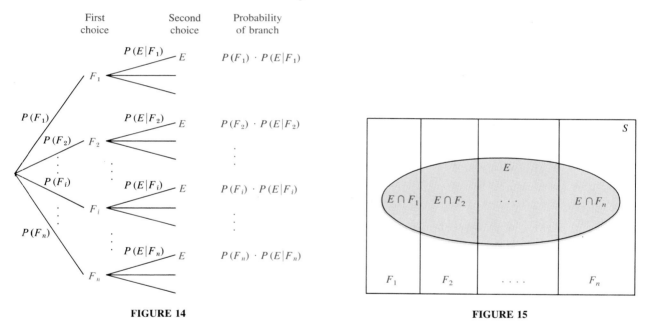

FIGURE 14

FIGURE 15

The probability $P(F_i|E)$, where $1 \le i \le n$, can be found by dividing the probability for the branch containing $P(E|F_i)$ by the sum of the probabilities of all the branches producing event E.

BAYES' FORMULA

$$P(F_i|E) = \frac{P(F_i) \cdot P(E|F_i)}{P(F_1) \cdot P(E|F_1) + P(F_2) \cdot P(E|F_2) + \cdots + P(F_n) \cdot P(E|F_n)}$$

This result is known as **Bayes' formula,** after the Reverend Thomas Bayes (1702–1761), whose paper on probability was published about 200 years ago.

The statement of Bayes' formula can be daunting. Actually, it is easier to remember the formula by thinking of the tree diagram that produced it. Go through the following steps.

USING BAYES' FORMULA

1. Start a tree diagram with branches representing events $F_1, F_2, \cdots,$ F_n. Label each branch with its corresponding probability.

2. From the end of each of these branches, draw a branch for event E. Label this branch with the probability of getting to it, $P(E|F_i)$.

3. You now have n different paths that result in event E. Next to each path, put its probability—the product of the probabilities that the first branch occurs, $P(F_i)$, and that the second branch occurs, $P(E|F_i)$; that is, the product $P(F_i) \cdot P(E|F_i)$.

4. The desired probability is given by the probability of the branch you want, divided by the sum of the probabilities of all the branches producing event E.

EXAMPLE 2

Based on past experience, a company knows that an experienced machine operator (one or more years of experience) will produce a defective item 1% of the time. Operators with some experience (up to one year) have a 2.5% defect rate, and new operators have a 6% defect rate. At any one time, the company has 60% experienced operators, 30% with some experience, and 10% new operators. Find the probability that a particular defective item was produced by a new operator.

Let E represent the event "item is defective" F_1 represent "item was made by an experienced operator," F_2 "item was made by an operator with some experience," and F_3 "item was made by a new operator." Then

$$P(F_1) = .60 \qquad P(E|F_1) = .01$$
$$P(F_2) = .30 \qquad P(E|F_2) = .025$$
$$P(F_3) = .10 \qquad P(E|F_3) = .06.$$

We need to find $P(F_3|E)$, the probability that an item was produced by a new operator, given that it is defective. First, draw a tree diagram using the given information, as in Figure 16. The steps leading to event E are shown in heavy type.

Find $P(F_3|E)$ with the bottom branch of the tree in Figure 16; divide the probability for this branch by the sum of the probabilities of all the branches leading to E, or

$$P(F_3|E) = \frac{.10(.06)}{.60(.01) + .30(.025) + .10(.06)} = \frac{.006}{.0195} = \frac{4}{13}.$$

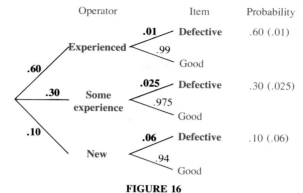

| Operator | | Item | Probability |

FIGURE 16

In a similar way,

$$P(F_2|E) = \frac{.30(.025)}{.60(.01) + .30(.025) + .10(.06)} = \frac{.0075}{.0195} = \frac{5}{13}.$$

Finally, $P(F_1|E) = 4/13$. Check that $P(F_1|E) + P(F_2|E) + P(F_3|E) = 1$. (That is, the defective item was made by *someone*.)   ▬

▬ EXAMPLE 3

A manufacturer buys items from 6 different suppliers. The fraction of the total number of items obtained from each supplier, along with the probability that an item purchased from that supplier is defective, are shown in the following chart.

Supplier	Fraction of Total Supplied	Probability of Defect
1	.05	.04
2	.12	.02
3	.16	.07
4	.23	.01
5	.35	.03
6	.09	.05

Find the probability that a defective item came from supplier 5.

Let F_1 be the event that an item came from supplier 1, with F_2, F_3, F_4, F_5, and F_6 defined in a similar manner. Let E be the event that an item is defective. We want to find $P(F_5|E)$. By Bayes' formula (draw a tree),

$$P(F_5|E) = \frac{(.35)(.03)}{(.05)(.04) + (.12)(.02) + (.16)(.07) + (.23)(.01) + (.35)(.03) + (.09)(.05)}$$

$$= \frac{.0105}{.0329} \approx .319.$$

There is about a 32% chance that a defective item came from supplier 5.   ▬

6.4 EXERCISES

For two events M and N, $P(M) = .4$, $P(N|M) = .3$, and $P(N|M') = .4$. Find each of the following.

1. $P(M|N)$

2. $P(M'|N)$

For mutually exclusive events R_1, R_2, R_3, we have $P(R_1) = .05$, $P(R_2) = .6$, and $P(R_3) = .35$. Also, $P(Q|R_1) = .40$, $P(Q|R_2) = .30$, and $P(Q|R_3) = .60$. Find each of the following.

3. $P(R_1|Q)$

4. $P(R_2|Q)$

5. $P(R_3|Q)$

6. $P(R_1'|Q)$

Suppose you have three jars with the following contents: 2 black balls and 1 white ball in the first, 1 black ball and 2 white balls in the second, and 1 black ball and 1 white ball in the third. One jar is to be selected, and then 1 ball is to be drawn from the selected jar. If the probabilities of selecting the first, second, or third jar are 1/2, 1/3, and 1/6 respectively, find the probabilities that if a white ball is drawn, it came from the following jars.

7. The second jar

8. The third jar

Let F, G, and H be nonempty events with $F \cap G = \emptyset$, $F \cap H = \emptyset$, and $G \cap H = \emptyset$. Let $S = F \cup G \cup H$. Let E be any event. Prove each of the following.

9. $P(E) = P(E \cap F) + P(E \cap G) + P(E \cap H)$

10. $P(E) = P(E|F) \cdot P(F) + P(E|G) \cdot P(G) + P(E|H) \cdot P(H)$

APPLICATIONS

BUSINESS AND ECONOMICS

Airline Usage *The following table shows the fraction of the population at various income levels, as well as the probability that a person from that income level will take an airline flight within the next year.*

Income Level	Proportion of Population	Probability of a Flight During the Next Year
$0–$5999	12.8%	.04
$6,000–$9999	14.6%	.06
$10,000–$14,999	18.5%	.07
$15,000–$19,999	17.8%	.09
$20,000–$24,999	13.9%	.12
$25,000 and over	22.4%	.13

If a person is selected at random from an airline flight, find the probability that the person has the given income level.

11. $10,000–$14,999

12. $25,000 and over

Job Qualifications   *Of all the people applying for a certain job, 70% are qualified, and 30% are not. The personnel manager claims that she approves qualified people 85% of the time; she approves an unqualified person 20% of the time. Find each of the following probabilities.*

13. A person is qualified if he or she was approved by the manager.

14. A person is unqualified if he or she was approved by the manager.

Quality Control   *A building contractor buys 70% of his cement from supplier A, and 30% from supplier B. A total of 90% of the bags from A arrive undamaged, while 95% of the bags from B arrive undamaged. Give the probabilities that a damaged bag is from each of the following sources.*

15. Supplier A   **16.** Supplier B

Credit   *The probability that a customer of a local department store will be a "slow pay" is .02. The probability that a "slow pay" will make a large down payment when buying a refrigerator is .14. The probability that a person who is not a "slow pay" will make a large down payment when buying a refrigerator is .50. Suppose a customer makes a large down payment on a refrigerator. Find the probabilities that the customer is in the following categories.*

17. A "slow pay"   **18.** Not a "slow pay"

Appliance Reliability   *Companies A, B, and C produce 15%, 40%, and 45%, respectively, of the major appliances sold in a certain area. In that area, 1% of the Company A appliances, 1 1/2% of the Company B appliances, and 2% of the Company C appliances need service within the first year. Suppose a defective appliance is chosen at random; find the probabilities that it was manufactured by the following companies.*

19. Company A   **20.** Company B

Television Advertising   *On a given weekend in the fall, a tire company can buy television advertising time for a college football game, a baseball game, or a professional football game. If the company sponsors the college game, there is a 70% chance of a high rating, a 50% chance if they sponsor a baseball game, and a 60% chance if they sponsor a professional football game. The probability of the company sponsoring these various games is .5, .2, and .3, respectively. Suppose the company does get a high rating; find the probabilities that it sponsored the following.*

21. A college game   **22.** A professional football game

Economy versus Investments   *According to readings in business publications, there is a 50% chance of a booming economy next summer, a 20% chance of a mediocre economy, and a 30% chance of a recession. The probabilities that a particular investment strategy will produce a huge profit under each of these possibilities are .1, .6, and .3, respectively. Suppose the strategy does produce huge profits; find the probabilities that the economy was in the following conditions.*

23. Booming   **24.** In recession

LIFE SCIENCES

Hepatitis Blood Test   **25.** The probability that a person with certain symptoms has hepatitis is .8. The blood test used to confirm this diagnosis gives positive results for 90% of people with the disease and 5% of those without the disease. What is the probability that an individual who has the symptoms and who reacts positively to the test actually has hepatitis?

Toxemia Test *A magazine article described a new test for toxemia, a disease that affects pregnant women. To perform the test, the woman lies on her left side and then rolls over on her back. The test is considered positive if there is a 20 mm rise in her blood pressure within one minute. The article gives the following probabilities, where T represents having toxemia at some time during the pregnancy, and N represents a negative test:*

$$P(T'|N) = .90 \quad \text{and} \quad P(T|N') = .75.$$

Assume that $P(N') = .11$, and find each of the following.

26. $P(N|T)$

27. $P(N'|T)$

SOCIAL SCIENCES

Political Affiliation **28.** In a certain county, the Democrats have 53% of the registered voters, 12% of whom are under 21. The Republicans have 47% of all registered voters, of whom 10% are under 21. If Kay is a registered voter who is under 21, what is the probability that she is a Democrat?

Voting *The following table shows the proportion of people over 18 who are in various age categories, along with the probability that a person in a given age category will vote in a general election.*

Age	Proportion of Voting Age Population	Probability of a Person of This Age Voting
18–21	11.0%	.48
22–24	7.6%	.53
25–44	37.6%	.68
45–64	28.3%	.64
65 or over	15.5%	.74

Suppose a voter is picked at random. Find the probabilities that the voter is in the following age categories.

29. 18–21

30. 65 or over

6.5 APPLICATIONS OF COUNTING

Chapter 5 discussed permutations and combinations—ways of counting numbers of outcomes for various kinds of experiments. In this section, methods of counting are used to help solve problems in probability. The problems worked with tree diagrams in earlier sections often can be worked with combinations instead. Combinations are used to solve probability problems involving two or more *dependent* events. In the next section, *Bernoulli trials* will be used for probability problems involving two or more *independent* events.

The solution to problems worked with combinations depends on the basic probability principle stated in Section 6.1:

> Let S be a sample space with n equally likely outcomes. Let event E contain m of these outcomes. Then the probability that event E occurs, written $P(E)$, is
>
> $$P(E) = \frac{m}{n}.$$

To compare the method of using combinations with the method of tree diagrams used in Section 6.3, the first example repeats Example 6 from that section.

▬ EXAMPLE 1

From a box containing 3 white, 2 green, and 1 red marble, 2 marbles are drawn one at a time without replacement. Find the probability that one white and one green marble are drawn.

In Example 6 of Section 6.3, it was necessary to consider the order in which the marbles were drawn. With combinations, it is not necessary. Simply count the number of ways in which 1 white and 1 green marble can be drawn from the given selection. The white marble can be drawn from the 3 white marbles in $\binom{3}{1}$ ways, and the green marble can be drawn from the 2 green marbles in $\binom{2}{1}$ ways. By the multiplication principle, both results can occur in

$$\binom{3}{1} \cdot \binom{2}{1} \text{ ways,}$$

giving the numerator of the probability fraction, $P(E) = m/n$. For the denominator, 2 marbles are to be drawn from a total of 6 marbles. This can occur in $\binom{6}{2}$ ways. The required probability is

$$P(1 \text{ white and 1 green}) = \frac{\binom{3}{1}\binom{2}{1}}{\binom{6}{2}} = \frac{\dfrac{3!}{2!1!} \cdot \dfrac{2!}{1!1!}}{\dfrac{6!}{4!2!}} = \frac{6}{15} = \frac{2}{5}.$$

This agrees with the answer found earlier. ▬

▬ EXAMPLE 2

From a group of 22 nurses, 4 are to be selected to present a list of grievances to management.

(a) One of the nurses is Jill Streitsel; the group agrees that she must be one of the 4 people chosen. Find the probability that Streitsel will be among the 4 chosen.

The probability that Streitsel will be selected is given by m/n, where m is the number of ways the chosen group includes her, and n is the total number of ways the group of 4 can be chosen. If Streitsel must be one of the 4 people, the problem reduces to finding the number of ways that the 3 additional nurses can be chosen. The 3 are chosen from 21 nurses; this can be done in

$$\binom{21}{3} = \frac{21!}{3!18!} = 1330$$

ways so $m = 1330$. Since n is the number of ways 4 nurses can be selected from 22,

$$n = \binom{22}{4} = 7315.$$

The probability that Streitsel will be one of the 4 chosen is

$$P(\text{Streitsel is chosen}) = \frac{1330}{7315} \approx .182.$$

(b) Find the probability that Streitsel will not be selected.

The probability that she will not be chosen is $1 - .182 = .818.$ ▬

▬ EXAMPLE 3

When shipping diesel engines abroad, it is common to pack 12 engines in one container that is then loaded on a rail car and sent to a port. Suppose that a company has received complaints from its customers that many of the engines arrive in nonworking condition. To help solve this problem, the company decides to make a spot check of containers after loading—the company will test 3 engines from a container at random; if any of the 3 are nonworking, the container will not be shipped until each engine in it is checked. Suppose a given container has 2 nonworking engines. Find the probability that the container will not be shipped.

The container will not be shipped if the sample of 3 engines contains 1 or 2 defective engines. If $P(1 \text{ defective})$ represents the probability of exactly 1 defective engine in the sample, then

$$P(\text{not shipping}) = P(1 \text{ defective}) + P(2 \text{ defectives}).$$

There are $\binom{12}{3}$ ways to choose the 3 engines for testing:

$$\binom{12}{3} = \frac{12!}{3!9!} = \frac{12(11)(10)}{3(2)(1)} = 220.$$

There are $\binom{2}{1}$ ways of choosing 1 defective engine from the 2 in the container, and for each of these ways, there are $\binom{10}{2}$ ways of choosing 2 good engines from among the 10 in the container. This makes

$$\binom{2}{1}\binom{10}{2} = \frac{2!}{1!1!} \cdot \frac{10!}{2!8!} = 2(45) = 90$$

ways of choosing a sample of 3 engines containing 1 defective, with

$$P(1 \text{ defective}) = \frac{90}{220}.$$

There are $\binom{2}{2}$ ways of choosing 2 defective engines from the 2 defective engines in the container, and $\binom{10}{1}$ ways of choosing 1 good engine from among the 10 good engines, for

$$\binom{2}{2}\binom{10}{1} = \frac{2!}{2!0!} \cdot \frac{10!}{1!9!} = 1(10) = 10$$

ways of choosing a sample of 3 engines containing 2 defectives. Finally,

$$P(2 \text{ defectives}) = \frac{10}{220}$$

and

$$P(\text{not shipping}) = P(1 \text{ defective}) + P(2 \text{ defectives})$$
$$= \frac{90}{220} + \frac{10}{220} = \frac{100}{220} \approx .455.$$

Notice that the probability is $1 - .455 = .545$ that the container will be shipped, even though it has 2 defective engines. The management must decide whether this probability is acceptable; if not, it may be necessary to test more than 3 engines from a container. ▬

Instead of finding the sum $P(1 \text{ defective}) + P(2 \text{ defectives})$, the result in Example 3 could be found as $1 - P(0 \text{ defectives})$.

$$P(\text{not shipping}) = 1 - P(0 \text{ defectives in sample})$$

$$= 1 - \frac{\binom{2}{0}\binom{10}{3}}{\binom{12}{3}}$$

$$= 1 - \frac{1(120)}{220}$$

$$= 1 - \frac{120}{220} = \frac{100}{220} \approx .455$$

■ EXAMPLE 4

In a common form of the card game *poker,* a hand of 5 cards is dealt to each player from a deck of 52 cards. There are a total of

$$\binom{52}{5} = \frac{52!}{5!47!} = 2{,}598{,}960$$

such hands possible. Find the probability of getting each of the following hands.

(a) A hand containing only hearts, called a *heart flush*

There are 13 hearts in a deck, with

$$\binom{13}{5} = \frac{13!}{5!8!} = \frac{13(12)(11)(10)(9)}{5(4)(3)(2)(1)} = 1287$$

different hands containing only hearts. The probability of a heart flush is

$$P(\text{heart flush}) = \frac{1287}{2{,}598{,}960} \approx .000495.$$

(b) A flush of any suit (5 cards of the same suit)

There are 4 suits in a deck, so

$$P(\text{flush}) = 4 \cdot P(\text{heart flush}) = 4 \cdot .000495 \approx .00198.$$

(c) A full house of aces and eights (3 aces and 2 eights)

There are $\binom{4}{3}$ ways to choose 3 aces from among the 4 in the deck, and $\binom{4}{2}$ ways to choose 2 eights.

$$P(\text{3 aces, 2 eights}) = \frac{\binom{4}{3} \cdot \binom{4}{2}}{2{,}598{,}960} = \frac{4 \cdot 6}{2{,}598{,}960} \approx .00000923.$$

(d) Any full house (3 cards of one value, 2 of another)

The 13 values in a deck give 13 choices for the first value, leaving 12 choices for the second value (order *is* important here, since a full house of aces and eights is not the same as a full house of eights and aces). From part (c), the probability for any *particular* full house is .00000923; the probability of *any* full house is

$$P(\text{full house}) = 13 \cdot 12 \cdot .00000923 \approx .00144. \quad ■$$

EXAMPLE 5

Suppose a group of n people is in a room. Find the probability that at least 2 of the people have the same birthday.

Here we refer to the month and the day, not necessarily the same year. Also, ignore leap years, and assume that each day in the year is equally likely as a birthday. Let us first find the probability that *no 2 people* among 5 people have the same birthday. There are 365 different birthdays possible for the first of the 5 people, 364 for the second (so that the people have different birthdays), 363 for the third, and so on. The number of ways the 5 people can have different birthdays is the number of permutations of 365 things (days) taken 5 at a time, or

$$P(365, 5) = 365 \cdot 364 \cdot 363 \cdot 362 \cdot 361.$$

The number of ways that 5 people can have the same or different birthdays is

$$365 \cdot 365 \cdot 365 \cdot 365 \cdot 365 = (365)^5.$$

Finally, the *probability* that none of the 5 people have the same birthday is

$$\frac{P(365, 5)}{(365)^5} = \frac{365 \cdot 364 \cdot 363 \cdot 362 \cdot 361}{365 \cdot 365 \cdot 365 \cdot 365 \cdot 365} \approx .973.$$

The probability that at least 2 of the 5 people *do* have the same birthday is $1 - .973 = .027$.

This result can be extended to more than 5 people. Generalizing, the probability that no 2 people among n people have the same birthday is

$$\frac{P(365, n)}{(365)^n}.$$

The probability that at least 2 of the n people *do* have the same birthday is

$$1 - \frac{P(365, n)}{(365)^n}.$$

The following table shows this probability for various values of n.

Number of People, n	Probability that Two Have the Same Birthday
5	.027
10	.117
15	.253
20	.411
22	.476
23	.507
25	.569
30	.706
35	.814
40	.891
50	.970
365	1

The probability that 2 people among 23 have the same birthday is .507, a little more than half. Many people are surprised at this result—it seems that a larger number of people should be required. ▬

▬ 6.5 EXERCISES

A shipment of 9 typewriters contains 2 defectives. Find the probabilities that a sample of each of the following sizes, drawn from the 9, will not contain a defective.

1. 1   **2.** 2   **3.** 3   **4.** 4

Refer to Example 3. The management feels that the probability of .545 that a container will be shipped even though it contains 2 defectives is too high. They decide to increase the sample size chosen. Find the probabilities that a container will be shipped even though it contains 2 defectives if the sample size is increased to the following.

5. 4   **6.** 5

A basket contains 6 red apples and 4 yellow apples. A sample of 3 apples is drawn. Find the probabilities that the sample contains the following.

7. All red apples   **8.** All yellow apples

9. 2 yellow and 1 red apple   **10.** More red than yellow apples

Two cards are drawn at random from an ordinary deck of 52 cards.

11. How many 2-card hands are possible?

Find the probability that the 2-card hand described above contains the following.

12. 2 aces   **13.** At least 1 ace

14. All spades   **15.** 2 cards of the same suit

16. Only face cards   **17.** No face cards

18. No card higher than 8 (Count ace as 1.)

Twenty-six slips of paper are each marked with a different letter of the alphabet and placed in a basket. A slip is pulled out, its letter recorded (in the order in which the slip was drawn), and the slip is replaced. This is done 5 times. Find the probabilities that the following "words" are formed.

19. "chuck"   **20.** A word that starts with p

21. A word with no repetition of letters   **22.** A word that contains no x, y, or z

Find the probabilities of the following hands at poker. Assume aces are either high or low.

23. Royal flush (5 highest cards of a single suit)

24. Straight flush (5 in a row in a single suit, but not a royal flush)

25. Four of a kind (4 cards of the same value)

26. Straight (5 cards in a row, not all of the same suit) with ace either high or low

A bridge hand is made up of 13 *cards from a deck of* 52. *Set up the probability that a hand chosen at random contains the following.*

27. Only hearts

28. 4 aces

29. Exactly 3 aces and exactly 3 kings

30. 6 of one suit, 5 of another, and 2 of another

31. At a conference of black writers in Detroit, special-edition books were selected to be given away in contests. There were 9 books written by Langston Hughes, 5 books by James Baldwin, and 7 books by Toni Morrison. The judge of one contest selected 6 books at random for prizes. Find the probabilities that the selection consisted of

(a) 3 Hughes and 3 Morrison books

(b) Exactly 4 Baldwin books

(c) 2 Hughes, 3 Baldwin, and 1 Morrison book

(d) At least 4 Hughes books

(e) Exactly 4 books written by males (Morrison is female.)

(f) No more than 2 books written by Baldwin

32. At the first meeting of a committee to plan a Northern California pow-wow, there were 3 women and 3 men from the Miwok tribe, 2 men and 3 women from the Hoopa tribe, and 4 women and 5 men from the Pomo tribe. If the ceremony subcouncil consists of 5 people, and is randomly selected, find the probabilities that the subcouncil contains the following.

(a) 3 men and 2 women

(b) Exactly 3 Miwoks and 2 Pomos

(c) 2 Miwoks, 2 Hoopas, and a Pomo

(d) 2 Miwoks, 2 Hoopas, and 2 Pomos

(e) More women than men

(f) Exactly 3 Hoopas

(g) At least 2 Pomos

33. Set up the probability that at least 2 of the 41 men who have served as president of the United States have had the same birthday.

34. Estimate the probability that at least 2 of the 100 U.S. Senators have the same birthday.

35. Give the probability that 2 of the 435 members of the House of Representatives have the same birthday.

36. Show that the probability that in a group of n people *exactly one* pair have the same birthday is

$$\binom{n}{2} \cdot \frac{P(365, n - 1)}{(365)^n}.$$

37. An elevator has 4 passengers and stops at 7 floors. It is equally likely that a person will get off at any one of the 7 floors. Find the probability that no 2 passengers leave at the same floor.

Exercises 38–44 involve the idea of a circular permutation: *the number of ways of arranging distinct objects in a circle. The number of ways of arranging n distinct objects in a line is n!, but there are fewer ways for arranging the n items in a circle since the first item could be placed in any of n locations.*

38. Show that the number of ways of arranging n distinct items in a circle is $(n - 1)!$.

Find the numbers of ways of arranging the following numbers of distinct items in a circle.

39. 4   **40.** 7   **41.** 10

42. Suppose that 8 people sit at a circular table. Find the probability that 2 selected people are sitting next to each other.

43. A key ring contains 7 keys; 1 black, 1 gold, and 5 silver. If the keys are arranged at random on the ring, find the probability that the black key is next to the gold key.

44. A circular table for a board of directors has 10 seats for the 10 attending members of the board. The chairman of the board always sits closest to the window. The vice president for sales, who is currently out of favor, will sit 3 positions to the chairman's left, since the chairman doesn't see well out of his left eye. The chairman's daughter-in-law will sit opposite him. All other members take seats at random. Find the probability that a selected other member will sit next to the chairman.

45. Rework Exercises 23–26 on a computer using the Monte Carlo method to approximate the answers with $n = 25$. Since each hand has 5 cards, you will need $25 \cdot 5 = 125$ random numbers to "look at" 25 hands. Compare these experimental results with the theoretical results.

46. Rework Exercises 27–30 on a computer using the Monte Carlo method to approximate the answers with $n = 20$. Since each hand has 13 cards, you will need $20 \cdot 13 = 260$ random numbers to "look at" 20 hands.

47. Use a computer to evaluate the answer for Exercise 33.

6.6 BERNOULLI TRIALS

Many probability problems are concerned with experiments in which an event is repeated many times. Some examples include finding the probability of getting 7 heads in 8 tosses of a coin, of hitting a target 6 times out of 6, and of finding 1 defective item in a sample of 15 items. Probability problems of this kind are called **repeated trials** problems, or **Bernoulli processes,** named after the Swiss mathematician Jakob Bernoulli (1654–1705), who is well known for his work in probability theory. In each case, some outcome is designated a success, and any other outcome is considered a failure. Thus, if the probability of a success in a single trial is p, the probability of failure will be $1 - p$. Repeated trials problems, also called **binomial trials,** must satisfy the following conditions.

BERNOULLI TRIALS

1. The same experiment is repeated several times.

2. There are only two possible outcomes, success and failure.

3. The repeated trials are independent, so that the probability of each outcome remains the same for each trial.

Consider a typical problem of this type: find the probability of getting five 1's on five rolls of a die. Here, the experiment, rolling a die, is repeated five times. If getting a 1 is designated as a success, then getting any other outcome is a failure. The five trials are independent; the probability of success (getting a 1) is 1/6 on each trial, while the probability of a failure (any other result) is 5/6. Thus, the required probability is

$$P(\text{five 1's on five rolls}) = P(1) \cdot P(1) \cdot P(1) \cdot P(1) \cdot P(1) = \left(\frac{1}{6}\right)^5$$

$$\approx .00013.$$

Now suppose the problem is changed to that of finding the probability of getting a 1 exactly four times in five rolls of the die. Again, a success on any roll of the die is defined as getting a 1, and the trials are independent, with the same probability of success each time. The desired outcome for this experiment can occur in more than one way, as shown below, where s represents a success (getting a 1), and f represents a failure (getting any other result), and each row represents a different outcome.

$$
\begin{array}{ccccc}
s & s & s & s & f \\
s & s & s & f & s \\
s & s & f & s & s \\
s & f & s & s & s \\
f & s & s & s & s
\end{array}
$$

The probability of each of these five outcomes is

$$\left(\frac{1}{6}\right)^4\left(\frac{5}{6}\right).$$

Since the five outcomes represent mutually exclusive events, add the probabilities to get

$$P(\text{four 1's in five rolls}) = 5\left(\frac{1}{6}\right)^4\left(\frac{5}{6}\right) = \frac{5^2}{6^5} \approx .0032$$

In the same way, we can compute the probability of rolling a 1 exactly three times in five rolls of a die. The probability of any one way of achieving 3 successes and 2 failures will be

$$\left(\frac{1}{6}\right)^3\left(\frac{5}{6}\right)^2.$$

Again, the desired outcome can occur in more than one way. To see this, let the set $\{1, 2, 3, 4, 5\}$ represent the first, second, third, fourth, and fifth tosses. The number of 3-element subsets of this set will correspond to the number of ways in which 3 successes and 2 failures can occur. Using combinations shows that there are $\binom{5}{3}$ such subsets. Since $\binom{5}{3} = 5!/(3!2!) = 10$,

$$P(\text{three 1's in five rolls}) = 10\left(\frac{1}{6}\right)^3\left(\frac{5}{6}\right)^2 = \frac{250}{6^5} \approx .032.$$

Suppose now that the probability of a success on one trial of a Bernoulli experiment is p, and the probability of exactly x successes in n repeated trials is needed. It is possible that the x successes could come first, followed by $n - x$ failures:

$$\underbrace{s\ s\ s\ \cdots\ s}_{x \text{ successes, then}}\ \ \underbrace{f f \cdots f.}_{n\ -\ x \text{ failures}} \tag{1}$$

The probability of this result is

$$P(s\ s\ s\ \cdots\ s\ s\ f f \cdots f)$$

$$= \underbrace{P(s) \cdot P(s) \cdot P(s) \cdots P(s)}_{x \text{ factors}} \cdot \underbrace{P(f) \cdot P(f) \cdots P(f)}_{n\ -\ x \text{ factors}}$$

$$= \underbrace{p \cdot p \cdot p \cdots p}_{x \text{ factors}} \cdot \underbrace{(1 - p) \cdot (1 - p) \cdots (1 - p)}_{n\ -\ x \text{ factors}}$$

$$= p^x(1 - p)^{n-x}.$$

The x successes could also be obtained by rearranging the letters in (1) above. There are $\binom{n}{x}$ ways of choosing the x places where the s's occur and the $n - x$ places where the f's occur, so the probability of exactly x successes in n trials is

$$\binom{n}{x} p^x(1 - p)^{n-x}.$$

A summary follows.

PROBABILITY IN A BERNOULLI EXPERIMENT

If p is the probability of success in a single trial of a Bernoulli experiment, the probability of x successes and $n - x$ failures in n independent repeated trials of the experiment is

$$\binom{n}{x} \cdot p^x \cdot (1 - p)^{n-x}.$$

■ EXAMPLE 1

The advertising agency that handles the Diet Supercola account believes that 40% of all consumers prefer this product over its competitors. Suppose a sample of 6 people is chosen. Assume that all responses are independent of each other. Find the probability of the following.

(a) Exactly 3 of the 6 people prefer Diet Supercola.

Think of the 6 responses as 6 independent trials. A success occurs if a person prefers Diet Supercola. Then this is a Bernoulli trials problem. In this example, P (success) $=$ P(prefer Diet Supercola) $=$ .4. The sample is made up of 6 people, so $n = 6$. To find the probability that exactly 3 people prefer this drink, let $x = 3$ and use the result in the box.

$$P(\text{exactly } 3) = \binom{6}{3}(.4)^3(1 - .4)^{6-3}$$
$$= 20(.4)^3(.6)^3$$
$$= 20(.064)(.216)$$
$$= .27648$$

(b) None of the 6 people prefer Diet Supercola.

Let $x = 0$.

$$P(\text{exactly } 0) = \binom{6}{0}(.4)^0(1 - .4)^6 = 1(1)(.6)^6 \approx .0467 \quad \blacksquare$$

■ EXAMPLE 2

At a certain school in northern Michigan, 80% of the students ski. Suppose 5 students at this school are selected, and each is asked if he or she skis. Assume that their responses are independent. Let success be a yes response, so that the probability of success on each of the 5 trials is .8. Since this is a Bernoulli trials problem, the probability that exactly 1 of the 5 students skis is

$$P(\text{exactly } 1) = \binom{5}{1}(.8)^1(.2)^4 = .0064,$$

while the probability that exactly 4 of the 5 students ski is

$$P(\text{exactly } 4) = \binom{5}{4}(.8)^4(.2)^1 = .4096. \quad \blacksquare$$

■ EXAMPLE 3

Find each of the following probabilities.

(a) The probability of getting exactly 7 heads in 8 tosses of a fair coin

Repeated tosses of a coin are independent events. Use the formula for Bernoulli trials. The probability of success (getting a head in a single toss) is 1/2. The probability of a failure (getting a tail) is $1 - 1/2 = 1/2$. Thus,

$$P(7 \text{ heads in } 8 \text{ tosses}) = \binom{8}{7}\left(\frac{1}{2}\right)^7\left(\frac{1}{2}\right)^1 = 8\left(\frac{1}{2}\right)^8 = .03125.$$

(b) The probability of two 4's in eight rolls of a die

Successive rolls of a die are independent events, so the Bernoulli trials formula can be used. If getting a 4 is designated a success, the probability of success is 1/6, while the probability of failure (a number other than 4), is 5/6.

$$P(\text{two 4's in eight rolls}) = \binom{8}{2}\left(\frac{1}{6}\right)^2\left(\frac{5}{6}\right)^6 \approx .2605 \quad \blacksquare$$

■ EXAMPLE 4

Assuming that selection of items for a sample can be treated as independent trials, and that the probability that any 1 item is defective is .01, find the following.

(a) The probability of 1 defective item in a random sample of 15 items from a production line

Here, a "success" is a defective item. Since selecting each item for the sample is assumed to be an independent trial, the Bernoulli trials formula applies. The probability of success (a defective item), is .01, while the probability of failure (an acceptable item) is .99. This makes

$$P(\text{1 defective in 15 items}) = \binom{15}{1}(.01)^1(.99)^{14}$$
$$= 15(.01)(.99)^{14}$$
$$\approx .130.$$

(b) The probability of at most 1 defective item in a random sample of 15 items from a production line

"At most 1" means 0 defective items or 1 defective item. Since 0 defective items is equivalent to 15 acceptable items,

$$P(\text{0 defective}) = (.99)^{15} \approx .860.$$

Use the union rule, noting that 0 defective and 1 defective are mutually exclusive events, to get

$$P(\text{at most 1 defective}) = P(\text{0 defective}) + P(\text{1 defective})$$
$$\approx .860 + .130$$
$$= .990. \quad \blacksquare$$

■ EXAMPLE 5

A new style of shoe is sweeping the country. In one area, 30% of all the shoes are of this type. Assume that these sales are independent events, and find the following probabilities using Bernoulli trials.

(a) Of 10 people who buy shoes, at least 8 buy the new shoe style.

Let success be "buy the new style," so that P(success) = .3. For at least 8 people out of 10 to buy the shoe, it must be sold to 8, 9, or 10 people, with

$$P(\text{at least } 8) = P(8) + P(9) + P(10)$$

$$= \binom{10}{8}(.3)^8(.7)^2 + \binom{10}{9}(.3)^9(.7)^1 + \binom{10}{10}(.3)^{10}(.7)^0$$

$$\approx .0014467 + .0001378 + .0000059$$

$$= .0015904.$$

(b) Of 10 people who buy shoes, no more than 7 buy the new shoe style.

"No more than 7" means 0, 1, 2, 3, 4, 5, 6, or 7 people buy the shoe. The required probability could be found by adding $P(0)$, $P(1)$, and so on, but it is easier to use the formula $P(E) = 1 - P(E')$. The complement of "no more than 7" is "8 or more" or "at least 8." Finally,

$$P(\text{no more than } 7) = 1 - P(\text{at least } 8)$$

$$= 1 - .0015904 \quad \text{Answer from part (a)}$$

$$= .9984096. \ \blacksquare$$

The Probability of k Trials for m Successes   In the rest of this section we show how to find the probability that k trials will be needed to guarantee m successes in a Bernoulli experiment. As an example, suppose that a salesperson in a very competitive business makes a sale in 1 client visit out of 5, so P(sale) = .2. The probability of a sale on the first call is .2. The probability that the *first* sale will be made on the *second* call is

$$P(\text{no sale on first}) \cdot P(\text{sale on second}) = .8(.2) = .16.$$

The probability that the *first* sale will be made on the *third* call is

$$(.8)^2(.2) = .128.$$

Generalizing, the probability that the first sale will be made on the kth call is

$$(.8)^{k-1}(.2).$$

▬ EXAMPLE 6

How many calls must this salesperson make to have an 80% chance of making a sale?

There is a .2 chance of making a sale on the first call, a .16 chance of making the first sale on the second call, a .128 chance of making the first sale on the third call, and so on. The probability of a sale by the kth call is the sum of all the probabilities of sales on calls 1, 2, 3, $\cdots$, k. A calculator gives the results in the following table.

Call Number	Probability That First Sale Is Made on This Call	Total of All Probabilities Up to and Including This Call
1	$(.8)^0(.2) = .2$	.2
2	$(.8)^1(.2) = .16$	.36
3	$(.8)^2(.2) = .128$	.488
4	$(.8)^3(.2) = .1024$	.5904
5	$(.8)^4(.2) \approx .082$	$\approx .672$
6	$(.8)^5(.2) \approx .066$	$\approx .738$
7	$(.8)^6(.2) \approx .052$	$\approx .790$
8	$(.8)^7(.2) \approx .042$	$\approx .832$

The salesperson must make 8 calls in order to have an 80% chance of making 1 sale. ▅

This result can be generalized: let p be the probability of success on 1 trial in a Bernoulli experiment. Then to find the probability that k trials will be needed to guarantee m successes, we assume the kth trial was a success and that $m - 1$ successes were distributed in some order among the other $k - 1$ trials. The desired probability is thus

$$\left[\binom{k-1}{m-1} p^{(m-1)} \cdot (1-p)^{(k-1)-(m-1)} \right] \cdot p$$

or

$$\binom{k-1}{m-1} p^m \cdot (1-p)^{k-m}.$$

▅ EXAMPLE 7

Find the probability that the salesperson in Example 6 will require 9 calls to make 3 sales.

Let $k = 9$ and $m = 3$, with $p = .2$. The desired probability is

$$\binom{9-1}{3-1}(.2)^3(1 - .2)^{9-3} = \binom{8}{2}(.2)^3(.8)^6 \approx .0587. \quad ▅$$

▬ 6.6 EXERCISES

Suppose that a family has 5 children. Also, suppose that the probability of having a girl is 1/2. Find the probabilities that the family will have the following children.

1. Exactly 2 girls and 3 boys

2. Exactly 3 girls and 2 boys

3. No girls

4. No boys

5. At least 4 girls

6. At least 3 boys

7. No more than 3 boys

8. No more than 4 girls

A die is rolled 12 *times. Find the probabilities of rolling the following.*

9. Exactly twelve 1's   **10.** Exactly six 1's   **11.** Exactly one 1

12. Exactly two 1's   **13.** No more than three 1's   **14.** No more than one 1

A coin is tossed 5 *times. Find the probabilities of getting the following.*

15. All heads   **16.** Exactly 3 heads

17. No more than 3 heads   **18.** At least 3 heads

Find the probabilities that the following numbers of tosses of a fair coin will be required to obtain 3 *heads.*

19. 6   **20.** 5   **21.** 10   **22.** 8

Find the probability that the following numbers of rolls of a fair die will be required to get four 5's.

23. 10   **24.** 6   **25.** 16   **26.** 12

27. Suppose we find the probability of r successes out of n trials for a Bernoulli experiment having probability p. Show that the result is the same as for the probability of $n - r$ successes out of n trials for a Bernoulli experiment having probability $1 - p$.

APPLICATIONS

BUSINESS AND ECONOMICS

Fast Food Industry   *According to a recent article in a business publication, only* 20% *of the population of the United States has never had a Big Burg hamburger at a major fast-food chain. Assume independence and find the probabilities that a random sample of* 10 *people produces the following data.*

28. Exactly 2 people have never had a Big Burg.

29. Exactly 5 people have never had a Big Burg.

30. Three people or fewer have never had a Big Burg.

31. Four people or more *have* had a Big Burg.

New Car Purchases   *An economist feels that the probability that a person at a certain income level will buy a new car this year is* .2. *Find the probabilities that among* 12 *such people, the following numbers will buy new cars.*

32. Exactly 4   **33.** Exactly 6   **34.** No more than 3   **35.** At least 3

Mining Exploration   *The probability that a given exploration team sent out by a mining company will find commercial quantities of iron ore is* .15. *How many such teams must the company send out to have the following probabilities of finding ore?*

36. 60%   **37.** 75%   **38.** 80%

Personnel Screening   *A company gives prospective workers a 6-question multiple-choice test. Each question has 5 possible answers, so that there is a 1/5 or 20% chance of answering a question correctly just by guessing. Find the probabilities of getting the following results by chance.*

39. Exactly 2 correct answers

40. No correct answers

41. At least 4 correct answers

42. No more than 3 correct answers

Customer Satisfaction   **43.** Over the last decade, 10% of all clients of J. K. Loss & Company have lost their life savings. Suppose a sample of 3 of the current clients of the firm is chosen. Assuming independence, find the probability that exactly one of the 3 clients will lose everything.

Quality Control   *A factory tests a random sample of 20 transistors for defectives. The probability that a particular transistor will be defective has been established by past experience to be .05.*

44. What is the probability that there are no defectives in the sample?

45. What is the probability that the number of defectives in the sample is at most 2?

LIFE SCIENCES

Drug Effectiveness   *A new drug cures 70% of the people taking it. Suppose 20 people take the drug; find the probabilities of the following.*

46. Exactly 18 people are cured.

47. Exactly 17 people are cured.

48. At least 17 people are cured.

49. At least 18 people are cured.

Surgery Survival Rates   *Assume that the probability that a person will die within a month after undergoing a certain operation is 20%. Find the probability that in 3 such operations, the following results are obtained.*

50. All 3 people survive.

51. Exactly 1 person survives.

52. At least 2 people survive.

53. No more than 1 person survives.

Vitamin A Deficiency   *Six mice from the same litter, all suffering from a vitamin A deficiency, are fed a certain dose of carrots. If the probability of recovery under such treatment is .70, find the probabilities of the following results.*

54. None of the mice recover.

55. Exactly 3 of the 6 mice recover.

56. All of the mice recover.

57. No more than 3 mice recover.

Effects of Radiation   **58.** In an experiment on the effects of a radiation dose on cells, a beam of radioactive particles is aimed at a group of 10 cells. Find the probability that 8 of the cells will be hit by the beam, if the probability that any single cell will be hit is .6. (Assume independence.)

Effects of Radiation   **59.** The probability of a mutation of a given gene under a dose of 1 roentgen of radiation is approximately 2.5×10^{-7}. What is the probability that in 10,000 genes, at least 1 mutation occurs?

Drug Side Effects   **60.** A new drug being tested causes a serious side effect in 5 out of 100 patients. What is the probability that no side effects occur in a sample of 10 patients taking the drug?

SOCIAL SCIENCES

Testing   *In a 10-question multiple-choice biology test with 5 choices for each question, an unprepared student guesses the answer to each item. Find the probabilities of the following results.*

61. Exactly 6 correct answers

62. Exactly 7 correct answers

63. At least 8 correct answers

64. Less than 8 correct answers

Community College Population

65. According to the state of California, 33% of all state community college students belong to ethnic minorities. Find the probabilities of the following results in a random sample of 10 California community college students.

(a) Exactly 2 belong to an ethnic minority.

(b) Three or fewer belong to an ethnic minority.

(c) Exactly 5 do not belong to an ethnic minority.

(d) Six or more do not belong to an ethnic minority.

GENERAL

I Ching

66. When the *I Ching* (mentioned in the last chapter) is used as an oracle, a hexagram is selected at random and the associated philosophy is applied to the question or circumstance of the person looking for insight. A common method of randomly selecting the hexagram is by tossing coins. Three coins are tossed and each "head" (yang) is given a value of 3 and each "tail" (yin) is given a value of 2. The values of the coins are then added. This sum determines the bottom line of the hexagram. The process is repeated for each line in the hexagram in ascending order. If the sum is 6, the line is considered a *"yin"* line; if the sum is 9, the line is a *"yang"* line; if the sum is 7, the line is a *"changing yang"* line; and if the sum is 8, the line is a *"changing yin"* line.

(a) Find the probabilities of obtaining each of the 4 possible lines.

(b) What is the probability that the hexagram contains exactly 2 changing lines?

(c) What is the probability that the hexagram contains 0 changing lines?

(d) What is the probability that the hexagram contains at least 3 changing lines?

FOR THE COMPUTER

Calculate each of the probabilities in Exercises 67–70.

Flu Inoculations

67. A flu vaccine has a probability of 80% of preventing a person who is inoculated from getting the flu. A county health office inoculates 134 people. Find the probabilities of the following.

(a) Exactly 10 of the people inoculated get the flu.

(b) No more than 10 of the people inoculated get the flu.

(c) None of the people inoculated get the flu.

Color Blindness

68. The probability that a male will be color-blind is .042. Find the probabilities that in a group of 53 men, the following will be true.

(a) Exactly 5 are color-blind.

(b) No more than 5 are color-blind.

(c) At least 1 is color-blind.

Quality Control   69. The probability that a certain machine turns out a defective item is .05. Find the probabilities that in a run of 75 items, the following results are obtained.
(a) Exactly 5 defectives   (b) No defectives   (c) At least 1 defective

Survey Results   70. A company is taking a survey to find out whether people like its product. Their last survey indicated that 70% of the population like the product. Based on that, of a sample of 58 people, find the probabilities of the following.
(a) All 58 like the product.
(b) From 28 to 30 (inclusive) like the product.

```
▀▀▀ KEY WORDS ▀▀▀
```

6.1	experiment		union rule
	trial		odds
	outcome		empirical probability
	sample space		probability distribution
	event	6.3	conditional probability
	simple event		product rule
	certain event		independent events
	impossible event	6.4	Bayes' Formula
	probability	6.6	Bernoulli trials
	addition principle		repeated trials
6.2	mutually exclusive events		binomial trials

CHAPTER 6   REVIEW EXERCISES

Write the sample space for each of the following experiments.

1. Rolling a die

2. Drawing a card from a deck containing only the 13 spades

3. Measuring the weight of a person to the nearest half pound (the scale will not measure more than 300 lb)

4. Tossing a coin 4 times

An urn contains 5 balls labeled 3, 5, 7, 9, and 11, respectively, while a second urn contains 4 red and 2 green balls. An experiment consists of pulling 1 ball from each urn, in turn. In Exercises 5–7 write each set using set notation.

5. The sample space

6. Event E, the number on the first ball is greater than 5

7. Event F, the second ball is green

8. Are the outcomes in the sample space in Exercise 5 equally likely?

A company sells typewriters and copiers. Let E be the event "a customer buys a typewriter," and let F be the event "a customer buys a copier." Write each of the following using ∩, ∪, or ' as necessary.

9. A customer buys neither machine.

10. A customer buys at least one of the machines.

When a single card is drawn from an ordinary deck, find the probabilities that it will be the following.

11. A heart

12. A red queen

13. A face card

14. Black or a face card

15. Red, given it is a queen

16. A jack, given that it is a face card

17. A face card, given it is a king

Find the odds in favor of a card drawn from an ordinary deck being each of the following.

18. A club

19. A black jack

20. A red face card or a queen

A sample shipment of five hair dryers is chosen at random. The probability of exactly 0, 1, 2, 3, 4, or 5 hair dryers being defective is given in the following table.

Number Defective	0	1	2	3	4	5
Probability	.31	.25	.18	.12	.08	.06

Find the probabilities that the following numbers of hair dryers are defective.

21. No more than 3

22. At least 3

Find the probabilities of getting the following sums when 2 fair dice are rolled.

23. 8

24. 0

25. At least 10

26. No more than 5

27. An odd number greater than 8

28. 12, given that the sum is greater than 10

29. 7, given that at least 1 die is a four

30. At least 9, given that at least 1 die is a five

Suppose $P(E) = .51$, $P(F) = .37$, and $P(E \cap F) = .22$. Find each of the following.

31. $P(E \cup F)$

32. $P(E \cap F')$

33. $P(E' \cup F)$

34. $P(E' \cap F')$

A basket contains 4 black, 2 blue, and 5 green balls. A sample of 3 balls is drawn. Find the probabilities that the sample contains the following.

35. All black balls

36. All blue balls

37. 2 black balls and 1 green ball

38. Exactly 2 black balls

39. 2 green balls and 1 blue ball

40. Exactly 1 blue ball

Suppose 2 cards are drawn without replacement from an ordinary deck of 52. Find the probabilities of the following results.

41. Both cards are red.

42. Both cards are spades.

43. At least 1 card is a spade.

44. The second card is red, given that the first card was a diamond.

45. The second card is a face card, given that the first card was not.

46. The second card is a five, given that the first card was the five of diamonds.

The table below shows the results of a survey of 1000 *buyers of new or used cars of a certain model.*

	Satisfied	Not Satisfied	Totals
New	300	100	400
Used	450	150	600
Totals	750	250	1000

Let S represent the event "satisfied," and N the event "bought a new car." Find each of the following.

47. $P(N \cap S)$　　　　**48.** $P(N \cup S')$　　　　**49.** $P(N|S)$

50. $P(N'|S)$　　　　**51.** $P(S|N')$　　　　**52.** $P(S'|N')$

Of the appliance repair shops listed in the phone book, 80% *are competent and* 20% *are not. A competent shop can repair an appliance correctly* 95% *of the time; an incompetent shop can repair an appliance correctly* 60% *of the time. Suppose an appliance was repaired correctly. Find the probabilities that it was repaired by the following.*

53. A competent shop　　　　　　　　**54.** An incompetent shop

Suppose an appliance was repaired incorrectly. Find the probabilities that it was repaired by the following.

55. A competent shop　　　　　　　　**56.** An incompetent shop

57. Box A contains 5 red balls and 1 black ball; box B contains 2 red balls and 3 black balls. A box is chosen, and a ball is selected from it. The probability of choosing box A is 3/8. If the selected ball is black, what is the probability that it came from box A?

58. Find the probability that the ball in Exercise 57 came from box B, given that it is red.

Suppose a family plans 6 *children, and the probability that a particular child is a girl is* 1/2. *Find the probability that the 6-child family will have the following children.*

59. Exactly 3 girls　　　**60.** All girls　　　**61.** At least 4 girls　　　**62.** No more than 2 boys

APPLICATIONS

BUSINESS AND ECONOMICS

Quality Control　*A certain machine that is used to manufacture screws produces a defective rate of* .01. *A random sample of* 20 *screws is selected. Find the probability that the sample contains the following.*

63. Exactly 4 defective screws　　　　**64.** Exactly 3 defective screws

65. No more than 4 defective screws

66. *Set up* the probability that the sample has 12 or more defective screws. (Do not evaluate.)

Locating Oil　**67.** An oil company finds oil with 14% of the wells that it drills. How many wells must the company drill to have the following probabilities of finding oil?

(a) 2/3　　(b) 3/4

LIFE SCIENCES

Sickle Cell Anemia   *The square shows the 4 possible (equally likely) combinations when both parents are carriers of the sickle cell anemia trait. Each carrier parent has normal cells (N) and trait cells (T).*

		2nd Parent	
		N_2	T_2
1st Parent	N_1		$N_1 T_2$
	T_1		

68. Complete the table.

69. If the disease occurs only when two trait cells combine, find the probability that a child born to these parents will have sickle cell anemia.

70. The child will carry the trait but not have the disease if a normal cell combines with a trait cell. Find this probability.

71. Find the probability that the child is neither a carrier nor has the disease.

SOCIAL SCIENCES

Randomized Response Method   **72.** *Randomized Response Method for Getting Honest Answers to Sensitive Questions.** Basically, this is a method to guarantee an individual who answers sensitive questions will be anonymous, thus encouraging a truthful response. This method is, in effect, an application of the formula for finding the probability of an intersection and operates as follows. Two questions A and B are posed, one of which is sensitive and the other not. The probability of receiving a "yes" to the nonsensitive question must be known. For example, one could ask

> A: Does your Social Security number end in an odd digit? (Nonsensitive)
> B: Have you ever intentionally cheated on your income taxes? (Sensitive)

We know that $P(\text{answer yes}|\text{answer } A) = 1/2$. We wish to approximate $P(\text{answer yes}|\text{answer } B)$. The subject is asked to flip a coin and answer A if the coin comes up heads and otherwise to answer B. In this way, the interviewer does not know which question the subject is answering. Thus, a "yes" answer is not incriminating. There is no way for the interviewer to know whether the subject is saying "Yes, my Social Security number ends in an odd digit" or Yes, I have intentionally cheated on my income taxes." The percentage of subjects in the group answering "yes" is used to approximate $P(\text{answer yes})$.

(a) Use the fact that the event "answer yes" is the union of the event "answer yes and answer A" with the event "answer yes and answer B" to prove that

$$P(\text{answer yes}|\text{answer } B)$$
$$= \frac{P(\text{answer yes}) - P(\text{answer yes}|\text{answer } A) \cdot P(\text{answer } A)}{P(\text{answer } B)}$$

(b) If this technique is tried on 100 subjects and 60 answered "yes," what is the approximate probability that a person randomly selected from the group has intentionally cheated on income taxes?

---

*From *Applied Statistics With Probability* by J. S. Milton and J. J. Corbet. Copyright © 1979 by Litton Educational Publishing, Inc. Reprinted by permission of Brooks/Cole Publishing Company, Monterey, California.

EXTENDED APPLICATION

MAKING A FIRST DOWN

A first down is desirable in football—it guarantees four more plays by the team making it, assuming no score or turnover occurs in the plays. After getting a first down, a team can get another by advancing the ball at least ten yards. During the four plays given by a first down, a team's position will be indicated by a phrase such as "third and 4," which means that the team has already had two of its four plays, and that 4 more yards are needed to get the 10 yards necessary for another first down. An article in a management journal* offers the following results for 189 games of a recent National Football League season. "Trials" represents the number of times a team tried to make a first down, given that it was currently playing either a third or a fourth down. Here n represents the number of yards still needed for a first down.

n	Trials	Successes	Probability of Making First Down with n Yards to Go
1	543	388	
2	327	186	
3	356	146	
4	302	97	
5	336	91	

EXERCISES

1. Complete the table.

2. Why is the sum of the answers in Exercise 1 not equal to 1?

EXTENDED APPLICATION

MEDICAL DIAGNOSIS

When a patient is examined, information, typically incomplete, is obtained about his state of health. Probability theory provides a mathematical model appropriate for this situation, as well as a procedure for quantitatively interpreting such partial information to arrive at a reasonable diagnosis.†

To develop a model, we list the states of health that can be distinguished in such a way that the patient can be in one and only one state at the time of the examination. For each state of health H, we associate a number $P(H)$ between 0 and 1 such that the sum of all these numbers is 1. This number $P(H)$ represents the probability, before examination, that a patient is in the state of health H, and $P(H)$ may be chosen subjectively from medical experience, using any information available prior to the examination. The probability may be most conveniently established from clinical records, that is, a mean probability is established for patients in general, although the number would vary from patient to patient. Of course, the more information that is brought to bear in establishing $P(H)$, the better the diagnosis.

---

*Reprinted by permission of Virgil Carter and Robert Machols, "Optimal Strategies on Fourth Down," *Management Science*, Vol. 24, No. 16, December 1978, copyright © 1978 The Institute of Management Sciences.

†This example is based on "Probabilistic Medical Diagnosis," Roger Wright, from *Some Mathematical Models in Biology*, Robert M. Thrall, ed., rev. ed., (The University of Michigan, 1967), by permission of Robert M. Thrall.

For example, limiting the discussion to the condition of a patient's heart, suppose there are exactly 3 states of health, with probabilities as follows:

State of Health, H		$P(H)$
H_1	patient has a normal heart	.8
H_2	patient has minor heart irregularities	.15
H_3	patient has a severe heart condition	.05

Having selected $P(H)$, the information from the examination is processed. First, the results of the examination must be classified. The examination itself consists of observing the state of a number of characteristics of the patient. Let us assume that the examination for a heart condition consists of a stethoscope examination and a cardiogram. The outcome of such an examination, C, might be one of the following:

C_1—stethoscope shows normal heart
and cardiogram shows normal heart;

C_2—stethoscope shows normal heart
and cardiogram shows minor irregularities,

and so on.

It remains to assess for each state of health H the conditional probability $P(C|H)$ of each examination outcome C using only the knowledge that a patient is in a given state of health. (This may be based on the medical knowledge and clinical experience of the doctor.) The conditional probabilities $P(C|H)$ will not vary from patient to patient, so that they may be built into a diagnostic system, although they should be reviewed periodically.

Suppose the result of the examination is C_1. Let us assume the following probabilities:

$$P(C_1|H_1) = .9$$
$$P(C_1|H_2) = .4$$
$$P(C_1|H_3) = .1.$$

Now, for a given patient, the appropriate probability associated with each state of health H, after examination, is $P(H|C)$, where C is the outcome of the examination. This can be calculated by using Bayes' formula. For example, to find $P(H_1|C_1)$—that is, the probability that the patient has a normal heart given that the examination showed a normal stethoscope examination and a normal cardiogram—we use Bayes' formula as follows:

$$P(H_1|C_1) = \frac{P(C_1|H_1)P(H_1)}{P(C_1|H_1)P(H_1) + P(C_1|H_2)P(H_2) + P(C_1|H_3)P(H_3)}$$

$$= \frac{(.9)(.8)}{(.9)(.8) + (.4)(.15) + (.1)(.05)} \approx .92.$$

Hence, the probability is about .92 that the patient has a normal heart on the basis of the examination results. This means that in 8 out of 100 patients, some abnormality will be present and not be detected by the stethoscope or the cardiogram.

EXERCISES

1. Find $P(H_2|C_1)$.

2. Assuming the following probabilities, find $P(H_1|C_2)$:

$P(C_2|H_1) = .2$, $P(C_2|H_2) = .8$, $P(C_2|H_3) = .3$.

3. Assuming the probabilities of Exercise 2, find $P(H_3|C_2)$.

7 Statistics and Probability Distributions

7.1 Basic Properties of Probability Distributions

7.2 Expected Value

7.3 Variance and Standard Deviation

7.4 The Normal Distribution

7.5 The Normal Curve Approximation to the Binomial Distribution

Review Exercises

Extended Application
Optimal Inventory for a Service Truck

Extended Application
Bidding on a Potential Oil Field – Signal Oil

Statistics is a branch of mathematics that deals with the collection and summarization of data. Methods of statistical analysis make it possible for us to draw conclusions about a population based on data from a sample of the population. Statistical models have become increasingly useful in manufacturing, government, agriculture, medicine, and the social sciences, and in all types of research. In this chapter we give a brief introduction to some of the key topics from statistical theory.

▦ 7.1   BASIC PROPERTIES OF PROBABILITY DISTRIBUTIONS

The concept of *probability distributions* was discussed briefly in Chapter 6. A probability distribution depends on the idea of a *random variable*.

Random Variables   Suppose a bank is interested in improving its services to the public. The manager decides to begin by finding the amount of time tellers spend on each transaction, rounded to the nearest minute. To each transaction, then, will be assigned one of the numbers 0, 1, 2, 3, 4, · · ·. That is, if t represents an outcome of the experiment of timing a transaction, then t may take on any of the values from the list 0, 1, 2, 3, 4, · · ·. Since the value that t takes on for a particular transaction is random, t is called a random variable.

RANDOM VARIABLE

> ▦ A **random variable** is a function that assigns a real number to each outcome of an experiment.

Table 1

Time	Frequency
1	3
2	5
3	9
4	12
5	15
6	11
7	10
8	6
9	3
10	1
Total:	75

Probability Distribution   Suppose that the bank manager in our example records the times for 75 different transactions, with results as shown in Table 1. As the table shows, the shortest transaction time was 1 minute, with 3 transactions of 1-minute duration. The longest time was 10 minutes. Only one transaction took that long.

In Table 1, the first column gives the ten values assumed by the random variable t, and the second column shows the number of occurrences corresponding to each of these values, the **frequency** of that value. Table 1 is an example of a **frequency distribution,** a table listing the frequencies for each value a random variable may assume.

Now suppose that several weeks after starting new procedures to speed up transactions, the manager takes another survey. This time she observes 57 transactions, and she records their times as shown in the frequency distribution in Table 2 on the next page. (Table 1 is repeated on that page to allow comparison of the two tables.)

Table 1		**Table 2**	
Time	*Frequency*	*Time*	*Frequency*
1	3	1	4
2	5	2	5
3	9	3	8
4	12	4	10
5	15	5	12
6	11	6	17
7	10	7	0
8	6	8	1
9	3	9	0
10	1	10	0
	Total: 75		Total: 57

Do the results in Table 2 indicate an improvement? It is hard to compare the two tables, since one is based on 75 transactions and the other on 57. To make them comparable, we can add a column to each table that gives the *relative frequency* of each transaction time. These results are shown in Tables 3 and 4. Where necessary, decimals are rounded to the nearest hundredth. To find a **relative frequency,** divide each frequency by the total of the frequencies. Here the individual frequencies are divided by 75 or 57.

Table 3			**Table 4**		
Time	*Frequency*	*Relative Frequency*	*Time*	*Frequency*	*Relative Frequency*
1	3	$3/75 = .04$	1	4	$4/57 \approx .07$
2	5	$5/75 \approx .07$	2	5	$5/57 \approx .09$
3	9	$9/75 = .12$	3	8	$8/57 \approx .14$
4	12	$12/75 = .16$	4	10	$10/57 \approx .18$
5	15	$15/75 = .20$	5	12	$12/57 \approx .21$
6	11	$11/75 \approx .15$	6	17	$17/57 \approx .30$
7	10	$10/75 \approx .13$	7	0	$0/57 = 0$
8	6	$6/75 = .08$	8	1	$1/57 \approx .02$
9	3	$3/75 = .04$	9	0	$0/57 = 0$
10	1	$1/75 \approx .01$	10	0	$0/57 = 0$

Whether the differences in relative frequency between the distributions in Tables 3 and 4 are interpreted as desirable or undesirable depends on management goals. If the manager wanted to eliminate the most time-consuming transactions, the results appear to be desirable. However, before the new procedures were followed, the largest relative frequency of transactions, .20, was for transactions of 5 minutes. With the new procedures, the largest relative frequency, .30, corresponds to a transaction of 6 minutes. At any rate, the results shown in the two tables are easier to compare using relative frequencies.

The relative frequencies in Tables 3 and 4 can be considered as probabilities. Such a table that lists all the possible values of a random variable, together with the corresponding probabilities, is called a **probability distribution.** The sum of the probabilities in a probability distribution must always be 1. (The sum in an actual distribution may vary slightly from 1 because of rounding.)

■ EXAMPLE 1

Many plants have seed pods with variable numbers of seeds. One variety of green beans has no more than 6 seeds per pod. Suppose that examination of 30 such bean pods gives the results shown in Table 5. Here the random variable x tells the number of seeds per pod. Give a probability distribution for these results.

The probabilities are found by computing the relative frequencies. A total of 30 bean pods were examined, so each frequency should be divided by 30 to get the probability distribution shown as Table 6. Some of the results have been rounded to the nearest hundredth.

Table 5

x	Frequency
0	3
1	4
2	6
3	8
4	5
5	3
6	1
	Total: 30

Table 6

x	Frequency	Probability
0	3	$3/30 = .10$
1	4	$4/30 \approx .13$
2	6	$6/30 = .20$
3	8	$8/30 \approx .27$
4	5	$5/30 \approx .17$
5	3	$3/30 = .10$
6	1	$1/30 \approx .03$
	Total: 30	

As shown in Table 6, the probability that the random variable x takes on the value 2 is 6/30, or .20. This is often written as

$$P(x = 2) = .20.$$

Also, $P(x = 5) = .10$, and $P(x = 6) \approx .03$.   ■

Instead of writing the probability distribution in Example 1 as a table, we could write the same information as a set of ordered pairs:

$\{(0, .10), (1, .13), (2, .20), (3, .27), (4, .17), (5, .10), (6, .03)\}.$

There is just one probability for each value of the random variable. Thus, a probability distribution defines a function, called a **probability distribution function** or simply a **probability function.** We shall use the terms "probability distribution" and "probability function" interchangeably. The function described in Example 1 is a **discrete probability function,** since a graph of the ordered pairs would produce unconnected dots. In a **continuous probability function** the values of the random variable correspond to an interval on the number line. The points of the graph are "connected" or "next to" one another. Continuous probability distribution functions are discussed in Section 7.4.

The information in a probability distribution is often displayed graphically as a special kind of bar graph called a **histogram.** The bars of a histogram all have the same width, usually 1. The heights of the bars are determined by the frequencies. A histogram for the data in Table 3 is given in Figure 1. A histogram shows important characteristics of a distribution that may not be readily apparent in tabular form, such as the relative sizes of the probabilities and any symmetry in the distribution.

The area of the bar above $t = 1$ in Figure 1 is the product of 1 and .04, or $.04 \times 1 = .04$. Since each bar has a width of 1, its area is equal to the probability that corresponds to that value of t. The probability that a particular value will occur is thus given by the area of the appropriate bar of the graph. For example, the probability that a transaction will take less than four minutes is the sum of the areas for $t = 1$, $t = 2$, and $t = 3$. This area, shown in color in Figure 2, corresponds to 23% of the total area, since

$$P(t < 4) = P(t = 1) + P(t = 2) + P(t = 3)$$
$$= .04 + .07 + .12 = .23.$$

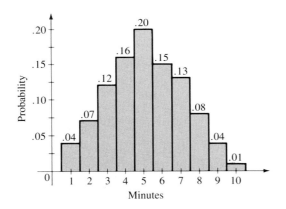

FIGURE 1

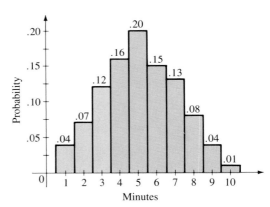

FIGURE 2

▬ EXAMPLE 2

Construct a histogram for the probability distribution in Example 1. Find the area that gives the probability that the number of seeds will be more than 4.

A histogram for this distribution is shown in Figure 3. The portion of the histogram in color represents

$$P(x > 4) = P(x = 5) + P(x = 6)$$
$$= .10 + .03 = .13,$$

or 13% of the total area. ▬

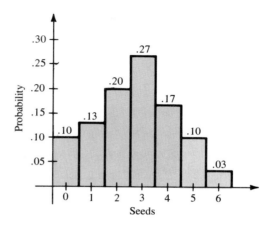

FIGURE 3

Table 7

x	$P(x)$
0	1/4
1	1/2
2	1/4

▬ **EXAMPLE 3**

(a) Give the probability distribution for the number of heads showing when two coins are tossed.

Let x represent the random variable "number of heads." Then x can take on the values 0, 1, or 2. Now find the probability of each outcome. The results are shown in Table 7.

(b) Draw a histogram for the distribution in Table 7. Find the probability that at least one coin comes up heads.

The histogram is shown in Figure 4. The portion in color represents

$$P(x \geq 1) = P(x = 1) + P(x = 2)$$
$$= \frac{3}{4}. \quad ▬$$

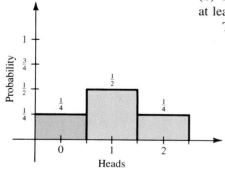

FIGURE 4

7.1 EXERCISES

In Exercises 1–6, **(a)** *give the probability distribution, and* **(b)** *sketch its histogram.*

1. In a seed viability test 50 seeds were placed in 10 rows of 5 seeds each. After a period of time, the number that germinated in each row were counted, giving the results in the table.

Number Germinated	Frequency
0	0
1	0
2	1
3	3
4	4
5	2
	Total: 10

2. At a large supermarket during the 5 P.M. rush, the manager counted the number of customers waiting in each of 10 checkout lines. The results are shown in the table.

Number Waiting	Frequency
2	1
3	2
4	4
5	2
6	0
7	1
	Total: 10

3. At a training program for police officers, each member of a class of 25 took 6 shots at a target. The numbers of bullseyes shot are shown in the table.

Number of Bullseyes	Frequency
0	0
1	1
2	0
3	4
4	10
5	8
6	2
	Total: 25

4. A class of 42 students took a 10-point quiz. The frequency of scores is given in the table.

Number of Points	Frequency
5	2
6	5
7	10
8	15
9	7
10	3
	Total: 42

5. Five mice are inoculated against a disease. After an incubation period, the number that contract the disease is noted. The experiment is repeated 20 times, with the results shown in the table.

Number With the Disease	Frequency
0	3
1	5
2	6
3	3
4	2
5	1
	Total: 20

6. The telephone company kept track of the calls for the correct time during a 24-hour period for two weeks. The results are shown in the table.

Number of Calls	Frequency
28	1
29	1
30	2
31	3
32	2
33	2
34	2
35	1
	Total: 14

For each of the experiments described below, let x determine a random variable, and use your knowledge of probability to prepare a probability distribution.

7. Four coins are tossed, and the number of heads is noted.

8. Two dice are rolled, and the total number of points is recorded.

9. Three cards are drawn from a deck. The number of aces is counted.

10. Two balls are drawn from a bag in which there are 4 white balls and 2 black balls. The number of black balls is counted.

11. A ballplayer with a batting average of .290 comes to bat 4 times in a game. The number of hits is counted.

12. Five cards are drawn from a deck. The number of black threes is counted.

Draw a histogram for each of the following, and shade the region that gives the indicated probability.

13. Exercise 7; $P(x \leq 2)$

14. Exercise 8; $P(x \geq 11)$

15. Exercise 9; $P($at least one ace$)$

16. Exercise 10; $P($at least one black ball$)$

17. Exercise 11; $P(x = 2 \text{ or } x = 3)$

18. Exercise 12; $P(1 \leq x \leq 2)$

The frequency with which letters occur in a large sample of any written language does not vary much. Therefore, determining the frequency of each letter in a coded message is usually the first step in deciphering it. The percent frequencies of the letters in the English language are as follows.

Letter	%	Letter	%	Letter	%
E	13	S, H	6	W, G, B	1.5
T	9	D	4	V	1
A, O	8	L	3.5	K, X, J	.5
N	7	C, U, M	3	Q, Z	.2
I, R	6.5	F, P, Y	2		

19. Use the preceding paragraph as a sample of the English language. Find the percent frequency for each letter in the sample. Compare your results with the frequencies given in the table.

20. The following message is written in a code in which the frequency of the symbols is the main key to the solution.

)? − − 8)) 6* + 8506 * 3 × 6 ; 4 ?* 7* & × * − 6.48 ( )6)985)?

(8 + 2: ;48) 81 & ?(;46 *3)6 *;48 & (+ 8(*598 + . 8 () 8 = 8

(5* − 8 − 5(81 ? 098 ;4 & +)& 15 * 50:)6)6 *; ? 6;6 & * 0? − 7

(a) Find the frequency of each symbol.

(b) By comparing the high-frequency symbols with the high-frequency letters in English, and the low-frequency symbols with the low-frequency letters, try to decipher the message. (*Hint:* Look for repeated two-symbol combinations and double letters for added clues. Try to identify vowels first.)

21. On the televised game show *Wheel of Fortune,* each contestant tries to discover a predetermined phrase by guessing letters that it contains. The game board shows boxes representing the letters in the various words, and the contestant is allowed to choose 5 consonants and 1 vowel. For each "correct" letter selected, the corresponding boxes are turned to show where the letters occur in the phrase. Most contestants pick the letters N, L, R, S, T, and E. Based on the table given above, are these the best choices? If not, what would be better?

7.2   EXPECTED VALUE

In working with experimental data, it is often useful to have a typical or "average" number that represents the entire set of data. For example, we compare our heights and weights to those of the typical or "average" person on weight charts. Students are familiar with the "class average" and their own "average" in a given course.

In a recent year, a citizen of the United States could expect to complete about 12 years of school, to be a member of a household earning $27,700 per year, and to live in a household of 2.7 people. What do we mean here by "expect"? Many people have completed less than 12 years of school; many others have completed more. Many households have less income than $27,700 per year; many others have more. The idea of a household of 2.7 people is a little hard to swallow. The numbers all refer to *averages*. When the term "expect" is used in this way, it refers to *mathematical expectation,* which is a kind of average.

Mean The **arithmetic mean,** or **average,** of a set of numbers is the sum of the numbers in the set, divided by the total number of numbers. To write the sum of the *n* numbers $x_1, x_2, x_3, \cdots, x_n$ in a compact way, use **summation notation:** using the Greek letter Σ (sigma), the sum $x_1 + x_2 + x_3 + \cdots + x_n$ is written

$$x_1 + x_2 + x_3 + \cdots + x_n = \sum_{i=1}^{n} x_i.$$

The symbol $\bar{x}$ (read *x*-bar) is used to represent the mean, so that the mean of the *n* numbers $x_1, x_2, x_3, \cdots, x_n$ is

$$\bar{x} = \frac{\sum_{i=1}^{n} x_i}{n}.$$

For example, the mean of the set of numbers 2, 3, 5, 6, 8 is

$$\frac{2 + 3 + 5 + 6 + 8}{5} = \frac{24}{5} = 4.8.$$

Expected Value What about an average value for a random variable? Can we use the mean to find it? As an example, let us find the average number of offspring for a certain species of pheasant, given the probability distribution in Table 8.

Table 8

Number of Offspring	Frequency	Probability
0	8	.08
1	14	.14
2	29	.29
3	32	.32
4	17	.17
	Total: 100	

We might be tempted to find the typical number of offspring by averaging the numbers 0, 1, 2, 3, and 4, which represent the numbers of offspring possible. This won't work, however, since the various numbers of offspring do not occur with equal probability: for example, a brood with 3 offspring is much more common than one with 0 or 1 offspring. The differing probabilities of occurrence can be taken into account with a **weighted average,** found by multiplying each of the possible numbers of offspring by its corresponding probability, as follows:

$$\text{Typical Number of Offspring} = 0(.08) + 1(.14) + 2(.29) + 3(.32) + 4(.17)$$
$$= 0 + .14 + .58 + .96 + .68$$
$$= 2.36.$$

Based on the data given above, a typical brood of pheasants includes 2.36 offspring.

It is certainly not possible for a pair of pheasants to produce 2.36 offspring. If the numbers of offspring produced by many different pairs of pheasants are found, however, the average (or the mean) of these numbers will be about 2.36.

We can use the results of this example to define the mean, or *expected value,* of a probability distribution as follows.

EXPECTED VALUE

Suppose the random variable x can take on the n values $x_1, x_2, x_3, \cdots, x_n$. Also, suppose the probabilities that these values occur are respectively $p_1, p_2, p_3, \cdots, p_n$. Then the **expected value** of the random variable is

$$E(x) = x_1p_1 + x_2p_2 + x_3p_3 + \cdots + x_np_n.$$

The symbol μ (the Greek letter *mu*) is used for the expected value of the random variable x. As in the example above, the expected value of a random variable may be a number that can never occur in any one trial of the experiment.

Physically, the expected value of a probability distribution represents a balance point. Figure 5 shows a histogram for the distribution of the pheasant offspring. If the histogram is thought of as a series of weights with magnitudes represented by the heights of the bars, then the system would balance if supported at the point corresponding to the expected value.

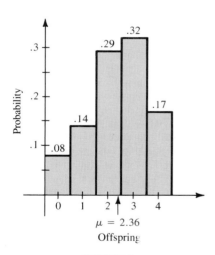

FIGURE 5

Table 9

Outcome (Net Winnings)	Probability
$399	$\dfrac{1}{2000}$
−$1	$\dfrac{1999}{2000}$

EXAMPLE 1

Suppose a local church decides to raise money by raffling a microwave oven worth $400. A total of 2000 tickets are sold at $1 each. Find the expected value of winning for a person who buys one ticket in the raffle.

Here the random variable represents the possible amounts of net winnings, where net winnings = amount won − cost of ticket. The net winnings of the person winning the oven are $400 (amount won) − $1 (cost of ticket) = $399. The net winnings for each losing ticket are $0 − $1 = −$1.

The probability of winning is 1 in 2000, or 1/2000, while the probability of losing is 1999/2000. See Table 9.

The expected winnings for a person buying one ticket are

$$399\left(\frac{1}{2000}\right) + (-1)\left(\frac{1999}{2000}\right) = \frac{399}{2000} - \frac{1999}{2000}$$

$$= -\frac{1600}{2000}$$

$$= -.80.$$

On the average, a person buying one ticket in the raffle will lose $.80, or 80¢.

It is not possible to lose 80¢ in this raffle—either you lose $1, or you win a $400 prize. If you bought tickets in many such raffles over a long period of time, however, you would lose 80¢ per ticket, on the average. ■

■ EXAMPLE 2

What is the expected number of girls in a family having exactly 3 children?

Some families with 3 children have 0 girls, others have 1 girl, and so on. We need to find the probabilities associated with 0, 1, 2, or 3 girls in a family of 3 children. To find these probabilities, first write the sample space S of all possible 3-child families: $S = \{ggg, ggb, bgg, gbg, gbb, bgb, bbg, bbb\}$. This sample space gives the probabilities shown in Table 10, assuming that the probability of a girl at each birth is 1/2.

Table 10

Outcome (Number of Girls)	Probability
0	1/8
1	3/8
2	3/8
3	1/8

The expected number of girls can now be found by multiplying each outcome (number of girls) by its corresponding probability and finding the sum of these values.

$$\text{Expected Number of Girls} = 0 \cdot \frac{1}{8} + 1 \cdot \frac{3}{8} + 2 \cdot \frac{3}{8} + 3 \cdot \frac{1}{8}$$

$$= \frac{3}{8} + \frac{6}{8} + \frac{3}{8}$$

$$= \frac{12}{8}$$

$$= \frac{3}{2} = 1.5$$

On the average, a 3-child family will have 1.5 girls. This result agrees with our intuition that, on the average, half the children born will be girls. ▬

▬ EXAMPLE 3

Each day Donna and Mary toss a coin to see who buys the coffee (40¢ a cup). One tosses and the other calls the outcome. If the person who calls the outcome is correct, the other buys the coffee; otherwise the caller pays. Find Donna's expected winnings.

Assume that an honest coin is used, that Mary tosses the coin, and that Donna calls the outcome. The possible results and corresponding probabilities are shown below.

	Possible Results			
Result of toss	Heads	Heads	Tails	Tails
Call	Heads	Tails	Heads	Tails
Caller wins?	Yes	No	No	Yes
Probability	1/4	1/4	1/4	1/4

Donna wins a 40¢ cup of coffee whenever the results and calls match, and she loses a 40¢ cup when there is no match. Here expected winnings are

$$(.40)\left(\frac{1}{4}\right) + (-.40)\left(\frac{1}{4}\right) + (-.40)\left(\frac{1}{4}\right) + (.40)\left(\frac{1}{4}\right) = 0.$$

On the average, over the long run, Donna neither wins nor loses. ▬

A game with an expected value of 0 (such as the one in Example 3) is called a **fair game.** Casinos do not offer fair games. If they did, they would win (on the average) $0, and have a hard time paying the help! Casino games have expected winnings for the house that vary from 1.5 cents per dollar to 60 cents per dollar. Exercises 30–33 at the end of the section ask you to find the expected winnings for certain games of chance.

The idea of expected value can be very useful in decision making, as shown by the next example.

▬ EXAMPLE 4

At age 50, you receive a letter from the Mutual of Mauritania Insurance Company. According to the letter, you must tell the company immediately which of the following two options you will choose: take $20,000 at age 60 (if you are alive, $0 otherwise) or $30,000 at age 70 (again, if you are alive, $0 otherwise). Based only on the idea of expected value, which should you choose?

Life insurance companies have constructed elaborate tables showing the probability of a person living a given number of years into the future. From a recent such table, the probability of living from age 50 to age 60 is .88, while the probability of living from age 50 to 70 is .64. The expected values of the two options are given below.

First Option: $(20,000)(.88) + (0)(.12) = 17,600$

Second Option: $(30,000)(.64) + (0)(.36) = 19,200$

Based strictly on expected values, choose the second option.   ▬

Numbers like the mean of a distribution are referred to as **measures of central tendency.** Two other (less important) measures of central tendency are the *median* and the *mode*.

Median   Asked by a reporter to give the average height of the players on his team, a Little League coach lined up his 15 players by increasing height. He picked the player in the middle and pronounced that player to be of average height. This kind of average, called the **median,** is defined as the middle entry in a set of data arranged in either increasing or decreasing order. If there is an even number of entries, the median is defined to be the mean of the two center entries.

Odd Number of Entries	*Even Number of Entries*
8	2
7	3
Median = 4	4
3	7 Median = $\frac{4+7}{2} = 5.5$
1	9
	12

▬ EXAMPLE 5

Find the median for each of the following lists of numbers.

(a) 11,   12,   17,   20,   23,   28,   29

The median is the middle number: in this case, 20. (Note that the numbers are already arranged in numerical order.) In this list, 3 numbers are smaller than 20 and 3 are larger.

(b) 15,   13,   7,   11,   19,   30,   39,   5   10

First arrange the numbers in numerical order, from smallest to largest.

5,   7,   10,   11,   13,   15,   19,   30,   39

The middle number, or median, can now be determined; it is 13.

(c) 47,   59,   32,   81,   74,   153

Write the numbers in numerical order.

32,   47,   59,   74,   81,   153

There are 6 numbers here; the median is the mean of the 2 middle numbers.

$$\text{Median} = \frac{59 + 74}{2} = \frac{133}{2} = 66\frac{1}{2} \quad \blacksquare$$

In some situations, the median gives a truer representative or typical element of the data than the mean. For example, suppose in an office there are 10 salespersons, 4 secretaries, the sales manager, and Ms. Daly, who owns the business. Their annual salaries are as follows: secretaries, $15,000 each; salespersons, $25,000 each; manager, $35,000; and owner, $200,000. The mean salary is

$$\bar{x} = \frac{(15,000)4 + (25,000)10 + 35,000 + 200,000}{16} = \$34,062.50.$$

However, since 14 people earn less than $34,062.50 and only 2 earn more, this does not seem very representative. The median salary is found by ranking the salaries by size: $15,000, $15,000, $15,000, $15,000, $25,000, $25,000, . . . , $200,000. Since there are 16 salaries (an even number) in the list, the mean of the eighth and ninth entries will give the value of the median. The eighth and ninth entries are both $25,000, so the median is $25,000. In this example, the median gives a truer average than the mean.

Mode   Sue's scores on ten class quizzes include one 7, two 8's, six 9's, and one 10. She claims that her average grade on quizzes is 9, because most of her scores are 9's. This kind of "average," found by selecting the most frequent entry, is called the **mode.**

■■ EXAMPLE 6
Find the mode for each list of numbers.

(a) 57,   38,   55,   55,   80,   87,   98

The number 55 occurs more often than any other, and is the mode. It is not necessary to place the numbers in numerical order when looking for the mode.

(b) 182,   185,   183,   185,   187,   187,   189

Both 185 and 187 occur twice. This list has *two* modes.

(c) 10,708,   11,519,   10,972,   17,546,   13,905,   12,182

No number occurs more than once. This list has no mode.   ■

The mode has the advantages of being easily found and not being influenced by data that are very large or very small compared to the rest of the data. It is often used in samples where the data to be "averaged" are not numerical. A major disadvantage of the mode is that we cannot always locate one mode for a set of values. There can be more than one mode, in case of ties, or there can be no mode if all entries occur with the same frequency.

The mean is the most commonly used measure of central tendency. Its advantages are that it is easy to compute, it takes all the data into consideration, and it is reliable—that is, repeated samples are likely to give very similar means. A disadvantage of the mean is that it is influenced by extreme values, as illustrated in the salary example above.

The median can be easy to compute and is influenced very little by extremes. Like the mode, the median can be found in situations where the data are not numerical. For example, in a taste test, people are asked to rank five soft drinks from the one they like best to the one they like least. The combined rankings then produce an ordered sample from which the median can be identified. A disadvantage of the median is the need to rank the data in order; this can be difficult when the number of items is large.

7.2 EXERCISES

Find the expected value for each random variable.

1.

x	2	3	4	5
$P(x)$	.1	.4	.3	.2

2.

y	4	6	8	10
$P(y)$	.4	.4	.05	.15

3.

z	9	12	15	18	21
$P(z)$	.14	.22	.36	.18	.10

4.

x	.30	32	36	38	44
$P(x)$	.31	.30	.29	.06	.04

Find the expected values for the random variables x having the probability functions graphed below.

5.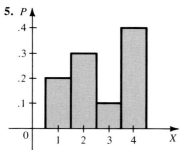

6.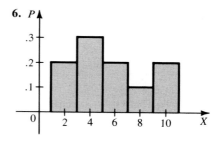

7.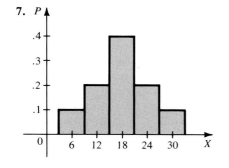

8.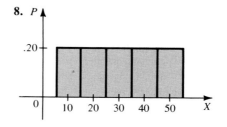

Find the median for each of the following lists of numbers. (Hint: Remember to first place the numbers in numerical order if necessary.)

9. 12, 18, 32, 51, 58, 92, 106

10. 596, 604, 612, 683, 719

11. 100, 114, 125, 135, 150, 172

12. 298, 346, 412, 501, 515, 521, 528, 621

13. 32, 58, 97, 21, 49, 38, 72, 46, 53

14. 1072, 1068, 1093, 1042, 1056, 1005, 1009

Find the mode or modes for each of the following lists of numbers.

15. 4, 9, 8, 6, 9, 2, 1, 3

16. 21, 32, 46, 32, 49, 32, 49

17. 80, 72, 64, 64, 72, 53, 64

18. 5, 9, 17, 3, 2, 8, 19, 1, 4, 20

19. 6.1, 6.8, 6.3, 6.3, 6.9, 6.7, 6.4, 6.1, 6.0

20. 12.75, 18.32, 19.41, 12.75, 18.30, 19.45, 18.33

21. A raffle offers a first prize of $100 and 2 second prizes of $40 each. One ticket costs $1, and 500 tickets are sold. Find the expected winnings for a person who buys 1 ticket. Is this a fair game?

22. A raffle offers a first prize of $1000, 2 second prizes of $300 each, and 20 third prizes of $10 each. If 10,000 tickets are sold at 50¢ each, find the expected winnings for a person buying 1 ticket. Is this a fair game?

Many of the following exercises involve combinations, which were discussed in Chapter 5.

23. If 3 marbles are drawn from a bag containing 3 yellow and 4 white marbles, what is the expected number of yellow marbles in the sample?

24. If 5 apples in a barrel of 25 apples are known to be rotten, what is the expected number of rotten apples in a sample of 2 apples?

25. A delegation of 3 is selected from a city council made up of 5 liberals and 4 conservatives.

 (a) What is the expected number of liberals in the delegation?

 (b) What is the expected number of conservatives in the delegation?

26. From a group of 2 women and 5 men, a delegation of 2 is selected. Find the expected number of women in the delegation.

27. In a club with 20 senior and 10 junior members, what is the expected number of junior members on a 3-member committee?

28. If 2 cards are drawn at one time from a deck of 52 cards, what is the expected number of diamonds?

29. Suppose someone offers to pay you $5 if you draw 2 diamonds in the game in Exercise 28. He says that you should pay 50¢ for the chance to play. Is this a fair game?

Find the expected winnings for the games of chance described in Exercises 30–33.

30. In one form of roulette, you bet $1 on "even." If one of the 18 even numbers comes up, you get your dollar back, plus another one. If one of the 20 noneven (18 odd, 0, and 00) numbers comes up, you lose your dollar.

31. In another form of roulette, there are only 19 noneven numbers (no 00).

32. Numbers is a game in which you bet $1 on any three-digit number from 000 to 999. If your number comes up, you get $500.

33. In one form of the game Keno, the house has a pot containing 80 balls, each marked with a different number from 1 to 80. You buy a ticket for $1 and mark one of the 80 numbers on it. The house then selects 20 numbers at random. If your number is among the 20, you get $3.20 (for a net winning of $2.20).

34. Use the assumptions of Example 3 to find Mary's expected winnings. If Mary tosses and Donna calls, it is still a fair game?

35. Suppose one day Mary brings a 2-headed coin and uses it to toss for the coffee. Since Mary tosses, Donna calls.

(a) Is this still a fair game?

(b) What is Donna's expected gain if she calls heads?

(c) What is Donna's expected gain if she calls tails?

36. Find the expected number of girls in a family of 4 children.

37. Find the expected number of boys in a family of 5 children.

38. Jack must choose at age 40 whether to inherit either $25,000 at age 50 (if he is still alive) or $30,000 at age 55 (if he is still alive.) If the probabilities for a person of age 40 living to be 50 and 55 are .90 and .85, respectively, which choice gives him the larger expected inheritance?

39. A magazine distributor offers a first prize of $100,000, two second prizes of $40,000 each, and two third prizes of $10,000 each. A total of 2,000,000 entries are received in the contest. Find the expected winnings if you submit one entry to the contest. If it would cost you 50¢ in time, paper, and stamps to enter, would it be worth it?

40. A local used-car dealer gets complaints about his cars as shown in the following table.

Number of Complaints per Day	0	1	2	3	4	5	6
Probability	.01	.05	.15	.26	.33	.14	.06

Find the expected number of complaints per day.

41. I can take one of two jobs. With job A, there is a 50% chance that I will make $60,000 per year after 5 years, and a 50% chance of making $30,000 after 5 years. With job B, there is a 30% chance that I will make $90,000 per year after 5 years and a 70% chance that I will make $20,000 after 5 years. Based stricly on expected value, which job should I take?

42. At the end of play in a major golf tournament, two players, an "old pro" and a "new kid," are tied. Suppose first prize is $80,000 and second prize is $20,000. Find the expected winnings for the old pro if

(a) both players are of equal ability,

(b) the new kid will freeze up, giving the old pro a 3/4 chance of winning.

43. In a certain animal species, the probability that a healthy adult female will have no offspring in a given year is .31, while the probabilities of 1, 2, 3, or 4 offspring are respectively .21, .19, .17, and .12. Find the expected number of offspring.

APPLICATIONS

BUSINESS AND ECONOMICS

Estimating Profit
44. A builder is considering a job that promises a profit of $30,000 with a probability of .7 or a loss (due to bad weather, strikes, and such) of $10,000 with a probability of .3. What is the expected profit?

Payout on Insurance Policies
45. An insurance company has written 100 policies of $10,000, 500 of $5000, and 1000 of $1000 on people of age 20. If experience shows that the probability that a person will die at age 20 is .001, how much can the company expect to pay out during the year the policies were written?

Estimating Occupancy
46. Experience has shown that a ski lodge will be full (160 guests) during the Christmas holidays if there is a heavy snow pack in December, while a light snowfall in December means that there will be only 90 guests. What is the expected number of guests if the probability for a heavy snow in December is .40? (Assume that there must be either a light snowfall or a heavy snowfall.)

Rating Sales Accounts
47. Levi Strauss and Company* uses expected value to help its salespeople rate their accounts. For each account, a salesperson estimates potential additional volume and the probability of getting it. The product of these gives the expected value of the potential, which is added to the existing volume. The totals are then classified as A, B, or C as follows: $40,000 or below, class C; from $40,000 up to and including $55,000, class B; above $55,000, class A. Complete the following chart for one salesperson.

Account Number	Existing Volume	Potential Additional Volume	Probability of Getting It	Expected Value of Potential	Existing Volume + Expected Value of Potential	Class
1	$15,000	$10,000	.25	$2,500	$17,500	C
2	$40,000	$0	—	—	$40,000	C
3	$20,000	$10,000	.20			
4	$50,000	$10,000	.10			
5	$5,000	$50,000	.50			
6	$0	$100,000	.60			
7	$30,000	$20,000	.80			

LIFE SCIENCES

Blood Pressure
48. According to an article in a magazine not known for its accuracy, a male decreases his life expectancy by 1 year, on the average, for every point that his blood pressure is above 120. The average life expectancy for a male is 76 years.
Find the life expectancies for males who have blood pressures as follows.

(a) 135   (b) 150   (c) 115   (d) 100

(e) Suppose a certain male has a blood pressure of 145. Find his life expectancy. How would you interpret the result to him?

---

*This example was supplied by James McDonald, Levi Strauss and Company, San Francisco.

PHYSICAL SCIENCES

Seeding Storms   **49.** One of the few methods that can be used in an attempt to cut the severity of a hurricane is to *seed* the storm. In this process, silver iodide crystals are dropped into the storm. Unfortunately, silver iodide crystals sometimes cause the storm to *increase* its speed. Wind speeds may also increase or decrease even with no seeding. The probabilities and amounts of property damage in the following tree diagram are from an article by R. A. Howard, J. E. Matheson, and D. W. North, "The Decision to Seed Hurricanes."*

   (a) Find the expected amount of damage under each option, "seed" and "do not seed."

   (b) To minimize total expected damage, what option should be chosen?

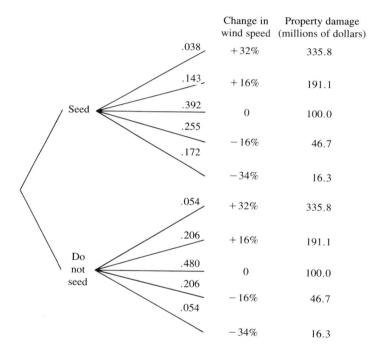

	Change in wind speed	Property damage (millions of dollars)
.038	+32%	335.8
.143	+16%	191.1
Seed .392	0	100.0
.255	−16%	46.7
.172	−34%	16.3
.054	+32%	335.8
.206	+16%	191.1
Do not seed .480	0	100.0
.206	−16%	46.7
.054	−34%	16.3

Contests **50.** A contest at a fast-food restaurant offered the following cash prizes and probabilities of winning on one visit.

Prize	Probability
$100,000	$\dfrac{1}{176,402,500}$
$25,000	$\dfrac{1}{39,200,556}$
$5000	$\dfrac{1}{17,640,250}$
$1000	$\dfrac{1}{1,568,022}$
$100	$\dfrac{1}{282,244}$
$5	$\dfrac{1}{7056}$
$1	$\dfrac{1}{588}$

Suppose you spend $1 to buy a bus pass that lets you go to 25 different restaurants in the chain and pick up entry forms. Find your expected value.

▰ 7.3 VARIANCE AND STANDARD DEVIATION

The mean of a distribution gives an average value of the distribution, but the mean tells nothing about the *spread* of the numbers in the distribution. For example, suppose seven measurements of the thickness (in centimeters) of a copper wire produced by one machine are

$$.010, \quad .010, \quad .009, \quad .008, \quad .007, \quad .009, \quad .010,$$

and seven measurements of the same type of wire produced by another machine are

$$.014, \quad .004, \quad .013, \quad .005, \quad .009, \quad .004, \quad .014.$$

The mean of both samples is .009, yet the samples are quite dissimilar; the amount of dispersion or variation within the samples is different. In addition to the mean, another kind of measure is needed to describe the variation of the numbers in a distribution.

Since the mean represents the center of the distribution, one way to measure the variation within a set of numbers is to find the average of their distances from the mean. That is, if the numbers are $x_1, x_2, \cdots, x_n$ and the mean is $\bar{x}$, we might first find the differences $x_1 - \bar{x}, x_2 - \bar{x}, \cdots, x_n - \bar{x}$, and then find the mean of the differences. The sum of these differences will always be 0, however, so their mean would also be 0. To see why, look at the four numbers x_1, x_2, x_3, x_4 having mean $\bar{x}$. The sum of the four differences is

$$\sum_{i=1}^{4} (x_i - \bar{x}) = (x_1 - \bar{x}) + (x_2 - \bar{x}) + (x_3 - \bar{x}) + (x_4 - \bar{x})$$

$$= x_1 + x_2 + x_3 + x_4 - 4(\bar{x}).$$

By definition, $\bar{x} = (x_1 + x_2 + x_3 + x_4)/4$, giving

$$\sum_{i=1}^{4} (x_i - \bar{x}) = x_1 + x_2 + x_3 + x_4 - 4\left(\frac{x_1 + x_2 + x_3 + x_4}{4}\right)$$

$$= 0.$$

While we proved this result only for four values, the proof could be extended to any finite number of values.

Since the sum of the differences from the mean is always 0, the mean of these differences also would be 0—not a good measure of the variability of a distribution. It turns out that a very useful measure of variability is found by *squaring* the differences from the mean.

For example, let us use the seven measurements given above,

.010,   .010,   .009,   .008,   .007,   .009,   .010.

The mean of these numbers is .009. Subtracting the mean from each value gives the differences

.001,   .001,   0, $-$.001, $-$.002,   0,   .001.

(Check that the sum of these differences is 0.) Now square each difference, getting

.000001,   .000001,   0,   .000001,   .000004,   0,   .000001.

Next, find the mean of these squares, which is .00000114 (rounded).

This number, the mean of the squares of the differences, is called the *variance* of the distribution. If x is the random variable for the distribution, then the variance is written $\text{Var}(x)$. The variance gives a measure of the variation of the numbers in the distribution, but since the squared differences were used to get it, the size of the variance does not reflect the actual amount of variation. To correct this problem, another measure of variation is used, the *standard deviation*, which is the square root of the variance. The symbol σ (the Greek lowercase sigma) is used for standard deviation. The standard deviation of the distribution discussed above is

$$\sigma = \sqrt{.00000114} \approx .001.$$

Variance and standard deviation of a set of numbers are defined as follows.

**VARIANCE AND
STANDARD DEVIATION**

The **variance** of a set of n numbers $x_1, x_2, x_3, \cdots x_n$, with mean $\bar{x}$, is

$$\text{Var}(x) = \frac{\Sigma(x - \bar{x})^2}{n}.$$

The **standard deviation** of the set is

$$\sigma = \sqrt{\frac{\Sigma(x - \bar{x})^2}{n}}.$$

EXAMPLE 1

Find the standard deviation of the seven measurements of copper wire produced by the second machine in the example above.

It is best to arrange the work in columns as in Table 11.

Table 11

x	$x - \bar{x}$	$(x - \bar{x})^2$
.014	.005	.000025
.004	−.005	.000025
.013	.004	.000016
.005	−.004	.000016
.009	0	0
.004	−.005	.000025
.014	.005	.000025
Totals	0	.000132

As mentioned above, the entries in the column $x - \bar{x}$ always should have a sum of 0. (This is a good way to check your work at that point.) To get the variance, divide the sum of the $(x - \bar{x})^2$ column by the number of values in the set: seven in this case. Then take the square root to get the standard deviation.

$$\text{Variance} = \text{Var}(x) = \frac{.000132}{7} \approx .0000189$$

$$\text{Standard Deviation} = \sigma = \sqrt{.0000189} \approx .004$$

Both measures of variation, the variance and the standard deviation, are larger for this sample than for the first sample, showing that the first machine produces copper wire with less variation than the second. ■

In the formulas for variance and standard deviation, sometimes $n - 1$ is used in the denominator instead of n. Usually $n - 1$ is used to calculate a sample standard deviation, denoted by s, and n is used for a population standard deviation, denoted by σ. For large values of n, the results from the two methods would be very close. Some calculators that are programmed to compute variance and standard deviation use n, and others use $n - 1$. Be sure to check how your calculator works before using it for the exercises. In this text, except for Example 5 in this section and the corresponding exercises (25 and 26), we will use n in the denominator.

For a probability distribution, the variance and standard deviation are found in a similar way, but the probabilities of the outcomes are again used as weights. (Recall that expected value is a weighted average, with the probabilities as weights.)

VARIANCE AND STANDARD DEVIATION FOR A PROBABILITY DISTRIBUTION

> If a random variable x takes on the n values $x_1, x_2, x_3, \cdots, x_n$ with respective probabilities $p_1, p_2, p_3, \cdots, p_n$, and if its expected value is $E(x) = \mu$, then the variance of x is
>
> $$\mathbf{Var}(x) = p_1(x_1 - \mu)^2 + p_2(x_2 - \mu)^2 + \cdots + p_n(x_n - \mu)^2.$$
>
> The standard deviation of x is
>
> $$\sigma = \sqrt{\mathbf{Var}(x)}.$$

EXAMPLE 2

Find the variance and the standard deviation of the number of pheasant offspring from Section 7.2, given the probability distribution shown in Table 12.

Table 12

x	p_i
0	.08
1	.14
2	.29
3	.32
4	.17

In Section 7.2, the mean of this distribution was found to be $\mu = 2.36$. When using the formula given above, it is easiest to work in columns as in Example 1.

Table 13

x	p_i	$x_i - \mu$	$(x_i - \mu)^2$	$p_i(x_i - \mu)^2$
0	.08	-2.36	5.57	.45
1	.14	-1.36	1.85	.26
2	.29	$-.36$	.13	.04
3	.32	.64	.41	.13
4	.17	1.64	2.69	.46
			Total:	1.34

The total of the last column in Table 13 gives the variance, 1.34. To find the standard deviation, take the square root of the variance.

$$\sigma = \sqrt{1.34} \approx 1.16$$

Chebyshev's Theorem Suppose the mean, or expected value, μ of an unknown distribution is known, along with the standard deviation σ. What then can be said about the values in the distribution? For example, if σ is very small, we would expect most of the values in the distribution to be close to μ, while a larger value of σ would suggest more spread in the values. One estimate of the fraction of values that lie within a specified distance of the mean is given by **Chebyshev's theorem,** named after the Russian mathematician P. L. Chebyshev (1821–1894).

CHEBYSHEV'S THEOREM

For any distribution of numbers with mean μ and standard deviation σ, the probability that a number will lie within k standard deviations of the mean is at least

$$1 - \frac{1}{k^2}.$$

That is,

$$P(\mu - k\sigma \leq x \leq \mu + k\sigma) \geq 1 - \frac{1}{k^2}.$$

■ EXAMPLE 3

By Chebyshev's theorem, at least

$$1 - \frac{1}{3^2} = 1 - \frac{1}{9} = \frac{8}{9},$$

or about 89%, of the numbers in any distribution lie within 3 standard deviations of the mean. Figure 6 shows a geometric interpretation of this result. ■

At least 89% of the distribution falls
in this interval.

FIGURE 6

Suppose a distribution has mean 52 and standard deviation 3.5. Then "3 standard deviations" is $3 \times 3.5 = 10.5$, and "3 standard deviations from the mean" is

$$52 - 10.5 \qquad \text{to} \qquad 52 + 10.5,$$

or $\qquad\qquad\qquad$ 41.5 $\qquad$ to $\qquad$ 62.5.

By Example 3, at least 89% of the values in this distribution will lie between 41.5 and 62.5.

Chebyshev's theorem gets much of its importance from the fact that it applies to *any* distribution—only the mean and the standard deviation must be known. Other results given later produce more accurate estimates, because more is known about the nature of the distribution.

■ EXAMPLE 4

The Forever Power Company claims that its batteries have a mean life of 26.2 hours, with a standard deviation of 4.1 hours. In a shipment of 100 batteries, about how many will have a life within 2 standard deviations of the mean—that is, between $26.2 - (4.1 \times 2) = 18$ and $26.2 + (4.1 \times 2) = 34.4$ hours?

Use Chebyshev's theorem with $k = 2$. At least

$$1 - \frac{1}{2^2} = 1 - \frac{1}{4} = \frac{3}{4},$$

or 75%, of the batteries should have a life within 2 standard deviations of the mean.

At least $75\% \times 100 = 75$ of the batteries can be expected to last between 18 and 34.4 hours. ■

■ EXAMPLE 5

Statistical process control is a method of determining when a manufacturing process is out of control, producing defective items. The procedure involves taking samples of a measurement on a product over a production run and calculating the mean and standard deviation of each sample. These results are used to determine when the manufacturing process is out of control. For example, three sample measurements from a manufacturing process on each of four days are given in Table 14. The mean $\bar{x}$ and standard deviation s are calculated for each sample. As mentioned earlier, the formula used for a sample standard deviation is

$$s = \sqrt{\frac{\Sigma(x_i - \bar{x})^2}{n - 1}},$$

and we will use this formula in this applied example.

Table 14

Day	1			2			3			4		
Sample Number	1	2	3	1	2	3	1	2	3	1	2	3
Measurements	-3	0	4	5	-2	4	3	-1	0	4	-2	1
	0	5	3	4	0	3	-2	0	0	3	0	3
	2	2	2	3	1	4	0	1	-2	3	-1	0
$\bar{x}$	$-1/3$	7/3	3	4	$-1/3$	11/3	1/3	0	$-2/3$	10/3	-1	4/3
s	2.5	2.5	1	1	1.5	.6	2.5	1	1.2	.6	1	1.5

Next, the mean of the 12 sample means and the mean of the 12 sample standard deviations are found (using the formula for $\bar{x}$). Here, these measures are

$$\mu = 1.3 \quad \text{and} \quad \bar{s} = 1.41.$$

The control limits for the sample means are given by

$$\mu \pm k_1 \bar{s},$$

where k_1 is a constant found from a manual. For samples of size 3, $k_1 = 1.954$, so the control limits for the sample means are

$$1.3 \pm (1.954)(1.41).$$

The upper control limit is 4.06, and the lower control limit is -1.46.

Similarly, the control limits for the sample standard deviations are given by $k_2 \cdot \bar{s}$ and $k_3 \cdot \bar{s}$, where k_2 and k_3 also are values given in the same manual. Here, $k_2 = 2.568$ and $k_3 = 0$, with the upper and lower control limits for the sample standard deviations equal to 2.568(1.41) and 0(1.41), or 3.62 and 0. As long as the sample means are between -1.46 and 4.06 and the sample standard deviations are between 0 and 3.62, the process is in control. ▬

7.3 EXERCISES

Find the standard deviation for each set of numbers.

1. 42, 38, 29, 74, 82, 71, 35

2. 122, 132, 141, 158, 162, 169, 180

3. 241, 248, 251, 257, 252, 287

4. 51, 58, 62, 64, 67, 71, 74, 78, 82, 93

5. 3, 7, 4, 12, 15, 18, 19, 27, 24, 11

6. 15, 42, 53, 7, 9, 12, 28, 47, 63, 14

Find the variance and standard deviation for each probability distribution.

7.

x_i	2	3	4	5
p_i	.1	.3	.4	.2

8.

x_i	10	20	30	40
p_i	.1	.5	.3	.1

9.

x_i	.01	.02	.03	.04	.05
p_i	.1	.5	.2	.1	.1

10.

x_i	100	105	110	115	120
p_i	.01	.08	.20	.50	.21

Find the standard deviation of the random variable in each of the following problems. (See Exercises 23–26 in Section 7.2.)

11. The number of yellow marbles in a sample of 3 marbles drawn from a bag containing 3 yellow and 4 white marbles

12. The number of rotten apples in a sample of 2 apples drawn from a barrel of 25 apples, 5 of which are known to be rotten

13. The number of liberals on a committee of 3 selected from a city council made up of 5 liberals and 4 conservatives

14. The number of women in a delegation of 2 selected from a group of 2 women and 5 men

Exercises 15 and 16 give histograms for pairs of probability distributions. Decide, using the graphs only, which distribution in each pair has the greater variance.

15.

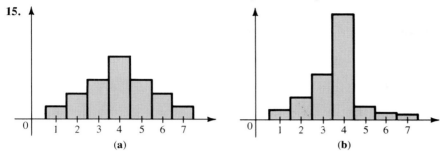

(a)   (b)

16.

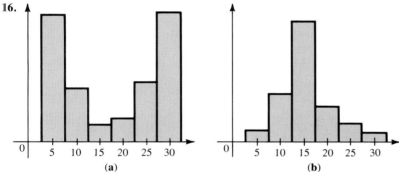

(a)   (b)

Calculate the variance of the probability distribution for each of the following histograms.

17.

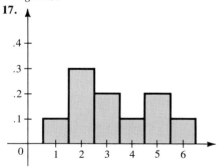

18.

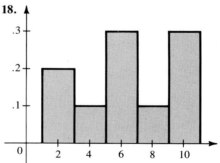

19. Use Chebyshev's theorem to find the fractions of a distribution that lie within the following numbers of standard deviations from the mean.

   (a) 2   **(b)** 4   **(c)** 5

20. A probability distribution has an expected value of 50 and a standard deviation of 6. Use Chebyshev's theorem to tell what percent of the numbers lie between the following pairs of values.

   (a) 38 and 62   **(b)** 32 and 68   **(c)** 26 and 74   **(d)** 20 and 80

   (e) Less than 38 or more than 62   **(f)** Less than 32 or more than 68

21. The weekly wages of the seven workers at Harold's Hardware Store are $180, $190, $240, $256, $300, $360, and $714.

   (a) Find the mean and standard deviation of this distribution.

   (b) How many of the workers earn wages within 1 standard deviation of the mean?

   (c) How many earn wages within 2 standard deviations of the mean?

   (d) What does Chebyshev's theorem give as the number earning wages within 2 standard deviations of the mean?

22. Show that the formula for variance given in the text can be rewritten as

$$\frac{1}{n^2}[n \cdot \Sigma(x^2) - (\Sigma x)^2].$$

 APPLICATIONS

BUSINESS AND ECONOMICS

Battery Life   **23.** The Forever Power Company conducted tests on the life of its batteries and those of a competitor (Brand X). They found that their batteries had a mean life in hours of 26.2, with a standard deviation of 4.1 (see Example 4). Their results for a sample of 10 Brand X batteries were as follows: 15, 18, 19, 23, 25, 25, 28, 30, 34, 38.

   (a) Find the mean and standard deviation for Brand X batteries.

   (b) Which batteries have a more uniform life in hours?

   (c) Which batteries have the highest average life in hours?

Sales Promotion   **24.** The Quaker Oats Company conducted a survey to determine whether a proposed premium, to be included in boxes of cereal, was appealing enough to generate new sales.* Four cities were used as test markets, where the cereal was distributed with the premium, and four cities as control markets, where the cereal was distributed without the premium. The eight cities were chosen on the basis of their similarity in terms of population, per capita income, and total cereal purchase volume. The results were as follows.

	City	Percent Change in Average Market Share Per Month
Test Cities	1	+18
	2	+15
	3	+7
	4	+10
Control Cities	1	+1
	2	-8
	3	-5
	4	0

---

*This example was supplied by Jeffery S. Berman, Senior Analyst, Marketing Information, Quaker Oats Company.

(a) Find the mean of the change in market share for the four test cities.

(b) Find the mean of the change in market share for the four control cities.

(c) Find the standard deviation of the change in market share for the test cities.

(d) Find the standard deviation of the change in market share for the control cities.

(e) Find the difference between the means of (a) and (b). This difference represents the estimate of the percent change in sales due to the premium.

(f) The two standard deviations from (c) and (d) were used to calculate an "error" of $\pm$ 7.95 for the estimate in (e). With this amount of error, what are the smallest and largest estimates of the increase in sales?

On the basis of the interval estimate of part (f), the company decided to mass-produce the premium and distribute it nationally.

Process Control   **25.** The following table gives 10 samples of three measurements each made during a production run.

				Sample Number					
1	2	3	4	5	6	7	8	9	10
2	3	-2	-3	-1	3	0	-1	2	0
-2	-1	0	1	2	2	1	2	3	0
1	4	1	2	4	2	2	3	2	2

Use the information in Example 5 to find the following.

(a) Find the mean $\bar{x}$ for each sample of 3 measurements.

(b) Find the standard deviation s for each sample of 3 measurements.

(c) Find the mean μ of the sample means.

(d) Find the mean $\bar{s}$ of the sample standard deviations.

(e) Using $k_1 = 1.954$, find the upper and lower control limits for the sample means.

(f) Using $k_2 = 2.568$ and $k_3 = 0$, find the upper and lower control limits for the sample standard deviations.

Process Control   **26.** Given the following measurements from later samples on the process in Exercise 25, decide whether the process is out of control.

	Sample Number				
1	2	3	4	5	6
3	-4	2	5	4	0
-5	2	0	1	-1	1
2	1	1	-4	-2	-6

FOR THE COMPUTER

Use a computer to solve the problems in Exercises 27–30.

27. The prices of pork bellies futures on the Chicago Mercantile Exchange over a period of several weeks are shown below.

48.25	48.50	47.75	48.45	46.85
47.10	46.50	46.90	46.60	47.00
46.35	46.65	46.85	47.20	46.60
47.00	45.00	45.15	44.65	45.15
46.25	45.90	46.10	45.82	45.70
47.05	46.95	46.90	47.15	47.10

Find the mean and standard deviation. ·

28. An assembly-line machine turns out washers with the following thicknesses in millimeters.

1.20	1.01	1.25	2.20	2.58	2.19
1.29	1.15	2.05	1.46	1.90	2.03
2.13	1.86	1.65	2.27	1.64	2.19
2.25	2.08	1.96	1.83	1.17	2.24

Find the mean and standard deviation of these thicknesses.

29. Twenty-five laboratory rats used in an experiment to test the food value of a new product made the following weight gains in grams.

5.25	5.03	4.90	4.97	5.03
5.12	5.08	5.15	5.20	4.95
4.90	5.00	5.13	5.18	5.18
5.22	5.04	5.09	5.10	5.11
5.23	5.22	5.19	4.99	4.93

Find the mean gain and the standard deviation of the gains.

30. A medical laboratory tested 21 samples of human blood for acidity on the pH scale with the following results.

7.1	7.5	7.3	7.4	7.6	7.2	7.3
7.4	7.5	7.3	7.2	7.4	7.3	7.5
7.5	7.4	7.4	7.1	7.3	7.4	7.4

Find the mean and standard deviation.

▃ 7.4 THE NORMAL DISTRIBUTION

The bank transactions in the example in Section 7.1 were timed to the nearest minute. Theoretically at least, they could have been timed to the nearest tenth of a minute, or hundredth of a minute, or even more accurately. Actually, it is possible for the transaction times to take on any real number value greater than 0. As mentioned earlier, a distribution in which the random variable can take any real-number value within some interval is a continuous distribution.

The distribution of heights (in inches) of female college freshmen is another example of a continuous distribution, since these heights include infinitely many possible measurements, such as 53, 58.5, 66.3, 72.666, . . . , and so on. Figure 7 shows the continuous distribution of heights of female college freshmen. Here the most frequent heights occur near the center of the interval shown.

Another continuous curve, which approximates the distribution of yearly incomes in the United States, is shown in Figure 8. From the graph, it can be seen that the most frequent incomes are grouped near the low end of the interval. This kind of distribution, where the peak is not at the center, is called **skewed.**

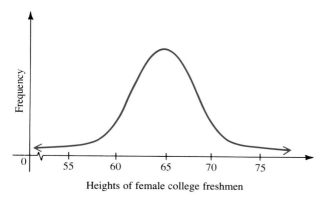

Heights of female college freshmen

FIGURE 7

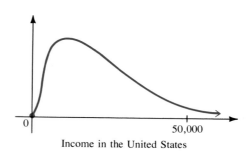

Income in the United States

FIGURE 8

Many different experiments produce probability distributions that come from a very important class of continuous distributions called **normal probability distributions.** Each normal probability distribution has associated with it a bell-shaped curve, such as the one in Figure 9. This curve, called a **normal curve,** is symmetric with respect to a vertical line drawn through the mean, μ. Vertical lines drawn at points $+ 1\sigma$ and $- 1\sigma$ from the mean show where the direction of ''curvature'' of the graph changes. (For those who have studied calculus, these points are the inflection points of the graph.)

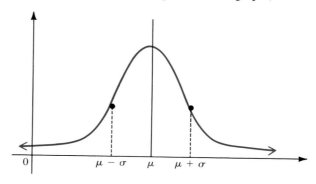

FIGURE 9

The distribution shown in Figure 7 is approximately a normal distribution, while the one in Figure 8 is not. The distribution of the lengths of the leaves of a certain tree would approximate a normal distribution, as would the distribution of the actual weights of cereal boxes that have an average weight of 16 ounces.

A normal curve never touches the *x*-axis—it extends indefinitely in both directions. The area under a normal curve is always the same: 1. If the value of the mean μ is fixed, changing the value of σ will change the shape of the normal curve. A larger value of σ produces a "flatter" normal curve, while smaller values of σ produce more values near the mean, which results in a "taller" normal curve. See Figure 10.

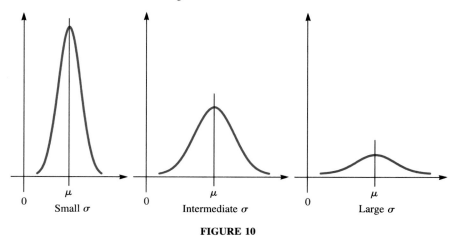

0 μ	0 μ	0 μ
Small σ	Intermediate σ	Large σ

FIGURE 10

For an experiment with normally distributed outcomes, the probability that an experiment produces a result between *a* and *b* is equal to the area under the associated normal curve from *a* to *b*. That is, the shaded area in Figure 11 gives the probability that the experimental outcome is between *a* and *b*. (Notice how the work under discussion in this section is related to the work with histograms in Section 7.1.)

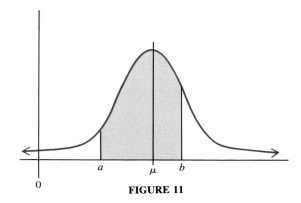

FIGURE 11

Since a normal curve is symmetric with respect to a vertical line through the mean, and since the total area under a normal curve is 1, the probability that a particular outcome is less than the mean is 1/2. A normal curve is the graph of a continuous distribution, with an infinite number of possible values, so the probability of the occurrence of any *particular* value is 0.

PROBABILITIES FOR A NORMAL PROBABILITY DISTRIBUTION

Let x be a random variable with a normal probability distribution. Then

1. $P(a \leq x \leq b)$ is the area under the associated normal curve between a and b;
2. $P(x < \mu) = 1/2$;
3. $P(x > \mu) = 1/2$;
4. $P(x) = 0$ for any real number x;
5. $P(x < a) = P(x \leq a)$ for any real number a.

Part (5) follows from part (4).

The equation of the normal curve having mean μ and standard deviation σ is given by

$$y = \frac{1}{\sigma\sqrt{2\pi}} e^{-[(x-\mu)/\sigma]^2/2},$$

where $e \approx 2.7182818$. This equation, along with calculus, can be used to find probabilities from normal curves. This approach produces an infinite number of different tables, however—one for each pair of values of μ and σ. This problem can be avoided by using just one table, the table for the normal curve where $\mu = 0$ and $\sigma = 1$, to find values for any normal curve, as shown below.

The normal curve having $\mu = 0$ and $\sigma = 1$ is called the **standard normal curve.** The normal curve table in the Appendix gives the areas under the standard normal curve, along with a sketch of the curve. The values in this table include the total area under the standard normal curve to the left of the number z.

■ EXAMPLE 1

Find the following areas from the table in the Appendix for the standard normal curve.

(a) To the left of $z = 1.25$

Look up 1.25 in the normal curve table. (Find 1.2 in the left-hand column and .05 at the top, then locate the intersection of the corresponding row and column.) The specified area is .8944, so the shaded area shown in Figure 12 is .8944. This area represents 89.44% of the total area under the normal curve. Thus, the probability that $z \leq 1.25$ is .8944.

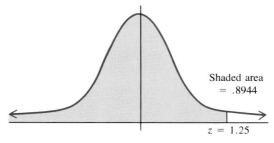

FIGURE 12

(b) To the right of $z = 1.25$

From part (a), the area to the left of $z = 1.25$ is .8944. The total area under the normal curve is 1, so the area to the right of $z = 1.25$ is

$$1 - .8944 = 1.0000 - .8944 = .1056.$$

See Figure 13, where the shaded area represents 10.56% of the total area under the normal curve, and the probability that $z \geq 1.25$ is .1056.

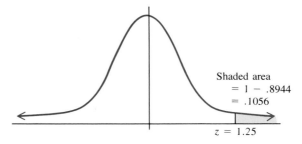

FIGURE 13

(c) Between $z = -1.02$ and $z = .92$

To find this area, which is shaded in Figure 14, start with the area to the left of $z = .92$ and subtract the area to the left of $z = -1.02$. See the two shaded regions in Figure 15. The result is $.8212 - .1539 = .6673$, which represents $P(-1.02 \leq z \leq .92)$. ■

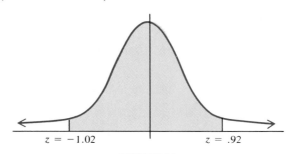

FIGURE 14

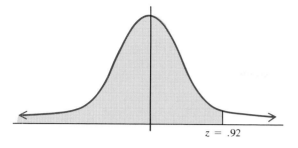

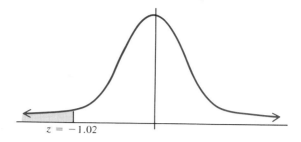

$z = .92$

$z = -1.02$

FIGURE 15

▰▰▰ EXAMPLE 2

Find a value of z satisfying the following conditions.

(a) 12.1% of the area is to the left of z.

Use the table backwards. Look in the body of the table for an area of .1210, and find the corresponding z using the left column and the top column of the table. You should find that $z = -1.17$ corresponds to an area of .1210.

(b) 20% of the area is to the right of z.

If 20% of the area is to the right, 80% is to the left. Find the z corresponding to an area of .8000. The closest value is $z = .84$. ▰▰▰

If a normal distribution does not have $\mu = 0$ and $\sigma = 1$, use the following theorem, which is stated without proof.

AREA UNDER A NORMAL CURVE

> ▰▰▰ Suppose a normal distribution has mean μ and standard deviation σ. The area under the associated normal curve to the left of the value x is exactly the same as the area to the left of
>
> $$z = \frac{x - \mu}{\sigma}$$
>
> for the standard normal curve.

Using this result, the normal curve table can be used for *any* normal curve with any values of μ and σ. The number z in the theorem is called a **z-score.**

▰▰▰ EXAMPLE 3

A normal distribution has mean 46 and standard deviation 7.2. Find the following areas under the associated normal curve.

(a) To the left of 50

Find the appropriate z-score using $x = 50$, $\mu = 46$, and $\sigma = 7.2$. Round to the nearest hundredth.

$$z = \frac{50 - 46}{7.2} = \frac{4}{7.2} \approx .56$$

From the table, the desired area is .7123.

Statistics and Probability Distributions

(b) To the right of 39

$$z = \frac{39 - 46}{7.2} = \frac{-7}{7.2} \approx -.97$$

The area to the *left* of $z = -.97$ is .1660, so the area to the *right* is

$$1 - .1660 = .8340.$$

(c) Between 32 and 43

Find z-scores for both values.

$$z = \frac{32 - 46}{7.2} = \frac{-14}{7.2} \approx -1.94 \quad \text{and} \quad z = \frac{43 - 46}{7.2} = \frac{-3}{7.2} \approx -.42$$

Start with the area to the left of $z = -.42$ and subtract the area to the left of $z = -1.94$, getting $.3372 - .0262 = .3110$.   ■

The z-scores are actually standard deviation multiples—that is, a z-score of 2.5 corresponds to a value 2.5 standard deviations above the mean. Looking up $z = 1.00$ and $z = -1.00$ in the table shows that

$$.8413 - .1587 = .6826,$$

or 68.26%, of the area under a normal curve lies within 1 standard deviation of the mean. Also,

$$.9772 - .0228 = .9544,$$

or 95.44% of the area lies within 2 standard deviations of the mean. These results, summarized in Figure 16, can be used to get a quick estimate of results when working with normal curves.

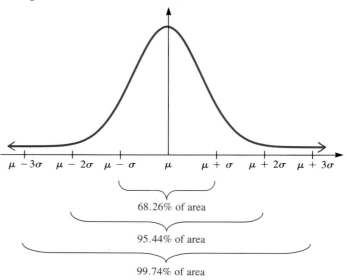

FIGURE 16

▬ EXAMPLE 4

A normal distribution has a mean of 23 and a standard deviation of 4.1. Find the following probabilities.

(a) $P(x \le 18)$

   This probability corresponds to the area under the normal curve to the left of x. Find the appropriate z-score as follows.

$$z = \frac{x - \mu}{\sigma} = \frac{18 - 23}{4.1} = -1.22$$

The table gives the area to the left of $z = -1.22$ as .1314, so

$$P(x \le 18) = .1314$$

(b) $P(20 \le x \le 25)$

   Find the two z-scores corresponding to 20 and 25.

For $x = 20$,

$$z = \frac{20 - 23}{4.1} = -.73.$$

For $x = 25$,

$$z = \frac{25 - 23}{4.1} = .49.$$

Now find the area under the standard normal curve between $-.73$ and $.49$. The area to the left of $-.73$ is .2327, and the area to the left of .49 is .6879. Then the area between $-.73$ and $.49$ is

$$.6879 - 2327 = .4552,$$

and $P(20 \le x \le 25) = .4552.$ ▬

▬ EXAMPLE 5

Suppose that the average salesperson for Dixie Office Supplies drives $\mu = 1200$ miles per month in a company car, with standard deviation $\sigma = 150$ miles. Assume that the number of miles driven is closely approximated by a normal curve. Find the percent of all drivers traveling the following distances.

(a) Between 1200 and 1600 miles per month

   First find the number of standard deviations above the mean that corresponds to 1600 miles. Do this by finding the z-score for 1600.

$$z = \frac{x - \mu}{\sigma}$$

$$= \frac{1600 - 1200}{150} \qquad \text{Let } x = 1600, \; \mu = 1200, \; \sigma = 150$$

$$= \frac{400}{150}$$

$$z \approx 2.67$$

From the table, the area to the left of $z = 2.67$ is .9962. Since $\mu = 1200$, the value 1200 corresponds to $z = 0$, the area to the left of $z = 0$ is .5000, and

$$.9962 - .5000 = .4962,$$

or 49.62%, of the drivers travel between 1200 and 1600 miles per month. See Figure 17.

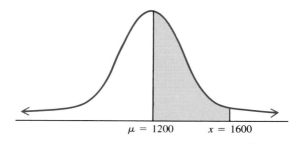

FIGURE 17

(b) Between 1000 and 1500 miles per month

As shown in Figure 18, z-scores for both $x = 1000$ and $x = 1500$ are needed.

For $x = 1000$,

$$z = \frac{1000 - 1200}{150}$$

$$= \frac{-200}{150}$$

$$z \approx -1.33.$$

For $x = 1500$,

$$z = \frac{1500 - 1200}{150}$$

$$= \frac{300}{150}$$

$$z = 2.00.$$

From the table, $z = -1.33$ leads to an area of .0918, while $z = 2.00$ corresponds to .9772. A total of $.9772 - .0918 = .8854$, or 88.54%, of the drivers travel between 1000 and 1500 miles per month. ■

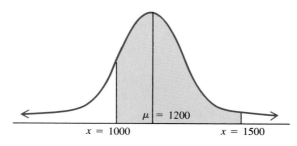

FIGURE 18

Suppose a normal distribution has $\mu = 1000$ and $\sigma = 150$. Then the method of this section can be used to show that 95.44% of all values lie within 2 standard deviations of the mean; that is, between

$$1000 - (2 \times 150) = 700 \quad \text{and} \quad 1000 + (2 \times 150) = 1300.$$

Chebyshev's theorem (Section 7.3) says that *at least*

$$1 - \frac{1}{2^2} = 1 - \frac{1}{4} = \frac{3}{4},$$

or 75%, of the values lie between 7000 and 1300. The difference between 95.44% and "at least 75%," from Chebyshev's theorem, comes from the fact that Chebyshev's theorem applies to *any* distribution, while the methods of this section apply only to *normal* distributions. Thus it should not be surprising that having more information (a normal distribution) should produce more accurate results (95.44% instead of "at least 75%.")

7.4 EXERCISES

Find the percent of the area under a normal curve between the mean and the given number of standard deviations from the mean.

1. 2.50

2. 1.68

3. .45

4. .81

5. -1.71

6. -2.04

7. 3.11

8. 2.80

Find the percent of the total area under the normal curve between each pair of z-scores.

9. $z = 1.41$ and $z = 2.83$

10. $z = .64$ and $z = 2.11$

11. $z = -2.48$ and $z = -.05$

12. $z = -1.74$ and $z = -1.02$

13. $z = -3.11$ and $z = 1.44$

14. $z = -2.94$ and $z = -.43$

15. $z = -.42$ and $z = .42$

16. $z = -1.98$ and $z = 1.98$

Find a z-score satisfying the following conditions.

17. 5% of the total area is to the left of z.

18. 1% of the total area is to the left of z.

19. 15% of the total area is to the right of z.

20. 25% of the total area is to the right of z.

Assume the distributions in Exercises 21–28 are normal, and use the areas under the normal curve given in the Appendix to answer the questions.

21. A machine produces bolts with an average diameter of .25 inches and a standard deviation of .02 inches. What is the probability that a bolt will be produced with a diameter greater than .3 inches?

22. The mean monthly income of the trainees of an engineering firm is $1200, with a standard deviation of $200. Find the probability that an individual trainee earns less than $1000 per month.

23. A machine that fills quart milk cartons is set up to average 32.2 oz per carton, with a standard deviation of 1.2 oz. What is the probability that a filled carton will contain less than 32 oz of milk?

24. The average contribution to the campaign of Polly Potter, a candidate for city council, was $50 with a standard deviation of $15. How many of the 200 people who contributed to Polly's campaign gave between $30 and $100?

25. At the Discount Market, the average weekly grocery bill is $32.25 with a standard deviation of $9.50. What are the largest and smallest amounts spent by the middle 50% of this market's customers?

26. The mean clotting time of blood is 7.45 sec, with a standard deviation of 3.6 sec. What is the probability that an individual's blood clotting time will be less than 7 sec or greater than 8 sec?

27. The average size of the fish in Lake Amotan is 12.3 inches, with a standard deviation of 4.1 inches. Find the probability of catching a fish longer than 18 inches in Lake Amotan.

28. To be graded extra large, an egg must weigh at least 2.2 oz. If the average weight for an egg is 1.5 oz, with a standard deviation of .4 oz, how many eggs in a sample of five dozen would you expect to grade extra large?

▋ APPLICATIONS

BUSINESS AND ECONOMICS

Life of Light Bulbs *A certain type of light bulb has an average life of 500 hr, with a standard deviation of 100 hr. The length of life of the bulb can be closely approximated by a normal curve. An amusement park buys and installs 10,000 such bulbs. Find the total number that can be expected to last for each of the following periods of time.*

29. At least 500 hr   30. Less than 500 hr

31. Between 500 and 650 hr   32. Between 300 and 500 hr

33. Between 650 and 780 hr   34. Between 290 and 540 hr

35. Less than 740 hr   36. More than 300 hr

37. More than 790 hr   38. Less than 410 hr

Quality Control *A box of oatmeal must contain 16 oz. The machine that fills the oatmeal boxes is set so that, on the average, a box contains 16.5 ounces. The boxes filled by the machine have weights that can be closely approximated by a normal curve. What fraction of the boxes filled by the machine are underweight if the standard deviation is as follows?*

39. .5 oz   40. .3 oz

41. .2 oz   42. .1 oz

Quality Control *The chickens at Colonel Thompson's Ranch have a mean weight of 1850 g, with a standard deviation of 150 g. The weights of the chickens are closely approximated by a normal curve. Find the percent of all chickens having weights in the following ranges.*

43. More than 1700 g   44. Less than 1800 g

45. Between 1750 and 1900 g   46. Between 1600 and 2000 g

47. Less than 1550 g   48. More than 2100 g

LIFE SCIENCES

Vitamin Requirements   *In nutrition, the Recommended Daily Allowance of vitamins is a number set by the government as a guide to an individual's daily vitamin intake. Actually, vitamin needs vary drastically from person to person, but the needs are very closely approximated by a normal curve. To calculate the Recommended Daily Allowance, the government first finds the average need for vitamins among people in the population, and the standard deviation. The Recommended Daily Allowance is then defined as the mean plus 2.5 times the standard deviation.*

49. What percent of the population will receive adequate amounts of vitamins under this plan?

Find the recommended daily allowance for each vitamin in Exercises 50–52.

50. Mean = 1800 units; standard deviation = 140 units

51. Mean = 159 units; standard deviation = 12 units

52. Mean = 1200 units; standard deviation = 92 units

SOCIAL SCIENCES

Education   *One professor uses the following system for assigning letter grades in a course.*

Grade	Total Points
A	Greater than $\mu + \frac{3}{2}\sigma$
B	$\mu + \frac{1}{2}\sigma$ to $\mu + \frac{3}{2}\sigma$
C	$\mu - \frac{1}{2}\sigma$ to $\mu + \frac{1}{2}\sigma$
D	$\mu - \frac{3}{2}\sigma$ to $\mu - \frac{1}{2}\sigma$
F	Below $\mu - \frac{3}{2}\sigma$

What percent of the students receive the following grades?

53. A   **54.** B   **55.** C

56. Do you think this system would be more likely to be fair in a large freshman class in psychology or in a graduate seminar of 5 students? Why?

Education   *A teacher gives a test to a large group of students. The results are closely approximated by a normal curve. The mean is 74, with a standard deviation of 6. The teacher wishes to give A's to the top 8% of the students and F's to the bottom 8%. A grade of B is given to the next 15%, with D's given similarly. All other students get C's. Find the bottom cutoff (rounded to the nearest whole number) for the following grades. (Hint: Use the table in the Appendix backwards.)*

57. A   **58.** B   **59.** C   **60.** D

FOR THE COMPUTER

Use a computer to find the following probabilities by finding the comparable area under a standard normal curve.

61. $P(1.372 \leq z \leq 2.548)$ **62.** $P(-2.751 \leq z \leq 1.693)$

63. $P(z > -2.476)$ **64.** $P(z < 1.692)$

65. $P(z < -.4753)$ **66.** $P(z > .2509)$

Use a computer to find the following probabilities for a distribution with a mean of 35.693 and a standard deviation of 7.104.

67. $P(12.275 < x < 28.432)$ **68.** $P(x > 38.913)$

69. $P(x < 17.462)$ **70.** $P(17.462 \leq x \leq 53.106)$

▤ 7.5 THE NORMAL CURVE APPROXIMATION TO THE BINOMIAL DISTRIBUTION

In many practical situations, an experiment can have one of only two possible outcomes: *success* and *failure*. Examples of such experiments, called *binomial trials,* or *Bernoulli trials,* include tossing a coin (perhaps *h* would be called a success and *t* a failure); rolling a die with the two outcomes being, for instance, five a success, and a result other than five a failure; or choosing a radio from a large batch and deciding whether the radio is defective or normal. (Bernoulli trials were first discussed in Section 6.7)

A **binomial distribution** must satisfy the following properties. The experiment is a series of independent trials with only two possible outcomes: success and failure. The trials are independent, and the probability of each outcome is constant from trial to trial.

As an example, suppose a die is tossed 5 times. Identify a result of a 1 or a 2 as a success and any other result as a failure. Since each trial (each toss) can result in a success or a failure, the result of the 5 tosses can be any number of successes from 0 through 5. These 6 possible outcomes are not equally likely. The various probabilities can be found with the formula from Section 6.7:

$$P(x) = \binom{n}{x} p^x (1 - p)^{n-x},$$

where *n* is the number of trials, *x* is the number of successes, *p* is the probability of success on a single trial, and $P(x)$ gives the probability that *x* of the *n* trials result in successes. In this example, $n = 5$ and $p = 1/3$, since either a 1 or a 2 results in a success. The probabilities for this experiment are tabulated in Table 15.

By definition, the mean μ of a probability distribution is given by the expected value of *x*. Expected value is found by summing the products of corresponding outcomes and probabilities. For the distribution in Table 15,

$$\mu = 0\left(\frac{32}{243}\right) + 1\left(\frac{80}{243}\right) + 2\left(\frac{80}{243}\right) + 3\left(\frac{40}{243}\right) + 4\left(\frac{10}{243}\right) + 5\left(\frac{1}{243}\right)$$

$$= \frac{405}{243} = 1\frac{2}{3}.$$

Table 15

x	$P(x)$
0	$\binom{5}{0}\left(\frac{1}{3}\right)^0\left(\frac{2}{3}\right)^5 = \dfrac{32}{243}$
1	$\binom{5}{1}\left(\frac{1}{3}\right)^1\left(\frac{2}{3}\right)^4 = \dfrac{80}{243}$
2	$\binom{5}{2}\left(\frac{1}{3}\right)^2\left(\frac{2}{3}\right)^3 = \dfrac{80}{243}$
3	$\binom{5}{3}\left(\frac{1}{3}\right)^3\left(\frac{2}{3}\right)^2 = \dfrac{40}{243}$
4	$\binom{5}{4}\left(\frac{1}{3}\right)^4\left(\frac{2}{3}\right)^1 = \dfrac{10}{243}$
5	$\binom{5}{5}\left(\frac{1}{3}\right)^5\left(\frac{2}{3}\right)^0 = \dfrac{1}{243}$

For a binomial distribution, which is a special kind of probability distribution, it can be shown that the method for finding the mean reduces to the formula

$$\mu = np,$$

where n is the number of trials and p is the probability of success on a single trial. Using this simplified formula, the computation of the mean in the example above is

$$\mu = np = 5\left(\frac{1}{3}\right) = 1\frac{2}{3},$$

which agrees with the result obtained using the expected value.

Like the mean, the variance Var(x) of a probability distribution is an expected value—the expected value of the squared deviations from the mean, $(x - \mu)^2$. To find the variance for the example given above, first use the mean $\mu = 5/3$ and find the quantities $(x - \mu)^2$. (See Table 16.)

Table 16

x	$P(x)$	$x - \mu$	$(x - \mu)^2$
0	$\dfrac{32}{243}$	$-\dfrac{5}{3}$	$\dfrac{25}{9}$
1	$\dfrac{80}{243}$	$-\dfrac{2}{3}$	$\dfrac{4}{9}$
2	$\dfrac{80}{243}$	$\dfrac{1}{3}$	$\dfrac{1}{9}$
3	$\dfrac{40}{243}$	$\dfrac{4}{3}$	$\dfrac{16}{9}$
4	$\dfrac{10}{243}$	$\dfrac{7}{3}$	$\dfrac{49}{9}$
5	$\dfrac{1}{243}$	$\dfrac{10}{3}$	$\dfrac{100}{9}$

Find Var(x) by finding the sum of the products $[(x - \mu)^2][P(x)]$.

$$\text{Var}(x) = \frac{25}{9}\left(\frac{32}{243}\right) + \frac{4}{9}\left(\frac{80}{243}\right) + \frac{1}{9}\left(\frac{80}{243}\right) + \frac{16}{9}\left(\frac{40}{243}\right)$$
$$+ \frac{49}{9}\left(\frac{10}{243}\right) + \frac{100}{9}\left(\frac{1}{243}\right)$$
$$= \frac{10}{9} = 1\frac{1}{9}$$

To find the standard deviation σ, find $\sqrt{10/9}$ or $\sqrt{10}/3$, which is approximately 1.05.

Just as with the mean, the variance of a binomial distribution can be found with a relatively simple formula. Again, it can be shown that

$$\textbf{Var}(x) = np(1 - p) \qquad \text{and} \qquad \sigma = \sqrt{np(1 - p)}.$$

By substituting the appropriate values for n and p from the example into this new formula, we get

$$\text{Var}(x) = 5\left(\frac{1}{3}\right)\left(\frac{2}{3}\right) = 10/9 = 1\frac{1}{9},$$

which agrees with the previous answer.

A summary of these results is given below.

BINOMIAL DISTRIBUTION

Suppose an experiment is a series of n independent repeated trials, where the probability of success in a single trial is always p. Let x be the number of successes in the n trials. Then the probability that exactly x successes will occcur in n trials is given by

$$\binom{n}{x}p^x(1 - p)^{n-x}.$$

The mean μ and variance $\text{Var}(x)$ of this binomial distribution are respectively

$$\mu = np \quad \text{and} \quad \text{Var}(x) = np(1 - p).$$

The standard deviation σ is

$$\sigma = \sqrt{np(1 - p)}.$$

■ EXAMPLE 1

The probability of picking a defective plate at random from a china factory's assembly line is .01. A sample of 3 plates is to be selected. Write the distribution for the number of defective plates in the sample, and give its mean and standard deviation.

Since 3 plates will be selected, the possible number of defective plates ranges from 0 to 3. Here, n (the number of trials) is 3, and p (the probability of selecting a defective plate on a single trial) is .01. The distribution and the probability of each outcome are shown in Table 17.

Table 17

x	$P(x)$
0	$\binom{3}{0}(.01)^0(.99)^3 = .970299$
1	$\binom{3}{1}(.01)(.99)^2 = .029403$
2	$\binom{3}{2}(.01)^2(.99) = .000297$
3	$\binom{3}{3}(.01)^3(.99)^0 = .000001$

The mean of the distribution is

$$\mu = np = 3(.01) = .03.$$

The standard deviation is

$$\sigma = \sqrt{np(1 - p)} = \sqrt{3(.01)(.99)} = \sqrt{.0297} = .17. \quad \blacksquare$$

The binomial distribution is extremely useful, but its use can lead to complicated calculations if n is large. However, the normal curve of the previous section can be used to get a good approximation to the binomial distribution. This approximation was first discovered in 1718 by Abraham DeMoivre (1667–1754) for the case $p = 1/2$. The result was generalized by the French mathematician Pierre–Simon Laplace (1749–1827) in a book published in 1812.

To see how the normal curve is used, look at the bar graph and normal curve in Figure 19. The histogram shows the expected number of heads if one coin is tossed 15 times, with the experiment repeated 32,768 times. Since the probability of heads on one toss is 1/2 and $n = 15$, the mean of this distribution is

$$\mu = np = 15\left(\frac{1}{2}\right) = 7.5.$$

The standard deviation is

$$\sigma = \sqrt{15\left(\frac{1}{2}\right)\left(1 - \frac{1}{2}\right)} = \sqrt{15\left(\frac{1}{2}\right)\left(\frac{1}{2}\right)} = \sqrt{3.75} \approx 1.94.$$

In Figure 19 we have superimposed the normal curve with $\mu = 7.5$ and $\sigma = 1.94$ over the histogram of the distribution.

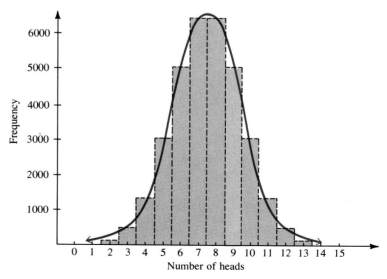

FIGURE 19

Suppose we need to know the fraction of the time that exactly 9 heads would be obtained in the 15 tosses. Using the binomial distribution formula, this can be calculated as

$$P(9) = \binom{15}{9}\left(\frac{1}{2}\right)^9\left(\frac{1}{2}\right)^6 \approx .153.$$

This answer is about the same fraction that would be found by dividing the area of the bar in color in Figure 19 by the total area of all 16 bars in the graph. (Some of the bars at the extreme left and right ends of the graph are too short to be visible.)

As the graph suggests, the area in color is approximately equal to the area under the normal curve from $x = 8.5$ to $x = 9.5$. The normal curve is higher than the top of the bar in the left half but lower in the right half.

To find the area under the normal curve from $x = 8.5$ to $x = 9.5$, first find z-scores, as in the last section. Do this with the mean and the standard deviation for the distribution given above to get z-scores for $x = 8.5$ and $x = 9.5$.

For $x = 8.5$,

$$z = \frac{8.5 - 7.5}{1.94}$$

$$= \frac{1.00}{1.94}$$

$$z \approx .52$$

For $x = 9.5$,

$$z = \frac{9.5 - 7.5}{1.94}$$

$$= \frac{2.00}{1.94}$$

$$z \approx 1.03$$

From the table in the Appendix, $z = .52$ gives an area of .6985, and $z = 1.03$ gives .8485. The difference between these two numbers is the desired result.

$$.8485 - .6985 = .1500$$

This answer (.1500) is not far from the exact answer, .153, found above.

■ EXAMPLE 2

About 6% of the bolts produced by a certain machine are defective.

(a) Find the probability that in a sample of 100 bolts, 3 or fewer are defective.

This problem satisfies the conditions of the definition of a binomial distribution, so the normal curve approximation can be used. First find the mean and the standard deviation using $n = 100$ and $p = 6\% = .06$.

$$\mu = 100(.06) \qquad \sigma = \sqrt{100(.06)(1 - .06)}$$
$$= 6 \qquad\qquad = \sqrt{100(.06)(.94)}$$
$$\qquad\qquad = \sqrt{5.64} \approx 2.37$$

As the graph in Figure 20 shows, we need to find the area to the left of $x = 3.5$ (since we want 3 or fewer defective bolts). The z-score corresponding to $x = 3.5$ is

$$z = \frac{3.5 - 6}{2.37} = \frac{-2.5}{2.37} \approx -1.05.$$

From the table, $z = -1.05$ leads to an area of .1469, so the probability of getting 3 or fewer defective bolts in a set of 100 bolts is .1469, or 14.69%.

(b) Find the probability of getting exactly 11 defective bolts in a sample of 100 bolts.

As Figure 21 shows, we need to find the area between $x = 10.5$ and $x = 11.5$.

$$\text{If } x = 10.5, \text{ then } z = \frac{10.5 - 6}{2.37} \approx 1.90.$$

$$\text{If } x = 11.5, \text{ then } z = \frac{11.5 - 6}{2.37} \approx 2.32.$$

Use the table to find that $z = 1.90$ gives an area of .9713, and $z = 2.32$ yields .9898. The final answer is the difference of these numbers, or

$$.9898 - .9713 = .0185.$$

The probability of having exactly 11 defective bolts is about .0185. ▬

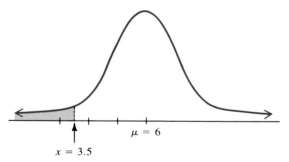

FIGURE 20

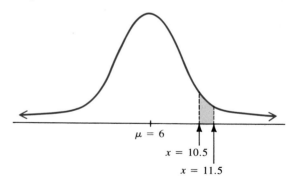

FIGURE 21

The normal curve approximation to the binomial distribution is usually quite accurate, especially for practical problems. For n up to say, 15 or 20, it is usually not too difficult to actually calculate the binomial probabilities directly. For larger values of n, a rule of thumb is that the normal curve approximation can be used as long as both np and $n(1 - p)$ are at least 5.

7.5 EXERCISES

In Exercises 1–6, several binomial experiments are described. For each one, give **(a)** *the distribution;* **(b)** *the mean; and* **(c)** *the standard deviation.*

1. A die is rolled six times and the number of 1's that come up is tallied. Write the distribution of 1's that can be expected to occur.

2. A 6-item multiple-choice test has 4 possible answers for each item. A student selects all his answers randomly. Give the distribution of correct answers.

3. To maintain quality control on its production line, the Bright Lite Company randomly selects 3 light bulbs each day for testing. Experience has shown a defective rate of .02. Write the distribution for the number of defectives in the daily samples.

4. In a taste test, each member of a panel of 4 is given 2 glasses of Supercola, one made using the old formula and one with the new formula, and asked to identify the new formula. Assuming the panelists operate independently, write the distribution of the number of successful identifications, if each judge actually guesses.

5. The probability that a radish seed will germinate is .7. Joe's mother gives him 4 seeds to plant. Write the distribution for the number of seeds that germinate.

6. Five patients in Ward 8 of Memorial Hospital have a disease with a known mortality rate of .1. Write the distribution of the number who survive.

Work the following problems involving binomial experiments.

7. The probability that an infant will die in the first year of life is about .025. In a group of 500 babies, what are the mean and standard deviation of the number of babies who can be expected to die in their first year of life?

8. The probability that a particular kind of mouse will have a brown coat is 1/4. In a litter of 8, assuming independence, how many could be expected to have a brown coat? With what standard deviation?

9. A certain drug is effective 80% of the time. Give the mean and standard deviation of the number of patients using the drug who recover, out of a group of 64 patients.

10. The probability that a newborn infant will be a girl is .49. If 50 infants are born on Susan B. Anthony's birthday, how many can be expected to be girls? With what standard deviation?

For the remaining exercises, use the normal curve approximation to the binomial distribution.

Suppose 16 coins are tossed. Find the probability of getting each of the following results.

11. Exactly 8 heads   12. Exactly 7 heads   13. Exactly 10 tails   14. Exactly 12 tails

Suppose 1000 coins are tossed. Find the probability of getting each of the following results.

15. Exactly 500 heads   16. Exactly 510 heads   17. 480 heads or more

18. Less than 470 tails   19. Less than 518 heads   20. More than 550 tails

A die is tossed 120 times. Find the probability of getting each of the following results. (*Hint:* $\sigma = 4.08$)

21. Exactly twenty 5's   22. Exactly twenty-four 6's   23. Exactly seventeen 3's

24. Exactly twenty-two 2's   **25.** More than eighteen 3's   **26.** Fewer than twenty-two 6's

27. An experimental drug causes a rash in 15% of all people taking it. If the drug is given to 12,000 people, find the probability that more than 1700 people will get the rash.

28. In one state, 55% of the voters expect to vote for Jones. Suppose 1400 people are asked the name of the person for whom they expect to vote. Find the probability that at least 700 people will say that they expect to vote for Jones.

▤ APPLICATIONS

BUSINESS AND ECONOMICS

Quality Control   *Two percent of the quartz heaters produced in a certain plant are defective. Suppose the plant produced 10,000 such heaters last month. Find the probability that among these heaters, the following numbers were defective.*

29. Fewer than 170   **30.** More than 222

LIFE SCIENCES

Drug Effectiveness   *A new drug cures 80% of the patients to whom it is administered. It is given to 25 patients. Find the probability that among these patients, the following results occur.*

31. Exactly 20 are cured.   **32.** Exactly 23 are cured.

33. All are cured.   **34.** No one is cured.

35. Twelve or fewer are cured.   **36.** Between 17 and 23 (inclusive) are cured.

FOR THE COMPUTER

37. Rework Exercises 67–70 from Section 6.6 using the normal curve to approximate the binomial probabilities. Compare your answers with the results found in Section 6.6 for the binomial distribution.

38. A coin is tossed 100 times. Find the probability of getting the following results.

   (a) Exactly 50 heads   **(b)** At least 55 heads   **(c)** No more than 40 heads

▤ KEY WORDS ▤

7.1 random variable
frequency distribution
probability distribution
probability distribution function
discrete probability function
continuous probability function
histogram

7.2 arithmetic mean
expected value
weighted average
fair game

median
mode

7.3 variance
standard deviation
Chebyshev's theorem

7.4 skewed distribution
normal distribution
standard normal curve
z-score

7.5 binomial distribution

CHAPTER 7   REVIEW EXERCISES

In Exercises 1–5, **(a)** *give a probability distribution, and* **(b)** *sketch its histogram.*

1.

x	1	2	3	4	5
Frequency	3	7	9	3	2

2.

x	8	9	10	11	12	13	14
Frequency	1	0	2	5	8	4	3

3. A coin is tossed 3 times and the number of heads is recorded.

4. A pair of dice are rolled and the sum of the results for each roll is recorded.

5. Patients in groups of 5 are given a new treatment for a fatal disease. The experiment is repeated 10 times with the following results.

Number Who Survived	Frequency
0	1
1	1
2	2
3	3
4	3
5	0
	Total: 10

In Exercises 6 and 7, give the probability that corresponds to the shaded region of each histogram.

6.

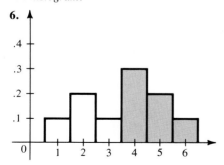

7.

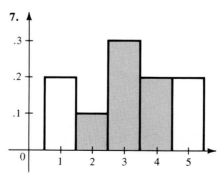

Solve the following problems.

8. You pay $6 to play in a game where you will roll a die, with payoffs as follows: $8 for a 6, $7 for a 5, and $4 for any other results. What are your expected winnings? Is the game fair?

9. A lottery has a first prize of $5000, two second prizes of $1000 each, and two $100 third prizes. A total of 10,000 tickets is sold, at $1 each. Find the expected winnings of a person buying 1 ticket.

10. Find the expected number of girls in a family of 5 children.

11. A developer can buy a piece of property that will produce a profit of $16,000 with probability .7, or a loss of $9000 with probability .3. What is the expected profit?

12. Game boards for a recent United Airlines contest could be obtained by sending a self-addressed stamped envelope to a certain address. The prize was a ticket for any city to which United flies. Assume that the value of the ticket was $1000 (we might as well go first-class), and that the probability that a particular game board would win was 1/4000. If the stamps to enter the contest cost 30¢ and envelopes cost 1¢ each, find the expected winnings for a person ordering 1 game board.

Find the expected value of the random variable in each of the following.

13. The data in Exercise 2   **14.** The experiment in Exercise 3   **15.** The experiment in Exercise 5

16. Three cards are drawn from a standard deck of 52 cards.
  (a) What is the expected number of aces?
  (b) What is the expected number of clubs?

17. Suppose someone offers to pay you $100 if you draw 3 cards from a standard deck of 52 cards and all the cards are clubs. What should you pay for the chance to win if it is a fair game?

Find the variance and standard deviation of the random variable in the following.

18. The experiment in Exercise 3   **19.** The experiment in Exercise 5

20. A probability distribution has an expected value of 28 and a standard deviation of 4. Use Chebyshev's theorem to decide what percent of the distribution is
  (a) between 20 and 36;   **(b)** less than 23.2 or greater than 32.8.

21. **(a)** Find the percent of the area under a normal curve within 2.5 standard deviations of the mean.
  (b) Compare your answer to part **(a)** with the result using Chebyshev's theorem.

22. Find the percent of the total area under the normal curve that corresponds to
  (a) $z \geq 2.2$;   **(b)** between $z = -1.3$ and $z = .2$.

23. Find a z-score such that 8% of the area under the curve is to the right of z.

24. Find the probability of getting the following number of heads in 15 tosses of a coin.
  (a) Exactly 7   **(b)** Between 7 and 10 (exclusive)   **(c)** At least 10

▮▮▮ APPLICATIONS

BUSINESS AND ECONOMICS

Stock Returns   **25.** The annual returns of two stocks for three years are given below.

Stock	1986	1987	1988
Stock I	11%	−1%	14%
Stock II	9%	5%	10%

(a) Find the mean and standard deviation for each stock over the 3-year period.

(b) If you are looking for security with an 8% return, which of these stocks should you choose?

Quality Control 26. A machine that fills quart milk cartons is set to fill them with 32.1 oz. If the actual contents of the cartons vary normally with a standard deviation of .1 oz, what percent of the cartons contain less than a quart (32 oz)?

Quality Control 27. The probability that a can of beer from a certain brewery is defective is .005. A sample of 4 cans is selected at random. Write a distribution for the number of defective cans in the sample, and give its mean and standard deviation.

Quality Control 28. About 6% of the frankfurters produced by a certain machine are overstuffed and thus defective. Find the probability that in a sample of 500 frankfurters,

(a) 25 or fewer are overstuffed (*Hint:* $\sigma = 5.3$);

(b) exactly 30 are overstuffed;

(c) more than 40 are overstuffed.

Bankruptcy 29. The probability that a small business will go bankrupt in its first year is .21. For 50 such small businesses, find the following probabilities.

(a) Exactly 8 go bankrupt.

(b) No more than 2 go bankrupt.

LIFE SCIENCES

Rat Diets 30. The weight gains of 2 groups of 10 rats fed different experimental diets were as follows.

Diet	Weight Gains									
Diet A	1	0	3	7	1	1	5	4	1	4
Diet B	2	1	1	2	3	2	1	0	1	0

Compute the mean and standard deviation for each group.

(a) Which diet produced the greatest mean gain?

(b) Which diet produced the most consistent gain?

SOCIAL SCIENCES

Commuting Times 31. The residents of a certain suburb average 42 min a day commuting to work, with a standard deviation of 12 min. Assume that commuting times are closely approximated by a normal curve and find the percent of the residents of this suburb who commute

(a) at least 50 min per day;   (b) no more than 35 min per day;

(c) between 32 and 40 min per day;   (d) between 38 and 60 min per day.

I.Q. Scores 32. On standard IQ tests, the mean is 100, with a standard deviation of 15. The results are very close to fitting a normal curve. Suppose an IQ test is given to a very large group of people. Find the percentage of those people whose IQ score is as follows:

(a) more than 130;   (b) less than 85;   (c) between 85 and 115.

EXTENDED APPLICATION

OPTIMAL INVENTORY FOR A SERVICE TRUCK

For many different items it is difficult or impossible to take the item to a central repair facility when service is required. Washing machines, large television sets, office copiers, and computers are only a few examples of such items. Service for items of this type is commonly performed by sending a repair person to the item, with the person driving to the item in a truck containing various parts that might be required in repairing the item. Ideally, the truck should contain all the parts that might be required in repairing the item. However, most parts would be needed only infrequently, so that inventory costs for the parts would be high.

An optimum policy for deciding on the parts to stock on a truck would require that the probability of not being able to repair an item without a trip back to the warehouse for needed parts be as low as possible, consistent with minimum inventory costs. An analysis similar to the one below was developed at the Xerox Corporation.*

To set up a mathematical model for deciding on the optimum truck stocking policy, let us assume that a broken machine might require one of 5 different parts (we could assume any number of different parts—we use 5 to simplify the notation). Suppose also that the probability that a particular machine requires part 1 is p_1, that it requires part 2 is p_2, and so on. Assume also that failures of different part types are independent, and that at most one part of each type is used on a given job.

Suppose that, on the average, a repair person makes N service calls per time period. If the repair person is unable to make a repair because at least one of the parts is unavailable, there is a penalty cost, L, corresponding to wasted time for the repair person, an extra trip to the parts depot, customer unhappiness, and so on. For each of the parts carried on the truck, an average inventory cost is incurred. Let H_i be the average inventory cost for part i, where $1 \leq i \leq 5$.

Let M_1 represent a policy of carrying only part 1 on the repair truck, M_{24} represent a policy of carrying only parts 2 and 4, with M_{12345} and M_0 representing policies of carrying all parts and no parts, respectively.

For policy M_{35}, carrying parts 3 and 5 only, the expected cost per time period per repair person, written $C(M_{35})$, is

$$C(M_{35}) = (H_3 + H_5) + NL[1 - (1 - p_1)(1 - p_2)(1 - p_4)].$$

(The expression in brackets represents the probability of needing at least one of the parts not carried, 1, 2, or 4 here.) As further examples,

$$C(M_{125}) = (H_1 + H_2 + H_5) + NL[1 - (1 - p_3)(1 - p_4)],$$

while

$$C(M_{12345}) = (H_1 + H_2 + H_3 + H_4 + H_5) + NL[1 - 1]$$
$$= H_1 + H_2 + H_3 + H_4 + H_5 \,,$$

and

$$C(M_0) = NL[1 - (1 - p_1)(1 - p_2)(1 - p_3)(1 - p_4)(1 - p_5)].$$

To find the best policy, evaluate $C(M_0)$, $C(M_1)$, $\cdots$, $C(M_{12345})$ and choose the smallest result. (A general solution method is in the *Management Science* paper.)

*Reprinted by permission of Stephen Smith, John Chambers, and Eli Shlifer, "Optimal Inventories Based on Job Completion Rate for Repairs Requiring Multiple Items," *Management Science*, Vol. 26, No. 8, August 1980, copyright © 1980 The Institute of Management Sciences.

EXAMPLE

Suppose that for a particular item, only 3 possible parts might need to be replaced. By studying past records of failures of the item, and finding necessary inventory costs, suppose that the following values have been found.

p_1	p_2	p_3	H_1	H_2	H_3
.09	.24	.17	\$15	\$40	\$9

Suppose $N = 3$ and L is \$54. Then, as an example,

$$\begin{aligned}
C(M_1) &= H_1 + NL[1 - (1 - p_2)(1 - p_3)]\\
&= 15 + 3(54)[1 - (1 - .24)(1 - .17)]\\
&= 15 + 3(54)[1 - (.76)(.83)]\\
&\approx 15 + 59.81\\
&= 74.81.
\end{aligned}$$

Thus, if policy M_1 is followed (carrying only part 1 on the truck), the expected cost per repair person per time period is \$74.81. Also,

$$\begin{aligned}
C(M_{23}) &= H_2 + H_3 + NL[1 - (1 - p_1)]\\
&= 40 + 9 + 3(54)[.09]\\
&= 63.58,
\end{aligned}$$

so that M_{23} is a better policy than M_1. By finding the expected values for all other possible policies (see the exercises below), the optimum policy may be chosen. ▬

EXERCISES

1. Refer to the example above and find each of the following.
   (a) $C(M_0)$ **(b)** $C(M_2)$ **(c)** $C(M_3)$ **(d)** $C(M_{12})$ **(e)** $C(M_{13})$ **(f)** $C(M_{123})$
2. Which policy leads to lowest expected cost?
3. In the example above, $p_1 + p_2 + p_3 = .09 + .24 + .17 = .50$. Why is it not necessary that the probabilities add up to 1?
4. Suppose an item to be repaired might need one of n different parts. How many different policies would then need to be evaluated?

EXTENDED APPLICATION

▬ **BIDDING ON A POTENTIAL OIL FIELD—SIGNAL OIL**

Signal Oil, with headquarters in Los Angeles, is a major petroleum company. In this example we use probability and expected values to help determine the best bid price for a new offshore oil field. The company has used all the modern methods of oil exploration to help interpret the economic potential of each tract.*

    Two uncontrollable (and therefore uncertain) variables dominate a problem of this type: (a) the amount of commercial oil reserves that might be found in a tract, and (b) the length of time that would be required to develop and begin commercial production using these reserves. Another important variable is the amount to be bid for the right to develop the tract. Although the bid is a variable, it is not subject to uncertainty, but is under the control of the company. The company must analyze the effects of bids of various sizes along with the variables involving uncertainty so that the proper bid can be made.

---

*This example was supplied by Kenneth P. King, Senior Planning Analyst, Signal Oil Company.

The following chart shows the probabilities of various events. Commercial production includes events B_2, B_3, and B_4. Note that commercial production is given a 20% chance of occurring, with an 80% chance of the occurrence of less than a commercially profitable level of oil reserves.

Oil Reserves		
Event B_j	Millions of Barrels	Chance of Occurrence
B_1	.0	.80
B_2	19.0	.06
B_3	25.5	.10
B_4	30.6	.04

Any delay in beginning the commercial development of the field adversely affects the overall profitability of the project. This delay can be caused by seasonal weather variation in the offshore area, together with its relative isolation and the uncertainty of drilling rig availability. Beginning development in a shorter-than-normal time would require a concerted speedup effort that would incur cost increases over the normal period of development. This additional cost, however, is somewhat offset by the fact that the income from the field would be received sooner. The chart below shows the probabilities of various lengths of time required for commercial development to begin.

Years From Bid to Start of Drilling		
Event A_i	Years	Chance of Occurrence
A_1	2	.75
A_2	1	.13
A_3	3	.12

The time required for drilling to begin is independent of the quantity of reserves in the field. Hence, for each possible value of i and j, we have $P(A_i \text{ and } B_j) = P(A_i) \cdot P(B_j)$. For example, $P(2 \text{ years' delay and } 25.5 \text{ million barrels}) = P(A_1 \text{ and } B_3) = P(A_1) \cdot P(B_3) = (.75)(.10) = .075$. The chart below shows the probabilities for all possible cases, along with the payoffs to the company for different bid levels.

		Payoff (in millions of dollars)				
Case	Event	Probability	$0 Bid	$2	$5	$10
1	A_1 and B_1	.600	− 1.1	− 2.3	− 4.0	− 6.9
2	A_1 and B_2	.045	10.4	8.4	5.4	.4
3	A_1 and B_3	.075	17.0	15.0	12.0	7.0
4	A_1 and B_4	.030	22.2	20.2	17.2	12.2
5	A_2 and B_1	.104	− 1.1	− 2.3	− 4.0	− 6.9
6	A_2 and B_2	.008	13.2	11.2	8.2	3.2
7	A_2 and B_3	.013	21.5	19.5	16.5	11.5
8	A_2 and B_4	.005	27.7	25.7	22.7	17.7
9	A_3 and B_1	.096	− 1.1	− 2.3	− 4.0	− 6.9
10	A_3 and B_2	.007	7.5	5.5	2.5	− 2.5
11	A_3 and B_3	.012	13.0	11.0	8.0	3.0
12	A_3 and B_4	.005	17.5	15.5	12.5	7.5
	Total:	1.000				

Now the company must calculate the expected value for each different bid level. For example, the expected value at a bid level of $2 million is given by

$$E(\text{bid of \$2 million}) = (-2.3)(.600) + (8.4)(.045) + (15.0)(.075) + \cdots + (15.5)(.005).$$

If the expected values for various possible bid levels are found in the same way and plotted, we get the graph in Figure 22. Using techniques from mathematics of finance (the payoffs above are actually present values), the company knows that the expected value of a profitable bid must be $0 or more. As shown in the figure, this means that $3.5 million is the most the company can bid for this particular tract.

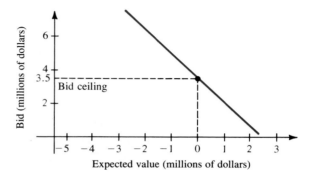

FIGURE 22

EXERCISES

1. Calculate $E(\text{bid of \$2 million})$.

2. Calculate $E(\text{bid of \$5 million})$.

8 Markov Chains

8.1   Basic Properties of Markov Chains

8.2   Regular Markov Chains

8.3   Absorbing Markov Chains

Review Exercises

In Chapter 6 we touched on *stochastic processes,* mathematical models in which the outcome of an experiment depends on the outcomes of previous experiments. In this chapter we study a special kind of stochastic process called a *Markov chain,* where the outcome of an experiment depends only on the outcome of the previous experiment. In other words, the next state of the system depends only on the present state, not on preceding states. Such experiments are common enough in applications to make their study worthwhile. Markov chains are named after the Russian mathematician A. A. Markov, 1856–1922, who started the theory of stochastic processes.

▤ 8.1 BASIC PROPERTIES OF MARKOV CHAINS

Transition Matrix   In sociology, it is convenient to classify people by income as *lower-class, middle-class,* and *upper-class.* Sociologists have found that the strongest determinant of the income class of an individual is the income class of the individual's parents. For example, if an individual in the lower-income class is said to be in *state 1,* an individual in the middle-income class is in *state 2,* and an individual in the upper-income class is in *state 3,* then the following probabilities of change in income class from one generation to the next might apply.

		Next Generation		
	State	*1*	*2*	*3*
Current	*1*	.65	.28	.07
Generation	*2*	.15	.67	.18
	3	.12	.36	.52

This table shows that if an individual is in state 1 (lower-income class) then there is a probability of .65 that any offspring will be in the lower-income class, a probability of .28 that offspring will be in the middle-income class, and a probability of .07 that offspring will be in the upper-income class.

The symbol p_{ij} will be used for the probability of transition from state i to state $j,$ in one generation. For example, p_{23} represents the probability that a person in state 2 will have offspring in state 3; from the table above,

$$p_{23} = .18.$$

Also from the table, $p_{31} = .12,$ $p_{22} = .67,$ and so on.

The information from the table can be written in other forms. Figure 1 is a **transition diagram** that shows the three states and the probabilities of going from one state to another.

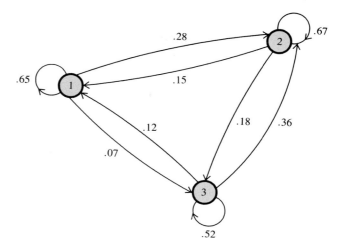

FIGURE 1

In a **transition matrix,** the states are indicated at the side and top. If *p* represents the transition matrix for the table above, then

$$
\begin{array}{c}
 \\
\begin{array}{ccc}
 & 1 & 2 & 3
\end{array}
\end{array}
$$

$$
\begin{array}{c}
1 \\
2 \\
3
\end{array}
\left[
\begin{array}{ccc}
.65 & .28 & .07 \\
.15 & .67 & .18 \\
.12 & .36 & .52
\end{array}
\right] = P.
$$

A transition matrix has several features:

1. It is square, since all possible states must be used both as rows and as columns.
2. All entries are between 0 and 1, inclusive; this is because all entries represent probabilities.
3. The sum of the entries in any row must be 1, since the numbers in the row give the probability of changing from the state at the left to one of the states indicated across the top.

Markov Chains   A transition matrix, such as matrix *P* above, also shows two key features of a Markov chain.

MARKOV CHAIN

A sequence of trials of an experiment is a **Markov chain** if

1. the outcome of each experiment is one of a set of discrete states;
2. the outcome of an experiment depends only on the present state, and not on any past states.

For example, in transition matrix *P,* a person is assumed to be in one of three discrete states (lower, middle, or upper income), with each offspring in one of these same three discrete states.

▬▬ EXAMPLE 1

A small town has only two dry cleaners, Johnson and NorthClean. Johnson's manager hopes to increase the firm's market share by conducting an extensive advertising campaign. After the campaign, a market research firm finds that there is a probability of .8 that a customer of Johnson's will bring his next batch of dirty clothes to Johnson, and a .35 chance that a NorthClean customer will switch to Johnson for his next batch. Write a transition matrix showing this information.

We must assume that the probability that a customer comes to a given dry cleaner depends only on where the last batch of clothes was taken. If there is an .8 chance that a Johnson customer will return to Johnson, then there must be a $1 - .8 = .2$ chance that the customer will switch to NorthClean. In the same way, there is a $1 - .35 = .65$ chance that a NorthClean customer will return to NorthClean. These probabilities give the following transition matrix.

$$
\begin{array}{cc}
 & \textit{Second Batch} \\
 & \begin{array}{cc} \text{Johnson} & \text{NorthClean} \end{array} \\
\textit{First Batch} \quad \begin{array}{c} \text{Johnson} \\ \text{NorthClean} \end{array} & \begin{bmatrix} .8 & .2 \\ .35 & .65 \end{bmatrix}
\end{array}
$$

We shall come back to this transition matrix later in this section (Example 4).

Figure 2 shows a transition diagram with the probabilities of using each dry cleaner for the second batch of dirty clothes. ▬▬

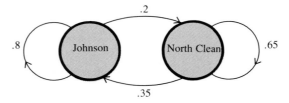

FIGURE 2

Look again at transition matrix *P* for income class changes.

$$
\begin{array}{c}
\begin{array}{ccc} 1 & 2 & 3 \end{array} \\
\begin{array}{c} 1 \\ 2 \\ 3 \end{array} \begin{bmatrix} .65 & .28 & .07 \\ .15 & .67 & .18 \\ .12 & .36 & .52 \end{bmatrix} = P
\end{array}
$$

This matrix shows the probability of change in income class from one generation to the next. Now let us investigate the probabilities for changes in income class over *two* generations. For example, if a parent is in state 3 (the upper-income class), what is the probability that a grandchild will be in state 2?

To find out, start with a tree diagram as shown in Figure 3; the various probabilities come from transition matrix P.

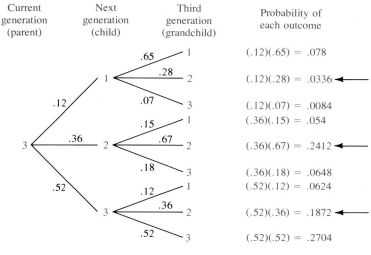

FIGURE 3

The arrows point to the outcomes ''grandchild in state 2''; the grandchild can get to state 2 after having had parents in either state 1, state 2, or state 3. The probability that a parent in state 3 will have a grandchild in state 2 is given by the sum of the probabilities indicated with arrows, or

$$.0336 + .2412 + .1872 = .4620.$$

We used p_{ij} to represent the probability of changing from state i to state j in one generation. This notation can be used to write the probability that a parent in state 3 will have a grandchild in state 2:

$$p_{31} \cdot p_{12} + p_{32} \cdot p_{22} + p_{33} \cdot p_{32}.$$

This sum of products of probabilities should remind you of matrix multiplication—it is nothing more than one step in the process of multiplying matrix P by itself. In particular, it is row 3 of P times column 2 of P. If P^2 represents the matrix product $P \cdot P$, then P^2 gives the probabilities of a transition from one state to another in *two* repetitions of an experiment. Generalizing,

P^k **gives the probabilities of a transition from one state to another in k repetitions of an experiment.**

EXAMPLE 2

For transition matrix P (income class changes),

$$P^2 = \begin{bmatrix} .65 & .28 & .07 \\ .15 & .67 & .18 \\ .12 & .36 & .52 \end{bmatrix} \begin{bmatrix} .65 & .28 & .07 \\ .15 & .67 & .18 \\ .12 & .36 & .52 \end{bmatrix} \approx \begin{bmatrix} .47 & .39 & .13 \\ .22 & .56 & .22 \\ .19 & .46 & .34 \end{bmatrix}.$$

(The numbers in the product have been rounded to the same number of decimal places as in matrix P.) The entry in row 3, column 2 of P^2 gives the probability that a person in state 3 will have a grandchild in state 2. This number, .46, is the result (rounded to 2 decimal places) found through use of the tree diagram.

Row 1, column 3 of P^2 gives the number .13, the probability that a person in state 1 will have a grandchild in state 3. How would the entry .47 be interpreted? ▬

EXAMPLE 3

In the same way that matrix P^2 gives the probabilities of income class changes after *two* generations, the matrix $P^3 = P \cdot P^2$ gives the probabilities of change after *three* generations.

For matrix P,

$$P^3 = P \cdot P^2 = \begin{bmatrix} .65 & .28 & .07 \\ .15 & .67 & .18 \\ .12 & .36 & .52 \end{bmatrix} \begin{bmatrix} .47 & .39 & .13 \\ .22 & .56 & .22 \\ .19 & .46 & .34 \end{bmatrix} \approx \begin{bmatrix} .38 & .44 & .17 \\ .25 & .52 & .23 \\ .23 & .49 & .27 \end{bmatrix}.$$

(The rows of P^3 don't necessarily total 1 exactly because of rounding errors.) Matrix P^3 gives a probability of .25 that a person in state 2 will have a great-grandchild in state 1. The probability is .52 that a person in state 2 will have a great-grandchild in state 2. ▬

EXAMPLE 4

Let us return to the transition matrix for the dry cleaners.

$$\begin{array}{cc} & \text{Second Batch} \\ & \begin{array}{cc} \text{Johnson} & \text{NorthClean} \end{array} \\ \textit{First Batch}\quad \begin{array}{c} \text{Johnson} \\ \text{NorthClean} \end{array} & \begin{bmatrix} .8 & .2 \\ .35 & .65 \end{bmatrix} \end{array}$$

As this matrix shows, there is a .8 chance that a person bringing his first batch to Johnson will also bring his second batch to Johnson, and so on. To find the probabilities for the third batch, the second stage of this Markov chain, find the square of the transition matrix. If C represents the transition matrix, then

$$C^2 = C \cdot C = \begin{bmatrix} .8 & .2 \\ .35 & .65 \end{bmatrix} \begin{bmatrix} .8 & .2 \\ .35 & .65 \end{bmatrix} = \begin{bmatrix} .71 & .29 \\ .51 & .49 \end{bmatrix}.$$

From C^2, the probability that a person bringing his first batch of clothes to Johnson will also bring his third batch to Johnson is .71; the probability that a person bringing his first batch to NorthClean will bring his third batch to NorthClean is .49.

The cube of matrix C gives the probabilities for the fourth batch, the third step in our experiment.

$$C^3 = C \cdot C^2 = \begin{bmatrix} .67 & .33 \\ .58 & .42 \end{bmatrix}$$

The probability is .58, for example, that a person bringing his first batch to NorthClean will bring his fourth batch to Johnson. ▬

Distribution of States   Look again at the transition matrix for income class changes:

$$P = \begin{bmatrix} .65 & .28 & .07 \\ .15 & .67 & .18 \\ .12 & .36 & .52 \end{bmatrix}.$$

Suppose the following table gives the initial distribution of people in the three income classes.

Class	State	Proportion
Lower	1	21%
Middle	2	68%
Upper	3	11%

To see how these proportions would change after one generation, use the tree diagram in Figure 4. For example, to find the proportion of people in state 2 after one generation, add the numbers indicated with arrows.

$$.0588 + .4556 + .0396 = .5540$$

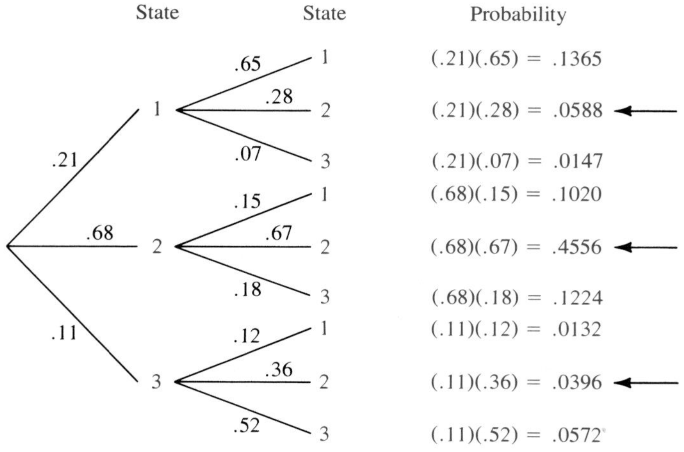

FIGURE 4

In a similar way, the proportion of people in state 1 after one generation is

$$.1365 + .1020 + .0132 = .2517,$$

and the proportion of people in state 3 after one generation is

$$.0147 + .1224 + .0572 = .1943.$$

The initial distribution of states, 21%, 68%, and 11%, becomes, after one generation, 25.17% in state 1, 55.4% in state 2, and 19.43% in state 3. These distributions can be written as *probability vectors* (where the percents have been changed to decimals rounded to the nearest hundredth)

$$[.21 \quad .68 \quad .11] \quad \text{and} \quad [.25 \quad .55 \quad .19],$$

respectively. A **probability vector** is a matrix of only one row, having non-negative entries, with the sum of the entries equal to 1.

The work with the tree diagram to find the distribution of states after one generation is exactly the work required to multiply the initial probability vector, $[.21 \quad .68 \quad .11]$, and the transition matrix P:

$$[.21 \quad .68 \quad .11] \begin{bmatrix} .65 & .28 & .07 \\ .15 & .67 & .18 \\ .12 & .36 & .52 \end{bmatrix} \approx [.25 \quad .55 \quad .19].$$

In a similar way, the distribution of income classes after two generations can be found by multiplying the initial probability vector and the square of P, the matrix P^2. Using P^2 from above,

$$[.21 \quad .68 \quad .11] \begin{bmatrix} .47 & .39 & .13 \\ .22 & .56 & .22 \\ .19 & .46 & .34 \end{bmatrix} \approx [.27 \quad .51 \quad .21].$$

In the next section we will develop a long-range prediction for the proportions of the population in each income class. The work in this section is summarized below.

> Suppose a Markov chain has initial probability vector
>
> $$I = [i_1 \quad i_2 \quad i_3 \quad \cdots \quad i_n]$$
>
> and transition matrix P. The probability vector after n repetitions of the experiment is
>
> $$I \cdot P^n.$$

8.1 EXERCISES

Decide whether each of the matrices in Exercises 1–9 could be a probability vector.

1. $[\frac{2}{3} \quad \frac{1}{2}]$

2. $[\frac{1}{2} \quad 1]$

3. $[0 \quad 1]$

4. $[.1 \quad .1]$

5. $[.4 \quad .2 \quad 0]$

6. $[\frac{1}{4} \quad \frac{1}{8} \quad \frac{5}{8}]$

7. [.07 .04 .37 .52]   **8.** [.3 −.1 .8]   **9.** [0 −.2 .6 .6]

Decide whether each of the matrices in Exercises 10–15 could be a transition matrix, by definition. Sketch a transition diagram for any transition matrices.

10. $\begin{bmatrix} .5 & 0 \\ 0 & .5 \end{bmatrix}$

11. $\begin{bmatrix} \frac{2}{3} & \frac{1}{3} \\ 1 & 0 \end{bmatrix}$

12. $\begin{bmatrix} \frac{1}{4} & \frac{3}{4} \\ \frac{1}{2} & \frac{1}{2} \end{bmatrix}$

13. $\begin{bmatrix} \frac{1}{4} & \frac{3}{4} & 0 \\ 2 & 0 & 1 \\ 1 & \frac{2}{3} & 3 \end{bmatrix}$

14. $\begin{bmatrix} \frac{1}{3} & \frac{1}{3} & \frac{1}{3} \\ 0 & 1 & 0 \\ \frac{1}{2} & 0 & \frac{1}{2} \end{bmatrix}$

15. $\begin{bmatrix} \frac{1}{3} & \frac{1}{2} & 1 \\ 0 & 1 & 0 \\ \frac{1}{2} & \frac{1}{2} & 1 \end{bmatrix}$

In Exercises 16–18 write any transition diagrams as transition matrices.

16.

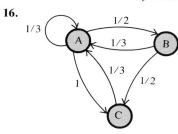

17.

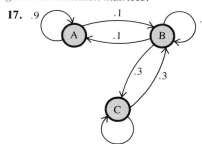

18.

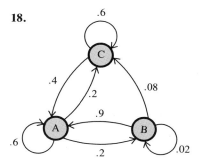

Find the first three powers of each of the transition matrices in Exercises 19–24 (for example, A, A², and A³ in Exercise 19). For each transition matrix, find the probability that state 1 changes to state 2 after three repetitions of the experiment.

19. $A = \begin{bmatrix} 1 & 0 \\ .8 & .2 \end{bmatrix}$

20. $B = \begin{bmatrix} .7 & .3 \\ 0 & 1 \end{bmatrix}$

21. $C = \begin{bmatrix} .5 & .5 \\ .72 & .28 \end{bmatrix}$

22. $D = \begin{bmatrix} .3 & .2 & .5 \\ 0 & 0 & 1 \\ .6 & .1 & .3 \end{bmatrix}$

23. $E = \begin{bmatrix} .8 & .1 & .1 \\ .3 & .6 & .1 \\ 0 & 1 & 0 \end{bmatrix}$

24. $F = \begin{bmatrix} .01 & .9 & .09 \\ .72 & .1 & .18 \\ .34 & 0 & .66 \end{bmatrix}$

▇ APPLICATIONS

LIFE SCIENCES

Genetics **25.** In Exercise 52 of Section 6.2 we discussed the effect on flower color of cross-pollinating pea plants. As shown in Exercise 52, since the gene for red is dominant and the gene for white is recessive, 75% of these pea plants have red flowers and 25% have white flowers, because plants with one red and one white gene appear red. If a red-flowered plant is crossed with a red-flowered plant known to have one red and one white gene, then 75% of the offspring will be red and 25% will be white. Crossing a red-flowered plant that has one red and one white gene with a white-flowered plant produces 50% red-flowered offspring and 50% white-flowered offspring.

(a) Write a transition matrix using this information.

(b) Write a probability vector for the initial distribution of colors.

Find the distribution of colors after each of the following numbers of generations.

(c) 1   **(d)** 2   **(e)** 3   **(f)** 4

BUSINESS AND ECONOMICS

Dry Cleaning **26.** The dry cleaning example in the text used the following transition matrix:

$$\begin{array}{cc} & \begin{array}{cc} \text{Johnson} & \text{NorthClean} \end{array} \\ \begin{array}{c} \text{Johnson} \\ \text{NorthClean} \end{array} & \left[\begin{array}{cc} .8 & .2 \\ .35 & .65 \end{array}\right] \end{array}.$$

Suppose now that each customer brings in one batch of clothes per week. Use various powers of the transition matrix to find the probability that a customer bringing a batch of clothes to Johnson initially also brings a batch to Johnson after

(a) 1 week;   **(b)** 2 weeks;   **(c)** 3 weeks;   **(d)** 4 weeks.

Dry Cleaning **27.** Suppose Johnson has a 40% market share initially, with NorthClean having a 60% share. Use this information to write a probability vector; use this vector along with the transition matrix above to find the share of the market for each firm after each of the following time periods. (As in Exercise 26, assume that customers bring in one batch of dry cleaning per week.)

(a) 1 week   **(b)** 2 weeks   **(c)** 3 weeks   **(d)** 4 weeks

Insurance **28.** An insurance company classifies its drivers into three groups: G_0 (no accidents), G_1 (one accident), and G_2 (more than one accident). The probability that a G_0 driver will remain a G_0 after one year is .85, that the driver will become a G_1 is .10, and that the driver will become a G_2 is .05. A G_1 driver cannot become a G_0 (this company has a long memory). There is a .8 probability that a G_1 driver will remain a G_1. A G_2 driver must remain a G_2. Write a transition matrix using this information.

Insurance **29.** Suppose that the company in Exercise 28 accepts 50,000 new policyholders, all of whom are G_0 drivers. Find the number in each group after

(a) 1 year;   **(b)** 2 years;   **(c)** 3 years;   **(d)** 4 years.

Insurance **30.** The difficulty with the mathematical model in Exercises 28 and 29 is that no "grace period" is provided; there should be a certain positive probability of moving from G_1 or G_2 back to G_0 (say, after four years with no accidents). A new system with this feature might produce the following transition matrix.

$$\left[\begin{array}{ccc} .85 & .10 & .05 \\ .15 & .75 & .10 \\ .10 & .30 & .60 \end{array}\right]$$

Suppose that when this new policy is adopted, the company has 50,000 policyholders, all in G_0. Find the number in each group after

(a) 1 year;   **(b)** 2 years;   **(c)** 3 years.

Market Share **31.** Research done by the Gulf Oil Corporation* produced the following transition matrix for the probabilities that during a given year a person with one system of home heating will keep the same system or switch to another.

---

*Reprinted by permission of Ali Ezzati, "Forecasting Market Shares of Alternative Home Heating Units," *Management Science,* Vol. 21, No. 4, December 1974, copyright © 1974 The Institute of Management Sciences.

Next Year's
Heating System

		Oil	Gas	Electric
This Year's	Oil	.825	.175	0
Heating System	Gas	.060	.919	.021
	Electric	.049	0	.951

The current share of the market held by these three heating systems is given by the vector [.26   .60   .14]. Assuming that these trends continue, find the share of the market held by each heating system after

(a) 1 year;   (b) 2 years;   (c) 3 years.

Land Use **32.** In one state, a Board of Realtors land use survey showed that 35% of all land was used for agricultural purposes, while 10% was urban. Ten years later, of the agricultural land, 15% had become urban and 80% had remained agricultural. (The remainder lay idle.) Of the idle land, 20% had become urbanized and 10% had been converted for agricultural use. Of the urban land, 90% remained urban and 10% was idle. Assume that these trends continue.

(a) Write a transition matrix using this information.

(b) Write a probability vector for the initial distribution of land.

Find the land use pattern after

(c) 10 years;   (d) 20 years.

SOCIAL SCIENCES

Housing Patterns **33.** In a survey investigating change in housing patterns in one urban area, it was found that 75% of the population lived in single-family dwellings and 25% in multiple housing of some kind. Five years later, in a follow-up survey, of those who had been living in single family dwellings, 90% still did so, but 10% had moved to multiple-family dwellings. Of those in multiple-family housing, 95% were still living in that type of housing, while 5% had moved to single-family dwellings. Assume that these trends continue.

(a) Write a transition matrix for this information.

(b) Write a probability vector for the initial distribution of housing.

What percent of the population can be expected in each category

(c) 5 years later?   (d) 10 years later?

Voting Trends **34.** At the end of June in a Presidential election year, 40% of the voters were registered as liberal, 45% as conservative, and 15% as independent. Over a one-month period, the liberals retained 80% of their constituency, while 15% switched to conservative and 5% to independent. The conservatives retained 70%, and lost 20% to the liberals. The independents retained 60% and lost 20% each to the conservatives and liberals. Assume that these trends continue.

(a) Write a transition matrix using this information.

(b) Write a probability vector for the initial distribution.

Find the percent of each type of voter at the end of

(c) July;   (d) August;   (e) September;   (f) October.

FOR THE COMPUTER

For each of the following transition matrices, find the first five powers of the matrix. Then find the probability that state 2 changes to state 4 after 5 repetitions of the experiment.

35.
$$\begin{bmatrix} .1 & .2 & .2 & .3 & .2 \\ .2 & .1 & .1 & .2 & .4 \\ .2 & .1 & .4 & .2 & .1 \\ .3 & .1 & .1 & .2 & .3 \\ .1 & .3 & .1 & .1 & .4 \end{bmatrix}$$

36.
$$\begin{bmatrix} .3 & .2 & .3 & .1 & .1 \\ .4 & .2 & .1 & .2 & .1 \\ .1 & .3 & .2 & .2 & .2 \\ .2 & .1 & .3 & .2 & .2 \\ .1 & .1 & .4 & .2 & .2 \end{bmatrix}$$

Company Training Program   **37.** A company with a new training program classified each worker into one of the following four states: s_1, never in the program; s_2, currently in the program; s_3, discharged; s_4, completed the program. The transition matrix for this company is given below.

$$
\begin{array}{c}
 \\ s_1 \\ s_2 \\ s_3 \\ s_4
\end{array}
\begin{array}{c}
\begin{array}{cccc} s_1 & s_2 & s_3 & s_4 \end{array} \\
\begin{bmatrix} .40 & .20 & .05 & .35 \\ 0 & .45 & .05 & .50 \\ 0 & 0 & 1.00 & 0 \\ 0 & 0 & 0 & 1.00 \end{bmatrix}
\end{array}
$$

(a) What proportion of workers who had never been in the program (state s_1) completed the program (state s_4) after the program had been offered 5 times?

(b) If the initial proportions of workers in each state were [.5   .5   0   0], find the corresponding proportions after the program had been offered 4 times.

▰▰ 8.2   REGULAR MARKOV CHAINS

If we start with a transition matrix P and an initial probability vector, we can use the nth power of P to find the probability vector for n repetitions of an experiment. In this section we try to decide what happens to an initial probability vector "in the long run," that is, as n gets larger and larger.

For example, let us use the transition matrix associated with the dry cleaning example in the previous section,

$$\begin{bmatrix} .8 & .2 \\ .35 & .65 \end{bmatrix}.$$

The initial probability vector, which gives the market share for each firm at the beginning of the experiment, is [.4   .6]. The market shares shown in the following table were found by using powers of the transition matrix. (See Exercise 27 in Section 8.1.)

Weeks After Start	Johnson	NorthClean
0	.4	.6
1	.53	.47
2	.59	.41
3	.62	.38
4	.63	.37
5	.63	.37
12	.64	.36

The results seem to approach the numbers in the probability vector [.64   .36].

What happens if the initial probability vector is different from [.4   .6]? Suppose [.75   .25] is used; the same powers of the transition matrix as above give the following results.

Weeks After Start	Johnston	NorthClean
0	.75	.25
1	.69	.31
2	.66	.34
3	.65	.35
4	.64	.36
5	.64	.36
6	.64	.36

The results again seem to be approaching the numbers in the probability vector [.64   .36], the same numbers approached with the initial probability vector [.4   .6]. In either case, the long-range trend is for a market share of about 64% for Johnson and 36% for NorthClean. The example above suggests that this long-range trend does not depend on the initial distribution of market shares.

Regular Transition Matrices   One of the many applications of Markov chains is in finding long-range predictions. It is not possible to make long-range predictions with all transition matrices, but for a large set of transition matrices, long-range predictions *are* possible. Such predictions are always possible with **regular transition matrices.** A transition matrix is **regular** if some power of the matrix contains all positive entries. A Markov chain is a **regular Markov chain** if its transition matrix is regular.

■ EXAMPLE 1
Decide whether the following transition matrices are regular.

(a) $A = \begin{bmatrix} .75 & .25 & 0 \\ 0 & .5 & .5 \\ .6 & .4 & 0 \end{bmatrix}$

Square A.

$$A^2 = \begin{bmatrix} .5625 & .3125 & .125 \\ .3 & .45 & .25 \\ .45 & .35 & .2 \end{bmatrix}$$

Since all entries in A^2 are positive, matrix A is regular.

(b) $B = \begin{bmatrix} .5 & 0 & .5 \\ 0 & 1 & 0 \\ 0 & 0 & 1 \end{bmatrix}$

Find various powers of B.

$$B^2 = \begin{bmatrix} .25 & 0 & .75 \\ 0 & 1 & 0 \\ 0 & 0 & 1 \end{bmatrix}; \ B^3 = \begin{bmatrix} .125 & 0 & .875 \\ 0 & 1 & 0 \\ 0 & 0 & 1 \end{bmatrix}; \ B^4 = \begin{bmatrix} .0625 & 0 & .9375 \\ 0 & 1 & 0 \\ 0 & 0 & 1 \end{bmatrix}$$

Further powers of B will still give the same zero entries, so no power of matrix B contains all positive entries. For this reason, B is not regular. ■

Suppose that v is a probability vector. It can be shown that for a regular Markov chain with a transition matrix P, there exists a single vector V such that $v \cdot P^n$ gets closer and closer to V as n gets larger and larger.

EQUILIBRIUM VECTOR OF A MARKOV CHAIN

> If a Markov chain with transition matrix P is regular, then there is a unique vector V such that for any probability vector v and for large values of n,
>
> $$v \cdot P^n \approx V.$$
>
> Vector V is called the **equilibrium vector** or the **fixed vector** of the Markov chain.

In the example with Johnson Cleaners, the equilibrium vector V is approximately [.64 .36]. Vector V can be determined by finding P^n for larger and larger values of n, and then looking for a vector that the product $v \cdot P^n$ approaches. Such an approach can be very tedious, however, and is prone to error. To find a better way, start with the fact that for a large value of n,

$$v \cdot P^n \approx V,$$

as mentioned above. From this result, $v \cdot P^n \cdot P \approx V \cdot P$, so that

$$v \cdot P^n \cdot P = v \cdot P^{n+1} \approx VP.$$

Since $v \cdot P^n \approx V$ for large values of n, it is also true that $v \cdot P^{n+1} \approx V$ for large values of n (the product $v \cdot P^n$ approaches V, so that $v \cdot P^{n+1}$ must also approach V). Thus, $v \cdot P^{n+1} \approx V$ and $v \cdot P^{n+1} \approx VP$, which suggests that

$$VP = V.$$

> If a Markov chain with transition matrix P is regular, then there exists a probability vector V such that
> $$VP = V.$$

This vector V gives the long-range trend of the Markov chain. Vector V is found by solving a system of linear equations, as shown in the next examples.

EXAMPLE 2

Find the long-range trend for the Markov chain in the dry cleaning example with transition matrix

$$\begin{bmatrix} .8 & .2 \\ .35 & .65 \end{bmatrix}.$$

This matrix is regular since all entries are positive. Let P represent this transition matrix, and let V be the probability vector $[v_1 \quad v_2]$. We want to find V such that

$$VP = V,$$

or

$$[v_1 \quad v_2]\begin{bmatrix} .8 & .2 \\ .35 & .65 \end{bmatrix} = [v_1 \quad v_2].$$

Use matrix multiplication on the left.

$$[.8v_1 + .35v_2 \quad .2v_1 + .65v_2] = [v_1 \quad v_2]$$

Set corresponding entries from the two matrices equal to get

$$.8v_1 + .35v_2 = v_1 \quad \text{and} \quad .2v_1 + .65v_2 = v_2.$$

Simplify each of these equations.

$$-.2v_1 + .35v_2 = 0 \quad \text{and} \quad .2v_1 - .35v_2 = 0$$

These last two equations are really the same. (The equations in the system obtained from $VP = V$ are always dependent.) To find the values of v_1 and v_2, recall that $V = [v_1 \quad v_2]$ is a probability vector, so that

$$v_1 + v_2 = 1.$$

To find v_1 and v_2, solve the system

$$-.2v_1 + .35v_2 = 0$$
$$v_1 + \quad v_2 = 1.$$

From the second equation, $v_1 = 1 - v_2$. Substitute $1 - v_2$ for v_1 in the first equation.

$$-.2(1 - v_2) + .35v_2 = 0$$

$$-.2 + .2v_2 + .35v_2 = 0$$

$$.55v_2 = .2$$

$$v_2 = \frac{4}{11} \approx .364$$

Since $v_1 = 1 - v_2$, $v_1 = 7/11 \approx .636$, and the equilibrium vector is $V = [7/11 \quad 4/11] \approx [.636 \quad .364]$. ■

Some powers of the transition matrix P in Example 1 (with entries rounded to two decimal places) are shown here.

$$P^2 = \begin{bmatrix} .71 & .29 \\ .51 & .49 \end{bmatrix} \qquad P^3 = \begin{bmatrix} .67 & .33 \\ .58 & .42 \end{bmatrix} \qquad P^4 = \begin{bmatrix} .65 & .35 \\ .62 & .38 \end{bmatrix}$$

$$P^5 = \begin{bmatrix} .65 & .35 \\ .63 & .37 \end{bmatrix} \qquad P^6 = \begin{bmatrix} .64 & .36 \\ .63 & .37 \end{bmatrix} \qquad P^{10} = \begin{bmatrix} .64 & .36 \\ .64 & .36 \end{bmatrix}$$

As these results suggest, higher and higher powers of the transition matrix P approach a matrix having all rows identical; these identical rows have as entries the entries of the equilibrium vector V. This agrees with the statement above: the initial state does not matter. Regardless of the initial probability vector, the system will approach a fixed vector V. This unexpected and remarkable fact is the basic property of regular Markov chains—*the limiting distribution is independent of the initial distribution.* This happens because some power of the transition matrix has all positive entries, so that all the initial probabilities are thoroughly mixed.

Let us summarize the results of this section.

PROPERTIES OF REGULAR MARKOV CHAINS

Suppose a regular Markov chain has a transition matrix P.

1. As n gets larger and larger, the product $v \cdot P^n$ approaches a unique vector V for any initial probability vector v. Vector V is called the *equilibrium vector* or *fixed vector*.
2. Vector V has the property that $VP = V$.
3. To find V, solve a system of equations obtained from the matrix equation $VP = V$, and from the fact that the sum of the entries of V is 1.
4. The powers P^n come closer and closer to a matrix whose rows are made up of the entries of the equilibrium vector V.

■ EXAMPLE 3

Find the equilibrium vector for the transition matrix

$$K = \begin{bmatrix} .2 & .6 & .2 \\ .1 & .1 & .8 \\ .3 & .3 & .4 \end{bmatrix}.$$

Matrix K has all positive entries and thus is regular. For this reason, an equilibrium vector V must exist such that $VK = V$. Let $V = [v_1 \quad v_2 \quad v_3]$. Then

$$[v_1 \quad v_2 \quad v_3] \begin{bmatrix} .2 & .6 & .2 \\ .1 & .1 & .8 \\ .3 & .3 & .4 \end{bmatrix} = [v_1 \quad v_2 \quad v_3].$$

Use matrix multiplication on the left.

$$[.2v_1 + .1v_2 + .3v_3 \quad .6v_1 + .1v_2 + .3v_3 \quad .2v_1 + .8v_2 + .4v_3]$$
$$= [v_1 \quad v_2 \quad v_3]$$

Setting corresponding entries equal gives the following equations:

$$.2v_1 + .1v_2 + .3v_3 = v_1$$
$$.6v_1 + .1v_2 + .3v_3 = v_2$$
$$.2v_1 + .8v_2 + .4v_3 = v_3.$$

Simplifying these equations gives

$$-.8v_1 + .1v_2 + .3v_3 = 0$$
$$.6v_1 - .9v_2 + .3v_3 = 0$$
$$.2v_1 + .8v_2 - .6v_3 = 0.$$

Since V is a probability vector,

$$v_1 + v_2 + v_3 = 1.$$

This gives a system of four equations in three unknowns:

$$v_1 + v_2 + v_3 = 1$$
$$-.8v_1 + .1v_2 + .3v_3 = 0$$
$$.6v_1 - .9v_2 + .3v_3 = 0$$
$$.2v_1 + .8v_2 - .6v_3 = 0.$$

This system can be solved with the Gauss-Jordan method presented earlier. Start with the augmented matrix

$$\begin{bmatrix} 1 & 1 & 1 & | & 1 \\ -.8 & .1 & .3 & | & 0 \\ .6 & -.9 & .3 & | & 0 \\ .2 & .8 & -.6 & | & 0 \end{bmatrix}.$$

The solution of this system is $v_1 = 5/23$, $v_2 = 7/23$, $v_3 = 11/23$, and

$$V = \begin{bmatrix} \frac{5}{23} & \frac{7}{23} & \frac{11}{23} \end{bmatrix} \approx [.22 \quad .30 \quad .48]. \quad ■$$

8.2 EXERCISES

Which of the transition matrices in Exercises 1–6 are regular?

1. $\begin{bmatrix} .2 & .8 \\ .9 & .1 \end{bmatrix}$

2. $\begin{bmatrix} .22 & .78 \\ .43 & .57 \end{bmatrix}$

3. $\begin{bmatrix} 1 & 0 \\ .6 & .4 \end{bmatrix}$

4. $\begin{bmatrix} .55 & .45 \\ 0 & 1 \end{bmatrix}$

5. $\begin{bmatrix} 0 & 1 & 0 \\ .4 & .2 & .4 \\ 1 & 0 & 0 \end{bmatrix}$

6. $\begin{bmatrix} .3 & .5 & .2 \\ 1 & 0 & 0 \\ .5 & .1 & .4 \end{bmatrix}$

Find the equilibrium vector for each transition matrix in Exercises 7–14.

7. $\begin{bmatrix} \frac{1}{4} & \frac{3}{4} \\ \frac{1}{2} & \frac{1}{2} \end{bmatrix}$

8. $\begin{bmatrix} \frac{2}{3} & \frac{1}{3} \\ \frac{1}{8} & \frac{7}{8} \end{bmatrix}$

9. $\begin{bmatrix} .3 & .7 \\ .4 & .6 \end{bmatrix}$

10. $\begin{bmatrix} .8 & .2 \\ .1 & .9 \end{bmatrix}$

11. $\begin{bmatrix} .1 & .1 & .8 \\ .4 & .4 & .2 \\ .1 & .2 & .7 \end{bmatrix}$

12. $\begin{bmatrix} .5 & .2 & .3 \\ .1 & .4 & .5 \\ .2 & .2 & .6 \end{bmatrix}$

13. $\begin{bmatrix} .25 & .35 & .4 \\ .1 & .3 & .6 \\ .55 & .4 & .05 \end{bmatrix}$

14. $\begin{bmatrix} .16 & .28 & .56 \\ .43 & .12 & .45 \\ .86 & .05 & .09 \end{bmatrix}$

Find the equilibrium vector for each transition matrix in Exercises 15–21. These matrices were first used in the exercises for Section 8.1. (Note: Not all of these transition matrices are regular, but equilibrium vectors still exist. Why doesn't this contradict the work of this section?)

15. Flower color (8.1 Exercise 25)

$\begin{bmatrix} .75 & .25 \\ .50 & .50 \end{bmatrix}$

16. Housing patterns (8.1 Exercise 33)

$\begin{bmatrix} .90 & .10 \\ .05 & .95 \end{bmatrix}$

17. Insurance categories (8.1 Exercise 28)

$\begin{bmatrix} .85 & .10 & .05 \\ 0 & .80 & .20 \\ 0 & 0 & 1 \end{bmatrix}$

18. "Modified" insurance categories (8.1 Exercise 30)

$\begin{bmatrix} .85 & .10 & .05 \\ .15 & .75 & .10 \\ .10 & .30 & .60 \end{bmatrix}$

19. Land use
(8.1 Exercise 32)

$\begin{bmatrix} .80 & .15 & .05 \\ 0 & .90 & .10 \\ .10 & .20 & .70 \end{bmatrix}$

20. Voter registration
(8.1 Exercise 34)

$\begin{bmatrix} .80 & .15 & .05 \\ .20 & .70 & .10 \\ .20 & .20 & .60 \end{bmatrix}$

21. Home heating systems
(8.1 Exercise 31)

$\begin{bmatrix} .825 & .175 & 0 \\ .060 & .919 & .021 \\ .049 & 0 & .951 \end{bmatrix}$

22. Find the equilibrium vector for the transition matrix

$$\begin{bmatrix} p & 1-p \\ 1-q & q \end{bmatrix},$$

where $0 < p < 1$ and $0 < q < 1$. Under what conditions is this matrix regular?

23. Show that the transition matrix

$$K = \begin{bmatrix} \frac{1}{4} & 0 & \frac{3}{4} \\ 0 & 1 & 0 \\ 0 & 0 & 1 \end{bmatrix}$$

has more than one vector V such that $VK = V$. Why does this not violate the statements of this section?

24. Let

$$P = \begin{bmatrix} a_{11} & a_{12} \\ a_{21} & a_{22} \end{bmatrix}$$

be a regular transition matrix having *column* sums of 1. Show that the equilibrium vector for P is [1/2 1/2].

▰ APPLICATIONS

BUSINESS AND ECONOMICS

Quality Control **25.** The probability that a complex assembly line works correctly depends on whether the line worked correctly the last time it was used. There is a .9 chance that the line will work correctly if it worked correctly the time before, and a .7 chance that it will work correctly if it did *not* work correctly the time before. Set up a transition matrix with this information and find the long-run probability that the line will work correctly.

Quality Control **26.** Suppose improvements are made in the assembly line of Exercise 25, so that the transition matrix becomes

	Works	Doesn't Work
Works	.95	.05
Doesn't Work	.80	.20

Find the new long-run probability that the line will work properly.

Sales Management **27.** Each month, a sales manager classifies her salespeople as low, medium, or high producers. There is a .4 chance that a low producer one month will become a medium producer the following month, and a .1 chance that a low producer will become a high producer. A medium producer will become a low or high producer, respectively, with probabilities .25 and .3. A high producer will become a low or medium producer, respectively, with probabilities .05 and .4. Find the long-range trend for the proportion of low, medium, and high producers.

LIFE SCIENCES

Research with Mice **28.** A large group of mice is kept in a cage having connected compartments A, B, and C. Mice in compartment A move to B with probability .3 and to C with probability .4. Mice in B move to A or C with probabilities of .15 and .55, respectively. Mice in C move to A or B with probabilities .3 and .6 respectively. Find the long-range prediction for the fraction of mice in each of the compartments.

Genetics **29.** Snapdragons with one red gene and one white gene produce pink-flowered offspring. If a red snapdragon is crossed with a pink snapdragon, the probabilities that the offspring will be red, pink, or white are 1/2, 1/2, and 0, respectively. If two pink snapdragons are crossed, the probabilities of red, pink, or white offspring are 1/4, 1/2, and 1/4, respectively. For a cross between a white and a pink snapdragon, the corresponding probabilities are 0, 1/2, and 1/2. Find the long-range prediction for the fraction of red, pink, and white snapdragons.

PHYSICAL SCIENCES

Weather   **30.** The weather in a certain spot is classified as fair, cloudy without rain, or rainy. A fair day is followed by a fair day 60% of the time, and by a cloudy day 25% of the time. A cloudy day is followed by a cloudy day 35% of the time, and by a rainy day 25% of the time. A rainy day is followed by a cloudy day 40% of the time, and by another rainy day 25% of the time. Find the long-range prediction for the proportion of fair, cloudy, and rainy days.

SOCIAL SCIENCES

Education   **31.** At one liberal arts college, students are classified as humanities majors, science majors, or undecideds. There is a 20% chance that a humanities major will change to a science major from one year to the next, and a 45% chance that a humanities major will change to undecided. A science major will change to humanities with probability .15, and to undecided with probability .35. An undecided will switch to humanities or science with probabilities of .5 and .3 respectively. Find the long-range prediction for the fraction of students in each of these three majors.

Rumors   **32.** The manager of the slot machines at a major casino makes a decision about whether or not to "loosen up" the slots so that the customers get a larger payback. The manager tells only one other person, a person whose word cannot be trusted. In fact, there is only a probability p, where $0 < p < 1$, that this person will tell the truth. Suppose this person tells several other people, each of whom tells several people, what the manager's decision is. Suppose there is always a probability p that the decision is passed on as heard. Find the long-range prediction for the fraction of the people who will hear the decision correctly. (*Hint:* Use a transition matrix; let the first row be $[p \quad 1 - p]$ and the second row be $[1 - p \quad p]$.)

FOR THE COMPUTER

Find the equilibrium vector for transition matrices 33 and 34 by taking powers of the matrix.

33.
$$\begin{bmatrix} .1 & .2 & .2 & .3 & .2 \\ .2 & .1 & .1 & .2 & .4 \\ .2 & .1 & .4 & .2 & .1 \\ .3 & .1 & .1 & .2 & .3 \\ .1 & .3 & .1 & .1 & .4 \end{bmatrix}$$

34.
$$\begin{bmatrix} .3 & .2 & .3 & .1 & .1 \\ .4 & .2 & .1 & .2 & .1 \\ .1 & .3 & .2 & .2 & .2 \\ .2 & .1 & .3 & .2 & .2 \\ .1 & .1 & .4 & .2 & .2 \end{bmatrix}$$

8.3   ABSORBING MARKOV CHAINS

Suppose a Markov chain has transition matrix

$$\begin{array}{c} \\ 1 \\ 2 \\ 3 \end{array} \begin{array}{ccc} 1 & 2 & 3 \\ \begin{bmatrix} .3 & .6 & .1 \\ 0 & 1 & 0 \\ .6 & .2 & .2 \end{bmatrix} \end{array} = P.$$

The matrix shows that p_{12}, the probability of going from state 1 to state 2, is .6, and that p_{22}, the probability of staying in state 2, is 1. Thus, once state 2 is entered, it is impossible to leave. For this reason, state 2 is called an *absorbing state*. Figure 5 shows a transition diagram for this matrix. The diagram shows that it is not possible to leave state 2.

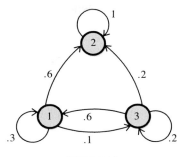

FIGURE 5

Generalizing from this example leads to the following definition.

ABSORBING STATE

> State i of a Markov chain is an **absorbing state** if $p_{ii} = 1$.

Using the idea of an absorbing state, we can define an absorbing Markov chain.

ABSORBING MARKOV CHAINS

> A Markov chain is an **absorbing chain** if and only if the following two conditions are satisfied:
>
> **1.** the chain has at least one absorbing state;
> **2.** it is possible to go from any nonabsorbing state to an absorbing state (perhaps in more than one step).

EXAMPLE 1

Identify all absorbing states in the Markov chains having the following matrices. Decide whether the Markov chain is absorbing.

$$\text{(a)} \quad \begin{array}{c} \\ 1 \\ 2 \\ 3 \end{array} \begin{array}{ccc} 1 & 2 & 3 \\ \begin{bmatrix} 1 & 0 & 0 \\ .3 & .5 & .2 \\ 0 & 0 & 1 \end{bmatrix} \end{array}$$

Since $p_{11} = 1$ and $p_{33} = 1$, both state 1 and state 3 are absorbing states. (Once these states are reached, they cannot be left.) The only nonabsorbing state is state 2. There is a .3 probability of going from state 2 to the absorbing state 1, so that it is possible to go from the nonabsorbing state to an absorbing state. This Markov chain is absorbing. The transition diagram is shown in Figure 6.

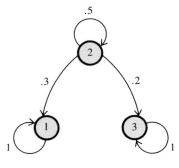

FIGURE 6

$$\text{(b)} \quad \begin{array}{c} \\ 1 \\ 2 \\ 3 \\ 4 \end{array} \begin{array}{cccc} 1 & 2 & 3 & 4 \\ \end{array} \\ \left[\begin{array}{cccc} .6 & 0 & .4 & 0 \\ 0 & 1 & 0 & 0 \\ .9 & 0 & .1 & 0 \\ 0 & 0 & 0 & 1 \end{array} \right]$$

States 2 and 4 are absorbing, with states 1 and 3 nonabsorbing. From state 1, it is possible to go only to states 1 or 3; from state 3 it is possible to go only to states 1 or 3. As the transition diagram in Figure 7 shows, neither nonabsorbing state leads to an absorbing state, so that this Markov chain is nonabsorbing. ■

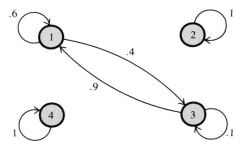

FIGURE 7

■ EXAMPLE 2

(*Gambler's Ruin*) Suppose players A and B have a coin tossing game going on—a fair coin is tossed and the player predicting the toss correctly wins $1 from the other player. Suppose the players have a total of $6 between them, and that the game goes on until one player has no money (is ruined.)

Let us agree that the states of this system are the amounts of money held by player A. There are seven possible states: A can have 0, 1, 2, 3, 4, 5, or 6 dollars. When either state 0 or state 6 is reached, the game is over. In any other state, the amount of money held by player A will increase by $1, or decrease by $1, with each of these events having probability 1/2 (since we assume a fair coin). For example, in state 3 (A has $3), there is 1/2 chance of changing to state 2 and a 1/2 chance of changing to state 4. Thus, $p_{32} = 1/2$ and $p_{34} = 1/2$. The probability of changing from state 3 to any other state is 0. Using this information gives the following 7×7 transition matrix.

$$\begin{array}{c} \\ 0 \\ 1 \\ 2 \\ 3 \\ 4 \\ 5 \\ 6 \end{array} \begin{array}{ccccccc} 0 & 1 & 2 & 3 & 4 & 5 & 6 \\ \end{array} \\ \left[\begin{array}{ccccccc} 1 & 0 & 0 & 0 & 0 & 0 & 0 \\ \frac{1}{2} & 0 & \frac{1}{2} & 0 & 0 & 0 & 0 \\ 0 & \frac{1}{2} & 0 & \frac{1}{2} & 0 & 0 & 0 \\ 0 & 0 & \frac{1}{2} & 0 & \frac{1}{2} & 0 & 0 \\ 0 & 0 & 0 & \frac{1}{2} & 0 & \frac{1}{2} & 0 \\ 0 & 0 & 0 & 0 & \frac{1}{2} & 0 & \frac{1}{2} \\ 0 & 0 & 0 & 0 & 0 & 0 & 1 \end{array} \right] = G$$

Based on the rules of the game given above, states 0 and 6 are absorbing—once these states are reached, they can never be left, and the game is over. It is possible to get from one of the nonabsorbing states, 1, 2, 3, 4, or 5, to one of the absorbing states, so the Markov chain is absorbing.

For the long-term trend of the game, find various powers of the transition matrix. A computer or a programmable calculator can be used to verify these results.

$$G^6 = \begin{bmatrix} 1.0000 & .0000 & .0000 & .0000 & .0000 & .0000 & .0000 \\ .6875 & .0781 & .0000 & .1406 & .0000 & .0625 & .0313 \\ .4531 & .0000 & .2188 & .0000 & .2031 & .0000 & .1250 \\ .2188 & .1406 & .0000 & .2813 & .0000 & .1406 & .2188 \\ .1250 & .0000 & .2031 & .0000 & .2188 & .0000 & .4531 \\ .0313 & .0625 & .0000 & .1406 & .0000 & .0781 & .6875 \\ .0000 & .0000 & .0000 & .0000 & .0000 & .0000 & 1.0000 \end{bmatrix}$$

$$G^{10} = \begin{bmatrix} 1.0000 & .0000 & .0000 & .0000 & .0000 & .0000 & .0000 \\ .7539 & .0400 & .0000 & .0791 & .0000 & .0391 & .0879 \\ .5479 & .0000 & .1191 & .0000 & .1182 & .0000 & .2148 \\ .3418 & .0791 & .0000 & .1582 & .0000 & .0791 & .3418 \\ .2148 & .0000 & .1182 & .0000 & .1191 & .0000 & .5479 \\ .0879 & .0391 & .0000 & .0791 & .0000 & .0400 & .7539 \\ .0000 & .0000 & .0000 & .0000 & .0000 & .0000 & 1.0000 \end{bmatrix}$$

As these results suggest, the system tends to one of the absorbing states, so that the probability is 1 that one of the two gamblers will eventually be wiped out. ▬

▬ EXAMPLE 3

Estimate the long-term trend for the transition matrix

$$P = \begin{bmatrix} .3 & .2 & .5 \\ 0 & 1 & 0 \\ 0 & 0 & 1 \end{bmatrix}.$$

Both states 2 and 3 are absorbing, and since it is possible to go from nonabsorbing state 1 to an absorbing state, the chain will eventually enter either state 2 or state 3. To find the long-term trend, let us find various powers of P.

$$P^2 = \begin{bmatrix} .09 & .26 & .65 \\ 0 & 1 & 0 \\ 0 & 0 & 1 \end{bmatrix} \qquad P^4 = \begin{bmatrix} .0081 & .2834 & .7085 \\ 0 & 1 & 0 \\ 0 & 0 & 1 \end{bmatrix}$$

$$P^8 = \begin{bmatrix} .0001 & .2857 & .7142 \\ 0 & 1 & 0 \\ 0 & 0 & 1 \end{bmatrix} \qquad P^{16} = \begin{bmatrix} .0000 & .2857 & .7142 \\ 0 & 1 & 0 \\ 0 & 0 & 1 \end{bmatrix}$$

Based on these powers, it appears that the transition matrix is getting closer and closer to the matrix

$$\begin{bmatrix} 0 & .29 & .71 \\ 0 & 1 & 0 \\ 0 & 0 & 1 \end{bmatrix}.$$

If the system is originally in state 1, there is no chance it will end up in state 1, but a .29 chance that it will end up in state 2 and a .71 chance it will end up in state 3. If the system was originally in state 2 it will end up in state 2; a similar statement can be made for state 3. ■

The examples suggest the following properties of absorbing Markov chains, which can be verified using more advanced methods.

1. Regardless of the original state of an absorbing Markov chain, in a finite number of steps the chain will enter an absorbing state and then stay in that state.

2. The powers of the transition matrix get closer and closer to some particular matrix.

3. The long-term trend depends on the initial state—changing the initial state can change the final result.

The third property distinguishes absorbing Markov chains from regular Markov chains, where the final result is independent of the initial state.

It would be preferable to have a method for finding the final probabilities of entering an absorbing state without finding all the powers of the transition matrix, as in Example 3. We do not really need to worry about the absorbing states (to enter an absorbing state is to stay there). Therefore, it is necessary only to work with the nonabsorbing states. To see how this is done, let us use as an example the transition matrix from the gambler's ruin problem in Example 2. Rewrite the matrix so that the rows and columns corresponding to the absorbing states come first.

	Absorbing		Nonabsorbing				
	0	6	1	2	3	4	5
0	1	0	0	0	0	0	0
6	0	1	0	0	0	0	0
1	$\frac{1}{2}$	0	0	$\frac{1}{2}$	0	0	0
2	0	0	$\frac{1}{2}$	0	$\frac{1}{2}$	0	0
3	0	0	0	$\frac{1}{2}$	0	$\frac{1}{2}$	0
4	0	0	0	0	$\frac{1}{2}$	0	$\frac{1}{2}$
5	0	$\frac{1}{2}$	0	0	0	$\frac{1}{2}$	0

$= G$

Let I_2 represent the 2×2 identity matrix in the upper left corner, let θ (the Greek letter *theta*) represent the matrix of zeros in the upper right, let R represent the matrix in the lower left, and let Q represent the matrix in the lower right. Using these symbols, G can be written as

$$G = \left[\begin{array}{c|c} I_2 & \theta \\ \hline R & Q \end{array}\right].$$

The **fundamental matrix** for an absorbing Markov chain is defined as matrix F, where

$$F = [I_n - Q]^{-1}.$$

Here I_n is the $n \times n$ identity matrix corresponding in size to matrix Q, so that the difference $I_n - Q$ exists.

For the gambler's ruin problem, using I_5 gives

$$F = \left[\begin{bmatrix} 1 & 0 & 0 & 0 & 0 \\ 0 & 1 & 0 & 0 & 0 \\ 0 & 0 & 1 & 0 & 0 \\ 0 & 0 & 0 & 1 & 0 \\ 0 & 0 & 0 & 0 & 1 \end{bmatrix} - \begin{bmatrix} 0 & \frac{1}{2} & 0 & 0 & 0 \\ \frac{1}{2} & 0 & \frac{1}{2} & 0 & 0 \\ 0 & \frac{1}{2} & 0 & \frac{1}{2} & 0 \\ 0 & 0 & \frac{1}{2} & 0 & \frac{1}{2} \\ 0 & 0 & 0 & \frac{1}{2} & 0 \end{bmatrix} \right]^{-1}$$

$$= \begin{bmatrix} 1 & -\frac{1}{2} & 0 & 0 & 0 \\ -\frac{1}{2} & 1 & -\frac{1}{2} & 0 & 0 \\ 0 & -\frac{1}{2} & 1 & -\frac{1}{2} & 0 \\ 0 & 0 & -\frac{1}{2} & 1 & -\frac{1}{2} \\ 0 & 0 & 0 & -\frac{1}{2} & 1 \end{bmatrix}^{-1}$$

$$= \begin{array}{c} \\ 1 \\ 2 \\ 3 \\ 4 \\ 5 \end{array} \begin{array}{ccccc} 1 & 2 & 3 & 4 & 5 \\ \begin{bmatrix} \frac{5}{3} & \frac{4}{3} & 1 & \frac{2}{3} & \frac{1}{3} \\ \frac{4}{3} & \frac{8}{3} & 2 & \frac{4}{3} & \frac{2}{3} \\ 1 & 2 & 3 & 2 & 1 \\ \frac{2}{3} & \frac{4}{3} & 2 & \frac{8}{3} & \frac{4}{3} \\ \frac{1}{3} & \frac{2}{3} & 1 & \frac{4}{3} & \frac{5}{3} \end{bmatrix} \end{array}$$

The inverse was found using techniques from Chapter 2.

Finally, use the fundamental matrix F along with matrix R found above to get the product FR.

$$FR = \begin{bmatrix} \frac{5}{3} & \frac{4}{3} & 1 & \frac{2}{3} & \frac{1}{3} \\ \frac{4}{3} & \frac{8}{3} & 2 & \frac{4}{3} & \frac{2}{3} \\ 1 & 2 & 3 & 2 & 1 \\ \frac{2}{3} & \frac{4}{3} & 2 & \frac{8}{3} & \frac{4}{3} \\ \frac{1}{3} & \frac{2}{3} & 1 & \frac{4}{3} & \frac{5}{3} \end{bmatrix} \begin{bmatrix} \frac{1}{2} & 0 \\ 0 & 0 \\ 0 & 0 \\ 0 & 0 \\ 0 & \frac{1}{2} \end{bmatrix} = \begin{array}{c} \\ 1 \\ 2 \\ 3 \\ 4 \\ 5 \end{array} \begin{array}{cc} 0 & 6 \\ \begin{bmatrix} \frac{5}{6} & \frac{1}{6} \\ \frac{2}{3} & \frac{1}{3} \\ \frac{1}{2} & \frac{1}{2} \\ \frac{1}{3} & \frac{2}{3} \\ \frac{1}{6} & \frac{5}{6} \end{bmatrix} \end{array}$$

The product matrix *FR* gives the probability that if the system was originally in a nonabsorbing state, it ended up in either of the two absorbing states. For example, the probability is 2/3 that if the system was originally in state 2, it ended up in state 0; the probability is 5/6 that if the system was originally in state 5 it ended up in state 6, and so on. Based on the original statement of the gambler's ruin problem, if player A starts with $2 (state 2), there is a 2/3 chance of ending in state 0 (player A is ruined); if player A starts with $5 (state 5) there is a 1/6 chance of player A being ruined, and so on.

In the fundamental matrix *F*, the sum of the elements in row *i* gives the expected number of steps for the matrix to enter an absorbing state from the *i*th nonabsorbing state. Thus, for $i = 2$, $4/3 + 8/3 + 2 + 4/3 + 2/3 = 8$ steps will be needed for the system to go from state 2 to an absorbing state (0 or 6).

Let us summarize what we have learned about absorbing Markov chains.

PROPERTIES OF ABSORBING MARKOV CHAINS

1. Regardless of the initial state, in a finite number of steps the chain will enter an absorbing state and then stay in that state.

2. The powers of the transition matrix get closer and closer to some particular matrix.

3. The long-term depends on the initial state.

4. Let *G* be the transition matrix for an absorbing Markov chain. Rearrange the rows and columns of *G* so that the absorbing states come first. Matrix *G* will have the form

$$G = \left[\begin{array}{c|c} I_n & \theta \\ \hline R & Q \end{array}\right],$$

where I_n is an identity matrix, with *n* equal to the number of absorbing states, and θ is a matrix of all zeros. The fundamental matrix is defined as

$$F = [I_m - Q]^{-1},$$

where I_m has the same order as *Q*.

5. The product *FR* gives the matrix of probabilities that a particular initial nonabsorbing state will lead to a particular absorbing state.

▬ EXAMPLE 4

Find the long-term trend for the transition matrix

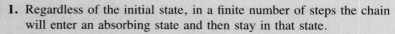

$$\begin{array}{c} \\ 1 \\ 2 \\ 3 \end{array} \begin{array}{ccc} 1 & 2 & 3 \end{array} \\ \left[\begin{array}{ccc} .3 & .2 & .5 \\ 0 & 1 & 0 \\ 0 & 0 & 1 \end{array}\right] = P$$

of Example 3.

Rewrite the matrix so that absorbing states 2 and 3 come first.

$$\begin{array}{c} \\ 2 \\ 3 \\ 1 \end{array} \begin{array}{ccc} 2 & 3 & 1 \\ \left[\begin{array}{cc|c} 1 & 0 & 0 \\ 0 & 1 & 0 \\ \hline .2 & .5 & .3 \end{array} \right] \end{array}$$

Here $R = [.2 \quad .5]$ and $Q = [.3]$. Find the fundamental matrix F.

$$F = [I_1 - Q]^{-1} = [1 - .3]^{-1} = [.7]^{-1} = [1/.7]$$

The product FR is

$$FR = [1/.7][.2 \quad .5] = [2/7 \quad 5/7] \approx [.286 \quad .714].$$

If the system starts in the nonabsorbing state 1, there is a 2/7 chance of ending up in the absorbing state 2 and a 5/7 chance of ending in the absorbing state 3. ■

8.3 EXERCISES

Find all absorbing states for the transition matrices in Exercises 1–8. Which are transition matrices for absorbing Markov chains?

1. $\begin{bmatrix} .15 & .05 & .8 \\ 0 & 1 & 0 \\ .4 & .6 & 0 \end{bmatrix}$
2. $\begin{bmatrix} .1 & .5 & .4 \\ .2 & .2 & .6 \\ 0 & 0 & 1 \end{bmatrix}$
3. $\begin{bmatrix} .4 & 0 & .6 \\ 0 & 1 & 0 \\ .9 & 0 & .1 \end{bmatrix}$
4. $\begin{bmatrix} .5 & .5 & 0 \\ .8 & .2 & 0 \\ 0 & 0 & 1 \end{bmatrix}$

5. $\begin{bmatrix} .2 & .5 & .1 & .2 \\ 0 & 1 & 0 & 0 \\ .9 & .02 & .04 & .04 \\ 0 & 0 & 0 & 1 \end{bmatrix}$
6. $\begin{bmatrix} 1 & 0 & 0 & 0 \\ .9 & .1 & 0 & 0 \\ 0 & 0 & 1 & 0 \\ .6 & 0 & .4 & 0 \end{bmatrix}$
7. $\begin{bmatrix} .1 & .8 & 0 & .1 \\ 0 & 1 & 0 & 0 \\ 1 & 0 & 0 & 0 \\ 0 & 0 & 0 & 1 \end{bmatrix}$
8. $\begin{bmatrix} .32 & .41 & .16 & .11 \\ .42 & .30 & 0 & .28 \\ 0 & 0 & 0 & 1 \\ 1 & 0 & 0 & 0 \end{bmatrix}$

Find the fundamental matrix F for the absorbing Markov chains with the matrices in Exercises 9–18. Also find the product matrix FR.

9. $\begin{bmatrix} 1 & 0 & 0 \\ 0 & 1 & 0 \\ .2 & .3 & .5 \end{bmatrix}$
10. $\begin{bmatrix} 1 & 0 & 0 \\ .6 & .1 & .3 \\ 0 & 0 & 1 \end{bmatrix}$
11. $\begin{bmatrix} .8 & .15 & .05 \\ 0 & 1 & 0 \\ 0 & 0 & 1 \end{bmatrix}$
12. $\begin{bmatrix} .42 & .37 & .21 \\ 0 & 1 & 0 \\ 0 & 0 & 1 \end{bmatrix}$

13. $\begin{bmatrix} 1 & 0 & 0 \\ 0 & 1 & 0 \\ \frac{1}{3} & \frac{1}{3} & \frac{1}{3} \end{bmatrix}$
14. $\begin{bmatrix} 1 & 0 & 0 \\ \frac{3}{8} & \frac{1}{8} & \frac{1}{2} \\ 0 & 0 & 1 \end{bmatrix}$
15. $\begin{bmatrix} 1 & 0 & 0 & 0 \\ \frac{1}{3} & 0 & \frac{2}{3} & 0 \\ 0 & 0 & 1 & 0 \\ \frac{1}{4} & \frac{1}{4} & \frac{1}{4} & \frac{1}{4} \end{bmatrix}$
16. $\begin{bmatrix} \frac{1}{4} & \frac{1}{2} & 0 & \frac{1}{4} \\ 0 & 1 & 0 & 0 \\ 0 & 0 & 1 & 0 \\ \frac{1}{2} & 0 & 0 & \frac{1}{2} \end{bmatrix}$

17. $\begin{bmatrix} 1 & 0 & 0 & 0 & 0 \\ 0 & 1 & 0 & 0 & 0 \\ .1 & .2 & .3 & .2 & .2 \\ .3 & .5 & .1 & 0 & .1 \\ 0 & 0 & 0 & 0 & 1 \end{bmatrix}$
18. $\begin{bmatrix} .4 & .2 & .3 & 0 & .1 \\ 0 & 1 & 0 & 0 & 0 \\ 0 & 0 & 1 & 0 & 0 \\ .1 & .5 & .1 & .1 & .2 \\ 0 & 0 & 0 & 0 & 1 \end{bmatrix}$

19. Write a transition matrix for a gambler's ruin problem when player A and player B start with a total of $4.

    (a) Find matrix F for this transition matrix, and find the product matrix FR.

    (b) Suppose player A starts with $1. What is the probability of ruin for A?

    (c) Suppose player A starts with $3. What is the probability of ruin for A?

20. Suppose player B (Exercise 19) slips in a coin that is slightly "loaded"—such that the probability that B wins a particular toss changes from 1/2 to 3/5. Suppose that A and B start the game with a total of $5.

    (a) If B starts with $3, find the probability that A will be ruined.

    (b) If B starts with $1, find the probability that A will be ruined.

It can be shown that the probability of ruin for player A in a game such as the one described in this section is

$$x_a = \frac{b}{a + b} \text{ if } r = 1, \quad \text{and} \quad x_a = \frac{r^a - r^{a+b}}{1 - r^{a+b}} \text{ if } r \neq 1,$$

where a is the initial amount of money that player A has, b is the initial amount that player B has, $r = (1 - p)/p$, and p is the probability that player A will win on a given play.

21. Find the probability that A will be ruined if $a = 10$, $b = 30$, and $p = .49$.

22. Find the probability in Exercise 21 if p changes to .50.

23. Complete the following chart, assuming $a = 10$ and $b = 10$.

p	.1	.2	.3	.4	.5	.6	.7	.8	.9
x_a									

▰ APPLICATIONS

SOCIAL SCIENCES

Student Retention 24. At a particular two-year college, a student has a probability of .25 of flunking out during a given year, a .15 probability of having to repeat the year, and a .6 probability of finishing the year. Use the states at the side.

(a) Write a transition matrix. Find F and FR.

(b) Find the probability that a freshman will graduate.

State	Meaning
1	Freshman
2	Sophomore
3	Has flunked out
4	Has graduated

Transportation 25. The city of Sacramento recently completed a new light rail system to bring commuters and shoppers into the downtown area and relieve freeway congestion. City planners estimate that each year 15% of those who drive or ride in an automobile will change to the light rail system, 80% will continue to use automobiles, and the rest will no longer go to the downtown area. Of those who use light rail, 5% will go back to using an automobile, 80% will continue to use light rail, and the rest will stay out of the downtown area. Assume those who do not go downtown will continue to stay out of the downtown area.

(a) Write a transition matrix. Find F and FR.

(b) Find the probability that a person who commuted by automobile ends up avoiding the downtown area.

Rat Maze 26. A rat is placed at random in one of the compartments of the maze pictured below. The probability that a rat in compartment 1 will move to compartment 2 is .3; to compartment 3 is .2; and to compartment 4 is .1. A rat in compartment 2 will move to compartments 1, 4, or 5 with probabilities of .2, .6, and .1 respectively. A rat in compartment 3 cannot leave that compartment. A rat in compartment 4 will move to 1, 2, 3, or 5 with probabilities of .1, .1, .4, and .3, respectively. A rat in compartment 5 cannot leave that compartment.

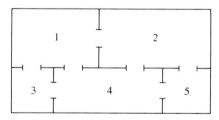

(a) Set up a transition matrix using this information. Find matrices F and FR. Find the probability that a rat ends up in compartment 5 if it was originally in compartment

(b) 1   (c) 2   (d) 3   (e) 4.

LIFE SCIENCES

Contagion 27. Under certain conditions, the probability that a person will get a particular contagious disease and die from it is .05, and the probability of getting the disease and surviving is .15. The probability that a survivor will pass the disease to another person who dies from it is also .05, that a survivor will pass the disease to another person who survives is .15, and so on. A transition matrix using the following states is given below. A person in state 1 is one who gets the disease and dies, a person in state 2 gets the disease and survives, and a person in state 3 does not get the disease.

$$
\begin{array}{cc}
 & \begin{array}{ccc} & \textit{Second Person} & \\ \;\;1 & \;\;2 & \;\;3 \end{array} \\
\textit{First Person} \begin{array}{c} 1 \\ 2 \\ 3 \end{array} &
\begin{bmatrix}
.05 & .15 & .8 \\
.05 & .15 & .8 \\
0 & 0 & 1
\end{bmatrix}
\end{array}
$$

(a) Find F and FR.

(b) Find the probability that an individual at the end of the chain gets the disease.

FOR THE COMPUTER

Company Training Program   **28.** **(a)** Find F and FR for the transition matrix from Section 8.1 for the company training program

$$
\begin{array}{c@{}c}
 & \begin{array}{cccc} s_1 & s_2 & s_3 & s_4 \end{array} \\
\begin{array}{c} s_1 \\ s_2 \\ s_3 \\ s_4 \end{array} &
\left[\begin{array}{cccc}
.4 & .2 & .05 & .35 \\
0 & .45 & .05 & .5 \\
0 & 0 & 1 & 0 \\
0 & 0 & 0 & 1
\end{array} \right].
\end{array}
$$

(b) Find the probability that a worker originally in the program is discharged.

(c) Find the probability that a worker not originally in the program goes on to complete the program.

Gambler's Ruin   **29.** Write a transition matrix for a gambler's ruin problem where players A and B start with a total of $10.

(a) Find the probability of ruin for A if A starts with $4.

(b) Find the probability of ruin for A if A starts with $5.

KEY WORDS

8.1 state
transition diagram
transition matrix
Markov chain
probability vector

8.2 regular transition matrix

regular Markov chain
equilibrium (or fixed) vector

8.3 absorbing state
absorbing chain
fundamental matrix

CHAPTER 8   REVIEW EXERCISES

Decide whether each of the matrices in Exercises 1–4 could be a transition matrix.

1. $\begin{bmatrix} .4 & .6 \\ 1 & 0 \end{bmatrix}$
    2. $\begin{bmatrix} -.2 & 1.2 \\ .8 & .2 \end{bmatrix}$
    3. $\begin{bmatrix} .8 & .2 & 0 \\ 0 & 1 & 0 \\ .1 & .4 & .5 \end{bmatrix}$
    4. $\begin{bmatrix} .6 & .2 & .3 \\ .1 & .5 & .4 \\ .3 & .3 & .4 \end{bmatrix}$

For each of the transition matrices in Exercises 5–8, (a) find the first three powers; and (b) find the probability that state 2 changes to state 1 after three repetitions of the experiment.

5. $C = \begin{bmatrix} .6 & .4 \\ 1 & 0 \end{bmatrix}$
    6. $D = \begin{bmatrix} .3 & .7 \\ .5 & .5 \end{bmatrix}$
    7. $E = \begin{bmatrix} .2 & .5 & .3 \\ .1 & .8 & .1 \\ 0 & 1 & 0 \end{bmatrix}$
    8. $F = \begin{bmatrix} .14 & .12 & .74 \\ .35 & .28 & .37 \\ .71 & .24 & .05 \end{bmatrix}$

In Exercises 9–12, use the transition matrices T, along with the given initial distribution D, to find the distribution after two repetitions of the experiment. Also predict the long-range distribution.

9. $D = [.3 \quad .7]; T = \begin{bmatrix} .4 & .6 \\ .5 & .5 \end{bmatrix}$
    10. $D = [.8 \quad .2]; T = \begin{bmatrix} .7 & .3 \\ .2 & .8 \end{bmatrix}$

11. $D = [.2 \quad .4 \quad .4]; T = \begin{bmatrix} .6 & .2 & .2 \\ .3 & .3 & .4 \\ .5 & .4 & .1 \end{bmatrix}$
    12. $D = [.1 \quad .1 \quad .8]; T = \begin{bmatrix} .2 & .3 & .5 \\ .1 & .1 & .8 \\ .7 & .1 & .2 \end{bmatrix}$

Decide whether each of the following transition matrices is regular.

13. $\begin{bmatrix} 0 & 1 \\ .2 & .8 \end{bmatrix}$

14. $\begin{bmatrix} .4 & .2 & .4 \\ 0 & 1 & 0 \\ .6 & .3 & .1 \end{bmatrix}$

15. $\begin{bmatrix} 1 & 0 & 0 \\ 0 & 1 & 0 \\ .3 & .5 & .2 \end{bmatrix}$

Find all absorbing states for the matrices in Exercises 16–18. Which are transition matrices for an absorbing Markov chain?

16. $\begin{bmatrix} 1 & 0 & 0 \\ .5 & .1 & .4 \\ 0 & 1 & 0 \end{bmatrix}$

17. $\begin{bmatrix} .2 & 0 & .8 \\ 0 & 1 & 0 \\ .7 & 0 & .3 \end{bmatrix}$

18. $\begin{bmatrix} .5 & .1 & .1 & .3 \\ 0 & 0 & 1 & 0 \\ 1 & 0 & 0 & 0 \\ .1 & .8 & .05 & .05 \end{bmatrix}$

In Exercises 19–22, find the fundamental matrix F for the absorbing Markov chains with matrices as follows. Also find the matrix FR.

19. $\begin{bmatrix} .2 & .5 & .3 \\ 0 & 1 & 0 \\ 0 & 0 & 1 \end{bmatrix}$

20. $\begin{bmatrix} 1 & 0 & 0 \\ 0 & 1 & 0 \\ .3 & .1 & .6 \end{bmatrix}$

21. $\begin{bmatrix} \frac{1}{5} & \frac{1}{5} & \frac{2}{5} & \frac{1}{5} \\ 0 & 1 & 0 & 0 \\ \frac{1}{2} & \frac{1}{4} & \frac{1}{8} & \frac{1}{8} \\ 0 & 0 & 0 & 1 \end{bmatrix}$

22. $\begin{bmatrix} .3 & .5 & .1 & .1 \\ .4 & .1 & .3 & .2 \\ 0 & 0 & 1 & 0 \\ 0 & 0 & 0 & 1 \end{bmatrix}$

▰ APPLICATIONS

BUSINESS AND ECONOMICS

Advertising   *Currently, 35% of all hot dogs sold in one area are made by Dogkins, and 65% are made by Long Dog. Suppose that Dogkins starts a heavy advertising campaign, with the campaign producing the following transition matrix.*

		After Campaign	
		Dogkins	Long Dog
Before	Dogkins	.8	.2
Campaign	Long Dog	.4	.6

23. Find the share of the market for each company

    (a) after one campaign;   **(b)** after three such campaigns.

24. Predict the long-range market share for Dogkins.

Credit Cards   *A credit card company classifies its customers in three groups: nonusers in a given month, light users, and heavy users. The transition matrix for these states is*

	Nonuser	Light	Heavy
Nonuser	.8	.15	.05
Light	.25	.55	.2
Heavy	.04	.21	.75

Suppose the initial distribution for the three states is $[.4 \quad .4 \quad .2]$. *Find the distribution after each of the following periods.*

25. 1 month   **26.** 2 months   **27.** 3 months

28. What is the long-range prediction for the distribution of users?

LIFE SCIENCES

Medical Research *A medical researcher is studying the risk of heart attack in men. She first divides men into three weight categories: thin, normal, and overweight. By studying the male ancestors, sons, and grandsons of these men, the researcher comes up with the following transition matrix.*

$$
\begin{array}{c}
\text{Thin} \\
\text{Normal} \\
\text{Overweight}
\end{array}
\begin{array}{ccc}
\text{Thin} & \text{Normal} & \text{Overweight} \\
\end{array}
\left[
\begin{array}{ccc}
.3 & .5 & .2 \\
.2 & .6 & .2 \\
.1 & .5 & .4
\end{array}
\right]
$$

Find the probabilities of the following for a man of normal weight.

29. Thin son   **30.** Thin grandson   **31.** Thin great-grandson

Find the probabilities of the following for an overweight man.

32. Overweight son   **33.** Overweight grandson   **34.** Overweight great-grandson

Suppose that the distribution of men by weight is initially given by [.2   .55   .25]. *Find each of the following distributions.*

35. After 1 generation   **36.** After 2 generations   **37.** After 3 generations

38. Find the long-range prediction for the distribution of weights.

Genetics *Researchers sometimes study the problem of mating offspring from the same two parents; two of these offspring are then mated, and so on. Let A be a dominant gene for some trait, and a the recessive gene. The original offspring can carry genes AA, Aa, or aa. There are 6 possible ways that these offspring can mate.*

State	Mating
1	AA and AA
2	AA and Aa
3	AA and aa
4	Aa and Aa
5	Aa and aa
6	aa and aa

Using these states gives the following transition matrix.

$$
\begin{array}{c}
1 \\ 2 \\ 3 \\ 4 \\ 5 \\ 6
\end{array}
\begin{array}{cccccc}
1 & 2 & 3 & 4 & 5 & 6
\end{array}
\left[
\begin{array}{cccccc}
1 & 0 & 0 & 0 & 0 & 0 \\
\frac{1}{4} & \frac{1}{2} & 0 & \frac{1}{4} & 0 & 0 \\
0 & 0 & 1 & 0 & 0 & 0 \\
\frac{1}{16} & \frac{1}{4} & \frac{1}{8} & \frac{1}{4} & \frac{1}{4} & \frac{1}{16} \\
0 & 0 & 0 & \frac{1}{4} & \frac{1}{2} & \frac{1}{4} \\
0 & 0 & 0 & 0 & 0 & 1
\end{array}
\right]
$$

39. Identify the absorbing states.

40. Find matrix Q.

41. Find F, and the product FR.

42. If the system starts in state 4, find the probability it will end in state 3.

9 Decision Theory

9.1   Decision Making

9.2   Strictly Determined Games

9.3   Mixed Strategies

Review Exercises

Extended Application
Decision Making in Life Insurance

Extended Application
Decision Making in the Military

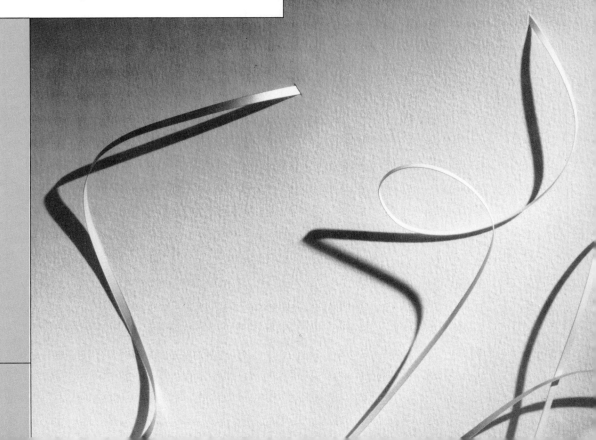

John F. Kennedy once remarked that he had assumed that as president he would find it difficult to choose between distinct, opposite alternatives when a decision needed to be made. He said that actually, however, such decisions were easy to make; the hard decisions came when he was faced with choices that were not as clear-cut. Most decisions that we must make fall in the second category—decisions that must be made under conditions of uncertainty. *Decision theory* is a mathematical model that provides a systematic way to attack problems of decision making when some alternatives are unclear or ambiguous.

9.1 DECISION MAKING

The idea of expected value, introduced earlier, is used in decision theory. The concepts are explained in the following example.

Freezing temperatures are endangering the orange crop in central California. A farmer can protect his crop by burning smudge pots—the heat from the pots keeps the oranges from freezing. Burning the pots is expensive, however; the cost is $4000. The farmer knows that if he burns smudge pots he will be able to sell his crop for a net profit (after smudge pot costs are deducted) of $10,000, provided that the freeze does develop and wipes out many of the other orange growers in California. If he does nothing he will either lose $2000 in planting costs if there is a freeze, or make a profit of $9600 if there is no freeze. (If there is no freeze, there will be a large supply of oranges, and thus his profit will be lower than if there was a small supply.)

What should the farmer do? He should begin by carefully defining the problem. First he must decide on the **states of nature,** the possible alternatives over which he has no control. Here there are two: freezing temperatures, or no freezing temperatures. Next, the farmer should list the things he can control—his actions or **strategies.** The farmer has two possible strategies: to use smudge pots or not to use smudge pots. The consequences of each action under each state of nature, called **payoffs,** can be summarized in a **payoff matrix,** as shown below. The payoffs in this case represent the profit for each possible combination of events.

		States of Nature	
		Freeze	No Freeze
Farmer's Strategies	Use Smudge Pots	$10,000	$5600
	Do Not Use Pots	−$2000	$9600

To get the $5600 entry in the payoff matrix, we took the profit if there is no freeze, $9600, and subtracted the $4000 cost of using the smudge pots.

Once the farmer makes the payoff matrix, what then? The farmer might be an optimist (some might call him a gambler); in this case he might assume that the best will happen and go for the biggest number on the matrix ($10,000). To get this profit, he must adopt the strategy "use smudge pots."

On the other hand, if the farmer were a pessimist, he would want to minimize the worst thing that could happen. If he uses smudge pots, the worst that can happen to him would be a profit of $5600, which will result if there is no freeze. If he does not use smudge pots, he might lose $2000. To minimize the worst, he once again should adopt the strategy "use smudge pots."

Suppose the farmer decides that he is neither an optimist nor a pessimist, but would like further information before choosing a strategy. For example, he might call the weather forecaster and ask for the probability of a freeze. Further, suppose the forecaster says that this probability is only .1. What should the farmer do? He should recall our earlier discussion of expected value and calculate the expected profit for each of his two possible strategies. If the probability of a freeze is .1, then the probability that there will be no freeze is .9. This information gives the following expected values:

$$\text{if smudge pots are used: } 10,000(.1) + 5600(.9) = \$6040;$$
$$\text{if no smudge pots are used: } -2000(.1) + 9600(.9) = \$8440.$$

Here the maximum expected profit, $8440, is obtained if smudge pots are *not* used. If the probability of a freeze is .6, the expected profit from the strategy "use pots" would be $8240 and from "use no pots," $2640. As the example shows, the farmer's beliefs about the probabilities of a freeze affect his choice of strategy.

▬ EXAMPLE 1

A small manufacturer of Christmas cards must decide in February what type of cards to emphasize in her fall line. She has three possible strategies: emphasize modern cards, emphasize old-fashioned cards, or emphasize a mixture of the two. Her success is dependent on the state of the economy in December. If the economy is strong, she will do well with her modern cards, while in a weak economy people long for the old days and buy old-fashioned cards. In an in-between economy, her mixture of lines would do the best. She first prepares a payoff matrix for all three possibilities. The numbers in the matrix represent her profits in thousands of dollars.

		States of Nature		
		Weak Economy	In-Between	Strong Economy
	Modern	40	85	120
Strategies	Old-Fashioned	106	46	83
	Mixture	72	90	68

(a) If the manufacturer is an optimist, she should aim for the biggest number on the matrix, 120 (representing $120,000 in profit). Her strategy in this case would be to produce modern cards.

(b) A pessimistic manufacturer wants to avoid the worst of all bad things that can happen. If she produces modern cards, the worst that can happen is a profit of $40,000. For old-fashioned cards, the worst is a profit of $46,000, while the worst that can happen from a mixture is a profit of $68,000. Her strategy here is to use a mixture.

(c) Suppose the manufacturer reads in a business magazine that leading experts feel there is a 50% chance of a weak economy at Christmas, a 20% chance of an in-between economy, and a 30% chance of a strong economy. The manufacturer can now find her expected profit for each possible strategy.

Modern: $40(.50) + 85(.20) + 120(.30) = 73$

Old-Fashioned: $106(.50) + 46(.20) + 83(.30) = 87.1$

Mixture: $72(.50) + 90(.20) + 68(.30) = 74.4$

Here the best strategy is old-fashioned cards; the expected profit is 87.1, or $87,100. ▬

Sometimes the numbers (or payoffs) in a payoff matrix do not represent money (profits or costs, for example), but *utility*. A **utility** is a number that measures the satisfaction (or lack of it) that results from a certain action. The numbers must be assigned by each individual, depending on how he or she feels about a situation. For example, one person might assign a utility of $+20$ to a week's vacation in San Francisco but only -6 if the vacation were moved to Sacramento. Matrices with utility payoffs are treated the same as those with dollar payoffs. (See Exercises 9 and 10 in this section.)

═ 9.1 EXERCISES   ═ APPLICATIONS

BUSINESS AND ECONOMICS

Investments 1. An investor has $20,000 to invest in stocks. She has two possible strategies: buy conservative blue-chip stocks or buy highly speculative stocks. There are two states of nature: the market goes up or the market goes down. The following payoff matrix shows the net amounts she will have under the various circumstances.

	Market Up	Market Down
Buy Blue-Chip	$25,000	$18,000
Buy Speculative	$30,000	$11,000

What should the investor do if she is

(a) an optimist?   (b) a pessimist?

(c) Suppose there is a .7 probability of the market going up. What is the best strategy? What is the expected profit?

(d) What is the best strategy if the probability of a market rise is .2?

Land Development 2. A developer has $100,000 to invest in land. He has a choice of two parcels (at the same price), one on the highway and one on the coast. With both parcels, his ultimate profit depends on whether he faces light opposition from environmental groups or heavy opposition. He estimates that the payoff matrix is as follows (the numbers represent his profit).

		Opposition	
		Light	Heavy
Parcels	Highway	$70,000	$30,000
	Coast	$150,000	$-$40,000

What should the developer do if he is

(**a**) an optimist?   (**b**) a pessimist?

(**c**) Suppose the probability of heavy opposition is .8. What is his best strategy? What is the expected profit?

(**d**) What is the best strategy if the probability of heavy opposition is only .4?

Concert Preparations　**3.** Hillsdale College has sold out all tickets for a jazz concert to be held in the stadium. If it rains, the show will have to be moved to the gym, which has a much smaller capacity. The dean must decide in advance whether to set up the seats and the stage in the gym or in the stadium, or both, just in case. The payoff matrix below shows the net profit in each case.

		States of Nature	
		Rain	No Rain
	Set Up in Stadium	−$1550	$1500
Strategies	Set Up in Gym	$1000	$1000
	Set Up in Both	$750	$1400

What strategy should the dean choose if she is

(**a**) an optimist?   (**b**) a pessimist?

(**c**) If the weather forecaster predicts rain with a probability of .6, what strategy should she choose to maximize expected profit? What is the maximum expected profit?

Machine Repairs　**4.** An analyst must decide what fraction of the items produced by a certain machine are defective. He has already decided that there are three possibilities for the fraction of defective items: .01, .10, and .20. He may recommend two courses of action: repair the machine or make no repairs. The payoff matrix below represents the *costs* to the company in each case.

		States of Nature		
		.01	.10	.20
	Repair	$130	$130	$130
Strategies	No Repair	$25	$200	$500

What strategy should the analyst recommend if he is

(**a**) an optimist?   (**b**) a pessimist?

(**c**) Suppose the analyst is able to estimate probabilities for the three states of nature as follows.

Fraction of Defectives	Probability
.01	.70
.10	.20
.20	.10

Which strategy should he recommend? Find the expected cost to the company if this strategy is chosen.

Marketing　**5.** The research department of the Allied Manufacturing Company has developed a new process that it believes will result in an improved product. Management must decide whether to go ahead and market the new product. The new product may or may not be better than the old one. If the new product is better and the company decides to market it, sales should increase by $50,000. If it is not better and they replace the old product with the new product on the market, they will lose $25,000 to competitors. If they decide not to market the new product they will lose a total of $40,000 if it is better, and just research costs of $10,000 if it is not.

(a) Prepare a payoff matrix.

(b) If management believes there is a probability of .4 that the new product is better, find the expected profits under each strategy and determine the best action.

Machinery Overhaul **6.** A businessman is planning to ship a used machine to his plant in Nigeria. He would like to use it there for the next four years. He must decide whether to overhaul the machine before sending it. The cost of overhaul is $2600. If the machine fails when in operation in Nigeria, it will cost him $6000 in lost production and repairs. He estimates the probability that it will fail at .3 if he does not overhaul it, and .1 if he does overhaul it. Neglect the possibility that the machine might fail more than once in the next four years.

(a) Prepare a payoff matrix.

(b) What should the businessman do to minimize his expected costs?

Contractor Bidding **7.** A contractor prepares to bid on a job. If all goes well, his bid should be $30,000, which will cover his costs plus his usual profit margin of $4500. If a threatened labor strike actually occurs, however, his bid should be $40,000 to give him the same profit. If there is a strike and he bids $30,000, he will lose $5500. If his bid is too high, he may lose the job entirely, while if it is too low, he may lose money.

(a) Prepare a payoff matrix.

(b) If the contractor believes that the probability of a strike is .6, how much should he bid?

LIFE SCIENCES

Anti-Smoking Campaign **8.** A community is considering an anti-smoking campaign.* The city council will choose one of three possible strategies: a campaign for everyone over age 10 in the community, a campaign for youths only, or no campaign at all. The two states of nature are a true cause-effect relationship between smoking and cancer and no cause-effect relationship. The costs to the community (including loss of life and productivity) in each case are as shown below.

		States of Nature	
		Cause-Effect Relationship	No Cause-Effect Relationship
	Campaign for All	$100,000	$800,000
Strategies	Campaign for Youth	$2,820,000	$20,000
	No Campaign	$3,100,100	$0

What action should the city council choose if it is

(a) optimistic?   **(b)** pessimistic?

(c) If the director of public health estimates that the probability of a true cause-effect relationship is .8, which strategy should the city council choose?

---

*This problem is based on an article by B. G. Greenberg in the September 1969 issue of the *Journal of the American Statistical Association*.

SOCIAL SCIENCES

Reelection Strategy   **9.** A politician must plan her reelection strategy. She can emphasize jobs or she can emphasize the environment. The voters may be concerned mainly about jobs or about the environment. A payoff matrix showing the utility of each possible outcome is shown below.

$$\begin{array}{cc} & \textit{Voters} \\ & \begin{array}{cc} \text{Jobs} & \text{Environment} \end{array} \\ \textit{Candidate} \begin{array}{c} \text{Jobs} \\ \text{Environment} \end{array} & \begin{bmatrix} +25 & -10 \\ -15 & +30 \end{bmatrix} \end{array}$$

The political analysts feel that there is a .35 chance that the voters will emphasize jobs. What strategy should the candidate adopt? What is its expected utility?

Education   **10.** In an accounting class, the instructor permits the students to bring a calculator or a reference book (but not both) to an examination. The examination itself can emphasize either numerical problems or definitions. In trying to decide which aid to take to an examination, a student first decides on the utilities shown in the following payoff matrix.

$$\begin{array}{cc} & \textit{Exam's Emphasis} \\ & \begin{array}{cc} \text{Numbers} & \text{Definitions} \end{array} \\ \textit{Student's Choice} \begin{array}{c} \text{Calculator} \\ \text{Book} \end{array} & \begin{bmatrix} +50 & 0 \\ +10 & +40 \end{bmatrix} \end{array}$$

(a) What strategy should the student choose if the probability that the examination will emphasize numbers is .6? What is the expected utility in this case?

(b) Suppose the probability that the examination emphasizes numbers is .4. What strategy should be chosen by the student?

9.2 STRICTLY DETERMINED GAMES

The word *game* in the title of this section may have led you to think of checkers or perhaps some card game. While **game theory** does have some application to these recreational games, it was developed in the 1940s to analyze competitive situations in business, warfare, and social situations. Game theory deals with how to make decisions when in competition with an aggressive opponent.

A game can be set up with a payoff matrix, such as the one shown below. This game involves the two players A and B, and is called a **two-person game.** Player A can choose either row 1 or row 2, while player B can choose either column 1 or column 2. A player's choice is called a **strategy,** just as before. The payoff is at the intersection of the row and column selected. As a general agreement, a positive number represents a payoff from B to A; a negative number represents a payoff from A to B. For example, if A chooses row 2 and B chooses column 2, then B pays $4 to A.

$$\begin{array}{cc} & \textbf{B} \\ & \begin{array}{cc} 1 & 2 \end{array} \\ \text{A} \begin{array}{c} 1 \\ 2 \end{array} & \begin{bmatrix} 2 & -1 \\ -3 & 4 \end{bmatrix} \end{array}$$

■ EXAMPLE 1

In the payoff matrix shown on the preceding page, suppose A chooses row 1 and B chooses column 2. Who gets what?

Row 1 and column 2 lead to the number -1. This number represents a payoff of \$1 from A to B.   ■

While the numbers in the payoff matrix on the preceding page represent money, they could just as easily represent goods or other property.

In the game above, no money enters the game from the outside; whenever one player wins, the other loses. Such a game is called a **zero-sum game.** The stock market is not a zero-sum game. Stocks can go up or down according to outside forces. Therefore, it is possible that all investors can make or lose money.

Only two-person zero-sum games are discussed in the rest of this chapter. Each player can have many different options. In particular, an $m \times n$ matrix game is one in which player A has m strategies (rows) and player B has n strategies (columns).

Dominant Strategies   In the rest of this section, the best possible strategy for each player is determined. Let us begin with the 3×3 game defined by the following matrix.

$$\begin{array}{c} \\ 1 \\ 2 \\ 3 \end{array} \begin{array}{ccc} 1 & 2 & 3 \\ \left[\begin{array}{ccc} -3 & -6 & 10 \\ 3 & 0 & -9 \\ 5 & -4 & -8 \end{array}\right] \end{array}$$

From B's viewpoint, strategy 2 is better than strategy 1 no matter which strategy A selects. This can be seen by comparing the two columns. If A chooses row 1, receiving \$6 from A is better than receiving \$3; in row 2 breaking even is better than paying \$3, and in row 3, getting \$4 from A is better than paying \$5. Therefore, B should never select strategy 1. Strategy 2 is said to *dominate* strategy 1, and strategy 1 (the dominated strategy) can be removed from consideration, producing the following reduced matrix.

$$\begin{array}{c} \\ 1 \\ 2 \\ 3 \end{array} \begin{array}{cc} 2 & 3 \\ \left[\begin{array}{cc} -6 & 10 \\ 0 & -9 \\ -4 & -8 \end{array}\right] \end{array}$$

Either player may have dominated strategies. In fact, after a dominated strategy for one player is removed, the other player may then have a dominated strategy where there was none before.

DOMINANT STRATEGIES

■ A row for A **dominates** another row if every entry in the first row is *larger* than the corresponding entry in the second row. For a column for B to dominate another, each entry must be *smaller*.

In the 3 × 2 matrix above, neither player now has a dominated strategy. From A's viewpoint strategy 1 is best if B chooses strategy 3, while strategy 2 is best if B chooses strategy 1. Verify that there are no dominated strategies for either player.

▬ EXAMPLE 2

Find any dominated strategies in the games with the given payoff matrices.

(a)
$$
\begin{array}{c}
\\
1\\
2
\end{array}
\begin{array}{cccc}
\;\;1 & \;\;2 & \;\;3 & \;\;4 \\
\end{array}
\left[
\begin{array}{cccc}
-8 & -4 & -6 & -9 \\
-3 & 0 & -9 & 12
\end{array}
\right]
$$

Here every entry in column 3 is smaller than the corresponding entry in column 2. Thus, column 3 dominates column 2. By removing the dominated column 2, the final game is as follows.

$$
\begin{array}{c}
1\\
2
\end{array}
\begin{array}{ccc}
\;\;1 & \;\;3 & \;\;4 \\
\end{array}
\left[
\begin{array}{ccc}
-8 & -6 & -9 \\
-3 & -9 & 12
\end{array}
\right]
$$

(b)
$$
\begin{array}{c}
1\\
2\\
3
\end{array}
\begin{array}{cc}
\;\;1 & \;\;2 \\
\end{array}
\left[
\begin{array}{cc}
3 & -2 \\
0 & 8 \\
6 & 4
\end{array}
\right]
$$

Each entry in row 3 is greater than the corresponding entry in row 1, so that row 3 dominates row 1. Removing row 1 gives the following game.

$$
\begin{array}{c}
2\\
3
\end{array}
\begin{array}{cc}
1 & 2 \\
\end{array}
\left[
\begin{array}{cc}
0 & 8 \\
6 & 4
\end{array}
\right]
\;\;▬
$$

Strictly Determined Games   Which strategies should the players choose in the following game?

$$
\begin{array}{cc}
 & \qquad\qquad\quad B \\
 &
\begin{array}{c}
\\
1\\
A\quad 2\\
3
\end{array}
\begin{array}{ccc}
\;\;1 & \;\;2 & \;\;3 \\
\end{array}
\left[
\begin{array}{ccc}
-9 & 11 & -4 \\
2 & 3 & 5 \\
-1 & -9 & 6
\end{array}
\right]
\end{array}
$$

The goal of game theory is to find **optimum strategies:** those that are the most profitable to the respective players. The payoff that results from each player's choosing the optimum strategy is called the **value** of the game.

The simplest strategy for a player is to consistently choose a certain row (or column). Such a strategy is called a **pure strategy,** in contrast to strategies requiring the random choice of a row (or column); these alternate strategies are discussed in the next section.*

---

*In this section we solve (find the optimum strategies for) only games that have optimum *pure* strategies.

To choose a pure strategy in the game above, player A could choose row 1, in hopes of getting the payoff of $11. Player B would quickly discover this, however, and start playing column 1. By playing column 1, B would receive $9 from A. If A were to choose row 2 consistently, then B would again minimize outgo by choosing column 1 (a payoff of $2 by B to A is better than paying $3 or $5, respectively, to A). By choosing row 3 consistently, A would cause B to choose column 2. The table shows what B will do when A chooses a given row consistently.

If A Chooses Pure Strategy:	Then B Will Choose:	With Payoff:
Row 1	Column 1	$9 to B
Row 2	Column 1	$2 to A
Row 3	Column 2	$9 to B

Based on these results, A's optimum strategy is to choose row 2; in this way A will guarantee a minimum payoff of $2 per play of the game, no matter what B does.

The optimum pure strategy in this game for A (the *row* player), is found by identifying the *smallest* number in each row of the payoff matrix; the row giving the *largest* such number gives the optimum strategy.

By going through a similar analysis for player B, we find that B should choose the column that will minimize the amount A can win. In the game above, B will pay $2 to A if B consistently chooses column 1. By choosing column 2 consistently, B will pay $11 to A, and by choosing column 3 player B will pay $6 to A. The optimum strategy for B is thus to choose column 1—with each play of the game B will pay $2 to A.

The optimum pure strategy in this game for B (the *column* player) is to identify the *largest* number in each column of the payoff matrix, and then choose the column producing the *smallest* such number.

In the game above, the entry 2 is both the *smallest* entry in its *row* and the *largest* entry in its *column*. Such an entry is called a **saddle point.** (The seat of a saddle is the maximum from one direction and the minimum from another direction. See Figure 1.) As Example 3(c) shows, there may be more than one such entry, but then the entries will have the same value.

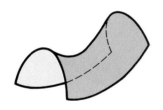

FIGURE 1

OPTIMUM PURE STRATEGY

In a game with a saddle point, the optimum pure strategy for player A is to choose the row containing the saddle point, while the optimum pure strategy for B is to choose the column containing the saddle point.

A game with a saddle point is called a **strictly determined game.** By using these optimum strategies, A and B will ensure that the same amount always changes hands with each play of the game; this amount, given by the saddle point, is the value of the game. The value of the game above is $2. A game having a value of 0 is a **fair game;** the game above is not fair.

EXAMPLE 3

Find the saddle points in the following games.

(a)
$$\begin{array}{c} & 1 & 2 \\ 1 & 2 & 2 \\ 2 & 0 & 4 \\ 3 & 1 & 6 \\ 4 & 3 & 7 \end{array}$$

The number that is both the smallest number in its row and the largest number in its column is 3. Thus, 3 is the saddle point, and the game has value 3. The strategies producing the saddle point can be written (4, 1). (Player A's strategy is written first.)

(b)
$$\begin{array}{c} & 1 & 2 \\ 1 & 6 & 5 \\ 2 & 2 & 3 \end{array}$$

The saddle point is 5, at strategies (1, 2).

(c)
$$\begin{array}{c} & 1 & 2 & 3 & 4 \\ 1 & 4 & 6 & 4 & 12 \\ 2 & -8 & -9 & 3 & 2 \end{array}$$

The saddle point, 4, occurs with either of two strategies, (1, 1), or (1, 3). The value of the game is 4. (None of the games in parts (a), (b), or (c) of this example are fair games, because none has a value of 0.)

(d)
$$\begin{array}{c} & 1 & 2 & 3 \\ 1 & 3 & 6 & -2 \\ 2 & 8 & -3 & 5 \end{array}$$

There is no number that is both the smallest number in its row and the largest number in its column, so the game has no saddle point. Since the game has no saddle point, it is not strictly determined. In the next section, methods are given for finding optimum strategies for such games.

9.2 EXERCISES

In the following game, decide on the payoff when the strategies of Exercises 1–6 are used.

$$\begin{array}{cc} & B \\ & \begin{array}{ccc} 1 & 2 & 3 \end{array} \\ A \begin{array}{c} 1 \\ 2 \\ 3 \end{array} & \begin{bmatrix} 6 & -4 & 0 \\ 3 & -2 & 6 \\ -1 & 5 & 11 \end{bmatrix} \end{array}$$

1. (1, 1) **2.** (1, 2) **3.** (2, 2)

4. (2, 3) **5.** (3, 1) **6.** (3, 2)

7. Does the game have any dominated strategies?

8. Does it have a saddle point?

Remove any dominated strategies in the games in Exercises 9–14. (From now on, we will save space by deleting the names of the strategies.)

9. $\begin{bmatrix} 0 & -2 & 8 \\ 3 & -1 & -9 \end{bmatrix}$

10. $\begin{bmatrix} 6 & 5 \\ 3 & 8 \\ -1 & -4 \end{bmatrix}$

11. $\begin{bmatrix} 1 & 4 \\ 4 & -1 \\ 3 & 5 \\ -4 & 0 \end{bmatrix}$

12. $\begin{bmatrix} 2 & 3 & 1 & -5 \\ -1 & 5 & 4 & 1 \\ 1 & 0 & 2 & -3 \end{bmatrix}$

13. $\begin{bmatrix} 8 & 12 & -7 \\ -2 & 1 & 4 \end{bmatrix}$

14. $\begin{bmatrix} 6 & 2 \\ -1 & 10 \\ 3 & 5 \end{bmatrix}$

When it exists, find the saddle point of each game and find the value of the game. Identify any games that are strictly determined.

15. $\begin{bmatrix} 3 & 5 \\ 2 & -5 \end{bmatrix}$

16. $\begin{bmatrix} 7 & 8 \\ -2 & 15 \end{bmatrix}$

17. $\begin{bmatrix} 3 & -4 & 1 \\ 5 & 3 & -2 \end{bmatrix}$

18. $\begin{bmatrix} -4 & 2 & -3 & -7 \\ 4 & 3 & 5 & -9 \end{bmatrix}$

19. $\begin{bmatrix} -6 & 2 \\ -1 & -10 \\ 3 & 5 \end{bmatrix}$

20. $\begin{bmatrix} 1 & 4 & -3 & 1 & -1 \\ 2 & 5 & 0 & 4 & 10 \\ 1 & -3 & 2 & 5 & 2 \end{bmatrix}$

21. $\begin{bmatrix} 2 & 3 & 1 \\ -1 & 4 & -7 \\ 5 & 2 & 0 \\ 8 & -4 & -1 \end{bmatrix}$

22. $\begin{bmatrix} 3 & 8 & -4 & -9 \\ -1 & -2 & -3 & 0 \\ -2 & 6 & -4 & 5 \end{bmatrix}$

23. $\begin{bmatrix} -6 & 1 & 4 & 2 \\ 9 & 3 & -8 & -7 \end{bmatrix}$

24. $\begin{bmatrix} 6 & -1 \\ 0 & 3 \\ 4 & 0 \end{bmatrix}$

25. Write a payoff matrix for the child's game *stone, scissors, paper*. Each of two children writes down one of these three words: *stone, scissors,* or *paper*. If the words are the same, the game is a tie. Otherwise, *stone* beats *scissors* (since stone ⌐an break scissors), *scissors* beats *paper* (since scissors can cut paper), and *paper* beats *stone* (since paper can hide stone). The winner receives $1 from the loser; no money changes hands in case of a tie. Is the game strictly determined?

26. John and Joann play a finger matching game—each shows one or two fingers, simultaneously. If the sum of the number of fingers showing is even, Joann pays John that number of dollars; for an odd sum, John pays Joann. Find the payoff matrix for this game. Is the game strictly determined?

27. Suppose the payoff matrix for a game has at least three rows. Also, suppose that row 1 dominates row 2, and row 2 dominates row 3. Show that row 1 must dominate row 3.

▤ APPLICATIONS

GENERAL

War Games   **28.** Two armies, *A* and *B,* are involved in a war game. Each army has available three different strategies, with payoffs as shown below. These payoffs represent square kilometers of land with positive numbers representing gains by *A*.

$$\begin{bmatrix} 3 & -8 & -9 \\ 0 & 6 & -12 \\ -8 & 4 & -10 \end{bmatrix}$$

Find the saddle point and the value of the game.

Football   **29.** When a football team has the ball and is planning its next play, it can choose one of several plays or strategies. The success of the chosen play depends largely on how well the other team "reads" the chosen play. Suppose a team with the ball (team *A*) can choose from three plays, while the opposition (team *B*) has four possible strategies. The numbers shown in the following payoff matrix represent yards of gain to team *A*.

$$\begin{bmatrix} 9 & -3 & -4 & 16 \\ 12 & 9 & 6 & 8 \\ -5 & -2 & 3 & 18 \end{bmatrix}$$

Find the saddle point. Find the value of the game.

SOCIAL SCIENCES

Border Patrol   **30.** A person attempting an illegal crossing of the border into the United States has two options. He can cross into unoccupied territory on foot at night, or he can cross at a regular entry point if he is hidden in a vehicle. His chances of detection depend on whether extra border guards are sent to the unoccupied territory or to the regular entry point. His chances of getting across the unoccupied territory are 50% if there are no extra patrols and 40% if there are extra patrols. His chances of crossing at a regular entry point are 30% if there are no extra border guards and 20% if there are extra guards. Find the saddle point and the value of the game.

BUSINESS AND ECONOMICS

Competition   **31.** Two merchants are planning competing stores to serve an area of three small cities. The fraction of the total population that live in each city is shown in the figure. If both merchants locate in the same city, merchant A will get 65% of the total business. If the merchants locate in different cities, each will get 80% of the business in the city it is in, and A will get 60% of the business from the city not containing B. Payoffs are measured by the number of percentage points above or below 50%. Write a payoff matrix for this game. Is this game strictly determined?

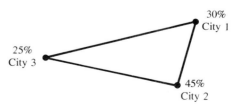

Competition   **32.** In Exercise 31, if merchant A gets 55% of the total business when both merchants locate in the same city, and the problem is unchanged otherwise, write the payoff matrix and find the saddle point and the value of the game.

▰▰▰ 9.3 MIXED STRATEGIES

As mentioned earlier, not every game has a saddle point. Two-person zero-sum games still have optimum strategies, however, even if the strategy is not as simple as the ones we saw earlier. In a game with a saddle point, the optimum strategy for player A is to pick the row containing the saddle point. Such a strategy is called a *pure strategy,* since the same row is always chosen.

If there is no saddle point, then it will be necessary for both players to mix their strategies. For example, A will sometimes play row 1, sometimes row 2, and so on. If this were done in some specific pattern, the competitor would soon guess it and play accordingly.

For this reason, it is best to mix strategies according to previously determined probabilities. For example, if a player has only two strategies and has decided to play them with equal probability, the random choice could be made by tossing a fair coin, letting heads represent one strategy and tails the other. This would result in the two strategies being used about equally over the long run. However, on a particular play it would not be possible to predetermine the strategy to be used. (Some other device, such as a spinner, is necessary if there are more than two strategies or if the probabilities are not 1/2.)

▰▰▰ EXAMPLE 1

Suppose a game has payoff matrix

$$\begin{bmatrix} -1 & 2 \\ 1 & 0 \end{bmatrix},$$

where the entries represent dollar winnings. Suppose player A chooses row 1 with probability 1/3 and row 2 with probability 2/3, and player B chooses each column with probability 1/2. Find the expected value of the game.

Assume that rows and columns are chosen independently, so that

$$P(\text{row 1, column 1}) = P(\text{row 1}) \cdot P(\text{column 1}) = \frac{1}{3} \cdot \frac{1}{2} = \frac{1}{6}$$

$$P(\text{row 1, column 2}) = P(\text{row 1}) \cdot P(\text{column 2}) = \frac{1}{3} \cdot \frac{1}{2} = \frac{1}{6}$$

$$P(\text{row 2, column 1}) = P(\text{row 2}) \cdot P(\text{column 1}) = \frac{2}{3} \cdot \frac{1}{2} = \frac{1}{3}$$

$$P(\text{row 2, column 2}) = P(\text{row 2}) \cdot P(\text{column 2}) = \frac{2}{3} \cdot \frac{1}{2} = \frac{1}{3}.$$

The table below lists the probability of each possible outcome, along with the payoff to player A.

Outcome	Probability of Outcome	Payoff for A
Row 1, column 1	1/6	−1
Row 1, column 2	1/6	2
Row 2, column 1	1/3	1
Row 2, column 2	1/3	0

The expected value of the game is given by the sum of the products of the probabilities and the payoffs, or

$$\text{Expected Value} = \frac{1}{6}(-1) + \frac{1}{6}(2) + \frac{1}{3}(1) + \frac{1}{3}(0) = \frac{1}{2}.$$

In the long run, for a great many plays of the game, the payoff to A will average 1/2 dollar per play of the game. It is important to note that as the mixed strategies used by A and B are changed, the expected value of the game may well change. (See Example 2 below.) ▬

To generalize the work of Example 1, let the payoff matrix for a 2 × 2 game be

$$M = \begin{bmatrix} a_{11} & a_{12} \\ a_{21} & a_{22} \end{bmatrix}.$$

Let player A choose row 1 with probability p_1 and row 2 with probability p_2, where $p_1 + p_2 = 1$. Write these probabilities as the row matrix

$$A = [p_1 \quad p_2].$$

Let player B choose column 1 with probability q_1 and column 2 with probability q_2, where $q_1 + q_2 = 1$. Write this as the column matrix

$$B = \begin{bmatrix} q_1 \\ q_2 \end{bmatrix}.$$

The probability of choosing row 1 and column 1 is

$$P(\text{Row 1, Column 1}) = P(\text{Row 1}) \cdot P(\text{Column 1}) = p_1 \cdot q_1.$$

In the same way, the probabilities of each possible outcome are shown in the table below, along with the payoff matrix for each outcome.

Outcome	Probability of Outcome	Payoff for A
Row 1, column 1	$p_1 \cdot q_1$	a_{11}
Row 1, column 2	$p_1 \cdot q_2$	a_{12}
Row 2, column 1	$p_2 \cdot q_1$	a_{21}
Row 2, column 2	$p_2 \cdot q_2$	a_{22}

The expected value for this game is

$$(p_1 \cdot q_1) \cdot a_{11} + (p_1 \cdot q_2) \cdot a_{12} + (p_2 \cdot q_1) \cdot a_{21} + (p_2 \cdot q_2) \cdot a_{22}.$$

This same result can be written as the matrix product

$$\text{Expected Value} = [p_1 \quad p_2] \begin{bmatrix} a_{11} & a_{12} \\ a_{21} & a_{22} \end{bmatrix} \begin{bmatrix} q_1 \\ q_2 \end{bmatrix} = AMB.$$

The same method works for games larger than 2×2: let the payoff matrix for a game have dimension $m \times n$; call this matrix $M = [a_{ij}]$. Let the mixed strategy for player A be given by the row matrix

$$A = [p_1 \quad p_2 \quad p_3 \cdots p_m]$$

and the mixed strategy for player B be given by the column matrix

$$B = \begin{bmatrix} q_1 \\ q_2 \\ \cdot \\ \cdot \\ \cdot \\ q_n \end{bmatrix}.$$

The expected value for this game is the product

$$AMB = [p_1 \quad p_2 \cdots p_m] \begin{bmatrix} a_{11} & a_{12} & \cdots & a_{1n} \\ a_{21} & a_{22} & \cdots & a_{2n} \\ \cdot & & & \cdot \\ \cdot & & & \cdot \\ a_{m1} & a_{m2} & \cdots & a_{mn} \end{bmatrix} \begin{bmatrix} q_1 \\ q_2 \\ \cdot \\ \cdot \\ \cdot \\ q_n \end{bmatrix}.$$

■ EXAMPLE 2

In the game in Example 1, having payoff matrix

$$M = \begin{bmatrix} -1 & 2 \\ 1 & 0 \end{bmatrix},$$

suppose player A chooses row 1 with probability .2, and player B chooses column 1 with probability .6. Find the expected value of the game.

If A chooses row 1 with probability .2, then row 2 is chosen with probability $1 - .2 = .8$, giving

$$A = [.2 \quad .8].$$

In the same way,

$$B = \begin{bmatrix} .6 \\ .4 \end{bmatrix}.$$

The expected value of this game is given by the product *AMB,* or

$$AMB = \begin{bmatrix} .2 & .8 \end{bmatrix} \begin{bmatrix} -1 & 2 \\ 1 & 0 \end{bmatrix} \begin{bmatrix} .6 \\ .4 \end{bmatrix}$$

$$= \begin{bmatrix} .6 & .4 \end{bmatrix} \begin{bmatrix} .6 \\ .4 \end{bmatrix}$$

$$= [.52].$$

On the average, these two strategies will produce a payoff of $.52, or 52¢, for A for each play of the game. This payoff is slightly better than the 50¢ in Example 1. ▬

In Example 2, player B could reduce the payoff to A by changing strategy. (Check this by choosing different matrices for B.) For this reason, player A needs to develop an *optimum strategy*—a strategy that will produce the best possible payoff no matter what B does. Just as in the previous section, this is done by finding the largest of the smallest possible amounts that can be won.

To find values of p_1 and p_2 so that the probability vector $[p_1 \quad p_2]$ produces an optimum strategy, start with the payoff matrix

$$M = \begin{bmatrix} -1 & 2 \\ 1 & 0 \end{bmatrix}$$

and assume that A chooses row 1 with probability p_1. If player B chooses column 1, then player A's expectation is given by E_1, where

$$E_1 = -1 \cdot p_1 + 1 \cdot p_2 = -p_1 + p_2.$$

Since $p_1 + p_2 = 1$, we have $p_2 = 1 - p_1$, and

$$E_1 = -p_1 + 1 - p_1 = 1 - 2p_1.$$

If B chooses column 2, then A's expected value is given by E_2, where

$$E_2 = 2 \cdot p_1 + 0 \cdot p_2 = 2p_1.$$

Draw graphs of $E_1 = 1 - 2p_1$ and $E_2 = 2p_1$; see Figure 2.

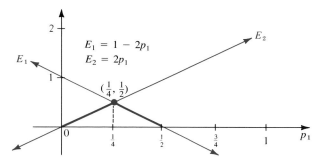

FIGURE 2

As mentioned above, A needs to maximize the smallest amounts that can be won. On the graph, the smallest amounts that can be won are represented by the points of E_2 up to the intersection point. To the right of the intersection point, the smallest amounts that can be won are represented by the points of the line E_1. Player A can maximize the smallest amounts that can be won by choosing the point of intersection itself, the peak of the heavily shaded line in Figure 2.

To find this point of intersection, find the simultaneous solution of the two equations. At the point of intersection, $E_1 = E_2$. Substitute $1 - 2p_1$ for E_1 and $2p_1$ for E_2.

$$E_1 = E_2$$
$$1 - 2p_1 = 2p_1$$
$$1 = 4p_1$$
$$\frac{1}{4} = p_1$$

By this result, player A should choose strategy 1 with probability 1/4, and strategy 2 with probability $1 - 1/4 = 3/4$. This will maximize A's expected winnings. To find the maximum winnings (which is also the value of the game), substitute 1/4 for p_1 in either E_1 or E_2. Choosing E_2 gives

$$E_2 = 2p_1 = 2\left(\frac{1}{4}\right) = \frac{1}{2},$$

that is, 1/2 dollar, or 50¢. Going through a similar argument for player B shows that the optimum strategy for player B is to choose each column with probability 1/2; in this case the value also turns out to be 50¢. In Example 2, A's winnings were 52¢; however, that was because B was not using his optimum strategy.

In the game above, player A can maximize expected winnings by playing row 1 with probability 1/4 and row 2 with probability 3/4. Such a strategy is called a **mixed strategy.** To actually decide which row to use on a given game, player A could use a spinner, such as the one in Figure 3.

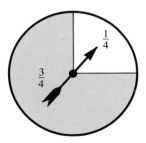

FIGURE 3

▰▰ EXAMPLE 3

Boll weevils threaten the cotton crop near Hattiesburg. Charles Dawkins owns a small farm; he can protect his crop by spraying with a potent (and expensive) insecticide. In trying to decide what to do, Dawkins first sets up a payoff matrix. The numbers in the matrix represent his profits.

		States of Nature	
		Boll Weevil Attack	No Attack
Strategies	Spray	$14,000	$7000
	Don't Spray	−$3000	$8000

Let p_1 represent the probability with which Dawkins chooses strategy 1, so that $1 - p_1$ is the probability with which he chooses strategy 2. If nature chooses strategy 1 (an attack), then Dawkins' expected value is

$$E_1 = 14{,}000p_1 - 3000(1 - p_1)$$
$$= 14{,}000p_1 - 3000 + 3000p_1$$
$$E_1 = 17{,}000p_1 - 3000.$$

For nature's strategy 2 (no attack), Dawkins has an expected value of

$$E_2 = 7000p_1 + 8000(1 - p_1)$$
$$= 7000p_1 + 8000 - 8000p_1$$
$$E_2 = 8000 - 1000p_1.$$

As suggested by the work above, to maximize his expected profit, Dawkins should find the value of p_1 for which $E_1 = E_2$.

$$E_1 = E_2$$
$$17{,}000p_1 - 3000 = 8000 - 1000p_1$$
$$18{,}000p_1 = 11{,}000$$
$$p_1 = 11/18$$

Thus, $p_2 = 1 - p_1 = 1 - 11/18 = 7/18$.

Dawkins will maximize his expected profit if he chooses strategy 1 with probability 11/18 and strategy 2 with probability 7/18. His expected profit from this mixed strategy, [11/18   7/18], can be found by substituting 11/18 for p_1 in either E_1 or E_2. If E_1 is chosen,

$$\text{Expected Profit} = 17{,}000\left(\frac{11}{18}\right) - 3000 = \frac{133{,}000}{18} \approx \$7400. \quad \blacksquare$$

To obtain a formula for the optimum strategy in a game that is not strictly determined, start with the matrix

$$M = \begin{bmatrix} a_{11} & a_{12} \\ a_{21} & a_{22} \end{bmatrix},$$

the payoff matrix of the game. Assume that A chooses row 1 with probability p_1. The expected value for A, assuming that B plays column 1, is E_1, where

$$E_1 = a_{11} \cdot p_1 + a_{21} \cdot (1 - p_1).$$

The expected value for A if B chooses column 2 is E_2, where

$$E_2 = a_{12} \cdot p_1 + a_{22} \cdot (1 - p_1).$$

As above, the optimum strategy for player A is found by letting $E_1 = E_2$.

$$a_{11} \cdot p_1 + a_{21} \cdot (1 - p_1) = a_{12} \cdot p_1 + a_{22} \cdot (1 - p_1)$$

Solve this equation for p_1.

$$a_{11} \cdot p_1 + a_{21} - a_{21} \cdot p_1 = a_{12} \cdot p_1 + a_{22} - a_{22} \cdot p_1$$

$$a_{11} \cdot p_1 - a_{21} \cdot p_1 - a_{12} \cdot p_1 + a_{22} \cdot p_1 = a_{22} - a_{21}$$

$$p_1(a_{11} - a_{21} - a_{12} + a_{22}) = a_{22} - a_{21}$$

$$p_1 = \frac{a_{22} - a_{21}}{a_{11} - a_{21} - a_{12} + a_{22}}$$

Since $p_2 = 1 - p_1$,

$$p_2 = 1 - \frac{a_{22} - a_{21}}{a_{11} - a_{21} - a_{12} + a_{22}}$$

$$= \frac{a_{11} - a_{21} - a_{12} + a_{22} - (a_{22} - a_{21})}{a_{11} - a_{21} - a_{12} + a_{22}}$$

$$= \frac{a_{11} - a_{12}}{a_{11} - a_{21} - a_{12} + a_{22}}.$$

This result is valid only if $a_{11} - a_{21} - a_{12} + a_{22} \neq 0$; this condition is satisfied if the game is not strictly determined.

There is a similar result for player B, which is included in the following summary.

OPTIMUM STRATEGIES IN A NON-STRICTLY-DETERMINED GAME

Let a non-strictly-determined game have payoff matrix

$$\begin{bmatrix} a_{11} & a_{12} \\ a_{21} & a_{22} \end{bmatrix}.$$

The optimum strategy for player A is $[p_1 \quad p_2]$, where

$$p_1 = \frac{a_{22} - a_{21}}{a_{11} - a_{21} - a_{12} + a_{22}} \quad \text{and} \quad p_2 = \frac{a_{11} - a_{12}}{a_{11} - a_{21} - a_{12} + a_{22}}.$$

The optimum strategy for player B is $\begin{bmatrix} q_1 \\ q_2 \end{bmatrix}$, where

$$q_1 = \frac{a_{22} - a_{12}}{a_{11} - a_{21} - a_{12} + a_{22}} \quad \text{and} \quad q_2 = \frac{a_{11} - a_{21}}{a_{11} - a_{21} - a_{12} + a_{22}}.$$

The value of the game is

$$\frac{a_{11}a_{22} - a_{12}a_{21}}{a_{11} - a_{21} - a_{12} + a_{22}}.$$

It is important to remember that the use of mixed strategies is meaningful only if a game is repeated many, many times. For one play of the game the mixed strategy has no meaning.

■ EXAMPLE 4

Suppose a game has payoff matrix

$$\begin{bmatrix} 5 & -2 \\ -3 & -1 \end{bmatrix}.$$

Here $a_{11} = 5$, $a_{12} = -2$, $a_{21} = -3$, and $a_{22} = -1$. To find the optimum strategy for player A, first find p_1.

$$p_1 = \frac{-1 - (-3)}{5 - (-3) - (-2) + (-1)} = \frac{2}{9}$$

Player A should play row 1 with probability 2/9 and row 2 with probability $1 - 2/9 = 7/9$.

For player B,

$$q_1 = \frac{-1 - (-2)}{5 - (-3) - (-2) + (-1)} = \frac{1}{9}.$$

Player B should choose column 1 with probability 1/9, and column 2 with probability 8/9. The value of the game is

$$\frac{5(-1) - (-2)(-3)}{5 - (-3) - (-2) + (-1)} = \frac{-11}{9}.$$

On the average, B will receive 11/9 dollars from A per play of the game. ■

9.3 EXERCISES

1. Suppose a game has payoff matrix

$$\begin{bmatrix} 3 & -4 \\ -5 & 2 \end{bmatrix}.$$

Suppose that player B uses the strategy $\begin{bmatrix} .3 \\ .7 \end{bmatrix}$. Find the expected value of the game if player A uses each of the following strategies.

(a) [.5 .5]   (b) [.1 .9]   (c) [.8 .2]   (d) [.2 .8]

2. Suppose a game has payoff matrix

$$\begin{bmatrix} 0 & -4 & 1 \\ 3 & 2 & -4 \\ 1 & -1 & 0 \end{bmatrix}.$$

Find the expected value of the game for the following strategies for players A and B.

(a) $A = [.1 \quad .4 \quad .5]$; $B = \begin{bmatrix} .2 \\ .4 \\ .4 \end{bmatrix}$ **(b)** $A = [.3 \quad .4 \quad .3]$; $B = \begin{bmatrix} .8 \\ .1 \\ .1 \end{bmatrix}$

Find the optimum strategies for player A and player B in the games in Exercises 3–14. Find the value of each game. (Be sure to look for a saddle point first.)

3. $\begin{bmatrix} 5 & 1 \\ 3 & 4 \end{bmatrix}$ **4.** $\begin{bmatrix} -4 & 5 \\ 3 & -4 \end{bmatrix}$ **5.** $\begin{bmatrix} -2 & 0 \\ 3 & -4 \end{bmatrix}$ **6.** $\begin{bmatrix} 6 & 2 \\ -1 & 10 \end{bmatrix}$

7. $\begin{bmatrix} 4 & -3 \\ -1 & 7 \end{bmatrix}$ **8.** $\begin{bmatrix} 0 & 6 \\ 4 & 0 \end{bmatrix}$ **9.** $\begin{bmatrix} -2 & \frac{1}{2} \\ 0 & -3 \end{bmatrix}$ **10.** $\begin{bmatrix} 6 & \frac{3}{4} \\ \frac{2}{3} & -1 \end{bmatrix}$

11. $\begin{bmatrix} \frac{8}{3} & -\frac{1}{2} \\ \frac{3}{4} & -\frac{5}{12} \end{bmatrix}$ **12.** $\begin{bmatrix} -\frac{1}{2} & \frac{2}{3} \\ \frac{7}{8} & -\frac{3}{4} \end{bmatrix}$ **13.** $\begin{bmatrix} -1 & 2 \\ 3 & 1 \end{bmatrix}$ **14.** $\begin{bmatrix} 8 & 18 \\ -4 & 2 \end{bmatrix}$

Remove any dominated strategies and then find the optimum strategy for each player and the value of the game.

15. $\begin{bmatrix} -4 & 9 \\ 3 & -5 \\ 8 & 7 \end{bmatrix}$ **16.** $\begin{bmatrix} 3 & 4 & -1 \\ -2 & 1 & 0 \end{bmatrix}$ **17.** $\begin{bmatrix} 8 & 6 & 3 \\ -1 & -2 & 4 \end{bmatrix}$

18. $\begin{bmatrix} -1 & 6 \\ 8 & 3 \\ -2 & 5 \end{bmatrix}$ **19.** $\begin{bmatrix} 9 & -1 & 6 \\ 13 & 11 & 8 \\ 6 & 0 & 9 \end{bmatrix}$ **20.** $\begin{bmatrix} 4 & 8 & -3 \\ 2 & -1 & 1 \\ 7 & 9 & 0 \end{bmatrix}$

21. In the game of matching coins, each of two players flips a coin. If both coins match (both show heads or both show tails), player A wins $1. If there is no match, player B wins $1, as in the payoff matrix below. Find the optimum strategies for the two players and the value of the game.

$$\begin{bmatrix} 1 & -1 \\ -1 & 1 \end{bmatrix}$$

22. Players A and B play a game in which they show either one or two fingers at the same time. If there is a match, A wins the amount of dollars equal to the total number of fingers shown. If there is no match, B wins the amount of dollars equal to the number of fingers shown.

(a) Write the payoff matrix.

(b) Find optimum strategies for A and B and the value of the game.

23. Repeat Exercise 22 if each player may show either 0 or 2 fingers with the same payoffs.

▬ APPLICATIONS

LIFE SCIENCES

Choosing Medication   **24.** The number of cases of African flu has reached epidemic levels. The disease is known to have two strains with similar symptoms. Doctor De Luca has two medicines available: the first is 60% effective against the first strain and 40% effective against the second. The second medicine is completely effective against the second strain but ineffective against the first. Use the matrix below to decide which medicine she should use and the results she can expect.

$$\text{Medicine} \begin{array}{c} \\ 1 \\ 2 \end{array} \begin{array}{c} \text{Strain} \\ \begin{array}{cc} 1 & 2 \end{array} \\ \begin{bmatrix} .6 & .4 \\ 0 & 1 \end{bmatrix} \end{array}$$

BUSINESS AND ECONOMICS

Advertising   **25.** Suppose Allied Manufacturing Company decides to put its new product on the market with a big television and radio advertising campaign. At the same time, the company finds out that its major competitor, Bates Manufacturing, also has decided to launch a big advertising campaign for a similar product. The payoff matrix below shows the increased sales (in millions) for Allied, as well as the decreased sales for Bates.

$$\text{Allied} \begin{array}{c} \\ \text{T.V.} \\ \text{Radio} \end{array} \begin{array}{c} \text{Bates} \\ \begin{array}{cc} \text{T.V.} & \text{Radio} \end{array} \\ \begin{bmatrix} 1.0 & -.7 \\ -.5 & .5 \end{bmatrix} \end{array}$$

Find the optimum strategy for Allied Manufacturing and the value of the game.

Pricing   **26.** The payoffs in the table below represent the differences between Boeing Aircraft Company's profit and its competitor's profit for two prices (in millions) on commercial jet transports, with positive payoffs being in Boeing's favor. What should Boeing's price strategy be?*

$$\text{Boeing's Strategy} \begin{array}{c} \\ 4.9 \\ 4.75 \end{array} \begin{array}{c} \text{Competitor's} \\ \text{Price Strategy} \\ \begin{array}{cc} 4.75 & 4.9 \end{array} \\ \begin{bmatrix} -4 & 2 \\ 2 & 0 \end{bmatrix} \end{array}$$

Purchasing   **27.** *The Huckster*† Merrill has a concession at Yankee Stadium for the sale of sunglasses and umbrellas. The business places quite a strain on him, the weather being what it is. He has observed that he can sell about 500 umbrellas when it rains, and about 100 when it is sunny; in the latter case he can also sell 1000 sunglasses. Umbrellas cost him 50 cents and sell for $1; sunglasses cost 20 cents and sell for 50 cents. He is willing to invest $250 in the project. Everything that is not sold is considered a total loss.

He assembles the facts regarding profit in a table.

$$\text{Buying for:} \begin{array}{c} \\ \text{Rain} \\ \text{Shine} \end{array} \begin{array}{c} \text{Selling during:} \\ \begin{array}{cc} \text{Rain} & \text{Shine} \end{array} \\ \begin{bmatrix} 250 & -150 \\ -150 & 350 \end{bmatrix} \end{array}$$

---

*From ''Pricing, Investment, and Games of Strategy,'' by Georges Brigham in *Management Sciences Models and Techniques,* Vol. 1. Copyright © 1960 Pergamon Press, Ltd. Reprinted with permission.

†From *The Compleat Strategyst* by J. D. Williams, Published 1966, by McGraw-Hill Book Company. Reprinted by permission of The Rand Corporation. This is an excellent nontechnical book on game theory.

He immediately takes heart, for this is a mixed-strategy game, and he should be able to find a stabilizing strategy that will save him from the vagaries of the weather. Find the best mixed strategy for Merrill.

SOCIAL SCIENCES

Law Enforcement   **28.** *The Squad Car†* This is a somewhat more harrowing example. A police dispatcher was conveying information and opinion, as fast as she could speak, to Patrol Car 2, cruising on the U.S. Highway: ". . . in a Cadillac; just left Hitch's Tavern on the old Country Road. Direction of flight unknown. Suspect Plesset is seriously wounded but may have an even chance if he finds a good doctor, like Doctor Haydon, soon—even Veterinary Paxson might save him, but his chances would be halved. Plesset shot Officer Flood, who has a large family."

Deputy Henderson finally untangled the microphone from the riot gun and his size 14 shoes. He replied: "Roger. We can cut him off if he heads for Haydon's and we have a fifty-fifty chance of cutting him off at the State Highway if he heads for the vet's. We must cut him off because we can't chase him—Deputy Root got this thing stuck in reverse a while ago, and our cruising has been a disgrace to the department ever since."

The headquarter's carrier-wave again hummed in the speaker, but the dispatcher's musical voice was now replaced by the grating tones of Sheriff Lipp. "If you know anything else, don't tell it. He has a hi-fi radio in that Cad. Get him."

Root suddenly was seized by an idea and stopped struggling with the gearshift. "Henderson, we may not need a gun tonight, but we need a pencil: this is just a two-by-two game. The dispatcher gave us all the dope we need." "You gonna use *her* estimates?" "You got better ones? She's got intuition; besides, that's information from headquarters. Now let's see . . . . Suppose we head for Haydon's. And suppose Plesset does too; then we rack up one good bandit, if you don't trip on that gun again. But if he heads for Paxson, the chances are three out of four that old doc will kill him."

"I don't get it." "Well, it didn't come easy. Remember, Haydon would have an even chance—one-half—of saving him. He'd have half as good a chance with Paxson; and half of one-half is one-quarter. So the chance he dies must be three-quarters—subtracting from one, you know."

"Yeah, it's obvious." "Huh. Now if we head for Paxson's it's tougher to figure. First of all, *he* may go to Haydon's, in which case we have to rely on the doc to kill him, of which the chance is only one-half."

"You ought to subtract that from one." "I did. Now suppose he too heads for Paxson's. Either of two things can happen. One is, we catch him, and the chance is one-half. The other is, we don't catch him—and again the chance is one-half—but there is a three-fourths chance that the doc will have a lethal touch. So the overall probability that he will get by us, but not by the doc, is one-half times three-fourths, or three-eighths. Add to that the one-half chance that he doesn't get by us, and we have seven-eighths."

---

†From *The Compleat Strategyst* by J. D. Williams, Published 1966, by McGraw-Hill Book Company. Reprinted by permission of The Rand Corporation. This is an excellent nontechnical book on game theory.

"I don't like this stuff. He's probably getting away while we're doodling."
"Relax. He has to figure it out too, doesn't he? And he's in worse shape than we are. Now let's see what we have."

Cad Goes to:

Haydon   Paxson

Patrol Car Goes to: Haydon $\begin{bmatrix} 1 & \frac{3}{4} \\ \frac{1}{2} & \frac{7}{8} \end{bmatrix}$
Paxson

"Fractions aren't so good in this light," Root continues. "Let's multiply everything by eight to clean it up. I hear it doesn't hurt anything."

Cad:

Haydon   Paxson

Patrol Car: Haydon $\begin{bmatrix} 8 & 6 \\ 4 & 7 \end{bmatrix}$
Paxson

"It is now clear that this is a very messy business . . ." "I know." "There is no single strategy which we can safely adopt. I shall therefore compute the best mixed strategy."

What mixed strategy should deputies Root and Henderson pursue?

KEY WORDS

9.1 states of nature
strategies
payoff matrix
utility

dominant strategy
optimum strategy
value of the game
pure strategy

9.2 game theory
two-person game
zero-sum game

saddle point
strictly determined game

9.3 mixed strategy

CHAPTER 9   REVIEW EXERCISES

APPLICATIONS

BUSINESS AND ECONOMICS

Labor Relations *In labor-management relations, both labor and management can adopt either a friendly or a hostile attitude. The results are shown in the following payoff matrix. The numbers give the wage gains made by an average worker.*

Management

Friendly   Hostile

Labor Friendly $\begin{bmatrix} \$600 & \$800 \\ \$400 & \$950 \end{bmatrix}$
Hostile

1. Suppose the chief negotiator for labor is an optimist. What strategy should he choose?

2. What strategy should he choose if he is a pessimist?

3. The chief negotiator for labor feels that there is a 70% chance that the company will be hostile. What strategy should he adopt? What is the expected payoff?

4. Just before negotiations begin, a new management is installed in the company. There is only a 40% chance that the new management will be hostile. What strategy should be adopted by labor?

SOCIAL SCIENCES

Politics *A candidate for city council can come out in favor of a new factory, be opposed to it, or waffle on the issue. The change in votes for the candidate depends on what her opponent does, with payoffs as shown.*

$$\text{Candidate} \quad \begin{array}{c} \\ \text{Favors} \\ \text{Waffles} \\ \text{Opposes} \end{array} \overset{\overset{\textit{Opponent}}{\begin{array}{ccc} \text{Favors} & \text{Waffles} & \text{Opposes} \end{array}}}{\begin{bmatrix} 0 & -1000 & -4000 \\ 1000 & 0 & -500 \\ 5000 & 2000 & 0 \end{bmatrix}}$$

5. What should the candidate do if she is an optimist?

6. What should she do if she is a pessimist?

7. Suppose the candidate's campaign manager feels there is a 40% chance that the opponent will favor the plant, and a 35% chance that he will waffle. What strategy should the candidate adopt? What is the expected change in the number of votes?

8. The opponent conducts a new poll that shows strong opposition to the new factory. This changes the probability he will favor the factory to 0 and the probability he will waffle to .7. What strategy should our candidate adopt? What is the expected change in the number of votes now?

Use the following payoff matrix to determine the payoff if each of the given strategies is used.

$$\begin{bmatrix} -2 & 5 & -6 & 3 \\ 0 & -1 & 7 & 5 \\ 2 & 6 & -4 & 4 \end{bmatrix}$$

9. (1, 1) **10.** (1, 4) **11.** (2, 3) **12.** (3, 4)

13. Are there any dominated strategies in this game? **14.** Is there a saddle point?

Remove any dominated strategies in the following games.

15. $\begin{bmatrix} -11 & 6 & 8 & 9 \\ -10 & -12 & 3 & 2 \end{bmatrix}$ **16.** $\begin{bmatrix} -1 & 9 & 0 \\ 4 & -10 & 6 \\ 8 & -6 & 7 \end{bmatrix}$ **17.** $\begin{bmatrix} -2 & 4 & 1 \\ 3 & 2 & 7 \\ -8 & 1 & 6 \\ 0 & 3 & 9 \end{bmatrix}$ **18.** $\begin{bmatrix} 3 & -1 & 4 \\ 0 & 4 & -1 \\ 1 & 2 & -3 \\ 0 & 0 & 2 \end{bmatrix}$

Find any saddle points for the following games. Give the value of the game. Identify any fair games.

19. $\begin{bmatrix} -2 & 3 \\ -4 & 5 \end{bmatrix}$ **20.** $\begin{bmatrix} -4 & 0 & 2 & -5 \\ 6 & 9 & 3 & 8 \end{bmatrix}$ **21.** $\begin{bmatrix} -4 & -1 \\ 6 & 0 \\ 8 & -3 \end{bmatrix}$

22. $\begin{bmatrix} 4 & -1 & 6 \\ -3 & -2 & 0 \\ -1 & -4 & 3 \end{bmatrix}$

23. $\begin{bmatrix} 8 & 1 & -7 & 2 \\ -1 & 4 & -3 & 3 \end{bmatrix}$

24. $\begin{bmatrix} 2 & -9 \\ 7 & 1 \\ 4 & 2 \end{bmatrix}$

Find the optimum strategies for each of the following games. Find the value of the game.

25. $\begin{bmatrix} 1 & 0 \\ -2 & 3 \end{bmatrix}$

26. $\begin{bmatrix} 2 & -3 \\ -3 & 5 \end{bmatrix}$

27. $\begin{bmatrix} -3 & 5 \\ 1 & 0 \end{bmatrix}$

28. $\begin{bmatrix} 8 & -3 \\ -6 & 2 \end{bmatrix}$

For each of the following games, remove any dominated strategies, then solve the game. Find the value of the game.

29. $\begin{bmatrix} -4 & 8 & 0 \\ -2 & 9 & -3 \end{bmatrix}$

30. $\begin{bmatrix} 1 & 0 & 3 & -3 \\ 4 & -2 & 4 & -1 \end{bmatrix}$

31. $\begin{bmatrix} 2 & -1 \\ -4 & 5 \\ -1 & -2 \end{bmatrix}$

32. $\begin{bmatrix} 8 & -6 \\ 4 & -8 \\ -9 & 9 \end{bmatrix}$

EXTENDED APPLICATION

DECISION MAKING IN LIFE INSURANCE

When a life insurance company receives an application from an agent requesting insurance on the life of an individual, it knows from experience that the applicant will be in one of three possible states of risk, with proportions as shown.*

States of Risk	Proportions
s_1 = standard risk	.90
s_2 = substandard risk (greater risk)	.07
s_3 = sub-substandard risk (greatest risk)	.03

A particular applicant could be correctly placed if all possible information about the applicant were known. This is not realistic in a practical situation; the company's problem is to obtain the maximum information at the lowest possible cost.

The company can take any of three possible strategies when it receives the application.

Strategies
a_1 = offer a standard policy
a_2 = offer a substandard policy (higher rates)
a_3 = offer a sub-substandard policy (highest rates)

---

*This example was supplied by Donald J. vanKeuren, actuary of Metropolitan Life Insurance Company, and Dave Halmstad, senior actuarial assistant. It is based on a paper by Donald Jones.

The payoff matrix in Table 1 below shows the payoffs associated with the possible strategies of the company and the states of the applicant. Here M represents the face value of the policy in thousands of dollars (for a $30,000 policy we have $M = 30$). For example, if the applicant is substandard (s_2) and the company offers him or her a standard policy (a_1), the company makes a profit of $13M$ (13 times the face value of the policy in thousands). Strategy a_2 would result in a larger profit of $20M$. However, if the prospective customer is a standard risk (s_1) but the company offers a substandard policy (a_2), the company loses $50 (the cost of preparing a policy) since the customer would reject the policy because it has higher rates than could be obtained elsewhere.

Table 1

		States of Nature		
		s_1	s_2	s_3
	a_1	20M	13M	3M
Strategies of Company	a_2	−50	20M	10M
	a_3	−50	−50	20M

Before deciding on the policy to be offered, the company can perform any of three experiments to help it decide.

$$e_0 = \text{no inspection report (no cost)}$$
$$e_1 = \text{regular inspection report (cost: \$5)}$$
$$e_2 = \text{special life report (cost: \$20)}$$

On the basis of this report, the company can classify the applicant as follows.

$$T_1 = \text{Applicant seems to be a standard risk.}$$
$$T_2 = \text{Applicant seems to be a substandard risk.}$$
$$T_3 = \text{Applicant seems to be a sub-substandard risk.}$$

Let $P(s|T)$ represent the probability that an applicant is in state s when the report indicates that he or she is in state T. For example, $P(s_1|T_2)$ represents the probability that an applicant is a standard risk (s_1) when the report indicates that he or she is a substandard risk (T_2). These probabilities, shown in Table 2, are based on Bayes' formula.

Table 2

True State	Regular Report			Special Report		
	$P(s_i|T_1)$	$P(s_i|T_2)$	$P(s_i|T_3)$	$P(s_i|T_1)$	$P(s_i|T_2)$	$P(s_i|T_3)$
s_1	.9695	.8411	.7377	.9984	.2081	.2299
s_2	.0251	.1309	.1148	.0012	.7850	.0268
s_3	.0054	.0280	.1475	.0004	.0069	.7433

Table 2 shows that $P(s_2|T_2)$, the probability that an applicant actually is substandard (s_2) if the regular report indicates substandard (T_2), is only .1309, while $P(s_2|T_2)$, using the special report, is .7850.

We now have probabilities and payoffs that can be used to find expected values for each possible strategy the company might adopt. There are many possibilities here: the company can use one of three experiments, the experiments can indicate one of three states, the company can offer one of three policies, and the applicant can be in one of three states. Figure 4 shows some of these possibilities in a *decision tree*.

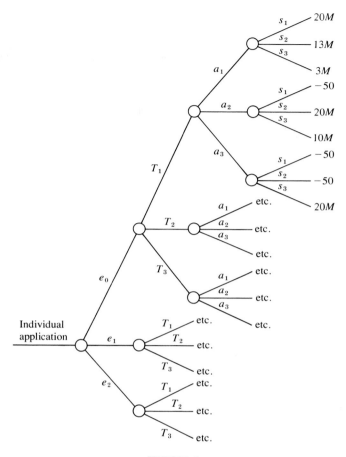

FIGURE 4

In order to find an optimum strategy for the company, consider an example. Suppose the company decides to perform experiment e_2 (special life report) with the report indicating a substandard risk, T_2. Then the expected values E_1, E_2, E_3 for the three possible actions a_1, a_2, a_3, respectively, are as shown below. (Recall that M is a variable representing the face amount of the policy in thousands.)

For action a_1 (offer standard policy):

$$E_1 = [P(s_1|T_2)](20M) + [P(s_2|T_2)](13M) + [P(s_3|T_2)](3M)$$
$$= (.2081)(20M) + (.7850)(13M) + (.0069)(3M)$$
$$= 4.162M + 10.205M + .0207M$$
$$\approx 14.388M.$$

For action a_2 (offer a substandard policy):

$$E_2 = [P(s_1|T_2)](-50) + [P(s_2|T_2)](20M) + [P(s_3|T_2)](10M)$$
$$= (.2081)(-50) + (.7850)(20M) + (.0069)(10M)$$
$$= 15.769M - 10.405.$$

For action a_3 (offer a sub-substandard policy):

$$E_2 = [P(s_1|T_2)](-50) + [P(s_2|T_2)](-50) + [P(s_3|T_2)](20M)$$
$$= .138M - 49.655.$$

Strategy a_2 is better than a_3 (for any positive M, $15.769M - 10.405 > .138M - 49.655$). The only choice is between strategies a_1 and a_2. Strategy a_2 is superior if it leads to a higher expected value than a_1. This happens for all values of M such that

$$15.769M - 10.405 > 14.388M$$
$$1.381M > 10.405$$
$$M > 7.535.$$

If the applicant applies for more than \$7535 of insurance, the company should use strategy a_2; otherwise it should use a_1.

Similar analyses can be performed for all possible strategies from the decision tree above to find the best strategy. It turns out that the company will maximize its expected profits if it offers a standard policy to all people applying for less than \$50,000 in life insurance, with a special report form required for all others.

EXERCISES

1. Find the expected values for each strategy a_1, a_2, and a_3 if the insurance company performs experiment e_1 (a regular report) with the report indicating that the applicant is a substandard risk.

2. Find the expected values for each action if the company performs e_1 with the report indicating that the applicant is a standard risk.

3. Find the expected values for each strategy if e_0 (no report) is selected. (*Hint:* Use the proportions given for the three states s_1, s_2, and s_3 as the probabilities.)

EXTENDED APPLICATION

DECISION MAKING IN THE MILITARY

The following article* by O. G. Haywood, Jr., entitled "Military Decision and Game Theory," is reprinted with only minor changes from the November 1954 issue of the journal of the Operations Research Society of America. The case is presented unedited to give you an idea of the type of article published in such a journal.

A military commander may approach decision with either of two philosophies. He may select his course of action on the basis of his estimate of what his enemy *is able to do* to oppose him. Or, he may make his selection on the basis of his estimate of what his enemy *is going to do*. The former is a doctrine of decision based on enemy capabilities; the latter, on enemy intentions.

The doctrine of decision of the armed forces of the United States is a doctrine based on enemy capabilities. A commander is enjoined to select the course of action which

*Reprinted by permission from *Operations Research,* Volume 3, Issue 6, 1954 pp. 365–69. Copyright 1954 Operations Research Society of America. No further reproduction permitted without the consent of the copyright owner.

offers the greatest promise of success in view of the enemy capabilities. The process of decision, as approved by the Joint Chiefs of Staff and taught in all service schools, is formalized in a five-step analysis called the *Estimate of the Situation*. These steps are illustrated in the following analysis of an actual World War II battle situation.

General Kenney was Commander of the Allied Air Forces in the Southwest Pacific Area. The struggle for New Guinea reached a critical stage in February 1943. Intelligence reports indicated a Japanese troop and supply convoy was assembling at Rabaul (see Figure 5). Lae was expected to be the unloading point. With this general background Kenney proceeded to make his five-step Estimate of the Situation.

STEP 1. THE MISSION

General MacArthur as Supreme Commander had ordered Kenney to intercept and inflict maximum destruction on the convoy. This then was Kenney's mission.

STEP 2. SITUATION AND COURSES OF ACTION

The situation as outlined above was generally known. One new critical factor was pointed out by Kenney's staff. Rain and poor visibility were predicted for the area north of New Britain. Visibility south of the island would be good.

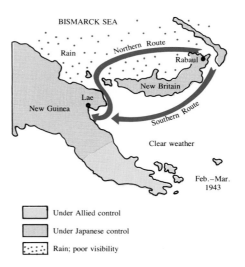

FIGURE 5   *The Rabaul-Lae Convoy Situation.* The problem is the distribution of reconnaissance to locate a convoy which may sail by either one of two routes.

The Japanese commander had two choices for routing his convoy from Rabaul to Lae. He could sail north of New Britain, or he could go south of that island. Either route required three days.

Kenney considered two courses of action, as he discusses in his memoirs. He could concentrate most of his reconnaissance aircraft either along the northern route where visibility would be poor, or along the southern route where clear weather was predicted. Mobility being one of the great advantages of air power, his bombing force could strike the convoy on either route once it was spotted.

STEP 3. ANALYSIS OF THE OPPOSING COURSES OF ACTION

With each commander having two alternative courses of action, four possible conflicts could ensue. These conflicts are pictured in Figure 6.

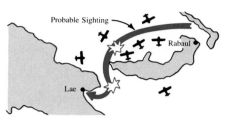

Kenney Strategy: Concentrate reconnaissance on northern route.
Japanese Strategy: Sail northern route.
Estimated Outcome: Although reconnaissance would be hampered by poor visibility, the convoy should be discovered by the second day, which would permit two days of bombing.
TWO DAYS OF BOMBING

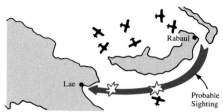

Kenney Strategy: Concentrate reconnaissance on northern route.
Japanese Strategy: Sail southern route.
Estimated Outcome: The convoy would be sailing in clear weather. However, with limited reconnaissance aircraft in this area, the convoy might be missed on the first day. Convoy should be sighted by second day, to permit two days of bombing.
TWO DAYS OF BOMBING

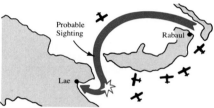

Kenney Strategy: Concentrate reconnaissance on southern route.
Japanese Strategy: Sail northern route.
Estimated Outcome: With poor visibility and limited reconnaissance, Kenney could not expect the convoy to be discovered until it broke out into clear weather on third day. This would permit only one day of bombing.
ONE DAY OF BOMBING

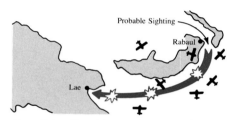

Kenney Strategy: Concentrate reconnaissance on southern route.
Japanese Strategy: Sail southern route.
Estimated Outcome: With good visibility and concentrated reconnaissance in the area, the convoy should be sighted almost as soon as it sailed from Rabaul. This would allow three days of bombing.
THREE DAYS OF BOMBING

FIGURE 6   *Possible Battles for the Rabaul-Lae Convoy Situation.* Four different engagements of forces may result from the interaction of Kenney's two strategies with the two Japanese strategies. Neither commander alone can determine which particular battle will result.

STEP 4. COMPARISON OF AVAILABLE COURSES OF ACTION

If Kenney concentrated on the northern route, he ensured one of the two battles of the top row of sketches. However, he alone could not determine which one of these two battles in the top row would result from his decision. Similarly, if Kenney concentrated on the southern route, he ensured one of the battles of the lower row. In the same manner, the Japanese commander could not select a particular battle, but could by his decision assure that the battle would be one of those pictured in the left column or one of those in the right column.

Kenney sought a battle which would provide the maximum opportunity for bombing the convoy. The Japanese commander desired the minimum exposure to bombing. But neither commander could determine the battle which would result from his own decision. Each commander had full and independent freedom to select either one of his alternative strategies. He had to do so with full realization of his opponent's freedom of choice. The particular battle which resulted would be determined by the two independent decisions.

The U.S. doctrine of decision—the doctrine that a commander base his action on his estimate of what the enemy is capable of doing to oppose him—dictated that Kenney select the course of action which offered the greatest promise of success in view of all of the enemy capabilities. If Kenney concentrated his reconnaissance on the northern route, he could expect two days of bombing regardless of his enemy's decision. If Kenney selected his other strategy, he must accept the possibility of a less favorable outcome.

STEP 5. THE DECISION
Kenney concentrated his reconnaissance aircraft on the northern route.

DISCUSSION   Let us assume that the Japanese commander used a similar philosophy of decision, basing his decision on his enemy's capabilities. Considering the four battles as sketched, the Japanese commander could select either the left or the right column, but could not select the row. If he sailed the northern route, he exposed the convoy to a maximum of two days of bombing. If he sailed the southern route, the convoy might be subjected to three days of bombing. Since he sought minimum exposure to bombing, he should select the northern route.

These two independent choices were the actual decisions which led to the conflict known in history as the Battle of the Bismarck Sea. Kenney concentrated his reconnaissance on the northern route; the Japanese convoy sailed the northern route; the convoy was sighted approximately one day after it sailed; and Allied bombing started shortly thereafter. Although the Battle of the Bismarck Sea ended in a disastrous defeat for the Japanese, we cannot say the Japanese commander erred in his decision. A similar convoy had reached Lae with minor losses two months earlier. The need was critical, and the Japanese were prepared to pay a high price. They did not know that Kenney had modified a number of his aircraft for low-level bombing and had perfected a deadly technique. The U.S. victory was the result of careful planning, thorough training, resolute execution, and tactical surprise of a new weapon—not of error in the Japanese decision.

EXERCISES

1. Use the results of Figure 6 to make a 2 × 2 game. Let the payoffs represent the number of days of bombing.
2. Find any saddle point and optimum strategy for the game.
3. Read the rest of the article used as the source for this example and prepare a discussion of the game theory aspects of the Avranches-Gap situation.

10 Mathematics of Finance

10.1 Simple Interest and Discount

10.2 Compound Interest

10.3 Sequences

10.4 Annuities

10.5 Present Value of an Annuity; Amortization

Review Exercises

Extended Application
A New Look at Athletes' Contracts

Extended Application
Present Value

Not too many years ago, some large corporations could borrow money at 3%, and home buyers could obtain loans at 4½%. Today, however, even the largest corporations must pay at least 10% for their money, and credit customers pay 18% or more to banks and retail outlets. Thus, it is important that both corporate managers and consumers understand **interest,** the cost of borrowing money. The formulas for interest are developed in this chapter.

10.1   SIMPLE INTEREST AND DISCOUNT

Interest on loans of a year or less is usually calculated as *simple interest,* a type of interest that is charged (or paid) only on the amount borrowed (or invested) and not on past interest. The amount borrowed is the **principal** P. The **rate** of interest r is given as a percent per year, and t is the **time** in years. Simple interest I is the product of the principal, rate, and time. (To use the formula for simple interest, write the rate r in decimal form.)

SIMPLE INTEREST

> The **simple interest** I on P dollars at a rate of interest r for a time t in years is given by
> $$I = Prt.$$

A deposit of P dollars today at a rate of interest r for a time t in years produces interest of $I = Prt$. This interest, added to the original principal P, gives

$$P + Prt = P(1 + rt).$$

This result, called the *future value* of P dollars at an interest rate r for time t in years, is summarized as follows. (When loans are involved, the future value is often called the *maturity value* of the loan.)

FUTURE OR MATURITY VALUE

> The **future** (or **maturity**) **value** A of P dollars at a rate of interest r for a time t in years is
> $$A = P(1 + rt).$$

EXAMPLE 1
Find the maturity value for each of the following loans at simple interest.

(a) A loan of $2500 to be repaid in 8 months with interest of 12.1%

The loan is for 8 months, or $8/12 = 2/3$ of a year. The maturity value is

$$A = P(1 + rt)$$
$$A = 2500\left[1 + .121\left(\frac{2}{3}\right)\right]$$
$$\approx 2500[1 + .08067] \approx 2701.67,$$

or $2701.67. (The answer is rounded to the nearest cent, as is customary in financial problems.) Of this maturity value,

$$\$2701.67 - \$2500 = \$201.67$$

represents interest.

(b) A loan of $11,280 for 85 days at 11% interest

It is common to assume 360 days in a year when working with simple interest. We shall usually make such an assumption in this book. The maturity value in this example is

$$A = 11,280\left[1 + .11\left(\frac{85}{360}\right)\right] \approx 11,280[1.0259722] \approx 11,572.97,$$

or $11,572.97. ▬

In part (b) of Example 1 we assumed 360 days in a year. Interest found using a 360-day year is called *ordinary interest,* and interest found using a 365-day year is *exact interest.*

Present Value   A sum of money that can be deposited today to yield some larger amount in the future is called the *present value* of that future amount. Let P be the present value of some amount A at some time t (in years) in the future. Assume a simple interest rate r. As above, the future value of this sum is

$$A = P(1 + rt).$$

Since P is the present value, divide both sides of this last result by $1 + rt$ to get

$$P = \frac{A}{1 + rt}.$$

A summary of present value follows.

PRESENT VALUE

> ▬
> The **present value** P of a future amount of A dollars at a simple interest rate r for t years is
>
> $$P = \frac{A}{1 + rt}.$$

▬ EXAMPLE 2

Find the present value of the following future amounts at simple interest.

(a) $10,000 in 1 year, if interest is 8%

Here $A = 10,000$, $t = 1$, and $r = .08$. Use the formula for present value.

$$P = \frac{10,000}{1 + (.08)(1)} = \frac{10,000}{1.08} \approx 9259.26$$

If \$9259.26 were deposited today at 8% interest, a total of \$10,000 would be in the account in 1 year. These two sums, \$9259.26 today, and \$10,000 in a year, are equivalent (at 8%) because the first amount becomes the other in a year.

(b) \$32,000 in 4 months at 9% interest

$$P = \frac{32,000}{1 + (.09)\left(\frac{4}{12}\right)} = \frac{32,000}{1.03} = \$31,067.96 \quad \blacksquare$$

■ EXAMPLE 3

Because of a court settlement, Charlie Dawkins owes \$5000 to Arnold Parker. The money must be paid in 10 months, with no interest. Suppose Dawkins wishes to pay the money today. What amount should Parker be willing to accept? Assume simple interest of 11%.

The amount that Parker should be willing to accept today is given by the present value:

$$P = \frac{5000}{1 + (.11)\left(\frac{10}{12}\right)} = \frac{5000}{1.09167} = 4580.14.$$

Parker should be willing to accept \$4580.14 in settlement.   ■

Simple Discount Notes   The loans discussed up to this point are called **simple interest notes,** where interest on the face value of the loan is added to the loan and paid at maturity. Another common type of note, called a **simple discount note,** has the interest deducted in advance from the amount of a loan before giving the *balance* to the borrower. The *full* value of the note must be paid at maturity. The money that is deducted is called the **bank discount** or just the **discount,** and the money actually received by the borrower is called the **proceeds.**

■ EXAMPLE 4

Elizabeth Thornton agrees to pay \$8500 to her banker in 9 months. The banker subtracts a discount of 12% and gives the balance to Thornton. Find the amount of the discount and the proceeds.

The discount is found in the same way that simple interest is found.

$$\text{Discount} = 8500(.12)\left(\frac{9}{12}\right) = 765.00$$

The proceeds are found by subtracting the discount from the original amount.

$$\text{Proceeds} = \$8500 - \$765.00 = \$7735.00 \quad \blacksquare$$

In Example 4, the borrower was charged a discount of 12%. However, 12% is *not* the interest rate paid, since 12% applies to the $8500, while the borrower actually received only $7735. To find the rate of interest actually paid by the borrower, called the **effective rate,** work as in the next example.

■ EXAMPLE 5

Find the actual rate of interest paid by Thornton in Example 4.

The rate of 12% stated in Example 4 is not the actual rate of interest since it applies to the total amount of $8500 and not to the amount actually borrowed, $7735. To find the rate of interest paid by Thornton, use the formula for simple interest, $I = Prt$, with r the unknown. Since the borrower received only $7735, $I = 765$, $P = 7735$, and $t = 9/12$. Substitute these values into $I = Prt$.

$$I = Prt$$

$$765 = 7735(r)\left(\frac{9}{12}\right)$$

$$\frac{765}{7735\left(\dfrac{9}{12}\right)} = r$$

$$.132 \approx r$$

The interest rate paid by the borrower is about 13.2%. ■

Let D represent the amount of discount on a loan. Then $D = Art$, where A is the maturity value of the loan (the amount borrowed plus interest), and r is the stated rate of interest. The amount actually received, the proceeds, can be written as $P = A - D$, or $P = A - Art$, from which $P = A(1 - rt)$.

The formulas for discount are summarized below.

DISCOUNT

> ■ If D is the discount on a loan having a maturity value A at a rate of interest r for t years, and if P represents the proceeds, then
> $$P = A - D \quad \text{or} \quad P = A(1 - rt).$$

One common use of discount is in *discounting a note,* a process by which a promissory note due at some time in the future can be converted to cash now.

■ EXAMPLE 6

Jim Levy owes $4250 to Jenny Chinn. The loan is payable in 1 year at 10% interest. Chinn needs cash to buy a new car, so 3 months before the loan is payable she goes to her bank to have the loan discounted. The bank will pay her the maturity value of the note less a 14% discount fee. Find the amount of cash she will receive from the bank.

First find the maturity value of the loan, the amount Levy must pay to Chinn. By the formula for maturity value,

$$A = P(1 + rt)$$
$$A = 4250[1 + (.10)(1)]$$
$$= 4250(1.10) = 4675,$$

or $4675.

The bank applies its discount rate to this total:

$$\text{Amount of Discount} = 4675(.14)\left(\frac{3}{12}\right) = 163.63.$$

(Remember that the loan was discounted 3 months before it was due.) Chinn actually receives

$$\$4675 - \$163.63 = \$4511.37$$

in cash from the bank. Three months later, the bank will get $4675 from Levy. ■

10.1 EXERCISES

Find the simple interest in Exercises 1–16.

1. $1000 at 5% for 1 year

2. $4500 at 10% for 1 year

3. $25,000 at 7% for 9 months

4. $3850 at 9% for 8 months

5. $1974 at 6.3% for 7 months

6. $3724 at 8.4% for 11 months

In Exercises 7–12, assume a 360-day year. Also, assume 30 days in each month.

7. $12,000 at 7% for 72 days

8. $38,000 at 9% for 216 days

9. $5147.18 at 10.1% for 58 days

10. $2930.42 at 11.9% for 123 days

11. $7980 at 10%; loan made on May 7 and due September 19

12. $5408 at 12%; loan made on August 16 and due December 30

In Exercises 13–16, assume 365 days in a year, and use the exact number of days in a month. (Assume 28 days in February.)

13. $7800 at 11%; made on July 7 and due October 25

14. $11,000 at 10%; made on February 19 and due May 31

15. $2579 at 9.6%; made on October 4 and due March 15

16. $37,098 at 11.2%; made on September 12 and due July 30

Find the present value of each of the future amounts in Exercises 17–22. Assume 360 days in a year.

17. $15,000 for 8 months; money earns 6%

18. $48,000 for 9 months; money earns 5%

19. $5276 for 3 months; money earns 5.4%

20. $6892 for 7 months; money earns 7.1%

21. $15,402 for 125 days; money earns 6.3%

22. $29,764 for 310 days; money earns 7.2%

Find the proceeds for the amounts in Exercises 23–26. Assume 360 *days in a year.*

23. $7150; discount rate 12%; length of loan 11 months

24. $9450; discount rate 13%; length of loan 7 months

25. $358; discount rate 11.6%; length of loan 183 days

26. $509; discount rate 9.2%; length of loan 238 days

▦ APPLICATIONS

BUSINESS AND ECONOMICS

Loan Repayment **27.** Donna Sharp borrowed $25,900 from her father to start a flower shop. She repaid him after 11 months, with interest of 8.4%. Find the total amount she repaid.

Delinquent Taxes **28.** An accountant for a corporation forgot to pay the firm's income tax of $725,896.15 on time. The government charged a penalty of 17.7% interest for the 34 days the money was late. Find the total amount (tax and penalty) that was paid. (Use a 365-day year.)

Savings **29.** Tuition of $1769 will be due when the spring term begins, in 4 months. What amount should a student deposit today, at 6.25%, to have enough to pay the tuition?

Savings **30.** A firm of attorneys has ordered 7 new IBM typewriters at a cost of $2104 each. The machines will not be delivered for 7 months. What amount could the firm deposit in an account paying 6.42% to have enough to pay for the machines?

Loan Repayment **31.** Hisao Kobayashi needs $5196 to pay for remodeling work on his house. He plans to repay the loan in 10 months. His bank loans money at a discount rate of 13%. Find the amount of his loan.

Loan Repayment **32.** Mary Collins decides to go back to college. To get to school she buys a small car for $6100. She decides to borrow the money from a bank that charges an 11.8% discount rate. If she will repay the loan in 7 months, find the amount of the loan.

Loan Interest **33.** Marge Prullage signs a $4200 note at the bank. The bank charges a 12.2% discount rate. Find the net proceeds if the note is for 10 months. Find the effective rate charged by the bank.

Loan Interest **34.** A bank charges a 13.1% discount rate on a $1000 note for 90 days. Find the effective rate.

Consumer Credit **35.** Maria Lopez owes $7000 to the Eastside Music Shop. She has agreed to pay the amount in 7 months at an interest rate of 10%. Two months before the loan is due to be paid, the store discounts it at the bank. The bank charges a 12.7% discount rate. How much money does the store receive?

Bank Discount **36.** A building contractor gives a $13,500 note to a plumber. The note is due in 9 months, with interest of 13%. Three months after the note is signed, the plumber discounts it at the bank. The bank charges a 14.1% discount rate. How much money does the plumber actually receive?

▇▇ 10.2   COMPOUND INTEREST

As mentioned earlier, simple interest is normally used for loans or investments of a year or less. For longer periods *compound interest* is used. With **compound interest,** interest is charged (or paid) on interest as well as on principal. To find a formula for compound interest, first suppose that P dollars is deposited at a rate of interest i per year. (Although r is used to represent simple interest rates, it is more common to use i for compound interest rates.) The interest earned during the first year is found by using the formula for simple interest:

$$\text{First-Year Interest} = P \cdot i \cdot 1 = Pi.$$

At the end of one year, the amount on deposit is the sum of the original principal and the interest earned, or

$$P + Pi = P(1 + i). \tag{1}$$

If the deposit earns compound interest, the interest earned during the second year is paid on the total amount on deposit at the end of the first year. Thus, the interest earned during the second year (again found by using the formula for simple interest) is given by

$$P(1 + i)(i)(1) = P(1 + i)i, \tag{2}$$

so that the total amount on deposit at the end of the second year is given by the sum of the amounts from (1) and (2) above, or

$$P(1 + i) + P(1 + i)i = P(1 + i) \cdot (1 + i)$$
$$= P(1 + i)^2.$$

In the same way, the total amount on deposit at the end of the third year is

$$P(1 + i)^3.$$

Generalizing, in t years the total amount on deposit is

$$A = P(1 + i)^t,$$

called the **compound amount.**

Compare this formula for compound interest with the formula for simple interest given in Section 1:

$$\begin{array}{ll} \textit{Compound Interest} & A = P(1 + i)^t \\ \textit{Simple Interest} & A = P(1 + rt). \end{array}$$

Both i and r represent the annual interest rate (in decimal form) and t represents time in years. The important distinction between the two formulas is that in the compound interest formula, the number of years, t, is an *exponent,* so that money grows much more rapidly when interest is compounded.

Interest can be compounded more than once a year. Suppose interest is compounded *m* times per year (*m periods* per year), at a rate *i* per year, so that i/m is the rate for each period. Suppose that interest is compounded for *n* years. Then the following formula for the compound amount can be derived in the same way as was the previous formula.

COMPOUND AMOUNT

If *P* dollars is deposited for *n* years with interest compounded *m* periods per year at a rate of interest *i* per year, the compound amount *A* is

$$A = P\left(1 + \frac{i}{m}\right)^{nm}.$$

EXAMPLE 1

Suppose $1000 is deposited for 6 years in an account paying 8% per year compounded annually.

(a) Find the compound amount.

In the formula above, $P = 1000$, $i = 8\% = .08$, $m = 1$, and $n = 6$. The compound amount is

$$A = P\left(1 + \frac{i}{m}\right)^{nm}$$

$$A = 1000\left(1 + \frac{.08}{1}\right)^{6(1)} = 1000(1.08)^6.$$

The quantity $(1.08)^6$ can be found by using a calculator with a y^x key or by using special compound interest tables. Such a table is given in the Appendix. To find $(1.08)^6$, look for 8% across the top and 6 (for 6 periods) down the side. You should find 1.58687; thus $(1.08)^6 \approx 1.58687$, and

$$A = 1000(1.58687) = 1586.87,$$

or $1586.87, which represents the final amount on deposit.

(b) Find the actual amount of interest earned.

From the compound amount, subtract the initial deposit.

$$\text{Amount of Interest} = \$1586.87 - \$1000 = \$586.87$$

EXAMPLE 2

Find the amount of interest earned by a deposit of $1000 for 6 years at 6% compounded quarterly.

Interest compounded quarterly is compounded 4 times a year. In 6 years, there are $6 \cdot 4 = 24$ quarters, or 24 periods. Interest of 6% per year is 6%/4, or 1.5% per quarter. The compound amount is

$$1000(1 + .015)^{24} = 1000(1.015)^{24}.$$

The value of $(1.015)^{24}$ can be found with a calculator or the table. Locate 1.5% across the top of the table and 24 periods at the left. You should find the number 1.42950, so that

$$A = 1000(1.42950) = 1429.50,$$

or $1429.50. The compound amount is $1429.50, and the interest earned is $1429.50 − $1000 = $429.50. ▬

▬ EXAMPLE 3

Find the compound amount if $900 is deposited at 8% compounded semiannually for 10 years.

In 10 years there are $10 \cdot 2 = 20$ semiannual periods. If interest is 8% per year, then 8%/2 = 4% is earned per semiannual period. Use a calculator, or look in the table for 4% and 20 periods, finding the number 2.19112. The compound amount is

$$A = 900(1.04)^{20} = 900(2.19112) = 1972.01,$$

or $1972.01. ▬

The more often interest is compounded within a given time period, the more interest will be earned. Using a calculator with a y^x key, together with the compound amount formula, we get the results shown in the following chart.

Interest on $1000 at 12% per Year for 10 Years

Type of Compounding	Number of Periods	Compound Amount	Interest
None at all (simple interest)	—	—	$1200.00
Annual	10	$1000(1 + .12)^{10} = \$3105.85$	$2105.85
Semiannual	20	$1000\left(1 + \dfrac{.12}{2}\right)^{20} = \$3207.14$	$2207.14
Quarterly	40	$1000\left(1 + \dfrac{.12}{4}\right)^{40} = \$3262.04$	$2262.04
Monthly	120	$1000\left(1 + \dfrac{.12}{12}\right)^{120} = \$3300.39$	$2300.39
Daily	3650	$1000\left(1 + \dfrac{.12}{365}\right)^{3650} = \$3319.46$	$2319.46
Hourly	87,600	$1000\left(1 + \dfrac{.12}{8760}\right)^{87,600} = \$3320.09$	$2320.09
Every Minute	5,256,000	$1000\left(1 + \dfrac{.12}{525,600}\right)^{5,256,000} = \$3320.11$	$2320.11

As suggested by the chart, the final amount on deposit is much greater when interest is compounded than when it is not. For example, $1000 deposited for 10 years at 12% simple interest produces $905.85 less than the same amount deposited for the same period at 12% compounded annually. As the frequency of compounding increases, however, there are smaller and smaller differences in the amount of interest earned. In fact, it can be shown that even if interest is compounded at intervals of time as small as one chooses (such as each hour, each minute, or each second), the total amount of interest earned will be only slightly more than for daily compounding. This is true even for a process called **continuous compounding,** which can be loosely described as compounding every instant. The interesting topic of continuous compounding is discussed in more detail in calculus.

▰▰ EXAMPLE 4

Suppose $24,000 is deposited at 8% interest for 9 years. Find the interest earned by (a) daily, and (b) hourly compounding.

(a) In 9 years there are $9 \times 365 = 3285$ days. The compound amount with daily compounding is

$$24,000\left(1 + \frac{.08}{365}\right)^{3285} = 49,302.51,$$

or $49,302.51. The interest earned is

$$\$49,302.51 - \$24,000 = \$25,302.51.$$

(b) In 1 year there are $365 \times 24 = 8760$ hours, and in 9 years there are $9 \times 8760 = 78,840$ hours. The compound amount is

$$24,000\left(1 + \frac{.08}{8760}\right)^{78,840} = 49,306.23,$$

or $49,306.23. This amount includes interest of

$$\$49,306.23 - \$24,000 = \$25,306.23,$$

only $3.72 more than when interest is compounded daily.   ▰▰

Effective Rate   If $1 is deposited at 4% compounded quarterly, a calculator or a table can be used to find that at the end of one year, the compound amount is $1.0406, an increase of 4.06% over the original $1. The actual increase of 4.06% in the money is somewhat higher than the stated increase of 4%. To differentiate between these two numbers, 4% is called the **nominal** or **stated** rate of interest, while 4.06% is called the **effective** rate.

■■ **EXAMPLE 5**

Find the effective rate corresponding to a nominal rate of 6% compounded semiannually.

Look in the compound interest table for an interest rate of 3% for 2 periods. You should find the number 1.06090. Alternatively, use a calculator to find $(1.03)^2$. By either method, $1 will increase to $1.06090, an actual increase of 6.09%. The effective rate is 6.09%. ■■

Generalizing from this example, the effective rate of interest is given by the following formula.

EFFECTIVE RATE

■■ The effective rate corresponding to a stated rate of interest i compounded m times per year is

$$\left(1 + \frac{i}{m}\right)^m - 1.$$

■■ **EXAMPLE 6**

A bank pays interest of 5% compounded monthly. Find the effective rate.

Use the formula given above, with $i = .05$ and $m = 12$. The effective rate is

$$\left(1 + \frac{.05}{12}\right)^{12} - 1.$$

Use a calculator with a y^x key to get

$$(1.0041667)^{12} - 1 = .0512,$$

or 5.12%. ■■

Present Value with Compound Interest   The formula for compound interest $A = P(1 + i/m)^{nm}$, has five variables: A, P, i, m, and n. Given the values of any four of these variables, the value of the fifth can be found. In particular, if A, the future amount, and i, m, and n are known, then P can be found. Here P is the amount that should be deposited today to produce A dollars in n years. The next example shows this.

■■ **EXAMPLE 7**

Joan Ailanjian must pay a lump sum of $6000 in 5 years. What amount deposited today at 8% compounded annually will amount to $6000 in 5 years?

Here $A = 6000$, $i = .08$, $m = 1$, $n = 5$, and P is unknown. Substituting these values into the formula for the compound amount gives

$$6000 = P\left(1 + \frac{.08}{1}\right)^{5(1)} = P(1.08)^5.$$

From a calculator or the compound interest table, $(1.08)^5 \approx 1.46933$, with

$$6000 \approx P(1.46933)$$

and
$$P = \frac{6000}{1.46933} \approx 4083.49,$$

or $4083.49. If Ailanjian leaves $4083.49 for 5 years in an account paying 8% compounded annually, she will have $6000 when she needs it. ▬

As Example 7 shows, $6000 in 5 years is the same as $4083.49 today (if money can be deposited at 8% compounded annually.) Recall from the first section that an amount that can be deposited today to yield a given sum in the future is called the *present value* of the future sum. Generalizing from Example 7, the present value of A dollars compounded m times a year at an interest rate i for n years is

$$P = \frac{A}{\left(1 + \dfrac{i}{m}\right)^{nm}}.$$

Compare this with the present value of an amount at simple interest i for t years, given in the previous section:

$$P = \frac{A}{1 + rt}.$$

The present value table in the Appendix gives values of

$$\frac{1}{(1 + i)^n} = (1 + i)^{-n}.$$

This table may be used instead of the compound interest table to find the present value P.

▬ EXAMPLE 8

Find the present value of $16,000 in 9 years if money can be deposited at 6% compounded semiannually.

In 9 years there are $2 \cdot 9 = 18$ semiannual periods. A rate of 6% per year is 3% in each semiannual period. Use a calculator, or look in the table (3% across the top and 18 periods down the side). You should find 1.70243. The present value is

$$\frac{16,000}{1.70243} \approx 9398.33.$$

A deposit of $9398.33 today, at 6% compounded semiannually, will produce a total of $16,000 in 9 years. ▬

We can solve the compound amount formula for n, also, as the following example shows.

▬ EXAMPLE 9

Suppose the general level of inflation in the economy averages 8% per year. Find the number of years it would take for the overall level of prices to double.

To find the number of years it will take for $1 worth of goods or services to cost $2, find n in the equation

$$2 = 1(1 + .08)^n,$$

where $A = 2$, $P = 1$, and $i = .08$. This equation simplifies to

$$2 = (1.08)^n.$$

The value of n could be found exactly by using logarithms, but we can get a reasonable approximation by reading down the 8% column of the table. Read down this column until you come to the number closest to 2, which is 1.99900. This number corresponds to 9 periods. The general level of prices will double in about 9 years. ▬

▬ 10.2 EXERCISES

Find the compound amount for each of the following deposits.

1. $1000 at 6% compounded annually for 8 years
2. $1000 at 7% compounded annually for 10 years
3. $470 at 10% compounded semiannually for 12 years
4. $15,000 at 6% compounded semiannually for 11 years
5. $6500 at 12% compounded quarterly for 6 years
6. $9100 at 8% compounded quarterly for 4 years

Find the amount of interest earned by each of the following deposits.

7. $6000 at 8% compounded annually for 8 years
8. $21,000 at 6% compounded annually for 5 years
9. $43,000 at 10% compounded semiannually for 9 years
10. $7500 at 8% compounded semiannually for 5 years
11. $2196.58 at 10.8% compounded quarterly for 4 years
12. $4915.73 at 11.6% compounded quarterly for 3 years

Find the present value of each of the following investments.

13. $4500 at 8% compounded annually for 9 years
14. $11,500 at 8% compounded annually for 12 years
15. $15,902.74 at 9.8% compounded annually for 7 years
16. $27,159.68 at 12.3% compounded annually for 11 years
17. $2000 at 9% compounded semiannually for 8 years
18. $2000 at 11% compounded semiannually for 8 years
19. $8800 at 10% compounded quarterly for 5 years
20. $7500 at 12% compounded quarterly for 9 years

Find the effective rate corresponding to each of the following nominal rates.

21. 4% compounded semiannually

22. 8% compounded quarterly

23. 8% compounded semiannually

24. 10% compounded semiannually

25. 12% compounded semiannually

26. 12% compounded quarterly

▦ APPLICATIONS

BUSINESS AND ECONOMICS

Savings

27. When Lindsay Branson was born, her grandfather made an initial deposit of $3000 in an account for her college education. Assuming an interest rate of 6% compounded quarterly, how much will the account be worth in 18 years?

Individual Retirement Account

28. Bill Poole has $10,000 in an Individual Retirement Account (IRA). Because of new tax laws, he decides to make no further deposits. The account earns 6% interest compounded semiannually. Find the amount on deposit in 15 years.

Inflation

29. Suppose the rate of inflation is steady at 5% per year compounded annually. How much will a car that costs $10,000 now cost in 10 years?

Inflation

30. Suppose the rate of inflation is 6% a year compounded annually. What will it cost in 10 years to buy a house currently valued at $70,000?

Loan Interest

31. A small business borrows $50,000 for expansion at 12% compounded monthly. The loan is due in 4 years. How much interest will the business pay?

Loan Interest

32. A developer needs $80,000 to buy land. He is able to borrow the money at 10% per year compounded quarterly. How much will the interest amount to if he pays off the loan in 5 years?

Savings

33. Linda Youngman wants to take a cruise in 3 years. She estimates she will need $5000 at that time. How much should she put in an account now, at 8% compounded semiannually, to have enough money for the cruise?

Savings

34. George Duda wants to have $20,000 available in 5 years for a down payment on a house. He has inherited $15,000. How much of the inheritance should he invest now to accumulate the $20,000, if he can get an interest rate of 8% compounded quarterly?

Comparing Investments

35. If money can be invested at 8% compounded quarterly, which is larger—$1000 now or the present value of $1210 left at 8% interest for 5 years?

Comparing Investments

36. If money can be invested at 6% compounded annually, which is larger—$10,000 now or the present value of $15,000 left at 6% interest for 10 years?

Doubling Time

Use the ideas from Example 9 to find the time it would take for the general level of prices in the economy to double at each of the following average annual inflation rates.

37. 4%   **38.** 6%

Doubling Time

39. The consumption of electricity has increased historically at 6% per year. If it continues to increase at this rate indefinitely, find the number of years before the electric utilities will need to double their generating capacity.

Doubling Time   **40.** Suppose a conservation campaign coupled with higher rates causes the demand for electricity to increase at only 2% per year, as it has recently. Find the number of years before the utilities will need to double generating capacity.

Negative Interest   *Under certain conditions, Swiss banks pay* negative *interest:* they charge you. (You didn't think all that secrecy was free?) *Suppose a bank "pays" —2.4% interest compounded annually. Use a calculator and find the compound amount for a deposit of $150,000 after each of the following periods.*

41. 2 years   **42.** 4 years

43. 8 years   **44.** 12 years

FOR THE COMPUTER

Find the compound amount for each of the following deposits.

45. $40,552 at 9.13% compounded quarterly for 5 years

46. $11,641.10 at 10.9% compounded monthly for 8 years

47. $673.27 at 8.6% compounded semiannually for 3.5 years

48. $2964.93 at 11.4% compounded monthly for 4.25 years

10.3   SEQUENCES

So far in this chapter on mathematics of finance, only lump sums have been discussed: a lump sum that is deposited today to produce a lump sum amount in the future, or the present value of a given lump sum amount in the future. In practice, however, many financial applications of mathematics deal with a sequence of *periodic payments,* such as car payments or house payments. Formulas for such payments will be developed in the next sections. To develop these formulas, we need a formula for the sum of the terms of a *sequence*.

A **sequence** is a function whose domain is the set of positive integers. For example,

$$a(n) = 2n, \qquad n = 1, 2, 3, 4, \ldots$$

is a sequence. The letter n is used as a variable instead of x to emphasize the fact that the domain includes only positive integers. For the same reason, a is used to name the function instead of f.

The range values of a sequence function, such as

$$a(1) = 2, \qquad a(2) = 4, \qquad a(3) = 6, \ldots$$

from the sequence given above, are called the **terms** of the sequence. Instead of writing $a(5)$ for the fifth term of the sequence, however, it is customary to write

$$a_5 = 10.$$

In the same way, for the sequence above, $a_1 = 2$, $a_2 = 4$, $a_8 = 16$, $a_{20} = 40$, and $a_{51} = 102$.

The symbol a_n is often used for the **general** or ***n*th term** of a sequence. For example, for the sequence 4, 7, 10, 13, 16, . . . the general term is given by $a_n = 1 + 3n$. This formula for a_n can be used to find any desired term of the sequence. For example, the first three terms of the sequence having $a_n = 1 + 3n$ are

$$a_1 = 1 + 3(1) = 4, \qquad a_2 = 1 + 3(2) = 7, \qquad a_3 = 1 + 3(3) = 10.$$

Also, $a_8 = 25$ and $a_{12} = 37$.

▬ EXAMPLE 1

Find the first four terms for the sequence having the general term $a_n = -4n + 2$.

Replace n, in turn, with 1, 2, 3, and 4. If $n = 1$,

$$a_1 = -4(\mathbf{1}) + 2 = -4 + 2 = -2. \qquad \text{Let } n = 1$$

Also, $\qquad a_2 = -4(\mathbf{2}) + 2 = -6.$ $\qquad\qquad$ Let $n = 2$

When $n = 3$, then $a_3 = -10$; also, $a_4 = -4(4) + 2 = -14$. The first four terms of this sequence are -2, -6, -10, and -14. ▬

Arithmetic Sequences   A sequence in which each term after the first is found by adding the same number to the preceding term is called an **arithmetic sequence.** The sequence in Example 1 is an arithmetic sequence; -4 is added to any term to get the next term.

The sequence

$$8, 13, 18, 23, 28, . . .$$

is an arithmetic sequence since each term after the first is found by adding 5 to the previous term. The number 5, the difference between any two adjacent terms, is called the **common difference.**

If a_1 is the first term of an arithmetic sequence and d is the common difference, then the second term can be found by adding the common difference d to the first term: $a_2 = a_1 + d$. The third term is found by adding d to the second term.

$$a_3 = a_2 + d = (a_1 + d) + d = a_1 + 2d$$

In the same way, $a_4 = a_1 + 3d$ and $a_5 = a_1 + 4d$. Generalizing, the *n*th term of an arithmetic sequence is given by $a_n = a_1 + (n - 1)d$.

***n*TH TERM**

> ▬ If an arithmetic sequence has first term a_1 and common difference d, then a_n, the *n*th term of the sequence, is given by
> $$a_n = a_1 + (n - 1)d.$$

■■ **EXAMPLE 2**

A company had sales of $50,000 during its first year of operation. If the sales increase by $6000 per year, find its sales in the eleventh year.

Since the sales for each year after the first are found by adding $6000 to the sales of the previous year, the sales form an arithmetic sequence with $a_1 = 50,000$ and $d = 6000$. Using the formula for the nth term of an arithmetic sequence, sales during the eleventh year are given by

$$a_{11} = 50,000 + (11 - 1)6000 = 50,000 + 60,000 = 110,000,$$

or $110,000. ■■

The formula above gives the nth term of an arithmetic sequence. To find a formula for the *sum* of the first n terms of an arithmetic sequence, let the sequence have first term a_1 and common difference d. Let S_n represent the sum of the first n terms of the sequence. Start by writing a formula for S_n as follows:

$$S_n = a_1 + [a_1 + d] + [a_1 + 2d] + \cdots + [a_1 + (n - 1)d].$$

Next, write this same sum in reverse order.

$$S_n = [a_1 + (n - 1)d] + [a_1 + (n - 2)d] + \cdots + [a_1 + d] + a_1$$

Now add the respective sides of these last two equations.

$$S_n + S_n = (a_1 + [a_1 + (n - 1)d]) + ([a_1 + d] + [a_1 + (n - 2)d])$$
$$+ \cdots + ([a_1 + (n - 1)d] + a_1)$$

From this,

$$2S_n = [2a_1 + (n - 1)d] + [2a_1 + (n - 1)d]$$
$$+ \cdots + [2a_1 + (n - 1)d].$$

There are n of the $[2a_1 + (n - 1)d]$ terms on the right, so that

$$2S_n = n[2a_1 + (n - 1)d],$$

$$S_n = \frac{n}{2}[2a_1 + (n - 1)d].$$

Since $a_n = a_1 + (n - 1)d$, and $S_n = \frac{n}{2}[a_1 + a_1 + (n - 1)d]$,

$$S_n = \frac{n}{2}(a_1 + a_n).$$

Our work with the sum of the first n terms of an arithmetic sequence is summarized on the next page.

SUM OF TERMS

Suppose an arithmetic sequence has first term a_1, common difference d, and nth term a_n. Then the sum S_n of the first n terms of the sequence is given by

$$S_n = \frac{n}{2}[2a_1 + (n - 1)d]$$

or

$$S_n = \frac{n}{2}(a_1 + a_n).$$

Either of these two formulas can be used to find the sum of the first n terms of an arithmetic sequence.

EXAMPLE 3
Find the sum of the first 25 terms of the arithmetic sequence

$$3, 8, 13, 18, 23, \ldots .$$

In this sequence, $a_1 = 3$ and $d = 5$. Using the first formula above gives

$$S_{25} = \frac{25}{2}[2(3) + (25 - 1)5] \qquad \text{Let } n = 25, a_1 = 3, d = 5$$

$$= \frac{25}{2}[6 + 120]$$

$$= 1575. \quad \blacksquare$$

EXAMPLE 4
Find the total sales of the company of Example 2 during its first 11 years.

From Example 2, $a_1 = 50{,}000$, $a_{11} = 110{,}000$, and $n = 11$. By the second formula above,

$$S_{11} = \frac{11}{2}(50{,}000 + 110{,}000) = 880{,}000.$$

The total sales for 11 years are \$880,000.   ■

Geometric Sequences   In an arithmetic sequence, each term after the first is found by adding the same number to the preceding term. In a **geometric sequence,** each term after the first is found by *multiplying* the preceding term by the same number. For example,

$$3, -6, 12, -24, 48, -96, \ldots$$

is a geometric sequence with each term after the first found by multiplying the preceding term by the number -2, which is called the **common ratio.**

If a_1 is the first term of a geometric sequence and r is the common ratio, then the second term is given by $a_2 = a_1 r$ and the third by $a_3 = a_2 r = a_1 r^2$. Also, $a_4 = a_1 r^3$, and $a_5 = a_1 r^4$. Generalizing from these results gives the following rule.

nTH TERM

> If a geometric sequence has first term a_1 and common ratio r, then
> $$a_n = a_1 r^{n-1}.$$

■ EXAMPLE 5

Find the indicated term for each of the following geometric sequences.

(a) 6, 24, 96, 384, . . . ; find a_7.

Here $a_1 = 6$. To find r, choose any term except the first and divide it by the preceding term. Choosing 96 gives

$$r = \frac{96}{24} = 4.$$

Now use the formula for the nth term.

$$a_7 = 6(4)^{7-1} \qquad \text{Let } n = 7, a_1 = 6, r = 4$$
$$= 6(4)^6$$
$$= 6(4096) \qquad 4^6 = 4096$$
$$= 24{,}576$$

(b) 8, -16, 32, -64, 128, . . . ; find a_6.

In this sequence $a_1 = 8$ and $r = -2$.

$$a_6 = 8(-2)^{6-1} \qquad \text{Let } n = 6, a_1 = 8, r = -2$$
$$= 8(-2)^5$$
$$= 8(-32)$$
$$= -256 \quad ■$$

In work with annuities, it is necessary to find the sum of the first n terms of a geometric sequence. To find a formula for this sum, let a geometric sequence have first term a_1 and common ratio r. Write the sum S_n of the first n terms as

$$S_n = a_1 + a_2 + a_3 + \cdots + a_n.$$

This sum also can be written as

$$S_n = a_1 + a_1 r + a_1 r^2 + \cdots + a_1 r^{n-1}. \tag{1}$$

If $r = 1$, then $S_n = na_1$, the correct result for this case. If $r \neq 1$, multiply both sides of equation (1) by r, obtaining

$$rS_n = a_1r + a_1r^2 + a_1r^3 + \cdots + a_1r^n. \tag{2}$$

Now subtract corresponding sides of equation (1) from equation (2):

$$rS_n - S_n = a_1r^n - a_1$$
$$S_n(r - 1) = a_1(r^n - 1),$$

giving
$$S_n = \frac{a_1(r^n - 1)}{r - 1}.$$

This result is summarized below.

SUM OF TERMS

> If a geometric sequence has first term a_1 and common ratio r, then the sum S_n of the first n terms is given by
> $$S_n = \frac{a_1(r^n - 1)}{r - 1}, \qquad r \neq 1.$$

EXAMPLE 6

Find the sum of the first six terms of the geometric sequence 3, 12, 48, ...
Here $a_1 = 3$ and $r = 4$. Find S_6 by the formula above.

$$S_6 = \frac{3(4^6 - 1)}{4 - 1} \qquad \text{Let } n = 6, a_1 = 3, r = 4$$

$$= \frac{3(4096 - 1)}{3}$$

$$= 4095$$

EXAMPLE 7

Suppose a child saves 1¢ on the first day of the week, 2¢ on the second day, 4¢ on the third day, and so on, doubling the amount each day. How much will the child have saved in one week?

This is a geometric sequence with $a_1 = 1$ and $r = 2$. Since a week has 7 days, find S_7.

$$S_7 = \frac{1(2^7 - 1)}{2 - 1}$$

$$= 127$$

In one week, the child will have saved 127¢, or $1.27.

10.3 EXERCISES

In Exercises 1–16, a formula for the general term of a sequence is given. Use the formula to find the first five terms of the sequence. Identify each sequence as arithmetic, geometric, *or* neither.

1. $a_n = 6n + 5$ **2.** $a_n = 12n - 3$ **3.** $a_n = 3n - 7$ **4.** $a_n = 5n - 12$

5. $a_n = -6n + 4$ **6.** $a_n = -11n + 10$ **7.** $a_n = 2^n$ **8.** $a_n = 3^n$

9. $a_n = (-2)^n$ **10.** $a_n = (-3)^n$ **11.** $a_n = 3(2^n)$ **12.** $a_n = -4(2^n)$

13. $a_n = \dfrac{n + 1}{n + 5}$ **14.** $a_n = \dfrac{2n}{n + 1}$ **15.** $a_n = \dfrac{1}{n + 1}$ **16.** $a_n = \dfrac{1}{n + 8}$

Identify each sequence as arithmetic, geometric, *or* neither. *For an arithmetic sequence, give the common difference. For a geometric sequence, give the common ratio.*

17. 6, 14, 22, 30, 38, 46, . . .   **18.** 40, 46, 52, 58, 64, . . .   **19.** 5, 8, 11, 14, 17, 20, 23, . . .

20. 23, 34, 45, 54, 63, 72, . . .   **21.** 4, 12, 36, 108, . . .   **22.** 7, 14, 28, 56, 112, . . .

23. 2, 5, 9, 14, 20, 27, . . .   **24.** 1, 4, 9, 16, 25, 36, . . .   **25.** 12, 9, 6, 3, 0, −3, −6, . . .

26. 37, 31, 25, 19, 13, 7, . . .   **27.** −18, −15, −12, −9, −6, . . .   **28.** −21, −17, −13, −9, −5, . . .

29. 3, −6, 12, −24, 48, −96, . . .   **30.** −5, 10, −20, 40, −80, . . .

31. −5, 6, −7, 8, −9, 10, −11, . . .   **32.** −12, 9, −6, 3, . . .

Find the indicated term for each arithmetic sequence.

33. $a_1 = 10$, $d = 5$; find a_{13} **34.** $a_1 = 6$, $d = 9$; find a_8

35. $a_1 = 8$, $d = 3$; find a_{20} **36.** $a_1 = 13$, $d = 7$; find a_{11}

37. 6, 9, 12, 15, 18, . . . ; find a_{25} **38.** 14, 17, 20, 23, 26, 29, . . . ; find a_{13}

39. −9, −13, −17, −21, −25, . . . ; find a_{15} **40.** −4, −11, −18, −25, −32, . . . ; find a_{18}

Find the sum of the first six terms of each arithmetic sequence.

41. 3, 6, 9, 12, . . .   **42.** 11, 13, 15, 17, . . .   **43.** 88, 98, 108, . . .   **44.** 92, 95, 98, . . .

45. $a_1 = 8$, $d = 9$ **46.** $a_1 = 12$, $d = 6$ **47.** $a_1 = 7$, $d = -4$ **48.** $a_1 = 13$, $d = -5$

Find a_5 for each geometric sequence.

49. $a_1 = 3$, $r = 2$ **50.** $a_1 = 5$, $r = 3$ **51.** $a_1 = -8$, $r = 3$

52. $a_1 = -6$, $r = 2$ **53.** $a_1 = 1$, $r = -3$ **54.** $a_1 = 12$, $r = -2$

55. $a_1 = 1024$, $r = 1/2$ **56.** $a_1 = 729$, $r = 1/3$

Find the sum of the first four terms of each geometric sequence.

57. $a_1 = 1$, $r = 2$ **58.** $a_1 = 3$, $r = 3$ **59.** $a_1 = 5$, $r = 1/5$

60. $a_1 = 6$, $r = 1/2$ **61.** $a_1 = 128$, $r = -3/2$ **62.** $a_1 = 81$, $r = -2/3$

Sums of the terms of a sequence are often written in sigma notation *(also used in statistics). For example, to evaluate*

$$\sum_{i=1}^{4}(3i + 7)$$

(where Σ is the capital Greek letter sigma*), first evaluate $3i + 7$ for $i = 1$, then for $i = 2$, $i = 3$, and finally, $i = 4$. Then add the results to get*

$$\sum_{i=1}^{4}(3i + 7) = [3(1) + 7] + [3(2) + 7] + [3(3) + 7] + [3(4) + 7]$$

$$= 10 + 13 + 16 + 19 = 58.$$

Evaluate each of the following sums.

63. $\displaystyle\sum_{i=1}^{5}(3 - 2i)$ **64.** $\displaystyle\sum_{i=1}^{4}(2i - 5)$ **65.** $\displaystyle\sum_{i=1}^{8}(2i + 1)$

66. $\displaystyle\sum_{i=1}^{4}(-6i + 8)$ **67.** $\displaystyle\sum_{i=1}^{5}i(2i + 1)$ **68.** $\displaystyle\sum_{i=1}^{4}(2i - 1)(i + 2)$

69. $\displaystyle\sum_{i=1}^{4}(3i + 1)(i + 1)$ **70.** $\displaystyle\sum_{i=1}^{5}(i - 3)(i + 5)$

A sum in the form

$$\sum_{i=1}^{n}(ai + b),$$

where a and b are real numbers, represents the sum of the first n terms of an arithmetic sequence. The first term is $a_1 = a(1) + b = a + b$ and the common difference is a. The nth term is $a_n = an + b$. Using these numbers, the sum can be found by using the formula for the sum of the first n terms. Use this method to find the following sums.

71. $\displaystyle\sum_{i=1}^{5}(2i + 8)$ **72.** $\displaystyle\sum_{i=1}^{6}(4i - 5)$ **73.** $\displaystyle\sum_{i=1}^{4}(-8i + 6)$

74. $\displaystyle\sum_{i=1}^{9}(2i - 3)$ **75.** $\displaystyle\sum_{i=1}^{500}i$ **76.** $\displaystyle\sum_{i=1}^{1000}i$

Sigma notation also can be used for geometric sequences. A sum in the form

$$\sum_{i=1}^{n}a(b^i)$$

represents the sum of the first n terms of a geometric sequence having first term $a_1 = ab$ and common ratio b. These sums can thus be found by applying the formula for the sum of the first n terms of a geometric sequence. Use this method to find the following sums.

77. $\displaystyle\sum_{i=1}^{4}3(2^i)$ **78.** $\displaystyle\sum_{i=1}^{3}2(3^i)$ **79.** $\displaystyle\sum_{i=1}^{4}\frac{1}{2}(4^i)$

80. $\displaystyle\sum_{i=1}^{4}\frac{3}{2}(2^i)$ **81.** $\displaystyle\sum_{i=1}^{4}\frac{4}{3}(3^i)$ **82.** $\displaystyle\sum_{i=1}^{4}\frac{5}{3}(3^i)$

APPLICATIONS

BUSINESS AND ECONOMICS

Salary **83.** Joy Watt is hired at a salary of $11,400 per year, with annual raises of $600. What will she earn during her eighth year with the company?

Income **84.** What will be the total income received by Watt (see Exercise 83) during her first 8 years with the company?

Depreciation **85.** A certain machine annually loses 30% of the value it had at the beginning of that year. If its initial value is $10,000, use a calculator to find its value

(a) at the end of the fifth year, and (b) at the end of the eighth year.

LIFE SCIENCES

Population Growth **86.** A certain colony of bacteria increases in number by 10% per hour. After 5 hours, what is the percentage increase in the population over the initial population?

10.4 ANNUITIES

In the first two sections of this chapter we discussed *lump sum* payments; now we will look at *sequences* of equal payments. For example, suppose $1500 is deposited at the end of each year for the next six years in an account paying 8% per year compounded annually. How much will be in the account after six years?

A sequence of equal payments made at equal periods of time is called an **annuity.** If the payments are made at the end of the time period, and if the frequency of payments is the same as the frequency of compounding, the annuity is called an **ordinary annuity.** The time between payments is the **payment period,** and the time from the beginning of the first payment period to the end of the last period is called the **term** of the annuity. The **future value of the annuity,** the final sum on deposit, is defined as the sum of the compound amounts of all the payments, compounded to the end of the term.

Figure 1 shows the annuity described above. To find the future value of this annuity, look separately at each of the $1500 payments. The first of these payments will produce a compound amount of

$$1500(1 + .08)^5 = 1500(1.08)^5$$

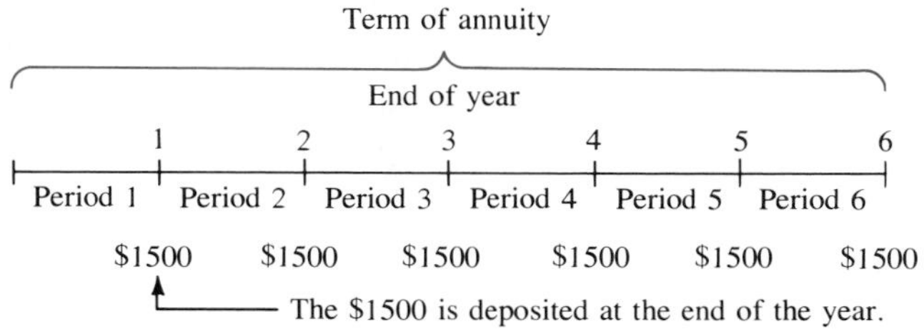

FIGURE 1

Use 5 as the exponent instead of 6 since the money is deposited at the *end* of the first year and earns interest for only five years. The second payment of $1500 will produce a compound amount of $1500(1.08)^4$. As shown in Figure 2, the future value of the annuity is

$$1500(1.08)^5 + 1500(1.08)^4 + 1500(1.08)^3 + 1500(1.08)^2$$
$$+ 1500(1.08)^1 + 1500.$$

(The last payment earns no interest at all.) From the compound interest table, or a calculator, this sum is

$$1500(1.46933) + 1500(1.36049) + 1500(1.25971) + 1500(1.16640)$$
$$+ 1500(1.08000) + 1500$$
$$\approx \$2204.00 + \$2040.74 + \$1889.57 + \$1749.60 + \$1620 + \$1500$$
$$= \$11{,}003.91.$$

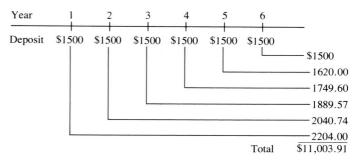

FIGURE 2

To generalize this result, suppose that payments of R dollars each are deposited into an account at the end of each period for n periods, at a rate of interest i per period. The first payment of R dollars will produce a compound amount of $R(1 + i)^{n-1}$ dollars, the second payment will produce $R(1 + i)^{n-2}$ dollars, and so on; the final payment earns no interest and contributes just R dollars to the total. If S represents the future value (or sum) of the annuity, then (as shown in Figure 3),

$$S = R(1 + i)^{n-1} + R(1 + i)^{n-2} + R(1 + i)^{n-3} + \cdots + R(1 + i) + R$$

or, written in reverse order,

$$S = R + R(1 + i)^1 + R(1 + i)^2 + \cdots + R(1 + i)^{n-1}.$$

This result is the sum of the first n terms of the geometric sequence having first term R and common ratio $1 + i$. Using the formula for the sum of the first n terms of a geometric sequence from the previous section,

$$S = \frac{R[(1 + i)^n - 1]}{(1 + i) - 1} = \frac{R[(1 + i)^n - 1]}{i} = R\left[\frac{(1 + i)^n - 1}{i}\right].$$

The quantity in brackets is commonly written $s_{\overline{n}|i}$ (read "*s-angle-n* at i"), so that

$$S = R \cdot s_{\overline{n}|i}.$$

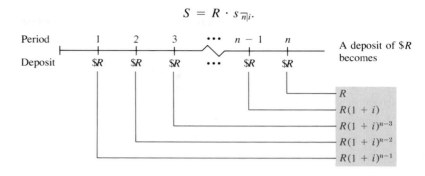

Period   1   2   3 $\cdots$ $n-1$ n A deposit of $\$R$ becomes

Deposit $\$R$ $\$R$ $\$R$ $\cdots$ $\$R$ $\$R$

R
$R(1 + i)$
$R(1 + i)^{n-3}$
$R(1 + i)^{n-2}$
$R(1 + i)^{n-1}$

The sum of these is the amount of the annuity.

FIGURE 3

Values of $s_{\overline{n}|i}$ can be found with a calculator or the amount of an annuity table in the Appendix.

To check the result above, go back to the annuity of \$1500 at the end of each year for 6 years; interest was 8% per year compounded annually. In the table, these values give the number 7.33593. Multiply this number and 1500:

$$\text{Future Value} = 1500(7.33593) \approx 11{,}003.90,$$

or \$11,003.90. The results differ by 1¢ due to rounding error.

We can summarize this work as follows.

FUTURE VALUE OF AN ANNUITY

The future value, S, of an annuity of n payments of R dollars each at the end of each consecutive interest period, with interest compounded at a rate i per period, is

$$S = R\left[\frac{(1 + i)^n - 1}{i}\right] \quad \text{or} \quad S = R \cdot s_{\overline{n}|i}.$$

■ EXAMPLE 1

Joe Kleine is an athlete who believes that his playing career will last 7 years. To prepare for his future, he deposits \$22,000 at the end of each year for 7 years in an account paying 6% compounded annually. How much will he have on deposit after 7 years?

His payments form an ordinary annuity with $R = 22{,}000$, $n = 7$, and $i = .06$. The future value of this annuity (by the formula above) is

$$S = 22{,}000\left[\frac{(1.06)^7 - 1}{.06}\right].$$

From the table or a calculator, the number in brackets, $s_{\overline{7}|.06}$, is 8.39384, so that

$$S = 22,000(8.39384) = 184,664.48,$$

or $184,664.48. ■

■ EXAMPLE 2

Suppose $1000 is deposited at the end of each six-month period for 5 years in an account paying 10% compounded semiannually. Find the future value of the annuity.

Interest of $10\%/2 = 5\%$ is earned semiannually. In 5 years there are $5 \cdot 2 = 10$ semiannual periods. Find $s_{\overline{10}|.05}$ with the table or a calculator. By looking in the 5% column and down 10 periods, we find $s_{\overline{10}|.05} = 12.57789$, with a future value of

$$S = 1000(12,57789) = 12,577.89,$$

or $12,577.89. ■

The formulas developed so far have been for *ordinary annuities*—those with payments made at the *end* of each time period. These results can be modified slightly to apply to **annuities due**—annuities in which payments are made at the *beginning* of each time period. To find the future value of an annuity due, treat each payment as if it were made at the *end* of the *preceding* period. That is, use the table or a calculator to find $s_{\overline{n}|i}$ for *one additional period;* to compensate for this, subtract the amount of one payment.

■ EXAMPLE 3

Find the future value of an annuity due if payments of $500 are made at the beginning of each quarter for 7 years, in an account paying 12% compounded quarterly.

In 7 years, there are 28 quarterly periods. Look in row 29 $(28 + 1)$ of the table. Use the $12\%/4 = 3\%$ column of the table. You should find the number 45.21885. Multiply this number by 500, the amount of each payment.

$$500(45.21885) \approx 22,609.43$$

Subtract the amount of one payment from this result.

$$\$22,609.43 - \$500 = \$22,109.43$$

The account will contain a total of $22,109.43 after 7 years. ■

Just as the formula $A = P(1 + i)^n$ can be solved for any of its variables, the formula for the future value of an annuity can be used to find the values of variables other than S. In Example 4 below we are given S, the amount of money wanted at the end of the time period, and we need to find R, the amount of each payment.

EXAMPLE 4

Bettina Martinelli wants to buy an expensive video camera three years from now. She plans to deposit an equal amount at the end of each quarter for three years in order to accumulate enough money to pay for the camera. The camera costs $2400, and the bank pays 6% interest compounded quarterly. Find the amount of each of the 12 deposits she will make.

This example describes an ordinary annuity with $S = 2400$, $i = .015$ (because $6\%/4 = 1.5\%$), and $n = 3 \cdot 4 = 12$ periods. The unknown here is the amount of each payment, R. By the formula for the amount of an annuity,

$$2400 = R \cdot s_{\overline{12}|.015}.$$

From the table or a calculator,

$$2400 = R(13.04121)$$
$$R \approx 184.03,$$

or $184.03 (dividing both sides by 13.04121).

Sinking Fund   The annuity in Example 4 is a *sinking fund:* a fund set up to receive periodic payments. These periodic payments ($184.03 in the example) together with the interest earned by the payments, are designed to produce a given sum at some time in the future. As another example, a sinking fund might be set up to receive money that will be needed to pay off the principal on a loan at some time in the future.

EXAMPLE 5

The Stockdales are close to retirement. They agree to sell an antique urn to the local museum for $17,000. Their tax adviser suggests that they defer receipt of this money until they retire, 5 years in the future. (At that time, they might well be in a lower tax bracket.) Find the amount of each payment the museum must make into a sinking fund so that it will have the necessary $17,000 in 5 years. Assume that the museum can earn 8% compounded annually on its money. Also, assume that the payments are made annually.

These payments are the periodic payments into an ordinary annuity. The annuity will amount to $17,000 in 5 years at 8% compounded annually. Using the formula,

$$17,000 = R \cdot s_{\overline{5}|.08}$$

or
$$R = \frac{17,000}{s_{\overline{5}|.08}}$$

$$R = \frac{17,000}{5.86660} \quad \text{From the table}$$

$$R \approx 2897.76,$$

or $2897.76. If the museum deposits $2897.76 at the end of each year for 5 years in an account paying 8% compounded annually, it will have the total amount that it needs. This result is shown in the following sinking fund table. In these tables, the last payment may differ slightly from the others because of rounding errors.

Payment Number	Amount of Deposit	Interest Earned	Total in Account
1	$2897.76	$0	$2897.76
2	$2897.76	$231.82	$6027.34
3	$2897.76	$482.19	$9407.29
4	$2897.76	$752.58	$13,057.63
5	$2897.76	$1044.61	$17,000.00

10.4 EXERCISES

Find each of the following.

1. $s_{\overline{12}|.05}$

2. $s_{\overline{20}|.06}$

3. $s_{\overline{16}|.04}$

4. $s_{\overline{40}|.02}$

5. $s_{\overline{20}|.01}$

6. $s_{\overline{18}|.015}$

7. $s_{\overline{15}|.04}$

8. $s_{\overline{30}|.015}$

Find the future value of each of the following ordinary annuities, if interest is compounded annually.

9. $R = 100$, $i = .06$, $n = 10$

10. $R = 1000$, $i = .06$, $n = 12$

11. $R = 10,000$, $i = .05$, $n = 19$

12. $R = 100,000$, $i = .08$, $n = 23$

13. $R = 8500$, $i = .06$, $n = 30$

14. $R = 11,200$, $i = .08$, $n = 25$

15. $R = 46,000$, $i = .06$, $n = 32$

16. $R = 29,500$, $i = .05$, $n = 15$

Find the future value of each of the following ordinary annuities, if payments are made and interest is compounded as given.

17. $R = 9200$, 10% interest compounded semiannually for 7 years

18. $R = 3700$, 8% interest compounded semiannually for 11 years

19. $R = 800$, 6% interest compounded semiannually for 12 years

20. $R = 4600$, 8% interest compounded quarterly for 9 years

21. $R = 15,000$, 12% interest compounded quarterly for 6 years

22. $R = 42,000$, 10% interest compounded semiannually for 12 years

Find the future value of each annuity due. Assume that interest is compounded annually.

23. $R = 600$, $i = .06$, $n = 8$

24. $R = 1400$, $i = .08$, $n = 10$

25. $R = 20,000$, $i = .08$, $n = 6$

26. $R = 4000$, $i = .06$, $n = 11$

Find the future value of each annuity due.

27. Payments of $1000 made at the beginning of each year for 9 years at 8% compounded annually

28. $750 deposited at the beginning of each year for 15 years at 6% compounded annually

29. $100 deposited at the beginning of each quarter for 9 years at 12% compounded quarterly

30. $1500 deposited at the beginning of each semiannual period for 11 years at 6% compounded semiannually

Find the periodic payment that will amount to each of the given sums under the given conditions.

31. $S = \$10,000$, interest is 5% compounded annually, payments are made at the end of each year for 12 years

32. $S = \$100,000$, interest is 8% compounded semiannually, payments are made at the end of each semiannual period for 9 years

33. $S = \$50,000$, interest is 12% compounded quarterly, payments are made at the end of each quarter for 8 years

Find the amount of each payment to be made into a sinking fund so that enough will be present to pay off the indicated loans.

34. $8500 loan, money earns 8% compounded annually, 7 annual payments

35. $2000 loan, money earns 6% compounded annually, 5 annual payments

36. $75,000 loan, money earns 6% compounded semiannually for 4 1/2 years

37. $11,000 loan, money earns 10% compounded semiannually for 6 years

38. $25,000 loan, money earns 12% compounded quarterly for 3 1/2 years

39. $50,000 loan, money earns 8% compounded quarterly for 2 1/2 years

40. $9000 loan, money earns 12% compounded monthly for 2 1/2 years

41. $6000 loan, money earns 12% compounded monthly for 3 years

━━ APPLICATIONS

BUSINESS AND ECONOMICS

Savings Annuity **42.** Pat Saivitzky deposits $12,000 at the end of each year for 9 years in an account paying 8% interest compounded annually. Find the final amount she will have on deposit.

Savings Annuity **43.** Pat's brother-in-law works in a bank that pays 6% compounded annually. If she deposits her money in this bank instead of the one of Exercise 42, how much will she have in her account?

44. How much would Pat lose over 9 years by using her brother-in-law's bank? (See Exercises 42 and 43.)

Savings Annuity   45. Pam Parker deposits $2435 at the beginning of each semiannual period for 8 years in an account paying 6% compounded semiannually. She then leaves that money alone, with no further deposits, for an additional 5 years. Find the final amount on deposit after the entire 13-year period.

Savings Annuity   46. Chuck deposits $10,000 at the beginning of each year for 12 years in an account paying 5% compounded annually. He then puts the total amount on deposit in another account paying 6% compounded semiannually for another 9 years. Find the final amount on deposit after the entire 21-year period.

Sinking Fund   47. Ray Berkowitz needs $10,000 in 8 years. What amount can he deposit at the end of each quarter at 8% compounded quarterly so that he will have his $10,000?

Sinking Fund   48. Find Berkowitz's quarterly deposit (see Exercise 47) if the money is deposited at 6% compounded quarterly.

Sinking Fund   49. Barb Silverman wants to buy an $18,000 car in 6 years. How much money must she deposit at the end of each quarter in an account paying 12% compounded quarterly, so that she will have enough to pay for her car?

Sinking Fund   50. The owner of Harv's Meats knows that he must buy a new deboner machine in 4 years. The machine costs $12,000. In order to accumulate enough money to pay for the machine, Harv decides to deposit a sum of money at the end of each six months in an account paying 6% compounded semiannually. How much should each payment be?

Individual Retirement Accounts   *Suppose a 40-year-old person deposits $2000 per year in an Individual Retirement Account until age 65. Find the total in the account with the following assumptions of interest rates. (Assume semiannual compounding, with payment made at the end of each semiannual period.)*

51. 6%   **52.** 8%   **53.** 4%   **54.** 10%

FOR THE COMPUTER

Find the final amount of each of the annuities in Exercises 55 and 56.

55. $892.17 per month for 2 years in an account paying 7.5% interest compounded monthly

56. $2476.32 each quarter for 5.5 years at 7.81% interest compounded quarterly

Sinking Fund   57. Jill Streitsel sells some land in Nevada. She will be paid a lump sum of $60,000 in 7 years. Until then, the buyer pays 8% simple interest, quarterly.

(a) Find the amount of each quarterly interest payment.

(b) The buyer sets up a sinking fund so that enough money will be present to pay off the $60,000. The buyer wants to make semiannual payments into the sinking fund; the account pays 6% compounded semiannually. Find the amount of each payment into the fund.

(c) Prepare a table showing the amount in the sinking fund after each deposit.

Sinking Fund   58. Jeff Reschke bought a rare stamp for his collection. He agreed to pay a lump sum of $4000 after 5 years. Until then, he pays 6% simple interest semiannually.

(a) Find the amount of each semiannual interest payment.

(b) Reschke sets up a sinking fund so that enough money will be present to pay off the $4000. He wants to make annual payments into the fund. The account pays 8% compounded annually. Find the amount of each payment.

(c) Prepare a table showing the amount in the sinking fund after each deposit.

≡ 10.5 PRESENT VALUE OF AN ANNUITY; AMORTIZATION

As shown in the previous section, if deposits of R dollars are made at the end of each period for n periods, at a rate of interest i per period, then the account will contain

$$S = R \cdot s_{\overline{n}|i} = R\left[\frac{(1 + i)^n - 1}{i}\right]$$

dollars after n periods. Let us now find the *lump sum P* that can be deposited today at a rate of interest i per period to become the same S dollars in n periods.

First recall that P dollars deposited today will amount to $P(1 + i)^n$ dollars after n periods at a rate of interest i per period. This amount, $P(1 + i)^n$, is to be the same as S, the future value of the annuity. Substituting $P(1 + i)^n$ for S in the formula above gives

$$P(1 + i)^n = R\left[\frac{(1 + i)^n - 1}{i}\right].$$

To solve this equation for P, multiply both sides by $(1 + i)^{-n}$.

$$P = R(1 + i)^{-n}\left[\frac{(1 + i)^n - 1}{i}\right]$$

Use the distributive property; also recall that $(1 + i)^{-n}(1 + i)^n = 1$.

$$P = R\left[\frac{(1 + i)^{-n}(1 + i)^n - (1 + i)^{-n}}{i}\right]$$

$$P = R\left[\frac{1 - (1 + i)^{-n}}{i}\right]$$

The amount P is called the **present value of the annuity.** The quantity in brackets is abbreviated as $a_{\overline{n}|i}$, so

$$a_{\overline{n}|i} = \frac{1 - (1 + i)^{-n}}{i}.$$

Values of $a_{\overline{n}|i}$ are given in the present value table in the Appendix. The formula for the present value of an annuity is summarized below.

PRESENT VALUE OF AN ANNUITY

≡ The present value P of an annuity of n payments of R dollars each at the end of consecutive interest periods with interest compounded at a rate of interest i per period is

$$P = R\left[\frac{1 - (1 + i)^{-n}}{i}\right] \qquad \text{or} \qquad P = R \cdot a_{\overline{n}|i}.$$

▬ EXAMPLE 1

What lump sum deposited today at 8% interest compounded annually will yield the same total amount as payments of $1500 at the end of each year for 12 years, also at 8% interest compounded annually?

We want to find the present value of an annuity of $1500 per year for 12 years at 8% compounded annually. Using the table or a calculator, $a_{\overline{12}|.08} = 7.53608$, so

$$P = 1500(7.53608) \approx 11{,}304.12,$$

or $11,304.12. A lump sum deposit of $11,304.12 today at 8% compounded annually will yield the same total after 12 years as deposits of $1500 at the end of each year for 12 years at 8% compounded annually.

Let's check this result. The compound amount in 12 years of a deposit of $11,304.12 today at 8% compounded annually can be found by the formula $A = P(1 + i)^n$. From the compound interest table, $11,304.12 will produce a total of

$$(11{,}304.12)(2.51817) \approx 28{,}465.70,$$

or $28,465.70. On the other hand, from the table for $s_{\overline{n}|i}$, deposits of $1500 at the end of each year for 12 years, at 8% compounded annually, give

$$1500(18.97713) \approx 28{,}465.70,$$

or $28,465.70, the same amount found above.

In summary, there are two ways to have $28,465.70 in 12 years at 8% compounded annually—a single deposit of $11,304.12 today, or payments of $1500 at the end of each year for 12 years.   ▬

▬ EXAMPLE 2

Mr. Jones and Ms. Gonsalez are both graduates of the Forestvire Institute of Technology. They both agree to contribute to the endowment fund of FIT. Mr. Jones says that he will give $500 at the end of each year for 9 years. Ms. Gonsalez prefers to give a lump sum today. What lump sum can she give that will be equivalent to Mr. Jones' annual gifts, if the endowment fund earns 8% compounded annually?

Here $R = 500$, $n = 9$, and $i = .08$. The necessary number from the present value table or a calculator is $a_{\overline{9}|.08} = 6.24689$. Ms. Gonsalez must therefore donate a lump sum of

$$500(6.24689) \approx 3123.45,$$

or $3123.45, today.   ▬

The formula for the present value of an annuity can also be used if we know the present value, the interest rate, and number of payments, and we want to find the size of a payment of the annuity. The next example shows how to do this.

▰ EXAMPLE 3

A car costs $6000. After a down payment of $1000, the balance will be paid off in 36 monthly payments with interest of 12% per year, compounded monthly. Find the amount of each payment.

A single lump sum payment of $5000 today would pay off the loan. Thus, $5000 is the present value of an annuity of 36 monthly payments with interest of 12%/12 $= 1\%$ per month. We need to find R, the amount of each payment. Start with $P = R \cdot a_{\overline{n}|i}$; replace P with 5000, n with 36, and i with .01. From the table or a calculator, $a_{\overline{36}|.01} = 30.10751$, so

$$5000 = R(30.10751)$$
$$R \approx 166.07,$$

or $166.07. A monthly payment of $166.07 will be needed. ▰

Amortization A loan is **amortized** if both the principal and interest are paid by a sequence of equal periodic payments. In Example 3 above, a loan of $5000 at 12% interest compounded monthly could be amortized by paying $166.07 per month for 36 months.

▰ EXAMPLE 4

A speculator agrees to pay $15,000 for a parcel of land; this amount, with interest, will be paid over 4 years, with semiannual payments, at an interest rate of 12% compounded semiannually.

(a) Find the amount of each payment.

If the speculator were to pay $15,000 immediately, there would be no need for any payments at all, making $15,000 the present value of an annuity of R dollars, $2 \cdot 4 = 8$ periods, and $i = 12\%/2 = 6\% = .06$ per period. Using $P = R \cdot a_{\overline{n}|i}$ with $P = 15,000$ gives

$$15,000 = R \cdot a_{\overline{8}|.06}$$

or
$$R = \frac{15,000}{a_{\overline{8}|.06}}.$$

From the table or a calculator, $a_{\overline{8}|.06} = 6.20979$, and

$$R = \frac{15,000}{6.20979} \approx 2415.54,$$

or $2415.54. Each payment is $2415.54.

(b) Find the part of the first payment that is applied to reduction of the debt.

Interest is 12% per year, compounded semiannually, or 6% per semiannual period. During the first period, the entire $15,000 is owed. Interest on this amount for 6 months (1/2 year) is found by the formula for simple interest:

$$I = Prt = 15,000(.12)\left(\frac{1}{2}\right) = 900,$$

or $900. At the end of 6 months, the speculator makes a payment of $2415.54; since $900 of this represents interest, a total of

$$\$2415.54 - \$900 = \$1515.54$$

is applied to the reduction of the original debt.

(c) Find the balance due after 6 months.

The original balance due is $15,000. After 6 months, $1515.54 is applied to reduction of the debt. The debt owed after 6 months is

$$\$15,000 - \$1515.54 = \$13,484.46.$$

(d) How much interest is owed for the second 6-month period?

A total of $13,484.46 is owed for the second 6 months. Interest on it is

$$I = 13,484.46 \ (.12)\left(\frac{1}{2}\right) = 809.07,$$

or $809.07. A payment of $2415.54 is made at the end of this period; a total of

$$\$2415.54 - \$809.07 = \$1606.47$$

is applied to reduction of the debt.

(e) Prepare an amortization schedule for this loan.

Continuing the process discussed in parts (a)–(d) gives the amortization schedule shown below. As the schedule shows, the payment is always the same, except perhaps for a small adjustment in the final payment. Payment 0 is the original amount of the loan. ▬

Amortization Schedule

Payment Number	Amount of Payment	Interest for Period	Portion to Principal	Principal at End of Period
0	—	—	—	$15,000.00
1	$2415.54	$900.00	$1515.54	$13,484.46
2	$2415.54	$809.07	$1606.47	$11,877.99
3	$2415.54	$712.68	$1702.86	$10,175.13
4	$2415.54	$610.51	$1805.03	$8370.10
5	$2415.54	$502.21	$1913.33	$6456.77
6	$2415.54	$387.41	$2028.13	$4428.64
7	$2415.54	$265.72	$2149.82	$2278.82
8	$2415.55	$136.73	$2278.82	$0

▬ EXAMPLE 5

The Millers buy a house for $74,000, making a down payment of $16,000. Interest is charged at 10.25% per year for 30 years. Find the amount of each monthly payment to amortize the loan.

Here, the present value, P, is 58,000 (or 74,000 − 16,000). Also, $i = .1025/12 = .0085416667$, and $n = 12 \cdot 30 = 360$. We need to find R. From the formula for the present value of an annuity,

$$58,000 = R\left[\frac{1 - (1 + .0085416667)^{-360}}{.0085416667}\right].$$

Use a financial calculator or a calculator with a y^x key to get

$$58,000 = R\left[\frac{1 - .0467967624}{.0085416667}\right]$$

$$58,000 = R\left[\frac{.9532032376}{.0085416667}\right]$$

$$58,000 = R[111.5945249]$$

or

$$R = 519.74.$$

Monthly payments of $519.74 will be required to amortize the loan. ▬

▬ 10.5 EXERCISES

Find each of the following.

1. $a\,\overline{_{15}|}.06$

2. $a\,\overline{_{10}|}.03$

3. $a\,\overline{_{18}|}.04$

4. $a\,\overline{_{30}|}.01$

5. $a\,\overline{_{16}|}.01$

6. $a\,\overline{_{32}|}.02$

7. $a\,\overline{_{6}|}.015$

8. $a\,\overline{_{18}|}.015$

Find the present value of each ordinary annuity.

9. Payments of $1000 each year for 9 years at 8% compounded annually

10. Payments of $5000 each year for 11 years at 6% compounded annually

11. Payments of $890 each year for 16 years at 8% compounded annually

12. Payments of $1400 each year for 8 years at 8% compounded annually

13. Payments of $10,000 semiannually for 15 years at 10% compounded semiannually

14. Payments of $50,000 quarterly for 10 years at 8% compounded quarterly

15. Payments of $15,806 quarterly for 3 years at 10.8% compounded quarterly

16. Payments of $18,579 every six months for 8 years at 9.4% compounded semiannually

In Exercises 17–22, find the lump sum deposited today that will yield the same total amount as payments of $10,000 at the end of each year for 15 years, at each of the given interest rates.

17. 4% compounded annually

18. 5% compounded annually

19. 6% compounded annually

20. 8% compounded annually

21. 10% compounded annually

22. 7% compounded annually

Find the payment necessary to amortize each of the following loans.

23. $1000, 8% compounded annually, 9 annual payments

24. $2500, 8% compounded quarterly, 6 quarterly payments

25. $41,000, 10% compounded semiannually, 10 semiannual payments

26. $90,000, 8% compounded annually, 12 annual payments

27. $140,000, 12% compounded quarterly, 15 quarterly payments

28. $7400, 8% compounded semiannually, 18 semiannual payments

29. $5500, 12% compounded monthly, 24 monthly payments

30. $45,000, 12% compounded monthly, 36 monthly payments

Find the monthly house payment necessary to amortize each of the loans in Exercises 31–34. You will need a financial calculator or one with a y^x key.

31. $49,560 at 10.75% for 25 years   **32.** $70,892 at 10.11% for 30 years

33. $53,762 at 11.45% for 30 years   **34.** $96,511 at 9.57% for 25 years

35. What sum deposited today at 5% compounded annually for 8 years will provide the same amount as $1000 deposited at the end of each year for 8 years at 6% compounded annually?

36. What lump sum deposited today at 8% compounded quarterly for 10 years will yield the same final amount as deposits of $4000 at the end of each six-month period for 10 years at 6% compounded semiannually?

▤ APPLICATIONS

BUSINESS AND ECONOMICS

Annuity   **37.** In his will the late Mr. Hudspeth said that each child in his family could have an annuity of $2000 at the end of each year for 9 years, or the equivalent present value. If money can be deposited at 8% compounded annually, what is the present value of the annuity?

Annuity   **38.** In the "Million-Dollar Lottery," a winner is paid a million dollars at the rate of $50,000 per year for 20 years. Assume that these payments form an ordinary annuity, and that the lottery managers can invest money at 6% compounded annually. Find the lump sum that the management must put away to pay off the "million-dollar" winner.

Car Payments   **39.** Um Sherif buys a new car costing $6000. She agrees to make payments at the end of each monthly period for 4 years. If she pays 12% interest, compounded monthly, what is the amount of each payment?

40. Find the total amount of interest Sherif will pay. (See Exercise 39.)

Amortization   *In Exercises 41–44, prepare an amortization schedule for each loan.*

41. An insurance firm pays $4000 for a new printer for its computer. It amortizes the loan for the printer in 4 annual payments at 8% compounded annually.

42. Large semitrailer trucks cost $72,000 each. Ace Trucking buys such a truck and agrees to pay for it by a loan that will be amortized with 9 semiannual payments at 10% compounded semiannually.

43. One retailer charges $1048 for a correcting electric typewriter. A firm of tax accountants buys 8 of these machines. They make a down payment of $1200 and agree to amortize the balance with monthly payments at 12% compounded monthly for 4 years. Prepare an amortization schedule showing the first six payments.

44. When Denise Sullivan opened her law office, she bought $14,000 worth of law books and $7200 worth of office furniture. She paid $1200 down and agreed to amortize the balance with semiannual payments for 5 years, at 12% compounded semiannually.

Inheritance **45.** Mrs. Yaroslavsky has died, leaving $25,000 to her husband. He deposits the money at 6% compounded annually. He wants to make annual withdrawals from the account so that the money (principal and interest) is gone in exactly 8 years.

(a) Find the amount of each withdrawal.

(b) Find the amount of each withdrawal if the money must last 12 years.

Charitable Trust **46.** The trustees of a college have accepted a gift of $150,000. The donor has directed the trustees to deposit the money in an account paying 6% per year, compounded semiannually. The trustees may withdraw money at the end of each 6-month period; the money must last 5 years.

(a) Find the amount of each withdrawal.

(b) Find the amount of each withdrawal if the money must last 6 years.

FOR THE COMPUTER

Amortization *Prepare an amortization schedule for each of the following loans.*

47. A loan of $37,947.50 with interest at 8.5% compounded annually, to be paid with equal annual payments over 10 years

48. A loan of $4835.80 at 9.25% interest compounded semiannually, to be repaid in 5 years in equal semiannual payments

KEY WORDS		
10.1	simple interest	terms
	principal	general term
	rate	arithmetic sequence
	time	common difference
	future value	geometric sequence
	maturity value	common ratio
	present value	**10.4** annuity
	bank discount	ordinary annuity
	proceeds	term of an annuity
	effective rate	annuity due
10.2	compound interest	sinking fund
	compound amount	**10.5** amortization
10.3	sequence	

CHAPTER 10   REVIEW EXERCISES

Many of these exercises will require a calculator with a y^x key.

Find the simple interest in each of the following.

1. $15,903 at 8% for 8 months

2. $4902 at 9.5% for 11 months

3. $42,368 at 5.22% for 5 months

4. $3478 at 7.4% for 88 days (assume a 360-day year)

5. $2390 at 8.7% from May 3 to July 28 (assume a 365-day year)

6. $69,056.12 at 5.5% from September 13 to March 25 of the following year
   (assume a 365-day year)

Find the present value of each of the future amounts in Exercises 7–9. Assume 360 days in a year; use simple interest.

7. 25,000 for 10 months, money earns 5%

8. $459.57 for 7 months, money earns 4.5%

9. $80,612 for 128 days, money earns 6.77%

Find the proceeds in Exercises 10–12. Assume 360 days in a year.

10 $56,882, discount rate 9%, length of loan 5 months

11. $802.34, discount rate 8.6%, length of loan 11 months

12. $12,000, discount rate 7.09%, length of loan 145 days

Find the effective annual rate for the following bank discount rates. (Assume a one-year loan of $1000.)

13. 11%   14. 12.5%

Find the compound amount in each of the following.

15. $1000 at 8% compounded annually for 9 years

16. $2800 at 6% compounded annually for 10 years

17. $19,456.11 at 12% compounded semiannually for 7 years

18. $312.45 at 6% compounded semiannually for 16 years

19. $1900 at 6% compounded quarterly for 9 years

20. $57,809.34 at 12% compounded quarterly for 5 years

21. $2500 at 12% compounded monthly for 3 years

22. $11,702.55 at 12% compounded monthly for 4 years

Find the amount of interest earned by each deposit.

23. $3954 at 8% compounded annually for 12 years

24. $12,699.36 at 10% compounded semiannually for 7 years

25. $7801.72 at 12% compounded quarterly for 5 years

26. $48,121.91 at 12% compounded monthly for 2 years

27. $12,903.45 at 10.37% compounded quarterly for 29 quarters

28. $34,677.23 at 9.72% compounded monthly for 32 months

Find the present value of each of the following amounts if money can be invested at the given rate.

29. $5000 in 9 years, 8% compounded annually

30. $12,250 in 5 years, 6% compounded semiannually

31. $42,000 in 7 years, 12% compounded monthly

32. $17,650 in 4 years, 8% compounded quarterly

33. $1347.89 in 3.5 years, 6.77% compounded semiannually

34. $2388.90 in 44 months, 5.93% compounded monthly

Write the first five terms for each of the sequences in Exercises 35–38. Identify any that are arithmetic or geometric.

35. $a_n = -4n + 2$ **36.** $a_n = (-2)^n$ **37.** $a_n = \dfrac{n + 2}{n + 5}$ **38.** $a_n = (n - 7)(n + 5)$

39. Find a_{12} for the arithmetic sequence having $a_1 = 6$ and $d = 5$.

40. Find the sum of the first 20 terms for the arithmetic sequence having $a_1 = -6$ and $d = 8$.

41. Find a_4 for the geometric sequence with $a_1 = -3$ and $r = 2$.

42. Find the sum of the first 6 terms for the geometric sequence with $a_1 = 8000$ and $r = -1/2$.

Find the value of each of the following annuities.

43. $500 deposited at the end of each six-month period for 8 years; money earns 6% compounded semiannually

44. $1288 deposited at the end of each year for 14 years; money earns 8% compounded annually

45. $4000 deposited at the end of each quarter for 7 years; money earns 6% compounded quarterly

46. $233 deposited at the end of each month for 4 years; money earns 12% compounded monthly

47. $672 deposited at the beginning of each quarter for 7 years; money earns 8% compounded quarterly

48. $11,900 deposited at the beginning of each month for 13 months; money earns 12% compounded monthly

Find the amount of each payment to be made into a sinking fund so that enough money will be available to pay off each of the following loans.

49. $6500 loan, money earns 8% compounded annually, 6 annual payments

50. $57,000 loan, money earns 6% compounded semiannually for 8 1/2 years

51. $233,188 loan, money earns 9.7% compounded quarterly for 7 3/4 years

52. $1,056,788 loan, money earns 8.12% compounded monthly for 4 1/2 years

Find the present value of each ordinary annuity.

53. Payments of $850 annually for 4 years at 8% compounded annually

54. Payments of $1500 quarterly for 7 years at 8% compounded quarterly

55. Payments of $4210 semiannually for 8 years at 8.6% compounded semiannually

56. Payments of $877.34 monthly for 17 months at 9.4% compounded monthly

Find the amount of the payment necessary to amortize each of the following loans.

57. $80,000 loan, 8% compounded annually, 9 annual payments

58. $3200 loan, 8% compounded quarterly, 10 quarterly payments

59. $32,000 loan, 9.4% compounded quarterly, 17 quarterly payments

60. $51,607 loan, 13.6% compounded monthly, 32 monthly payments

Find the monthly house payments for the following mortgages.

61. $56,890 at 10.74% for 25 years

62. $77,110 at 11.45% for 30 years

Prepare amortization schedules for the following loans.

63. $5000 at 10% compounded semiannually for 3 years

64. $12,500 at 12% compounded quarterly for 2 years

▬ APPLICATIONS

BUSINESS AND ECONOMICS

Bank Discount   **65.** Tom Wilson owes $5800 to his mother. He has agreed to pay back the money in 10 months, at an interest rate of 10%. Three months before the loan is due, the mother discounts the loan at the bank. The bank charges a 13.45% discount rate. How much money does Tom's mother receive?

Bank Discount   **66.** Larry DiCenso needs $9812 to buy new equipment for his business. The bank charges a discount of 12%. Find the amount of DiCenso's loan if he borrows the money for 7 months.

Sinking Fund   **67.** In 4 years, Mr. Heeren must pay a pledge of $5000 to his church's building fund. What lump sum can he invest today, at 8% compounded semiannually, so that he will have enough to pay his pledge?

Sinking Fund   **68.** Joann Tong must make an alimony payment of $1500 in 15 months. What lump sum can she invest today, at 12% compounded monthly, so that she will have enough to make the payment?

Savings Annuity   **69.** Georgette Dahl deposits $491 at the end of each quarter for 9 years. If the account pays 9.4% compounded quarterly, find the final amount in the account.

Savings Annuity   **70.** J. Euclid deposits $1526.38 at the beginning of each 6-month period in an account paying 7.6% compounded semiannually. How much will be in the account after 5 years?

Loan Payments   **71.** Toni Tsakoupolis borrows $20,000 from the bank to help her expand her business. She agrees to repay the money in equal payments at the end of each year for 9 years. Interest is at 10.9% compounded annually. Find the amount of each payment.

Loan Payments   **72.** Ken Murrill wants to expand his pharmacy. To do this, he takes out a loan of $49,275 from the bank, and he agrees to repay the money at 12.4% compounded monthly over 48 months. Find the amount of each payment necessary to amortize this loan.

**EXTENDED
APPLICATION**

A NEW LOOK AT ATHLETES' CONTRACTS*

The following article appeared in The Sacramento Bee *several years ago. Although the USFL is long gone and there have been many other changes in the sports world, the article still presents a valid and interesting way to compare sports contracts.*

There has been a lot of controversy lately about big-buck contracts with athletes and how this affects the nature of the sports business.

Indeed, the recent headlines of a $40-million contract with Steve Young from Brigham Young University to play football for the USFL has opened Pandora's box. Club owners in all sports are wondering when all of this is going to stop.

To put the issue in its proper perspective, however, it seems to me that the $40-million contract is misleading. Just mentioning dollars misses several points. For one, the USFL has got to be around long enough to pay off. Two, the present value of million-dollar payments that stretch 43 years into the future is reduced considerably.

In order to assess the effective value of these kinds of contracts, two important questions must be answered. First, what is the risk of the company to pay off on the periodic principal payments; and second, what is the present value of the return when parlayed into future cash benefits?

Recently, I conducted a survey among a number of sports-minded educators and students here on the campus. The survey was similar to the rating survey conducted by Wall Street rating firms when they evaluate risk of a corporation.

Specifically, the question was: "On a scale of 1 to 10, give me your estimate of the team's ability to continue in business and to meet the commitments of its contracts."

I was surprised to find that the baseball teams were rated with the lowest risk; the National Football League teams next; then the National Basketball Association teams; and then the United States Football League teams coming in last.

The cash flows which these athletes receive in the future have a present value today in total dollars based upon about an 8 percent discount factor each year for expected inflation. By determining the expected present value, the average per annum return of Athlete A who receives megabucks in the future can be compared with Athlete B who receives megabucks not so far into the future.

When risk of paying these megabucks is considered, we arrive at a completely different picture than just the dollar signs on the contract portray.

The following chart of the highly paid athletes, recently taken from a newspaper article, shows the athletes' names in sequence according to the best deal when risk and expected yearly return are assessed. A lesser coefficient of variation provides a statistical ratio of lesser risk to the per annum return for each athlete for each dollar of present value.

Notice that these risk assumptions show George Foster of the New York Mets with the best deal. Dave Winfield and Gary Carter are close behind. Steve Young is last, yet it is claimed that he received the most attractive sports contract ever signed.

---

*From "A New Look at Athletes' Contracts" by Richard F. Kaufman in *The Sacramento Bee,* April 30, 1984. Reprinted by permission of the author.

Some Highly Paid Athletes

	Present Value		Per Annum	
	of Contract (In Millions)	Expected Value	Risk Assessment	Coefficient of Variation
George Foster NL (Mets) $10 million, 5 years	$ 7.985	$1.597	2.3	1.44
Dave Winfield AL (Yankees) $21 million, 10 years	$14.091	$1.409	2.6	1.85
Gary Carter NL (Expos) $15 million, 8 years	$10.775	$1.347	3.0	2.23
Moses Malone NBA (76ers) $13.2 million, 6 years	$10.170	$1.695	5.4	3.19
Larry Bird NBA (Celtics) $15 million, 7 years	$11.157	$1.594	5.2	3.26
Wayne Gretzky NHL (Edmonton Oilers) $21 million, 21 years	$10.170	$ .477	5.8	12.16
Magic Johnson NBA (Lakers) $25 million, 25 years	$10.675	$ .427	5.6	13.12
Steve Young USFL (Express) $40 million, 43 years	$11.267	$ .262	7.2	27.48

It is true that risk is only our perception of the uncertainties of future events, and many times our judgment is wrong. But let's face it. It is the only game in town, unless you are gifted with extrasensory perception or have a crystal ball. In that case, I would appreciate your contacting me because I have work for you.

Anytime we purchase a bond or a like security, we make judgments about the future ability of the firm or municipality to meet its interest and principal payments. We employ expert rating agencies like Moody's or Standard & Poor's to professionally rate these securities. Indeed, the market price of the bond depends upon the judgments of these agencies.

It would seem to me that the time is long overdue for a similar rating system to be implemented to determine the risk of sports organizations—if for no other reason than to help the athlete decide where he would like to hang his hat. But more importantly, to provide a truer financial picture to the fans and broadcasting companies who are asked to pick up the tab. What do you think?

EXTENDED
APPLICATION

PRESENT VALUE*

The Southern Pacific Railroad, with lines running from Oregon to Louisiana, is one of the country's most profitable railroads. The railroad has vast landholdings (granted by the government in the last half of the nineteenth century) and is diversified into trucking, pipelines, and data transmission.

The railroad was recently faced with a decision on the fate of an old bridge that crosses the Kings River in central California. The bridge is on a minor line that carries very little traffic. Just north of the Southern Pacific bridge is a bridge owned by the Santa Fe Railroad; it too carries little traffic. The Southern Pacific had two alternatives: it could replace its own bridge or it could negotiate with the Santa Fe for the rights to run trains over its bridge. In the second alternative a yearly fee would be paid to the Santa Fe, and new connecting tracks would be built. The situation is shown in Figure 4.

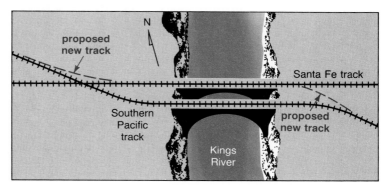

FIGURE 4

To find the better of these two alternatives, the railroad used the following approach.

1. Calculate estimated expenses for each alternative.

2. Calculate annual cash flows in after-tax dollars. At a 48% corporate tax rate, $1 of expenses actually costs the railroad $.52, and $1 of revenue can bring a maximum of $.52 in profit. Cash flow for a given year is found by the following formula.

$$\text{Cash Flow} = -.52 \text{ (Operating and Maintenance Expenses)}$$
$$+ .52 \text{ (Savings and Revenue)} + .48 \text{ (Depreciation)}$$

3. Calculate the net present values of all cash flows for future years. The formula used is

$$\text{Net Present Value} = \Sigma(\text{Cash Flow in Year } i)(1 + k)^{1-i},$$

where i is a particular year in the life of the project and k is the assumed annual rate of interest. (Recall that Σ indicates a sum.) The net present value is found for interest rates from 0% to 20%.

4. The interest rate that leads to a net present value of $0 is called the **rate of return** on the project.

---

*Based on information supplied by The Southern Pacific Transportation Company, San Francisco.

Let us now see how these steps worked out in practice.

Alternative 1: Operate Over the Santa Fe bridge. First, estimated expenses were calculated.

1976	Work done by Southern Pacific on Santa Fe track	$27,000
1976	Work by Southern Pacific on its own track	11,600
1976	Undepreciated portion of cost of current bridge	97,410
1976	Salvage value of bridge	12,690
1977	Annual maintenance of Santa Fe track	16,717
1977	Annual rental fee to Santa Fe	7,382

From these figures and others not given here, annual cash flows and net present values were calculated. The following table was then prepared.

Interest Rate, %	Net Present Value, $
0	85,731
4	67,566
8	53,332
12	42,008
16	32,875
20	25,414

Although the table does not show a net present value of $0, the interest rate that leads to that value is 44%. This is the rate of return for this alternative.

Alternative 2: Build a new bridge. Again, estimated expenses were calculated.

1976	Annual maintenance	$2870
1976	Annual inspection	$120
1976	Repair trestle	$17,920
1977	Install bridge	$189,943
1977	Install walks and handrails	$15,060
1978	Repaint steel (every 10 years)	$10,000
1978	Repair trusses (every 10 years, increases by $200)	$2000
1981	Replace ties	$31,000
1988	Repair concrete (every 10 years)	$400
2021	Replace ties	$31,000

After cash flows and net present values were calculated, the following table was prepared.

Interest Rate, %	Net Present Value, $
0	399,577
4	96,784
7.8	0
8	−3,151
12	−43,688
16	−62,615
20	−72,126

In this alternative the net present value is $0 at 7.8%, the rate of return.

Based on this analysis, the first alternative, renting from the Sante Fe, is clearly preferable.

EXERCISES

Find the cash flow in each of the following years. Use the formula given above.

1. Alternative 1, year 1977, operating expenses $6228, maintenance expenses $2976, savings $26,251, depreciation $10,778

2. Alternative 1, year 1984, same as Exercise 1, but depreciation is only $1347

3. Alternative 2, year 1976, maintenance $2870, operating expenses $6386, savings $26,251, no depreciation

4. Alternative 2, year 1980, operating expenses $6228, maintenance expenses $2976, savings $10,618, depreciation $6736

11

Digraphs and Networks

11.1   Graphs and Digraphs

11.2   Dominance Digraphs

11.3   Communication Digraphs

11.4   Networks

Review Exercises

One fairly new area of mathematics has found increasing application in social sciences and management. A brief introduction to this branch of mathematics, called **graph theory,** is given in this chapter. The use of the word *graph* in this chapter has no relationship to the word *graph* used earlier, as in "the graph of the equation $y = 2x + 5$." The alternate definition of a graph for this chapter is given in the first section.

11.1 GRAPHS AND DIGRAPHS

In each of the past few presidential election years, one party or the other has had a large number of candidates competing in the presidential primaries held in the various states. Not all the candidates compete in each state. For example, candidates A and B might be the only candidates in one state primary, while another state race might see candidates A, C, and D on the ballot.

Suppose that in one year six candidates, A, B, C, D, E, and F, compete for the presidential nomination. Suppose also that after a few state primaries, the situation is as shown in the following table.

Candidate	Opponents
A	B, D, and F
B	A and D
C	No one
D	A, B, E, and F
E	D
F	A and D

The information in this table can also be given in a diagram, as in Figure 1, where each candidate is represented by a point, and each line connecting a pair of points represents a primary contest between the two candidates.

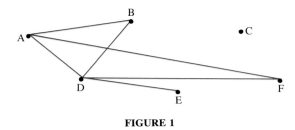

FIGURE 1

The diagram in Figure 1 is called a *graph*; each of the points A, B, C, D, E, and F is a *vertex,* and the lines AB, AD, BD, and so on, are called *edges*. While the edges in Figure 1 are straight lines, this is not at all necessary by definition. (Notice that the point where DB and AF cross is *not* a vertex.)

GRAPH

A **graph** is made up of a finite number of **vertices** A, B, C, . . ., N, together with a finite number of **edges** AB, AC, BC, . . ., each joining a pair of vertices.

As shown in Figure 1, not all vertices must have edges; there are no edges ending at vertex C, for example.

In our example of the presidential candidates, it is certainly possible that candidates A and B might run against each other in more than one state primary, so that more than one edge could connect vertices A and B.

▬ EXAMPLE 1

Identify all vertices and edges in Figure 2.

(a) The graph shown in Figure 2(a) has four vertices, A, B, C, and D, with edges AB, BC, CD, and DB.

(b) The four vertices in Figure 2(b) are M, N, P, and Q; the edges are MQ, QP, PN and NM. ▬

It could be very useful to modify the graph of Figure 1 so that the *winner* of each election were indicated. Suppose that

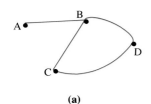

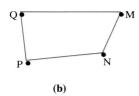

(a)

(b)

FIGURE 2

A   defeated B and D, but lost to F;

B   defeated D and F, but lost to A;

C   did not run;

D   defeated E, but lost to A, B, and F;

E   lost to D;

F   defeated A and D, but lost to B.

This information can be added to Figure 1 by making each edge into a **directed edge,** an edge with an arrowhead. The arrowhead points to the loser of each election. Figure 3 shows the modified graph.

The graph in Figure 3, with directed edges, is a *directed graph,* or *digraph.*

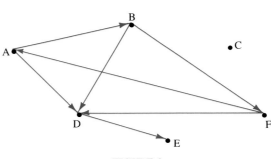

FIGURE 3

DIGRAPH

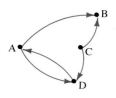

(a)

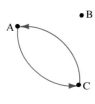

(b)

FIGURE 4

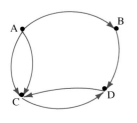

(a)

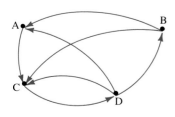

(b)

FIGURE 5

> A **directed graph,** or **digraph,** is made up of a finite number of vertices, A, B, C, . . ., N, together with a finite number of directed edges, AB, AC, BC, . . ., each joining a pair of vertices.

In a graph, the edge AB is the same as the edge BA. In a digraph, however, the order AB is used to show that the arrowhead points from A to B, so that AB is not the same as BA.

▬ EXAMPLE 2
Identify all vertices and edges in each digraph in Figure 4.

(a) The vertices in Figure 4(a) are A, B, C, and D. The directed edges are AD, DA, AB, CB, and CD.

(b) The digraph in Figure 4(b) has vertices A, B, and C, with directed edges AC and CA. ▬

Matrix Representation of Digraphs Recall the relationship between transition matrices and transition diagrams in Chapter 8. As shown there, a matrix is a very useful way to represent the information in a graph or digraph. After a matrix is obtained from a digraph, the results of Chapter 2 can be used to obtain further information. A square matrix is obtained from a digraph by placing in row A and column B the number of directed edges from A to B. By the definition of a digraph, there can be only 0 entries in the row 1 column 1 position, the row 2 column 2 position, and so on.

▬ EXAMPLE 3
Write a matrix corresponding to each digraph in Figure 5.

(a) In Figure 5(a), there are two directed edges from A to C, so the number 2 is placed in row 1, column 3. Place 1 in row 1, column 2, because of the one directed edge from A to B; put 0 in row 2, column 1. Complete the matrix as follows.

$$
\begin{array}{c}
 \\
A \\
B \\
C \\
D
\end{array}
\begin{array}{cccc}
A & B & C & D \\
\end{array}
\left[
\begin{array}{cccc}
0 & 1 & 2 & 0 \\
0 & 0 & 0 & 1 \\
0 & 0 & 0 & 1 \\
0 & 0 & 1 & 0
\end{array}
\right]
$$

(b) The matrix for the digraph in Figure 5(b) is as follows.

$$
\begin{array}{c}
 \\
A \\
B \\
C \\
D
\end{array}
\begin{array}{cccc}
A & B & C & D \\
\end{array}
\left[
\begin{array}{cccc}
0 & 0 & 1 & 0 \\
1 & 0 & 1 & 0 \\
0 & 0 & 0 & 1 \\
1 & 1 & 1 & 0
\end{array}
\right] \quad ▬
$$

Note that the matrix in Example 3(b) contains only 0 and 1 as entries. Such a matrix is called an **adjacency matrix.**

11.1 EXERCISES

In Exercises 1–12, write a matrix for each digraph. Identify any adjacency matrices.

1.

2.

3.

4.
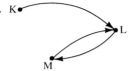

5.
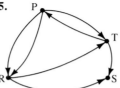

6.
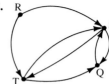

7.
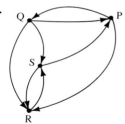

8.
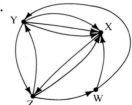

9.
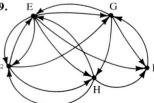

10.
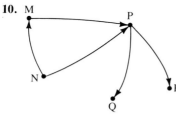

11.
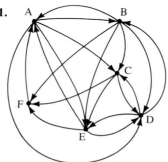

12.
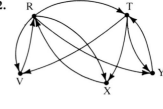

Sketch a digraph for each of the following matrices.

13. $\begin{bmatrix} 0 & 1 \\ 2 & 0 \end{bmatrix}$

14. $\begin{bmatrix} 0 & 1 \\ 1 & 0 \end{bmatrix}$

15. $\begin{bmatrix} 0 & 1 & 1 \\ 0 & 0 & 1 \\ 0 & 0 & 0 \end{bmatrix}$

16. $\begin{bmatrix} 0 & 0 & 1 \\ 1 & 0 & 1 \\ 1 & 1 & 0 \end{bmatrix}$

17. $\begin{bmatrix} 0 & 2 & 0 \\ 2 & 0 & 1 \\ 1 & 2 & 0 \end{bmatrix}$

18. $\begin{bmatrix} 0 & 1 & 2 \\ 0 & 0 & 1 \\ 2 & 0 & 0 \end{bmatrix}$ **19.** $\begin{bmatrix} 0 & 0 & 1 & 1 \\ 1 & 0 & 0 & 1 \\ 0 & 1 & 0 & 1 \\ 0 & 1 & 1 & 0 \end{bmatrix}$ **20.** $\begin{bmatrix} 0 & 1 & 0 & 0 \\ 0 & 0 & 1 & 1 \\ 1 & 1 & 0 & 0 \\ 1 & 0 & 1 & 0 \end{bmatrix}$ **21.** $\begin{bmatrix} 0 & 2 & 0 & 1 \\ 0 & 0 & 0 & 2 \\ 2 & 1 & 0 & 1 \\ 2 & 0 & 0 & 0 \end{bmatrix}$ **22.** $\begin{bmatrix} 0 & 1 & 1 & 2 \\ 0 & 0 & 1 & 0 \\ 2 & 0 & 0 & 1 \\ 0 & 1 & 0 & 0 \end{bmatrix}$

APPLICATIONS

SOCIAL SCIENCES

Draw a digraph and write a matrix for each of the following situations.

Interpersonal Relationships

23. Among the four people Al, Bill, Cynthia, and Donna, Al can't stand Bill or Cynthia, but likes Donna; Bill can't stand Al, but likes the other two; Cynthia doesn't like Bill or Donna, but likes Al, and Donna doesn't like anyone. No one dislikes himself or herself.

Election Results

24. In a series of elections, A defeated B and C but lost to D, B defeated D but lost to A and C, C defeated D and B but lost to A, and D defeated A but lost to B and C.

11.2 DOMINANCE DIGRAPHS

The study of *dominance* is an important aspect of sociology and political science. In the study of **dominance,** it is assumed that given a pair A and B (A and B might represent people, or teams, for example) then either A dominates B, or B dominates A, but not both. Dominance would be of interest when studying gang violence or attending a tournament in which each team in a league must play every other team exactly once, with no ties allowed.

As an example, the following chart shows the results of a tournament among five teams A, B, C, D, and E, where, for example, AB indicates a game between teams A and B.

Game	AB	AC	AD	AE	BC	BD	BE	CD	CE	DE
Winner	A	C	D	A	C	B	E	D	E	D

A digraph for this tournament is shown in Figure 6.

Notice that there is no transitive property for dominance. (Recall the transitivity property for "less than," for example: if $a < b$ and $b < c$, then $a < c$.) As shown previously, A dominates (won the game with) B, and B dominates D, but D dominates A.

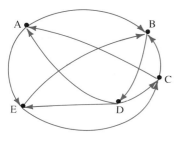

FIGURE 6

The matrix corresponding to the digraph in Figure 6 is as follows.

$$
\begin{array}{c c} & \begin{array}{c c c c c} A & B & C & D & E \end{array} \\ \begin{array}{c} A \\ B \\ C \\ D \\ E \end{array} & \left[\begin{array}{c c c c c} 0 & 1 & 0 & 0 & 1 \\ 0 & 0 & 0 & 1 & 0 \\ 1 & 1 & 0 & 0 & 0 \\ 1 & 0 & 1 & 0 & 1 \\ 0 & 1 & 1 & 0 & 0 \end{array} \right] \end{array}
$$

As mentioned at the end of Section 11.1, this matrix is an adjacency matrix. A matrix obtained from a dominance situation will always be an adjacency matrix, since either A dominates B, or B dominates A, but not both. Also, if the entry in row i and column j is 0 (where $i \neq j$), then the entry in row j and column i must be 1; if the entry in row i and column j is 1 (where $i \neq j$), then the entry that is in row j and column i must be 0. Such a matrix is called **asymmetric.**

> A digraph obtained from a dominance situation leads to a matrix that is an asymmetric adjacency matrix.

EXAMPLE 1

Each of the following matrices comes from a dominance situation. Complete the matrix.

(a) $\begin{bmatrix} 0 & \\ 1 & 0 \end{bmatrix}$

Because the matrix obtained from a dominance situation must be asymmetric, the 1 in row 2, column 1 leads to a 0 in row 1, column 2. Complete the matrix as follows.

$$
\begin{bmatrix} 0 & 0 \\ 1 & 0 \end{bmatrix}
$$

(b) $\begin{bmatrix} 0 & 1 & 1 \\ & 0 & 0 \\ & & 0 \end{bmatrix}$

The 1 in row 1, column 3 leads to 0 in row 3, column 1, and the 0 in row 2, column 3 leads to a 1 in row 3, column 2. Complete the matrix as follows.

$$
\begin{bmatrix} 0 & 1 & 1 \\ 0 & 0 & 0 \\ 0 & 1 & 0 \end{bmatrix}
$$

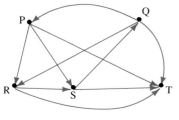

FIGURE 7

EXAMPLE 2

Write a matrix for the dominance situation in Figure 7. List all dominances.

The directed edge from Q to P shows that Q dominates P; also, P dominates R, and Q dominates R. Entering 1 for such dominances and 0 for a lack of dominance leads to the following matrix.

$$
\begin{array}{c}
 \\
P \\ Q \\ R \\ S \\ T
\end{array}
\begin{array}{c}
\begin{array}{ccccc} P & Q & R & S & T \end{array} \\
\left[\begin{array}{ccccc}
0 & 0 & 1 & 1 & 1 \\
1 & 0 & 1 & 0 & 1 \\
0 & 0 & 0 & 1 & 1 \\
0 & 1 & 0 & 0 & 1 \\
0 & 0 & 0 & 0 & 0
\end{array}\right]
\end{array}
$$

The matrix shows that P dominates R, S, and T; Q dominates P, R, and T; R dominates S and T; S dominates Q and T; and T dominates no one.

Since P dominates R, S, and T directly, as seen in the first row, P is said to have three **one-stage dominances.** Adding the numbers in the second row of the matrix shows that Q has three one-stage dominances; from the third and fourth rows, R and S have two each; from the fifth row, T has none. ■

In Example 2, Q dominated R and R dominated S, but Q did not dominate S. However, Q does have an indirect dominance over S, through R. Because of this, Q is said to have a **two-stage dominance** over S. Also, S has two-stage dominance over P.

Although we shall not prove it, the number of two-stage dominances can be found by *squaring* the adjacency matrix. If we use M for the matrix above, then

$$
M^2 =
\begin{bmatrix}
0 & 0 & 1 & 1 & 1 \\
1 & 0 & 1 & 0 & 1 \\
0 & 0 & 0 & 1 & 1 \\
0 & 1 & 0 & 0 & 1 \\
0 & 0 & 0 & 0 & 0
\end{bmatrix}
\begin{bmatrix}
0 & 0 & 1 & 1 & 1 \\
1 & 0 & 1 & 0 & 1 \\
0 & 0 & 0 & 1 & 1 \\
0 & 1 & 0 & 0 & 1 \\
0 & 0 & 0 & 0 & 0
\end{bmatrix}
=
\begin{array}{c}
\begin{array}{ccccc} P & Q & R & S & T \end{array} \\
\begin{bmatrix}
0 & 1 & 0 & 1 & 2 \\
0 & 0 & 1 & 2 & 2 \\
0 & 1 & 0 & 0 & 1 \\
1 & 0 & 1 & 0 & 1 \\
0 & 0 & 0 & 0 & 0
\end{bmatrix}
\begin{array}{c} P \\ Q \\ R \\ S \\ T \end{array}
\end{array}
$$

The 1 in row 1, column 2 of this result shows that P has two-stage dominance over Q (through S), while P has two-stage dominance over T in *two* ways (through R or through S). Adding the numbers in each row of this matrix shows that P has a total of $0 + 1 + 0 + 1 + 2 = 4$ two-stage dominances, while Q has 5, R has 2, S has 3, and T has none. (Notice that matrix M^2 is not an adjacency matrix.)

Adding matrices M and M^2 gives the total number of one-stage or two-stage dominances.

$$
M + M^2 =
\begin{bmatrix}
0 & 0 & 1 & 1 & 1 \\
1 & 0 & 1 & 0 & 1 \\
0 & 0 & 0 & 1 & 1 \\
0 & 1 & 0 & 0 & 1 \\
0 & 0 & 0 & 0 & 0
\end{bmatrix}
+
\begin{bmatrix}
0 & 1 & 0 & 1 & 2 \\
0 & 0 & 1 & 2 & 2 \\
0 & 1 & 0 & 0 & 1 \\
1 & 0 & 1 & 0 & 1 \\
0 & 0 & 0 & 0 & 0
\end{bmatrix}
=
\begin{array}{c}
\begin{array}{ccccc} P & Q & R & S & T \end{array} \\
\begin{bmatrix}
0 & 1 & 1 & 2 & 3 \\
1 & 0 & 2 & 2 & 3 \\
0 & 1 & 0 & 1 & 2 \\
1 & 1 & 1 & 0 & 2 \\
0 & 0 & 0 & 0 & 0
\end{bmatrix}
\begin{array}{c} P \\ Q \\ R \\ S \\ T \end{array}
\end{array}
$$

The entry 3 in the upper right-hand corner of the result shows that P has a total of 3 one-stage or two-stage dominances of T. Also, Q has a total of 2 one-stage or two-stage dominances over both R and S.

The sum of the entries in the P-column of matrix $M + M^2$ gives the number of ways that P is dominated. For example, from the first column, P is dominated in $0 + 1 + 0 + 1 + 0 = 2$ ways, while Q is dominated in 3 ways, R in 4 ways, S in 5 ways, and T in 10 ways. (This dominance is assumed to be in one stage or two stages.)

This process could be continued; finding M^3 would give the number of **three-stage dominances,** M^4 would give the number of **four-stage dominances,** and so on. A generalization of this idea follows.

Let M be an adjacency matrix and let k be a positive integer. Then the entry a_{ij} of M^k gives the number of k-stage dominances of i over j. Also, the sum of the entries in row i of the matrix

$$M + M^2 + M^3 + \cdots + M^k$$

is the total number of ways that i is dominant in one, two, . . ., k stages.

■ EXAMPLE 3

A league is made up of six teams, A, B, C, D, E, and F. In a recent tournament, the teams played each other exactly once, with results as shown in the digraph in Figure 8. (No ties were allowed.)

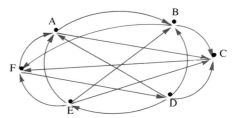

FIGURE 8

The adjacency matrix for the tournament follows.

$$
M = \begin{array}{c} \\ A \\ B \\ C \\ D \\ E \\ F \end{array}
\begin{array}{c}
\begin{array}{cccccc} A & B & C & D & E & F \end{array} \\
\left[\begin{array}{cccccc}
0 & 1 & 1 & 0 & 0 & 0 \\
0 & 0 & 1 & 0 & 0 & 1 \\
0 & 0 & 0 & 0 & 0 & 1 \\
1 & 1 & 1 & 0 & 1 & 0 \\
1 & 1 & 1 & 0 & 0 & 1 \\
1 & 0 & 0 & 1 & 0 & 0
\end{array} \right]
\end{array}
$$

It is not possible to use this matrix to decide on a winner, since both teams D and E won four games, while all other teams won fewer games. To help choose a winner, we could find M^2 to get two-stage wins for each team. Find M^2 as follows.

$$M^2 = \begin{bmatrix} 0 & 1 & 1 & 0 & 0 & 0 \\ 0 & 0 & 1 & 0 & 0 & 1 \\ 0 & 0 & 0 & 0 & 0 & 1 \\ 1 & 1 & 1 & 0 & 1 & 0 \\ 1 & 1 & 1 & 0 & 0 & 1 \\ 1 & 0 & 0 & 1 & 0 & 0 \end{bmatrix} \begin{bmatrix} 0 & 1 & 1 & 0 & 0 & 0 \\ 0 & 0 & 1 & 0 & 0 & 1 \\ 0 & 0 & 0 & 0 & 0 & 1 \\ 1 & 1 & 1 & 0 & 1 & 0 \\ 1 & 1 & 1 & 0 & 0 & 1 \\ 1 & 0 & 0 & 1 & 0 & 0 \end{bmatrix} = \begin{bmatrix} 0 & 0 & 1 & 0 & 0 & 2 \\ 1 & 0 & 0 & 1 & 0 & 1 \\ 1 & 0 & 0 & 1 & 0 & 0 \\ 1 & 2 & 3 & 0 & 0 & 3 \\ 1 & 1 & 2 & 1 & 0 & 2 \\ 1 & 2 & 2 & 0 & 1 & 0 \end{bmatrix}$$

The sum $M + M^2$ gives the total number of one-stage and two-stage wins for each team.

$$M + M^2 = \begin{bmatrix} 0 & 1 & 1 & 0 & 0 & 0 \\ 0 & 0 & 1 & 0 & 0 & 1 \\ 0 & 0 & 0 & 0 & 0 & 1 \\ 1 & 1 & 1 & 0 & 1 & 0 \\ 1 & 1 & 1 & 0 & 0 & 1 \\ 1 & 0 & 0 & 1 & 0 & 0 \end{bmatrix} + \begin{bmatrix} 0 & 0 & 1 & 0 & 0 & 2 \\ 1 & 0 & 0 & 1 & 0 & 1 \\ 1 & 0 & 0 & 1 & 0 & 0 \\ 1 & 2 & 3 & 0 & 0 & 3 \\ 1 & 1 & 2 & 1 & 0 & 2 \\ 1 & 2 & 2 & 0 & 1 & 0 \end{bmatrix} = \begin{array}{c} \\ \\ \\ \\ \\ \\ \end{array} \begin{array}{cccccc} A & B & C & D & E & F \\ \end{array}$$

$$\begin{bmatrix} 0 & 1 & 2 & 0 & 0 & 2 \\ 1 & 0 & 1 & 1 & 0 & 2 \\ 1 & 0 & 0 & 1 & 0 & 1 \\ 2 & 3 & 4 & 0 & 1 & 3 \\ 2 & 2 & 3 & 1 & 0 & 3 \\ 2 & 2 & 2 & 1 & 1 & 0 \end{bmatrix} \begin{array}{c} A \\ B \\ C \\ D \\ E \\ F \end{array}$$

The results shows that team D had a total of $2 + 3 + 4 + 0 + 1 + 3 = 13$ one- or two-stage wins, while all other teams had fewer wins. Based on this result, team D could be declared the winner of the tournament. ▬

11.2 EXERCISES

For each matrix in Exercises 1–10 that is the matrix of a dominance digraph, find the matrix representing two-stage dominance. Assume the matrices represent the result of 2-, 3-, 4-, or 5-person games. Find the total number of people dominated by each person in one or two stages.

1.

	A	B
A	0	1
B	0	0

2.

	A	B
A	0	0
B	1	0

3.

	P	Q	R
P	0	1	0
Q	0	0	1
R	1	0	0

4.

	Z	W	T
Z	0	0	0
W	1	0	0
T	1	1	0

5.

	X	Y	Z
X	0	0	1
Y	0	0	1
Z	0	0	0

6.

	M	N	P
M	0	1	0
N	0	0	1
P	1	1	0

7.

	R	S	T	V
R	0	1	0	1
S	0	0	1	0
T	1	0	0	0
V	0	1	1	0

8.

	P	Q	R	S
P	0	0	1	1
Q	1	0	1	1
R	0	0	0	1
S	0	0	0	0

9.

	A	B	C	D	E
A	0	1	1	0	0
B	0	0	1	0	1
C	0	0	0	1	1
D	1	1	0	0	0
E	1	0	0	1	0

10.

	X	Y	Z	W	T
X	0	0	1	0	1
Y	1	0	0	0	0
Z	0	1	0	1	0
W	1	1	0	0	0
T	0	1	1	1	0

For each digraph in Exercises 11–14, write the corresponding matrix. Find the number of two-stage dominances for A, B, C, *and* D. *List all two-stage dominances.*

11.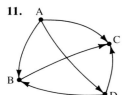

12.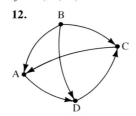

13.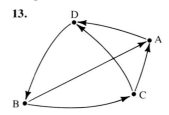

14.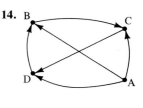

▆ APPLICATIONS

SOCIAL SCIENCES

Determining a Winner   *Decide on the winner of the tournaments in Exercises 15 and 16, using one- or two-stage dominances. Here* AB *represents a game between teams* A *and* B.

15.

Game	AB	AC	AD	AE	BC	BD	BE	CD	CE	DE
Winner	A	C	A	A	C	B	B	D	C	D

16.

Game	AB	AC	AD	AE	BC	BD	BE	CD	CE	DE
Winner	B	C	A	A	B	B	E	C	E	E

Preference Tests   *Another application of dominance comes from taste- or color-preference tests. For example, a consumer is shown several different colors for a new car, two at a time. The consumer then indicates a preference for one in each pair, for all possible pairs of colors. A "favorite" color can then be selected, using the methods above. In Exercises 17–20, find one- and two-stage dominances and then choose the favorite.*

17. The colors red, orange, yellow, and green are under consideration for a new cereal. A consumer is shown samples of the cereal in each color, two at a time. The results are as follows.

Pairs of Colors	RO	RY	RG	OY	OG	YG
Choice	R	R	G	O	G	Y

18. A perfume company is considering four possible fragrances for a new perfume: lilac, rose, apple blossom, and gardenia. The results of one consumer preference survey were as follows.

Pair of Fragrances	LR	LA	LG	RA	RG	AG
Choice	R	A	L	R	G	A

19. In a wine-tasting session a consumer compared five wines, A, B, C, D, and E, with the following results.

Wines	AB	AC	AD	AE	BC	BD	BE	CD	CE	DE
Choice	A	A	D	E	B	B	B	C	E	E

20. A manager made a comparison among five workers: A, B, C, D, and E. The workers were compared two at a time, with the following results.

Workers	AB	AC	AD	AE	BC	BD	BE	CD	CE	DE
Choice	A	A	D	A	C	B	B	C	C	D

FOR THE COMPUTER

Find the number of four-stage dominances for each person in the following matrices.

21. The matrix in Exercise 9   **22.** The matrix in Exercise 10

11.3 COMMUNICATION DIGRAPHS

In our study of dominance digraphs, we found that the associated matrix was asymmetric; that is, if the entry in row i and column j was a 1, say, then the entry in row j and column i had to be 0. With *communication digraphs,* however, the associated matrices are symmetric. The symmetry follows from the nature of communication: if A can communicate with B, then (normally) B can also communicate with A. Communication digraphs arise in studies of highway networks, pipelines, airline routes, and telephone links between cities.

A **communication digraph** has the following properties.

COMMUNICATION DIGRAPH

For any two vertices A and B:

1. The fact that A can communicate with B implies that B can communicate with A.

2. There can be more than one edge between A and B.

3. A can always communicate with B, although not necessarily directly.

4. The associated matrix is **symmetric;** that is, $a_{ij} = a_{ji}$ for all possible values of i and j.

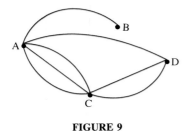

FIGURE 9

Figure 9 shows the possible routes of communication between cities A, B, C, and D. Since communication is two-way, no arrowheads are needed. The associated matrix M for the digraph in Figure 9 is

$$
\begin{array}{c}
 \\
A \\
B \\
C \\
D
\end{array}
\begin{array}{cccc}
A & B & C & D \\
\end{array}
\left[
\begin{array}{cccc}
0 & 1 & 3 & 1 \\
1 & 0 & 0 & 0 \\
3 & 0 & 0 & 2 \\
1 & 0 & 2 & 0
\end{array}
\right].
$$

Just as in the last section, the matrix M^2 gives the number of communication routes between two cities that go through exactly one other city. Matrix M^2 is

$$
M^2 =
\begin{bmatrix}
0 & 1 & 3 & 1 \\
1 & 0 & 0 & 0 \\
3 & 0 & 0 & 2 \\
1 & 0 & 2 & 0
\end{bmatrix}
\begin{bmatrix}
0 & 1 & 3 & 1 \\
1 & 0 & 0 & 0 \\
3 & 0 & 0 & 2 \\
1 & 0 & 2 & 0
\end{bmatrix}
=
\begin{bmatrix}
11 & 0 & 2 & 6 \\
0 & 1 & 3 & 1 \\
2 & 3 & 13 & 3 \\
6 & 1 & 3 & 5
\end{bmatrix}.
$$

The entry 3 in the fourth row of M^2 shows that there are 3 routes connecting C and D that pass through exactly one other city (in this case, all three routes go through A). The entry 5 in the fourth row shows that there are 5 routes connecting city D with itself that go through exactly one other city. (One of these 5 routes goes through city A; the other four go through C.)

As before, the matrix M^3 would give the number of routes that pass through exactly two other cities. Also, the sum $M + M^2$ gives the number of routes that pass through no cities or one city.

Organizational Communication   In some communication networks any two members can communicate directly. This is not always the case with *organizational communication* digraphs, however. (Mailroom clerks at Exxon probably don't chat with the chairman of the board, for example.) Assuming that only one method of communication exists between members of an organization, an **organizational communication digraph** has the following properties.

ORGANIZATIONAL COMMUNICATION

For any two vertices A and B:

1. If A communicates with B, then B communicates with A.
2. No more than one edge connects A and B.
3. A does not communicate with A.
4. The associated matrix is a symmetric adjacency matrix.

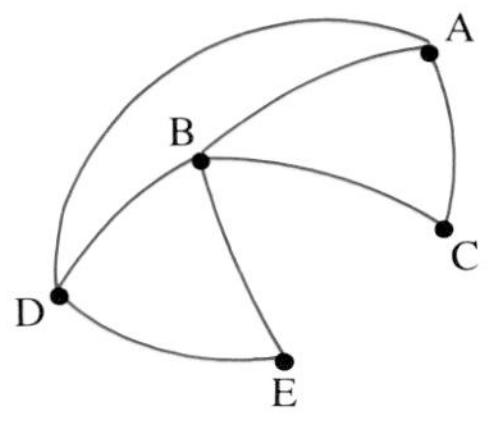

FIGURE 10

A typical organizational digraph is shown in Figure 10. The associated matrix for this figure is

$$\begin{array}{c} \\ A \\ B \\ C \\ D \\ E \end{array}\begin{array}{c} \begin{array}{ccccc} A & B & C & D & E \end{array} \\ \begin{bmatrix} 0 & 1 & 1 & 1 & 0 \\ 1 & 0 & 1 & 1 & 1 \\ 1 & 1 & 0 & 0 & 0 \\ 1 & 1 & 0 & 0 & 1 \\ 0 & 1 & 0 & 1 & 0 \end{bmatrix} \end{array}.$$

While A and E do not communicate directly in the digraph of Figure 10, they can communicate by going through B as an intermediary; alternate communication routes between A and E go through D, or through C and then B, among others. A route from one person to another, perhaps through intermediaries, is a *path*.

PATH

A collection of vertices and edges leading from one vertex to another, with no vertex repeated, is called a **path.** If a path exists between each pair of vertices, the digraph is **connected.**

EXAMPLE 1

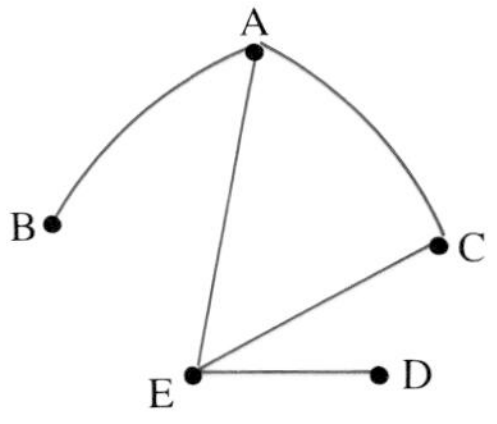

FIGURE 11

Use the organizational communication digraph in Figure 11 to answer the following questions.

(a) What are two possible paths from A to D?

Two possible paths from A to D are shown in Figure 12. Note that a route that "doubles back," such as going from A to E to C to A to E to D is *not* a path, since vertices A and E are repeated.

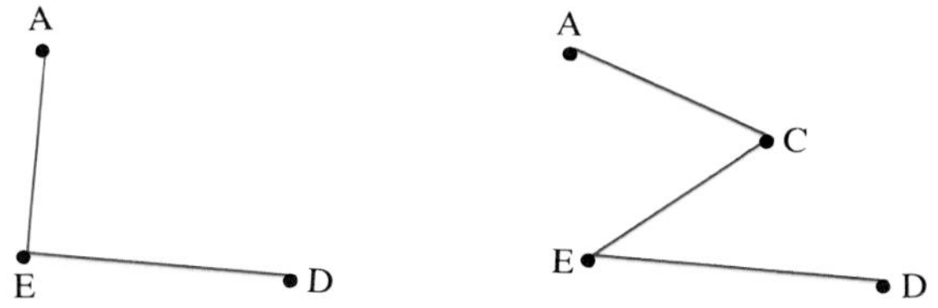

FIGURE 12

(b) Is this digraph connected?

There is a path connecting each pair of vertices, so the digraph is connected.

(c) What is the matrix for the digraph in Figure 11?

The associated matrix for the digraph follows.

$$
\begin{array}{c c}
 & \begin{array}{c c c c c} \text{A} & \text{B} & \text{C} & \text{D} & \text{E} \end{array} \\
\begin{array}{c} \text{A} \\ \text{B} \\ \text{C} \\ \text{D} \\ \text{E} \end{array} &
\left[\begin{array}{c c c c c}
0 & 1 & 1 & 0 & 1 \\
1 & 0 & 0 & 0 & 0 \\
1 & 0 & 0 & 0 & 1 \\
0 & 0 & 0 & 0 & 1 \\
1 & 0 & 1 & 1 & 0
\end{array} \right]
\end{array}
$$

We found by inspection that the digraph in Example 1 was connected. This process works well for most simple digraphs, but it is impractical for large ones. For larger digraphs, the following theorem (which we do not prove) is used.

THEOREM

To decide whether an organizational communication digraph with n vertices is connected, first find the associated adjacency matrix M, and then form the sum

$$M + M^2 + M^3 + \ldots + M^{n-1}.$$

The digraph is connected if the matrix sum has only positive entries.

EXAMPLE 2

Use the theorem to decide whether the following digraphs in Figure 13 are connected.

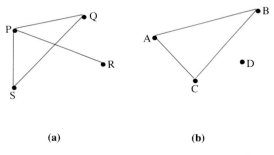

(a)   (b)

FIGURE 13

(a) For Figure 13(a), first write the associated adjacency matrix M.

$$
M = \begin{array}{c c}
 & \begin{array}{c c c c} \text{P} & \text{Q} & \text{R} & \text{S} \end{array} \\
\begin{array}{c} \text{P} \\ \text{Q} \\ \text{R} \\ \text{S} \end{array} &
\left[\begin{array}{c c c c}
0 & 1 & 1 & 1 \\
1 & 0 & 0 & 1 \\
1 & 0 & 0 & 0 \\
1 & 1 & 0 & 0
\end{array} \right]
\end{array}
$$

Since the digraph involves four vertices, the first $4 - 1 = 3$ powers of M must be found. Calculation of M^2 and M^3 leads to the sum

$$M + M^2 + M^3 = \begin{bmatrix} 0 & 1 & 1 & 1 \\ 1 & 0 & 0 & 1 \\ 1 & 0 & 0 & 0 \\ 1 & 1 & 0 & 0 \end{bmatrix} + \begin{bmatrix} 3 & 1 & 0 & 1 \\ 1 & 2 & 1 & 1 \\ 0 & 1 & 1 & 1 \\ 1 & 1 & 1 & 2 \end{bmatrix} + \begin{bmatrix} 2 & 4 & 3 & 4 \\ 4 & 2 & 1 & 3 \\ 3 & 1 & 0 & 1 \\ 4 & 3 & 1 & 2 \end{bmatrix}$$

$$= \begin{bmatrix} 5 & 6 & 4 & 6 \\ 6 & 4 & 2 & 5 \\ 4 & 2 & 1 & 2 \\ 6 & 5 & 2 & 4 \end{bmatrix}$$

All the entries in this matrix are positive, showing that the given digraph is connected.

(b) For Figure 13(b), the associated matrix M is

$$\begin{array}{c} \\ A \\ B \\ C \\ D \end{array} \begin{array}{cccc} A & B & C & D \\ \end{array} \\ \begin{bmatrix} 0 & 1 & 1 & 0 \\ 1 & 0 & 1 & 0 \\ 1 & 1 & 0 & 0 \\ 0 & 0 & 0 & 0 \end{bmatrix}.$$

The sum $M + M^2 + M^3$ is

$$M + M^2 + M^3 = \begin{bmatrix} 0 & 1 & 1 & 0 \\ 1 & 0 & 1 & 0 \\ 1 & 1 & 0 & 0 \\ 0 & 0 & 0 & 0 \end{bmatrix} + \begin{bmatrix} 2 & 1 & 1 & 0 \\ 1 & 2 & 1 & 0 \\ 1 & 1 & 2 & 0 \\ 0 & 0 & 0 & 0 \end{bmatrix} + \begin{bmatrix} 2 & 3 & 3 & 0 \\ 3 & 2 & 3 & 0 \\ 3 & 3 & 2 & 0 \\ 0 & 0 & 0 & 0 \end{bmatrix}$$

$$= \begin{bmatrix} 4 & 5 & 5 & 0 \\ 5 & 4 & 5 & 0 \\ 5 & 5 & 4 & 0 \\ 0 & 0 & 0 & 0 \end{bmatrix}.$$

The final matrix does not have all positive entries (some entries are 0), so the digraph is not connected. ▬

11.3 EXERCISES

Write the associated matrix for each digraph. Identify any communication digraphs or organizational communication digraphs.

1.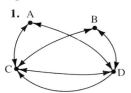

2.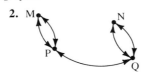

3.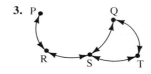

4.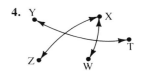

5.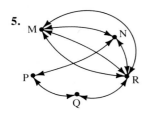

6.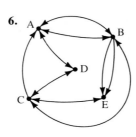

7.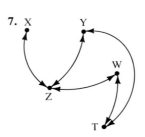

8.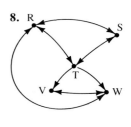

Sketch a digraph for each matrix of a communication digraph in Exercises 9–16. Find all two-stage paths of communication.

9. $\begin{bmatrix} 0 & 1 & 0 \\ 1 & 0 & 1 \\ 0 & 1 & 0 \end{bmatrix}$

10. $\begin{bmatrix} 0 & 2 & 1 \\ 2 & 0 & 0 \\ 1 & 0 & 0 \end{bmatrix}$

11. $\begin{bmatrix} 0 & 0 & 2 \\ 0 & 0 & 1 \\ 2 & 1 & 0 \end{bmatrix}$

12. $\begin{bmatrix} 0 & 1 & 0 \\ 1 & 0 & 2 \\ 0 & 2 & 0 \end{bmatrix}$

13. $\begin{bmatrix} 0 & 1 & 0 & 1 \\ 1 & 0 & 2 & 0 \\ 0 & 2 & 0 & 1 \\ 1 & 0 & 1 & 0 \end{bmatrix}$

14. $\begin{bmatrix} 0 & 0 & 2 & 1 \\ 0 & 0 & 1 & 0 \\ 2 & 1 & 0 & 2 \\ 1 & 0 & 2 & 0 \end{bmatrix}$

15. $\begin{bmatrix} 0 & 1 & 0 & 2 \\ 1 & 0 & 0 & 1 \\ 0 & 0 & 0 & 1 \\ 2 & 1 & 1 & 0 \end{bmatrix}$

16. $\begin{bmatrix} 0 & 1 & 1 & 0 \\ 0 & 0 & 1 & 0 \\ 1 & 1 & 0 & 2 \\ 1 & 0 & 2 & 0 \end{bmatrix}$

Identify all paths from A *to* D *in the following organizational communication digraphs.*

17.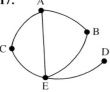

18.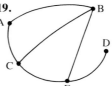

19.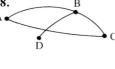

20.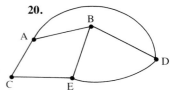

Use the theorem given at the end of the section to show that the following organizational communication digraphs are connected.

21.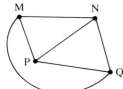

22.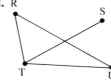

23.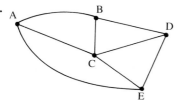

24.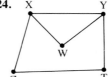

Show that the following organizational communication digraphs are not connected.

25.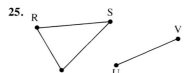

26.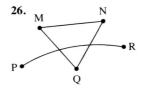

11.4 NETWORKS

Perhaps the single most useful application of graphs comes from the study of *networks*. The interstate highway system is an example of a network: given any city on the system, it is possible to find a route or path from that city to any other city on the system. Formally, as this example suggests, a network is a graph in which a path can be found between any two vertices.

NETWORK

> A **network** is a connected graph.

(a)

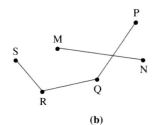

(b)

FIGURE 14

EXAMPLE 1

The graph in Figure 14(a) is a network. The one in Figure 14(b) is not, since there is no path between M and P, for example.

A **cycle** is a route that goes from one vertex back to the same vertex without passing along any edge more than once. For example, the graph in Figure 15(a) has five cycles: PNQP, NMPN, MQNM, QPMQ, and PNMQP. The graph in Figure 15(b) has only one cycle, WRSTW.

Graph theory is used to produce the minimum total cost when designing a network, such as a pipeline. For example, in the graph in Figure 15(b), the connection between W and T isn't really necessary since T can be reached from W by going through R and S. (Alternatively, the connection between W and R could be omitted, with W reached from R by going through T and S.) A network with no cycles is called a **tree.** In a tree, there is only one path between any two vertices.

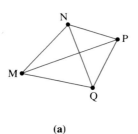

(a)

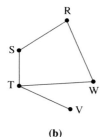

(b)

FIGURE 15

■ EXAMPLE 2

The graph in Figure 15(b) is not a tree because of the cycle WRSTW. However, the network becomes a tree if edge WR or edge WT is omitted, as seen in Figure 16. ■

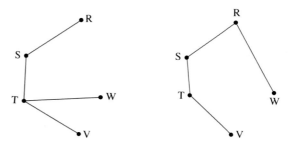

FIGURE 16

A tree that contains all the vertices of a network is called a **spanning tree.** Think of a spanning tree as a sort of "skeleton" for the graph: the spanning tree includes all the vertices of the graph; any edges that are not really necessary for each vertex to be reachable from all other vertices are omitted.

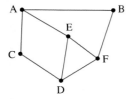

FIGURE 17

■ EXAMPLE 3

Find a spanning tree for the network in Figure 17.

This network contains several cycles, such as ACDEA, ACDFEA, AEFBA, ACDFBA, and so on. A spanning tree must contain no cycles, so it is necessary to remove at least one edge from each cycle. Two possible spanning trees are shown in Figure 18. Many others are possible. ■

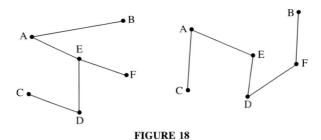

FIGURE 18

In an application, a length or cost is assigned to each edge of a network. For example, lengths might be assigned when designing a new microwave relay network, and costs might be assigned for a pipeline project where some parts go through relatively flat land but other parts involve expensive construction through mountains. In a network, a spanning tree for which the sum of numbers assigned to edges is a minimum is called a **minimum spanning tree.**

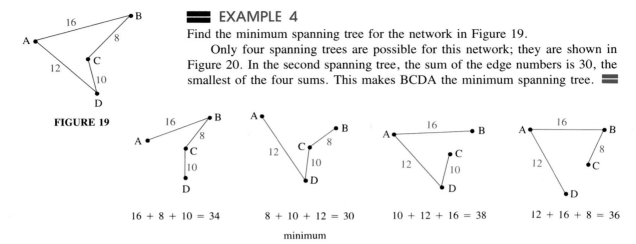

FIGURE 19

FIGURE 20

EXAMPLE 4

Find the minimum spanning tree for the network in Figure 19.

Only four spanning trees are possible for this network; they are shown in Figure 20. In the second spanning tree, the sum of the edge numbers is 30, the smallest of the four sums. This makes BCDA the minimum spanning tree. ▬

The minimum spanning tree for Example 4 was found by considering all possible spanning trees. This approach would be much too time-consuming in practice. Instead, we find the minimum spanning tree for a network by using the following procedure.

FINDING THE MINIMUM SPANNING TREE

> To find a minimum spanning tree for a network with numbers assigned to each edge:
>
> **1.** Find the edge with the smallest number (if more than one edge has this number, choose any one of them).
> **2.** Consider any edge with the next smallest number. If it would form a cycle with the edges already chosen, reject it; otherwise, choose it.
> **3.** Repeat Step 2 until the number of edges is one less than the number of vertices.
>
> Since several edges might have equal numbers, there may be more than one minimum spanning tree.

EXAMPLE 5

A company has branches in Washington, D.C., Charolotte, Nashville, Louisville, and Kansas City. The company wishes to build a private communications line between all these branches and Chicago headquarters. Find the line of minimum distance that will connect all the cities.

The first step is to draw a network showing all possible connections and the corresponding distances. This is done in Figure 21. (The numbers represent straight-line distances, but it is not convenient to draw all the edges as straight lines.)

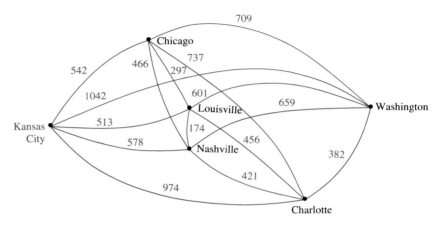

FIGURE 21

To find a minimum spanning tree, start with the edge with the smallest number,

Louisville–Nashville   174.

The next smaller number is associated with the edge

Chicago–Louisville   297.

Next, use the edge

Washington–Charlotte   382.

Now use

Charlotte–Nashville   421.

The next smaller number is Louisville–Charlotte, at 456 miles, but this edge would complete a cycle, so it cannot be used. The next usable edge is

Louisville–Kansas City   513.

The minimum spanning tree is shown in Figure 22. The length of this minimum spanning tree is

$$513 + 297 + 174 + 421 + 382 = 1787 \text{ miles.}$$

Any other spanning tree would produce a greater total mileage.∎

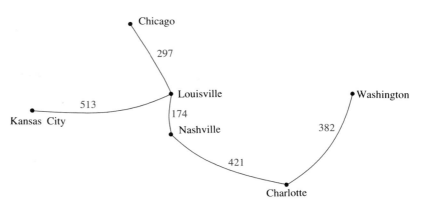

Chicago

297

Louisville

Washington

513

Kansas City

174

Nashville

382

421

Charlotte

FIGURE 22

11.4 EXERCISES

Identify all cycles in the following networks.

1.
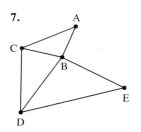

2.
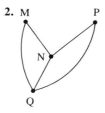

3.
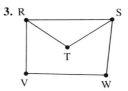

4.
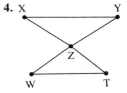

Find all spanning trees for each network.

5.
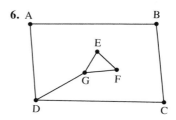

6.
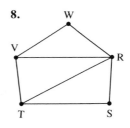

7.

8.

Find a minimum spanning tree for each network.

9.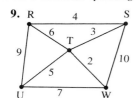

10.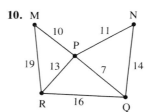

11. art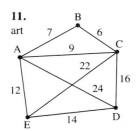

12.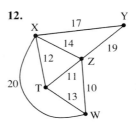

═══ APPLICATIONS

BUSINESS AND ECONOMICS

Minimum Cost   **13.** Find the minimum cost for the pipeline below, in which the numbers represent costs in suitable units.

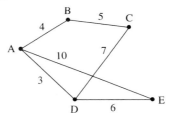

Minimum Cost   **14.** Find the minimum cost for the internal telephone system represented below, in which the numbers represent costs in appropriate units.

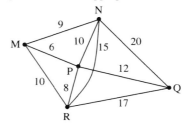

SOCIAL SCIENCES

Distances Between Cities   **15.** The distances between cities A, B, C, D, and E are given in the following mileage chart.

	A	B	C	D	E
A		168	352	152	317
B	168		177	172	296
C	352	177		235	152
D	152	172	235		164
E	317	296	152	164	

Find the minimum spanning tree for this network of cities.

Distances Between Cities   **16.** The distances between cities, R, S, T, V, W, and X are given in the following mileage chart.

	R	S	T	V	W	X
R		106	154	194	202	91
S	106		205	274	301	137
T	154	205		82	116	201
V	194	274	82		192	195
W	202	301	116	192		274
X	91	137	201	195	274	

Find the minimum spanning tree for this network of cities.

KEY WORDS

11.1   **graph theory**
graph
vertex
edge
directed edge
directed graph
digraph
adjacency matrix
11.2   **dominance**
asymmetric matrix

11.3   **communication digraph**
symmetric matrix
organizational communication digraph
path
connected digraph
11.4   **network**
cycle
tree
spanning tree
minimum spanning tree

CHAPTER 11   REVIEW EXERCISES

Write a matrix for each graph that is a digraph. Identify any adjacency matrices.

1.
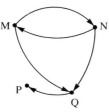

2.
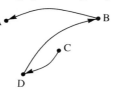

3.
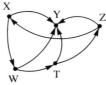

4.
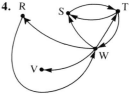

Sketch a digraph for each of the following matrices.

5.
$$\begin{array}{c} \\ A \\ B \\ C \end{array} \begin{array}{ccc} A & B & C \\ \left[\begin{array}{ccc} 0 & 2 & 1 \\ 1 & 0 & 2 \\ 0 & 1 & 0 \end{array}\right] \end{array}$$

6.
$$\begin{array}{c} \\ M \\ N \\ P \end{array} \begin{array}{ccc} M & N & P \\ \left[\begin{array}{ccc} 0 & 1 & 0 \\ 1 & 0 & 0 \\ 0 & 1 & 0 \end{array}\right] \end{array}$$

7.
$$\begin{array}{c} \\ R \\ S \\ T \\ V \end{array} \begin{array}{cccc} R & S & T & V \\ \left[\begin{array}{cccc} 0 & 1 & 0 & 1 \\ 0 & 0 & 1 & 0 \\ 1 & 0 & 0 & 1 \\ 0 & 1 & 1 & 0 \end{array}\right] \end{array}$$

8.
$$\begin{array}{c} \\ X \\ Y \\ Z \\ W \end{array} \begin{array}{cccc} X & Y & Z & W \\ \left[\begin{array}{cccc} 0 & 1 & 0 & 0 \\ 1 & 0 & 0 & 1 \\ 1 & 1 & 0 & 1 \\ 1 & 0 & 0 & 0 \end{array}\right] \end{array}$$

For each matrix in Exercises 9–14 that is the matrix of a dominance digraph, find the matrix representing two-stage dominance. Find the total number of vertices dominated by each vertex in one or two stages.

9.
$$\begin{array}{c} \\ A \\ B \\ C \end{array} \begin{array}{ccc} A & B & C \\ \left[\begin{array}{ccc} 0 & 1 & 0 \\ 0 & 0 & 1 \\ 1 & 0 & 0 \end{array}\right] \end{array}$$

10.
$$\begin{array}{c} \\ M \\ N \\ P \end{array} \begin{array}{ccc} M & N & P \\ \left[\begin{array}{ccc} 0 & 0 & 0 \\ 1 & 0 & 0 \\ 1 & 1 & 0 \end{array}\right] \end{array}$$

11.
$$\begin{array}{c} \\ X \\ Y \\ Z \end{array} \begin{array}{ccc} X & Y & Z \\ \left[\begin{array}{ccc} 0 & 1 & 1 \\ 0 & 0 & 1 \\ 1 & 0 & 0 \end{array}\right] \end{array}$$

12.
$$\begin{array}{c} \\ R \\ S \\ T \end{array} \begin{array}{ccc} R & S & T \\ \left[\begin{array}{ccc} 0 & 1 & 0 \\ 1 & 0 & 0 \\ 1 & 1 & 0 \end{array}\right] \end{array}$$

13.
$$\begin{array}{c} \\ P \\ Q \\ R \\ S \end{array} \begin{array}{cccc} P & Q & R & S \\ \left[\begin{array}{cccc} 0 & 1 & 0 & 1 \\ 0 & 0 & 1 & 1 \\ 1 & 0 & 0 & 1 \\ 0 & 0 & 0 & 0 \end{array}\right] \end{array}$$

14.
$$\begin{array}{c} \\ A \\ B \\ C \\ D \end{array} \begin{array}{cccc} A & B & C & D \\ \left[\begin{array}{cccc} 0 & 0 & 0 & 1 \\ 1 & 0 & 0 & 0 \\ 1 & 1 & 0 & 1 \\ 0 & 1 & 0 & 0 \end{array}\right] \end{array}$$

Write the corresponding matrix for each digraph. Find the number of two-stage dominances for A, B, C, and D.

15.
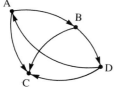

16.
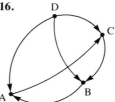

Decide on the winner of each of the following tournaments, using one- or two-stage dominances. Here AB means a game between teams A and B.

17.

Game	AB	AC	AD	AE	BC	BD	BE	CD	CE	DE
Winner	A	C	D	A	B	B	B	D	C	D

18.

Game	AB	AC	AD	AE	BC	BD	BE	CD	CE	DE
Winner	A	A	D	A	B	B	B	D	C	E

Write the associated matrix for each digraph. Identify any communication digraphs or organizational communication digraphs.

19.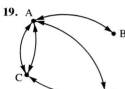

20.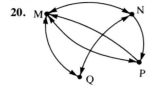

21.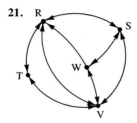

22.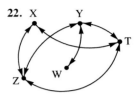

Sketch a digraph for each matrix of a communication digraph. Find all two-stage paths of communication.

23. $\begin{bmatrix} 0 & 0 & 2 \\ 0 & 0 & 1 \\ 2 & 1 & 0 \end{bmatrix}$

24. $\begin{bmatrix} 0 & 1 & 2 \\ 1 & 0 & 1 \\ 2 & 1 & 0 \end{bmatrix}$

25. $\begin{bmatrix} 0 & 2 & 1 & 2 \\ 2 & 0 & 0 & 1 \\ 1 & 0 & 0 & 1 \\ 2 & 1 & 1 & 0 \end{bmatrix}$

26. $\begin{bmatrix} 0 & 2 & 1 & 1 \\ 2 & 0 & 0 & 0 \\ 1 & 0 & 0 & 2 \\ 1 & 0 & 2 & 0 \end{bmatrix}$

Identify all paths from A to D in the following digraphs.

27.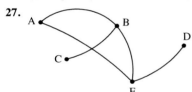

28.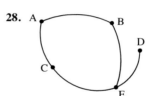

Show that each of the following digraphs is connected.

29.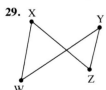

30.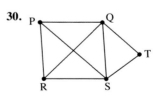

Identify all cycles in the following networks.

31.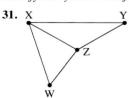

32.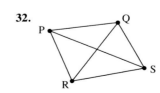

Find all spanning trees for each network.

33.

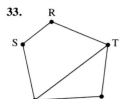

34.

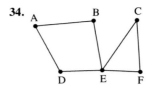

Find a minimum spanning tree for each network.

35.

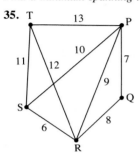

36.

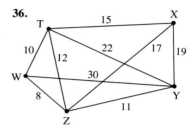

Tables

Combinations

Area Under a Normal Curve

Compound Interest

Present Value

Amount of an Annuity

Present Value of an Annuity

Combinations

n	$\binom{n}{0}$	$\binom{n}{1}$	$\binom{n}{2}$	$\binom{n}{3}$	$\binom{n}{4}$	$\binom{n}{5}$	$\binom{n}{6}$	$\binom{n}{7}$	$\binom{n}{8}$	$\binom{n}{9}$	$\binom{n}{10}$
0	1										
1	1	1									
2	1	2	1								
3	1	3	3	1							
4	1	4	6	4	1						
5	1	5	10	10	5	1					
6	1	6	15	20	15	6	1				
7	1	7	21	35	35	21	7	1			
8	1	8	28	56	70	56	28	8	1		
9	1	9	36	84	126	126	84	36	9	1	
10	1	10	45	120	210	252	210	120	45	10	1
11	1	11	55	165	330	462	462	330	165	55	11
12	1	12	66	220	495	792	924	792	495	220	66
13	1	13	78	286	715	1287	1716	1716	1287	715	286
14	1	14	91	364	1001	2002	3003	3432	3003	2002	1001
15	1	15	105	455	1365	3003	5005	6435	6435	5005	3003
16	1	16	120	560	1820	4368	8008	11440	12870	11440	8008
17	1	17	136	680	2380	6188	12376	19448	24310	24310	19448
18	1	18	153	816	3060	8658	18564	31824	43758	48620	43758
19	1	19	171	969	3876	11628	27132	50388	75582	92378	92378
20	1	20	190	1140	4845	15504	38760	77520	125970	167960	184756

For $r > 10$, it may be necessary to use the identity

$$\binom{n}{r} = \binom{n}{n-r}$$

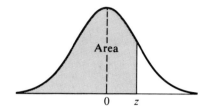

Area Under a Normal Curve to the Left of z, Where $z = \dfrac{x - \mu}{\sigma}$

z	.00	.01	.02	.03	.04	.05	.06	.07	.08	.09
− 3.4	.0003	.0003	.0003	.0003	.0003	.0003	.0003	.0003	.0003	.0002
− 3.3	.0005	.0005	.0005	.0004	.0004	.0004	.0004	.0004	.0004	.0003
− 3.2	.0007	.0007	.0006	.0006	.0006	.0006	.0006	.0005	.0005	.0005
− 3.1	.0010	.0009	.0009	.0009	.0008	.0008	.0008	.0008	.0007	.0007
− 3.0	.0013	.0013	.0013	.0012	.0012	.0011	.0011	.0011	.0010	.0010
− 2.9	.0019	.0018	.0017	.0017	.0016	.0016	.0015	.0015	.0014	.0014
− 2.8	.0026	.0025	.0024	.0023	.0023	.0022	.0021	.0021	.0020	.0019
− 2.7	.0035	.0034	.0033	.0032	.0031	.0030	.0029	.0028	.0027	.0026
− 2.6	.0047	.0045	.0044	.0043	.0041	.0040	.0039	.0038	.0037	.0036
− 2.5	.0062	.0060	.0059	.0057	.0055	.0054	.0052	.0051	.0049	.0048
− 2.4	.0082	.0080	.0078	.0075	.0073	.0071	.0069	.0068	.0066	.0064
− 2.3	.0107	.0104	.0102	.0099	.0096	.0094	.0091	.0089	.0087	.0084
− 2.2	.0139	.0136	.0132	.0129	.0125	.0122	.0119	.0116	.0113	.0110
− 2.1	.0179	.0174	.0170	.0166	.0162	.0158	.0154	.0150	.0146	.0143
− 2.0	.0228	.0222	.0217	.0212	.0207	.0202	.0197	.0192	.0188	.0183
− 1.9	.0287	.0281	.0274	.0268	.0262	.0256	.0250	.0244	.0239	.0233
− 1.8	.0359	.0352	.0344	.0336	.0329	.0322	.0314	.0307	.0301	.0294
− 1.7	.0446	.0436	.0427	.0418	.0409	.0401	.0392	.0384	.0375	.0367
− 1.6	.0548	.0537	.0526	.0516	.0505	.0495	.0485	.0475	.0465	.0455
− 1.5	.0668	.0655	.0643	.0630	.0618	.0606	.0594	.0582	.0571	.0559
− 1.4	.0808	.0793	.0778	.0764	.0749	.0735	.0722	.0708	.0694	.0681
− 1.3	.0968	.0951	.0934	.0918	.0901	.0885	.0869	.0853	.0838	.0823
− 1.2	.1151	.1131	.1112	.1093	.1075	.1056	.1038	.1020	.1003	.0985
− 1.1	.1357	.1335	.1314	.1292	.1271	.1251	.1230	.1210	.1190	.1170
− 1.0	.1587	.1562	.1539	.1515	.1492	.1469	.1446	.1423	.1401	.1379

Area Under a Normal Curve (continued)

z	.00	.01	.02	.03	.04	.05	.06	.07	.08	.09
−.9	.1841	.1814	.1788	.1762	.1736	.1711	.1685	.1660	.1635	.1611
−.8	.2119	.2090	.2061	.2033	.2005	.1977	.1949	.1922	.1894	.1867
−.7	.2420	.2389	.2358	.2327	.2296	.2266	.2236	.2206	.2177	.2148
−.6	.2743	.2709	.2676	.2643	.2611	.2578	.2546	.2514	.2483	.2451
−.5	.3085	.3050	.3015	.2981	.2946	.2912	.2877	.2843	.2810	.2776
−.4	.3446	.3409	.3372	.3336	.3300	.3264	.3228	.3192	.3156	.3121
−.3	.3821	.3783	.3745	.3707	.3669	.3632	.3594	.3557	.3520	.3483
−.2	.4207	.4168	.4129	.4090	.4052	.4013	.3974	.3936	.3897	.3859
−.1	.4602	.4562	.4522	.4483	.4443	.4404	.4364	.4325	.4286	.4247
−.0	.5000	.4960	.4920	.4880	.4840	.4801	.4761	.4721	.4681	.4641
.0	.5000	.5040	.5080	.5120	.5160	.5199	.5239	.5279	.5319	.5359
.1	.5398	.5438	.5478	.5517	.5557	.5596	.5636	.5675	.5714	.5753
.2	.5793	.5832	.5871	.5910	.5948	.5987	.6026	.6064	.6103	.6141
.3	.6179	.6217	.6255	.6293	.6331	.6368	.6406	.6443	.6480	.6517
.4	.6554	.6591	.6628	.6664	.6700	.6736	.6772	.6808	.6844	.6879
.5	.6915	.6950	.6985	.7019	.7054	.7088	.7123	.7157	.7190	.7224
.6	.7257	.7291	.7324	.7357	.7389	.7422	.7454	.7486	.7517	.7549
.7	.7580	.7611	.7642	.7673	.7704	.7734	.7764	.7794	.7823	.7852
.8	.7881	.7910	.7939	.7967	.7995	.8023	.8051	.8078	.8106	.8133
.9	.8159	.8186	.8212	.8238	.8264	.8289	.8315	.8340	.8365	.8389
1.0	.8413	.8438	.8461	.8485	.8508	.8531	.8554	.8577	.8599	.8621
1.1	.8643	.8665	.8686	.8708	.8729	.8749	.8770	.8790	.8810	.8830
1.2	.8849	.8869	.8888	.8907	.8925	.8944	.8962	.8980	.8997	.9015
1.3	.9032	.9049	.9066	.9082	.9099	.9115	.9131	.9147	.9162	.9177
1.4	.9192	.9207	.9222	.9236	.9251	.9265	.9278	.9292	.9306	.9319
1.5	.9332	.9345	.9357	.9370	.9382	.9394	.9406	.9418	.9429	.9441
1.6	.9452	.9463	.9474	.9484	.9495	.9505	.9515	.9525	.9535	.9545
1.7	.9554	.9564	.9573	.9582	.9591	.9599	.9608	.9616	.9625	.9633
1.8	.9641	.9649	.9656	.9664	.9671	.9678	.9686	.9693	.9699	.9706
1.9	.9713	.9719	.9726	.9732	.9738	.9744	.9750	.9756	.9761	.9767
2.0	.9772	.9778	.9783	.9788	.9793	.9798	.9803	.9808	.9812	.9817
2.1	.9821	.9826	.9830	.9834	.9838	.9842	.9846	.9850	.9854	.9857
2.2	.9861	.9864	.9868	.9871	.9875	.9878	.9881	.9884	.9887	.9890
2.3	.9893	.9896	.9898	.9901	.9904	.9906	.9909	.9911	.9913	.9916
2.4	.9918	.9920	.9922	.9925	.9927	.9929	.9931	.9932	.9934	.9936
2.5	.9938	.9940	.9941	.9943	.9945	.9946	.9948	.9949	.9951	.9952
2.6	.9953	.9955	.9956	.9957	.9959	.9960	.9961	.9962	.9963	.9964
2.7	.9965	.9966	.9967	.9968	.9969	.9970	.9971	.9972	.9973	.9974
2.8	.9974	.9975	.9976	.9977	.9977	.9978	.9979	.9979	.9980	.9981
2.9	.9981	.9982	.9982	.9983	.9984	.9984	.9985	.9985	.9986	.9986
3.0	.9987	.9987	.9987	.9988	.9988	.9989	.9989	.9989	.9990	.9990
3.1	.9990	.9991	.9991	.9991	.9992	.9992	.9992	.9992	.9993	.9993
3.2	.9993	.9993	.9994	.9994	.9994	.9994	.9994	.9995	.9995	.9995
3.3	.9995	.9995	.9995	.9996	.9996	.9996	.9996	.9996	.9996	.9997
3.4	.9997	.9997	.9997	.9997	.9997	.9997	.9997	.9997	.9997	.9998

Compound Interest

$$(1 + i)^n$$

$\dfrac{i}{n}$	1%	$1\frac{1}{2}$%	2%	3%	4%	5%	6%	8%
1	1.01000	1.01500	1.02000	1.03000	1.04000	1.05000	1.06000	1.08000
2	1.02010	1.03023	1.04040	1.06090	1.08160	1.10250	1.12360	1.16640
3	1.03030	1.04568	1.06121	1.09273	1.12486	1.15763	1.19102	1.25971
4	1.04060	1.06136	1.08243	1.12551	1.16986	1.21551	1.26248	1.36049
5	1.05101	1.07728	1.10408	1.15927	1.21665	1.27628	1.33823	1.46933
6	1.06152	1.09344	1.12616	1.19405	1.26532	1.34010	1.41852	1.58687
7	1.07214	1.10984	1.14869	1.22987	1.31593	1.40710	1.50363	1.71382
8	1.08286	1.12649	1.17166	1.26677	1.36857	1.47746	1.59385	1.85093
9	1.09369	1.14339	1.19509	1.30477	1.42331	1.55133	1.68948	1.99900
10	1.10462	1.16054	1.21899	1.34392	1.48024	1.62889	1.79085	2.15892
11	1.11567	1.17795	1.24337	1.38423	1.53945	1.71034	1.89830	2.33164
12	1.12683	1.19562	1.26824	1.42576	1.60103	1.79586	2.01220	2.51817
13	1.13809	1.21355	1.29361	1.46853	1.66507	1.88565	2.13293	2.71962
14	1.14947	1.23176	1.31948	1.51259	1.73168	1.97993	2.26090	2.93719
15	1.16097	1.25023	1.34587	1.55797	1.80094	2.07893	2.39656	3.17217
16	1.17258	1.26899	1.37279	1.60471	1.87298	2.18287	2.54035	3.42594
17	1.18430	1.28802	1.40024	1.65285	1.94790	2.29202	2.69277	3.70002
18	1.19615	1.30734	1.42825	1.70243	2.02582	2.40662	2.85434	3.99602
19	1.20811	1.32695	1.45681	1.75351	2.10685	2.52695	3.02560	4.31570
20	1.22019	1.34686	1.48595	1.80611	2.19112	2.65330	3.20714	4.66096
21	1.23239	1.36706	1.51567	1.86029	2.27877	2.78596	3.39956	5.03383
22	1.24472	1.38756	1.54598	1.91610	2.36992	2.92526	3.60354	5.43654
23	1.25716	1.40838	1.57690	1.97359	2.46472	3.07152	3.81975	5.87146
24	1.26973	1.42950	1.60844	2.03279	2.56330	3.22510	4.04893	6.34118
25	1.28243	1.45095	1.64061	2.09378	2.66584	3.38635	4.29187	6.84848
26	1.29526	1.47271	1.67342	2.15659	2.77247	3.55567	4.54938	7.39635
27	1.30821	1.49480	1.70689	2.22129	2.88337	3.73346	4.82235	7.98806
28	1.32129	1.51722	1.74102	2.28793	2.99870	3.92013	5.11169	8.62711
29	1.33450	1.53998	1.77584	2.35657	3.11865	4.11614	5.41839	9.31727
30	1.34785	1.56308	1.81136	2.42726	3.24340	4.32194	5.74349	10.06266
31	1.36133	1.58653	1.84759	2.50008	3.37313	4.53804	6.08810	10.86767
32	1.37494	1.61032	1.88454	2.57508	3.50806	4.76494	6.45339	11.73708
33	1.38869	1.63448	1.92223	2.65234	3.64838	5.00319	6.84059	12.67605
34	1.40258	1.65900	1.96068	2.73191	3.79432	5.25335	7.25103	13.69013
35	1.41660	1.68388	1.99989	2.81386	3.94609	5.51602	7.68609	14.78534
36	1.43077	1.70914	2.03989	2.89828	4.10393	5.79182	8.14725	15.96817
37	1.44508	1.73478	2.08069	2.98523	4.26809	6.08141	8.63609	17.24563
38	1.45953	1.76080	2.12230	3.07478	4.43881	6.38548	9.15425	18.62528
39	1.47412	1.78721	2.16474	3.16703	4.61637	6.70475	9.70351	20.11530
40	1.48886	1.81402	2.20804	3.26204	4.80102	7.03999	10.28572	21.72452
41	1.50375	1.84123	2.25220	3.35990	4.99306	7.39199	10.90286	23.46248
42	1.51879	1.86885	2.29724	3.46070	5.19278	7.76159	11.55703	25.33948
43	1.53398	1.89688	2.34319	3.56452	5.40050	8.14967	12.25045	27.36664
44	1.54932	1.92533	2.39005	3.67145	5.61652	8.55715	12.98548	29.55597
45	1.56481	1.95421	2.43785	3.78160	5.84118	8.98501	13.76461	31.92045
46	1.58046	1.98353	2.48661	3.89504	6.07482	9.43426	14.59049	34.47409
47	1.59626	2.01328	2.53634	4.01190	6.31782	9.90597	15.46592	37.23201
48	1.61223	2.04348	2.58707	4.13225	6.57053	10.40127	16.39387	40.21057
49	1.62835	2.07413	2.63881	4.25622	6.83335	10.92133	17.37750	43.42742
50	1.64463	2.10524	2.69159	4.38391	7.10668	11.46740	18.42015	46.90161

Present Value

$$\frac{1}{(1 + i)^n}$$

i / n	1%	1½%	2%	3%	4%	5%	6%	8%
1	.99010	.98522	.98039	.97087	.96154	.95238	.94340	.92593
2	.98030	.97066	.96117	.94260	.92456	.90703	.89000	.85734
3	.97059	.95632	.94232	.91514	.88900	.86384	.83962	.79383
4	.96098	.94218	.92385	.88849	.85480	.82270	.79209	.73503
5	.95147	.92826	.90573	.86261	.82193	.78353	.74726	.68058
6	.94205	.91454	.88797	.83748	.79031	.74622	.70496	.63017
7	.93272	.90103	.87056	.81309	.75992	.71068	.66506	.58349
8	.92348	.88771	.85349	.78941	.73069	.67684	.62741	.54027
9	.91434	.87459	.83676	.76642	.70259	.64461	.59190	.50025
10	.90529	.86167	.82035	.74409	.67556	.61391	.55839	.46319
11	.89632	.84893	.80426	.72242	.64958	.58468	.52679	.42888
12	.88745	.83639	.78849	.70138	.62460	.55684	.49697	.39711
13	.87866	.82403	.77303	.68095	.60057	.53032	.46884	.36770
14	.86996	.81185	.75788	.66112	.57748	.50507	.44230	.34046
15	.86135	.79985	.74301	.64186	.55526	.48102	.41727	.31524
16	.85282	.78803	.72845	.62317	.53391	.45811	.39365	.29189
17	.84438	.77639	.71416	.60502	.51337	.43630	.37136	.27027
18	.83602	.76491	.70016	.58739	.49363	.41552	.35034	.25025
19	.82774	.75361	.68643	.57029	.47464	.39573	.33051	.23171
20	.81954	.74247	.67297	.55368	.45639	.37689	.31180	.21455
21	.81143	.73150	.65978	.53755	.43883	.35894	.29416	.19866
22	.80340	.72069	.64684	.52189	.42196	.34185	.27751	.18394
23	.79544	.71004	.63416	.50669	.40573	.32557	.26180	.17032
24	.78757	.69954	.62172	.49193	.39012	.31007	.24698	.15770
25	.77977	.68921	.60953	.47761	.37512	.29530	.23300	.14602
26	.77205	.67902	.59758	.46369	.36069	.28124	.21981	.13520
27	.76440	.66899	.58586	.45019	.34682	.26785	.20737	.12519
28	.75684	.65910	.57437	.43708	.33348	.25509	.19563	.11591
29	.74934	.64936	.56311	.42435	.32065	.24295	.18456	.10733
30	.74192	.63976	.55207	.41199	.30832	.23138	.17411	.09938
31	.73458	.63031	.54125	.39999	.29646	.22036	.16425	.09202
32	.72730	.62099	.53063	.38834	.28506	.20987	.15496	.08520
33	.72010	.61182	.52023	.37703	.27409	.19987	.14619	.07889
34	.71297	.60277	.51003	.36604	.26355	.19035	.13791	.07305
35	.70591	.59387	.50003	.35538	.25342	.18129	.13011	.06763
36	.69892	.58509	.49022	.34503	.24367	.17266	.12274	.06262
37	.69200	.57644	.48061	.33498	.23430	.16444	.11579	.05799
38	.68515	.56792	.47119	.32523	.22529	.15661	.10924	.05369
39	.67837	.55953	.46195	.31575	.21662	.14915	.10306	.04971
40	.67165	.55126	.45289	.30656	.20829	.14205	.09722	.04603
41	.66500	.54312	.44401	.29763	.20028	.13528	.09172	.04262
42	.65842	.53509	.43530	.28896	.19257	.12884	.08653	.03946
43	.65190	.52718	.42677	.28054	.18517	.12270	.08163	.03654
44	.64545	.51939	.41840	.27237	.17805	.11686	.07701	.03383
45	.63905	.51171	.41020	.26444	.17120	.11130	.07265	.03133
46	.63273	.50415	.40215	.25674	.16461	.10600	.06854	.02901
47	.62646	.49670	.39427	.24926	.15828	.10095	.06466	.02686
48	.62026	.48936	.38654	.24200	.15219	.09614	.06100	.02487
49	.61412	.48213	.37896	.23495	.14634	.09156	.05755	.02303
50	.60804	.47500	.37153	.22811	.14071	.08720	.05429	.02132

Amount of an Annuity

$$s_{\overline{n}|i} = \frac{(1 + i)^n - 1}{i}$$

i \ n	1%	$1\frac{1}{2}$%	2%	3%	4%	5%	6%	8%
1	1.00000	1.00000	1.00000	1.00000	1.00000	1.00000	1.00000	1.00000
2	2.01000	2.01500	2.02000	2.03000	2.04000	2.05000	2.06000	2.08000
3	3.03010	3.04523	3.06040	3.09090	3.12160	3.15250	3.18360	3.24640
4	4.06040	4.09090	4.12161	4.18363	4.24646	4.31013	4.37462	4.50611
5	5.10101	5.15227	5.20404	5.30914	5.41632	5.52563	5.63709	5.86660
6	6.15202	6.22955	6.30812	6.46841	6.63298	6.80191	6.97532	7.33593
7	7.21354	7.32299	7.43428	7.66246	7.89829	8.14201	8.39384	8.92280
8	8.28567	8.43284	8.58297	8.89234	9.21423	9.54911	9.89747	10.63663
9	9.36853	9.55933	9.75463	10.15911	10.58280	11.02656	11.49132	12.48756
10	10.46221	10.70272	10.94972	11.46388	12.00611	12.57789	13.18079	14.48656
11	11.56683	11.86326	12.16872	12.80780	13.48635	14.20679	14.97164	16.64549
12	12.68250	13.04121	13.41209	14.19203	15.02581	15.91713	16.86994	18.97713
13	13.80933	14.23683	14.68033	15.61779	16.62684	17.71298	18.88214	21.49530
14	14.94742	15.45038	15.97394	17.08632	18.29191	19.59863	21.01507	24.21492
15	16.09690	16.68214	17.29342	18.59891	20.02359	21.57856	23.27597	27.15211
16	17.25786	17.93237	18.63929	20.15688	21.82453	23.65749	25.67253	30.32428
17	18.43044	19.20136	20.01207	21.76159	23.69751	25.84037	28.21288	33.75023
18	19.61475	20.48938	21.41231	23.41444	25.64541	28.13238	30.90565	37.45024
19	20.81090	21.79672	22.84056	25.11687	27.67123	30.53900	33.75999	41.44626
20	22.01900	23.12367	24.29737	26.87037	29.77808	33.06595	36.78559	45.76196
21	23.23919	24.47052	25.78332	28.67649	31.96920	35.71925	39.99273	50.42292
22	24.47159	25.83758	27.29898	30.53678	34.24797	38.50521	43.39229	55.45676
23	25.71630	27.22514	28.84496	32.45288	36.61789	41.43048	46.99583	60.89330
24	26.97346	28.63352	30.42186	34.42647	39.08260	44.50200	50.81558	66.76476
25	28.24320	30.06302	32.03030	36.45926	41.64591	47.72710	54.86451	73.10594
26	29.52563	31.51397	33.67091	38.55304	44.31174	51.11345	59.15638	79.95442
27	30.82089	32.98668	35.34432	40.70963	47.08421	54.66913	63.70577	87.35077
28	32.12910	34.48148	37.05121	42.93092	49.96758	58.40258	68.52811	95.33883
29	33.45039	35.99870	38.79223	45.21885	52.96629	62.32271	73.63980	103.96594
30	34.78489	37.53868	40.56808	47.57542	56.08494	66.43885	79.05819	113.28321
31	36.13274	39.10176	42.37944	50.00268	59.32834	70.76079	84.80168	123.34587
32	37.49407	40.68829	44.22703	52.50276	62.70147	75.29883	90.88978	134.21354
33	38.86901	42.29861	46.11157	55.07784	66.20953	80.06377	97.34316	145.95062
34	40.25770	43.93309	48.03380	57.73018	69.85791	85.06696	104.18375	158.62667
35	41.66028	45.59209	49.99448	60.46208	73.65222	90.32031	111.43478	172.31680
36	43.07688	47.27597	51.99437	63.27594	77.59831	95.83632	119.12087	187.10215
37	44.50765	48.98511	54.03425	66.17422	81.70225	101.62814	127.26812	203.07032
38	45.95272	50.71989	56.11494	69.15945	85.97034	107.70955	135.90421	220.31595
39	47.41225	52.48068	58.23724	72.23423	90.40915	114.09502	145.05846	238.94122
40	48.88637	54.26789	60.40198	75.40126	95.02552	120.79977	154.76197	259.05652
41	50.37524	56.08191	62.61002	78.66330	99.82654	127.83976	165.04768	280.78104
42	51.87899	57.92314	64.86222	82.02320	104.81960	135.23175	175.95054	304.24352
43	53.39778	59.79199	67.15947	85.48389	110.01238	142.99334	187.50758	329.58301
44	54.93176	61.68887	69.50266	89.04841	115.41288	151.14301	199.75803	356.94965
45	56.48107	63.61420	71.89271	92.71986	121.02939	159.70016	212.74351	386.50562
46	58.04589	65.56841	74.33056	96.50146	126.87057	168.68516	226.50812	418.42607
47	59.62634	67.55194	76.81718	100.39650	132.94539	178.11942	241.09861	452.90015
48	61.22261	69.56522	79.35352	104.40840	139.26321	188.02539	256.56453	490.13216
49	62.83483	71.60870	81.94059	108.54065	145.83373	198.42666	272.95840	530.34274
50	64.46318	73.68283	84.57940	112.79687	152.66708	209.34800	290.33590	573.77016

Present Value of an Annuity

$$a_{\overline{n}|i} = \frac{1 - (1 + i)^{-n}}{i}$$

n \ i	1%	$1\frac{1}{2}$%	2%	3%	4%	5%	6%	8%
1	.99010	.98522	.98039	.97087	.96154	.95238	.94340	.92593
2	1.97040	1.95588	1.94156	1.91347	1.88609	1.85941	1.83339	1.78326
3	2.94099	2.91220	2.88388	2.82861	2.77509	2.72325	2.67301	2.57710
4	3.90197	3.85438	3.80773	3.71710	3.62990	3.54595	3.46511	3.31213
5	4.85343	4.78264	4.71346	4.57971	4.45182	4.32948	4.21236	3.99271
6	5.79548	5.69719	5.60143	5.41719	5.24214	5.07569	4.91732	4.62288
7	6.72819	6.59821	6.47199	6.23028	6.00205	5.78637	5.58238	5.20637
8	7.65168	7.48593	7.32548	7.01969	6.73274	6.46321	6.20979	5.74664
9	8.56602	8.36052	8.16224	7.78611	7.43533	7.10782	6.80169	6.24689
10	9.47130	9.22218	8.98259	8.53020	8.11090	7.72173	7.36009	6.71008
11	10.36763	10.07112	9.78685	9.25262	8.76048	8.30641	7.88687	7.13896
12	11.25508	10.90751	10.57534	9.95400	9.38507	8.86325	8.38384	7.53608
13	12.13374	11.73153	11.34837	10.63496	9.98565	9.39357	8.85268	7.90378
14	13.00370	12.54338	12.10625	11.29607	10.56312	9.89864	9.29498	8.24424
15	13.86505	13.34323	12.84926	11.93794	11.11839	10.37966	9.71225	8.55948
16	14.71787	14.13126	13.57771	12.56110	11.65230	10.83777	10.10590	8.85137
17	15.56225	14.90765	14.29187	13.16612	12.16567	11.27407	10.47726	9.12164
18	16.39827	15.67256	14.99203	13.75351	12.65930	11.68959	10.82760	9.37189
19	17.22601	16.42617	15.67846	14.32380	13.13394	12.08532	11.15812	9.60360
20	18.04555	17.16864	16.35143	14.87747	13.59033	12.46221	11.46992	9.81815
21	18.85698	17.90014	17.01121	15.41502	14.02916	12.82115	11.76408	10.01680
22	19.66038	18.62082	17.65805	15.93692	14.45112	13.16300	12.04158	10.20074
23	20.45582	19.33086	18.29220	16.44361	14.85684	13.48857	12.30338	10.37106
24	21.24339	20.03041	18.91393	16.93554	15.24696	13.79864	12.55036	10.52876
25	22.02316	20.71961	19.52346	17.41315	15.62208	14.09394	12.78336	10.67478
26	22.79520	21.39863	20.12104	17.87684	15.98277	14.37519	13.00317	10.80998
27	23.55961	22.06762	20.70690	18.32703	16.32959	14.64303	13.21053	10.93516
28	24.31644	22.72672	21.28127	18.76411	16.66306	14.89813	13.40616	11.05108
29	25.06579	23.37608	21.84438	19.18845	16.98371	15.14107	13.59072	11.15841
30	25.80771	24.01584	22.39646	19.60044	17.29203	15.37245	13.76483	11.25778
31	26.54229	24.64615	22.93770	20.00043	17.58849	15.59281	13.92909	11.34980
32	27.26959	25.26714	23.46833	20.38877	17.87355	15.80268	14.08404	11.43500
33	27.98969	25.87895	23.98856	20.76579	18.14765	16.00255	14.23023	11.51389
34	28.70267	26.48173	24.49859	21.13184	18.41120	16.19290	14.36814	11.58693
35	29.40858	27.07559	24.99862	21.48722	18.66461	16.37419	14.49825	11.65457
36	30.10751	27.66068	25.48884	21.83225	18.90828	16.54685	14.62099	11.71719
37	30.79951	28.23713	25.96945	22.16724	19.14258	16.71129	14.73678	11.77518
38	31.48466	28.80505	26.44064	22.49246	19.36786	16.86789	14.84602	11.82887
39	32.16303	29.36458	26.90259	22.80822	19.58448	17.01704	14.94907	11.87858
40	32.83469	29.91585	27.35548	23.11477	19.79277	17.15909	15.04630	11.92461
41	33.49969	30.45896	27.79949	23.41240	19.99305	17.29437	15.13802	11.96723
42	34.15811	30.99405	28.23479	23.70136	20.18563	17.42321	15.22454	12.00670
43	34.81001	31.52123	28.66156	23.98190	20.37079	17.54591	15.30617	12.04324
44	35.45545	32.04062	29.07996	24.25427	20.54884	17.66277	15.38318	12.07707
45	36.09451	32.55234	29.49016	24.51871	20.72004	17.77407	15.45583	12.10840
46	36.72724	33.05649	29.89231	24.77545	20.88465	17.88007	15.52437	12.13741
47	37.35370	33.55319	30.28658	25.02471	21.04294	17.98102	15.58903	12.16427
48	37.97396	34.04255	30.67312	25.26671	21.19513	18.07716	15.65003	12.18914
49	38.58808	34.52468	31.05208	25.50166	21.34147	18.16872	15.70757	12.21216
50	39.19612	34.99969	31.42361	25.72976	21.48218	18.25593	15.76186	12.23348

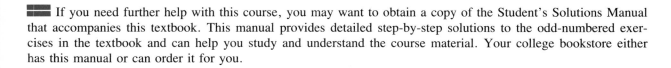
ANSWERS TO SELECTED EXERCISES

■■ If you need further help with this course, you may want to obtain a copy of the Student's Solutions Manual that accompanies this textbook. This manual provides detailed step-by-step solutions to the odd-numbered exercises in the textbook and can help you study and understand the course material. Your college bookstore either has this manual or can order it for you.

> ■■ Note that answers to computer problems may vary slightly, because different computers and software may be used to solve the problems.

■■ CHAPTER 1 SECTION 1.1 (PAGE 6)

1. $x = 4$ **3.** $m = 5/4$ **5.** $m = 12$ **7.** $k = -2/7$ **9.** $x = 5$ **11.** $r = -7/8$ **13.** $x = 3$ **15.** $x = 84$
17. $x \leq 10$ **19.** $m \leq -3/2$ **21.** $k < -5/3$ **23.** $z < -3/11$ **25.** $x \geq 16$ **27.** $t \geq -3/2$ **29.** $p \geq -3/5$
31. $k < 2$ **33.** $x = b - 3a$ **35.** $x = (3a + b)/(3 - a)$ **37.** $x = (3 - 3a)/(a^2 - a - 1)$ **39.** $x = 2a^2/(a^2 + 3)$
41. \$8000 **43.** \$6000 **45.** \$70,000 **47.** **(a)** 196 **(b)** $108 + .14x$ **(c)** $196 < 108 + .14x$; 628.6 mi **49.** 4
51. 400/3 or 133 1/3 liters

■■ SECTION 1.2 (PAGE 18)

1. Function **3.** Not a function **5.** Function **7.** **(a)** -12 **(b)** 2 **(c)** -4 **(d)** $-2a - 4$ **9.** **(a)** 6 **(b)** 6 **(c)** 6
(d) 6 **11.** **(a)** 48 **(b)** 6 **(c)** 0 **(d)** $2a^2 + 4a$ **13.** **(a)** 30 **(b)** 2 **(c)** 2 **(d)** $(a + 1)(a + 2)$, or $a^2 + 3a + 2$
15.

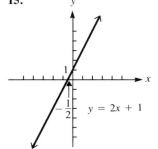

$y = 2x + 1$

17.

$y = 4x$

19.

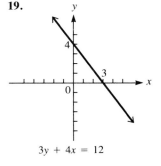

$3y + 4x = 12$

21.

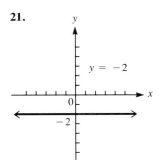

$y = -2$

23.

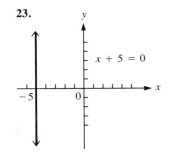

$x + 5 = 0$

25.

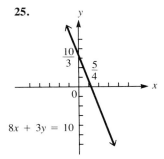

$8x + 3y = 10$

27.

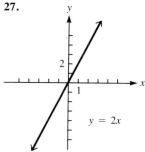

$y = 2x$

29.

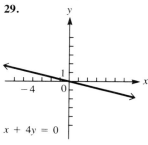

$x + 4y = 0$

31. (a) $1050 thousand, or $1,050,000
(b) $1100 thousand, or $1,100,000
(c) $1200 thousand, or $1,200,000
(d) $1300 thousand, or $1,300,000

33. (a) $11 **(b)** $11 **(c)** $18 **(d)** $32 **(e)** $32 **(f)** $39 **(g)** $39

35. (a) $16 **(b)** $11 **(c)** $6 **(d)** 8
(e) 4 **(f)** 0
(g) *p*

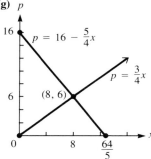

(h) 0 **(i)** 40/3 **(j)** 80/3
(k) See part (g). **(l)** 8 **(m)** $6

37. (a) *p*

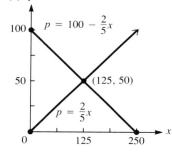

(b) 125 **(c)** $50

39. (a) $135 **(b)** $205
(c) $275 **(d)** $345
(e) *y*

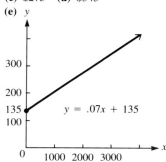

$y = .07x + 135$

SECTION 1.3 (PAGE 30)

1. $-1/5$ **3.** $2/3$ **5.** $-3/2$ **7.** Undefined slope **9.** 0 **11.** 3; 4 **13.** -4; 8 **15.** $-3/4$; 5/4
17. -3; 0 **19.** $-2/5$; 0 **21.** 0; 8 **23.** 0; -2 **25.** Undefined slope; no *y*-intercept
27.

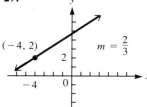

$m = \frac{2}{3}$

29.

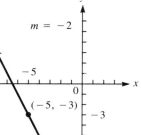

$m = -2$

31.

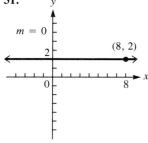

$m = 0$

33.

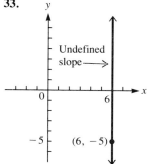

35.

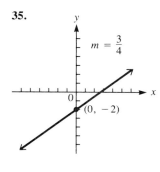

37.

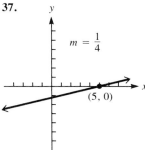

39. $y = (-3/4)x + 4$ **41.** $y = (-1/2)x - 2$ **43.** $y = (3/2)x + 5/4$ **45.** $2x - y = -9$ **47.** $3x + y = 3$
49. $x - 4y = -5$ **51.** $4x - 3y = -7$ **53.** $2x + 3y = 0$ **55.** $x = -8$ **57.** $y = 3$
63. $m = 640$; $y = 640x + 1100$ **65. (a)** $h = 3.5r + 83$ **(b)** About 163.5 cm; about 177.5 cm   **(c)** About 25 cm
67. (a) $y = 2.5x - 70$ **(b)** 55%   **69. (a)** $y = 5346.8x + 86.821$ **(b)** About 231,185

SECTION 1.4 (PAGE 39)

1. If $C(x)$ is the cost of renting a saw for x hours, then $C(x) = 12 + x$.   **3.** If $P(x)$ is the cost in cents of parking for x half-hours, then $P(x) = 35x + 50$.   **5.** $C(x) = 30x + 100$ **7.** $C(x) = 25x + 1000$ **9.** $C(x) = 50x + 500$
11. $C(x) = 90x + 2500$ **13.** $2000; $32,000   **15.** $2000; $52,000   **17.** $150,000; $600,000
19. (a) $C(x) = 3.50x + 90$ **(b)** 17   **(c)** 108   **21. (a)** $2600   **(b)** $2900   **(c)** $3200   **(d)** $2000   **(e)** $300
23. (a) $y = 82,500x + 850,000$ **(b)** $1,097,500   **(c)** $1,510,000   **25. (a)** $97   **(b)** $97.10   **(c)** $.097, or 9.7¢
(d) $.097, or 9.7¢   **27. (a)** $100   **(b)** $36   **(c)** $24   **29. (a)** $20,000 per year   **(b)** $80,000

31.

Year	Depreciation	Accumulated Depreciation at End of Year
1	$2000	$2000
2	$2000	$4000
3	$2000	$6000
4	$2000	$8000
5	$2000	$10,000
6	$2000	$12,000

33.

Year	Depreciation	Accumulated Depreciation at End of Year
1	$3750	$3750
2	$3750	$7500
3	$3750	$11,250
4	$3750	$15,000

35. (a) 2 units   **(b)** $980   **(c)** 52 units   **37. (a)** 500 units; $30,000   **(b)** $3000   **(c)** 1000 units   **39.** Break-even point is 45 units; don't produce   **41.** Break-even point is -50 units; impossible to make a profit here since $C(x) > R(x)$ for all positive x
43. (a) *y*

```
90          •
80        •
75      •
70    •
60  •
50 •
  0  1  2  3  4  5   x
```

(b) *y*

```
90            •
80         •
75       •
70     •
60   •
50 •
  0  1  2  3  4  5   x
```

(c) $y = 8x + 50$

(d)

Year	Actual Sales	Predicted Sales	Difference
0	48	50	-2
1	59	58	1
2	66	66	0
3	75	74	1
4	80	82	-2
5	90	90	0

(e) 106 thousand dollars, or $106,000
(f) 122 thousand dollars, or $122,000

45. (a) 100 thousand   **(b)** 70 thousand   **(c)** 0   **(d)** −5 thousand; the number of bacteria is decreasing
47. (a) 32.5 min   **(b)** 70 min   **(c)** 145 min   **(d)** 220 min   **(e)** 13.95 min   **(f)** 26.7 min   **(g)** 52.2 min   **(h)** 103.2 min
(i) About 69.2 min   **(j)** About 104.5 min

▬▬ SECTION 1.5   (PAGE 51)

1. Year 1: $5090.91; year 4: $3563.64   **3.** Year 1: $4171.43; year 4: $2085.71   **5. (a)** Year 1: $4000; year 2: $3000;
year 3: $2000; year 4: $1000   **(b)** $2500   **7. (a)** $60,000   **(b)** $50,000   **(c)** $40,000   **(d)** $30,000

9.

Year	Straight-Line Depreciation	Sum-of-the-Years'-Digits Depreciation
1	$375.00	$562.50
2	$375.00	$375.00
3	$375.00	$187.50
Totals	$1125.00	$1125.00

19. (a) $y' = 1.19x - .84$ **(b)** 2.4   **(c)** 3.3
(d) 2.8   **(e)** .94

21. (a)

Year	Straight-Line Depreciation	Sum-of-the-Years'-Digits Depreciation
1	$4833.33	$9354.84
2	$4833.33	$9043.01
3	$4833.33	$8731.18
4	$4833.33	$8419.35
5	$4833.33	$8107.53
6	$4833.33	$7795.70
7	$4833.33	$7483.87
8	$4833.33	$7172.04
9	$4833.33	$6860.22
10	$4833.33	$6548.39
11	$4833.33	$6236.56
12	$4833.33	$5924.73
13	$4833.33	$5612.90
14	$4833.33	$5301.08
15	$4833.33	$4989.25
16	$4833.33	$4677.42
17	$4833.33	$4365.59
18	$4833.33	$4053.76
19	$4833.33	$3741.94
20	$4833.33	$3430.11
21	$4833.33	$3118.28
22	$4833.33	$2806.45
23	$4833.33	$2494.62
24	$4833.33	$2182.80
25	$4833.33	$1870.97
26	$4833.33	$1559.14
27	$4833.33	$1247.31
28	$4833.33	$ 935.48
29	$4833.33	$ 623.66
30	$4833.33	$ 311.82

11. (a) $y' = .027x + 2.93$ **(b)** 13.7   **(c)** 447,000   **(d)** .97
13. (a) $y' = .95x + 5.3$ **(b)** 12.9 million   **(c)** 1992   **(d)** .90
15. (a) $y' = .3x + 1.5$ **(b)** .20   **(c)** 2.4
17. (a)

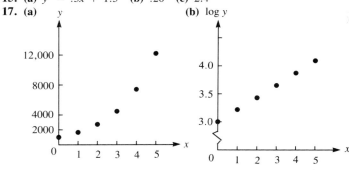

(c) $y' = .22x + 3.0$ **(d)** 4.54; about 35,000

21. (b)

Year	Straight-Line Depreciation	Sum-of-the-Years'-Digits Depreciation
1	$17,200.00	$32,250.00
2	$17,200.00	$30,100.00
3	$17,200.00	$27,950.00
4	$17,200.00	$25,800.00
5	$17,200.00	$23,650.00
6	$17,200.00	$21,500.00
7	$17,200.00	$19,350.00
8	$17,200.00	$17,200.00
9	$17,200.00	$15,050.00
10	$17,200.00	$12,900.00
11	$17,200.00	$10,750.00
12	$17,200.00	$ 8,600.00
13	$17,200.00	$ 6,450.00
14	$17,200.00	$ 4,300.00
15	$17,200.00	$ 2,150.00

23. $y' = 1.003077x + 8.486716$; .879913

CHAPTER 1   REVIEW EXERCISES   (PAGE 58)

1. $x = 2$ **3.** $k = 3$ **5.** $n = 7$ **7.** $z = 7/5$ **9.** $m \leq 5/2$ **11.** $m < -1$ **13.** $k \geq 7/3$ **15.** $x \leq 13$
17. $k > 3$ **19. (a)** 23   **(b)** -9 **(c)** -17 **(d)** $4r + 3$ **21. (a)** -28 **(b)** -12 **(c)** -28 **(d)** $-r^2 - 3$
23.

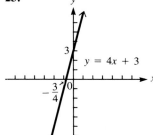

25.

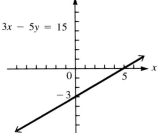

27.

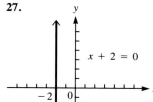

29.

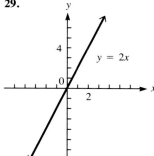

31. 1/3
32. $-2/11$
35. $-2/3$
37. Undefined slope
39. $2x - 3y = 13$
41. $5x + 4y = 17$
43. $x = -1$
45. \$18,000
47. (a) 7/6; 9/2 **(b)** 2; 2 **(c)** 5/2; 1/2
     (d) p **(e)** \$15   **(f)** 2; 2

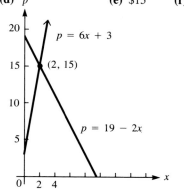

49. $C(x) = 30x + 60$ **51.** $C(x) = 30x + 85$
53. (a) 5 units **(b)** \$200   **55. (a)** \$2714.29 **(b)** \$3392.86

57.

Year	Straight-Line Depreciation	Sum-of-the-Years'-Digits Depreciation
1	\$17,000	\$27,200
2	\$17,000	\$20,400
3	\$17,000	\$13,600
4	\$17,000	\$ 6800

59.

Year	Straight-Line Depreciation	Sum-of-the-Years'-Digits Depreciation
1	\$6000	\$9000
2	\$6000	\$6000
3	\$6000	\$3000

61. $y' = 3.9 x - 7.9$ **63.** .998

65. **(a)** $1.11   **(b)** $.57   **(c)** $1.65   **(d)** $2.73

    (e) $C(x)$ **(f)** Domain: $\{x \mid x > 0\}$ (at least in theory); range: $\{.30, .57, .84, 1.11, 1.38, 1.65, \ldots\}$

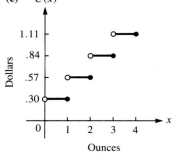

▬▬ EXTENDED APPLICATION   (PAGE 61)

1. $.1A + 200$ **2.** $.0001A - .3$ **3.** 8000   **4.** 8000 acres   **5.** 800 tons   **6.** $8,000,000

▬▬ EXTENDED APPLICATION   (PAGE 62)

1. 4.8 million units   **2.**

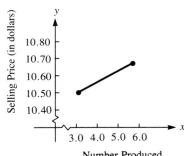

3. In the interval under discussion (3.1 to 5.7 million units), the marginal cost always exceeds the selling price.

4. (a) $9.87; $10.22   **(b)**

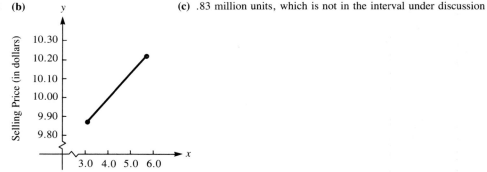

(c) .83 million units, which is not in the interval under discussion

CHAPTER 2   SECTION 2.1   (PAGE 77)

1. $(3, 6)$ **3.** $(-1, 4)$ **5.** $(-2, 0)$ **7.** $(1, 3)$ **9.** $(4, -2)$ **11.** $(2, -2)$ **13.** No solution
15. $(a, 4a - 9)$ for any real number a **17.** $(12, 6)$ **19.** $(7, -2)$ **21.** $(1, 2, -1)$ **23.** $(2, 0, 3)$
25. No solution   **27.** $(0, 2, 4)$ **29.** $(1, 2, 3)$ **31.** $(-1, 2, 1)$ **33.** $(4, 1, 2)$ **35.** $((1 - a)/2, (11 - a)/2, a)$ for
any real number a **37.** $(3 - a, 4 - a, a)$ for any real number a **39.** $((a - 7)/3, (4a - 49)/3, a, (-a - 2)/3)$
for any real number a **41.** $(3, -4)$ **43.** No solution   **45.** No solution   **47.** $k = 3$; solution is $(-1, 0, 2)$
49. Wife: 40 days; husband: 32 days   **51.** 5 model 201; 8 model 301   **53.** \$10,000 at 16%;
\$7000 at 20%; \$8000 at 18%

SECTION 2.2   (PAGE 86)

1. $\begin{bmatrix} 2 & 3 & | & 11 \\ 1 & 2 & | & 8 \end{bmatrix}$ **3.** $\begin{bmatrix} 1 & 5 & | & 6 \\ 0 & 1 & | & 1 \end{bmatrix}$ **5.** $\begin{bmatrix} 2 & 1 & 1 & | & 3 \\ 3 & -4 & 2 & | & -7 \\ 1 & 1 & 1 & | & 2 \end{bmatrix}$ **7.** $\begin{bmatrix} 1 & 1 & 0 & | & 2 \\ 0 & 2 & 1 & | & -4 \\ 0 & 0 & 1 & | & 2 \end{bmatrix}$ **9.** $\begin{bmatrix} 1 & 0 & 0 & | & 5 \\ 0 & 1 & 0 & | & -2 \\ 0 & 0 & 1 & | & 3 \end{bmatrix}$

11. $x = 2$ **13.** $2x + y = 1$ **15.** $x = 2$ **17.** $x + 4y = 10$ **19.** $x + 5z = 15$
$\quad y = 3$ $\quad\ 3x - 2y = -9$ $\quad y = 3$ $\qquad\ 0 = 0$ $\qquad y + 3z = 12$
$\qquad\qquad\qquad\qquad\qquad\qquad\qquad z = -2$ $\qquad\qquad\qquad\qquad\qquad\qquad\qquad 0 = 0$

21. $(2, 3)$ **23.** $(-3, 0)$ **25.** $(7/2, -1)$ **27.** $(5/2, -1)$ **29.** No solution
31. $(a, (6a - 1)/3)$ for any real number a **33.** $(-2, 1, 3)$ **35.** $(-1, 23, 16)$ **37.** $(3, 2, -4)$
39. No solution   **41.** $(-1, 3, 2)$ **43.** $(2, 4, 5)$ **45.** $(0, 2, -2, 1)$ The answers are given in the order x, y, z, w.

47. (a) $\begin{bmatrix} 1 & 0 & 0 & 1 & | & 1000 \\ 1 & 1 & 0 & 0 & | & 1100 \\ 0 & 1 & 1 & 0 & | & 700 \\ 0 & 0 & 1 & 1 & | & 600 \end{bmatrix}; \begin{bmatrix} 1 & 0 & 0 & 1 & | & 1000 \\ 0 & 1 & 0 & -1 & | & 100 \\ 0 & 0 & 1 & 1 & | & 600 \\ 0 & 0 & 0 & 0 & | & 0 \end{bmatrix}$ **(b)** $x_1 + x_4 = 1000; x_2 - x_4 = 100; x_3 + x_4 = 600$

(c) $x_4 = 1000 - x_1; x_4 = x_2 - 100; x_4 = 600 - x_3$ **(d)** $x_1 = 1000; x_4 = 1000$ **(e)** 100 **(f)** $x_3 = 600; x_4 = 600$
(g) $x_4 = 600; x_3 = 600; x_2 = 700; x_1 = 1000$ **49.** 12 cars from I to A, 8 cars from II to A, 16 cars from I to B, and no
cars from II to B   **51.** $(4.12062, 1.66866, .117699)$ **53.** $(30.7209, 39.6513, 31.386, 50.3966)$ **55.** 81 kg of
the first chemical, 382.286 kg of the second, and 286.714 kg of the third   **57.** 243 of A, 38 of B, and 101 of C (rounded)

SECTION 2.3   (PAGE 95)

1. False; not all corresponding elements are equal.   **3.** True   **5.** True   **7.** 2×2; square   **9.** 3×4 **11.** 2×1;
column   **13.** $x = 2, y = 4, z = 8$ **15.** $x = -15, y = 5, k = 3$ **17.** $z = 18, r = 3, s = 3, p = 3, a = 3/4$

19. $\begin{bmatrix} 9 & 12 & 0 & 2 \\ 1 & -1 & 2 & -4 \end{bmatrix}$ **21.** $\begin{bmatrix} 5 & 13 & 0 \\ 3 & 1 & 8 \end{bmatrix}$ **23.** Not possible   **25.** $\begin{bmatrix} 1 & 5 & 6 & -9 \\ 5 & 7 & 2 & 1 \\ -7 & 2 & 2 & -7 \end{bmatrix}$ **27.** $\begin{bmatrix} 3 & 4 \\ 4 & 8 \end{bmatrix}$

29. $\begin{bmatrix} 3 & 12 \\ -6 & 3 \end{bmatrix}$ **31.** $\begin{bmatrix} -12x + 8y & -x + y \\ x & 8x - y \end{bmatrix}$ **39.** $\begin{bmatrix} 7 & 2 \\ 9 & 0 \\ 8 & 6 \end{bmatrix}; \begin{bmatrix} 7 & 9 & 8 \\ 2 & 0 & 6 \end{bmatrix}$

41. (a) $\begin{bmatrix} 2 & 1 & 2 & 1 \\ 3 & 2 & 2 & 1 \\ 4 & 3 & 2 & 1 \end{bmatrix}$ **(b)** $\begin{bmatrix} 5 & 0 & 7 \\ 0 & 10 & 1 \\ 0 & 15 & 2 \\ 10 & 12 & 8 \end{bmatrix}$ **(c)** $\begin{bmatrix} 8 \\ 4 \\ 5 \end{bmatrix}$

■ SECTION 2.4   (PAGE 104)

1. 2×2; 2×2 **3.** 4×4; 2×2 **5.** 3×2; BA does not exist   **7.** AB does not exist; 3×2 **9.** $\begin{bmatrix} -4 & 8 \\ 0 & 6 \end{bmatrix}$

11. $\begin{bmatrix} 24 & -8 \\ -16 & 0 \end{bmatrix}$ **13.** $\begin{bmatrix} -22 & -6 \\ 20 & -12 \end{bmatrix}$ **15.** $\begin{bmatrix} 13 \\ 25 \end{bmatrix}$ **17.** $\begin{bmatrix} 5 \\ -21 \end{bmatrix}$ **19.** $\begin{bmatrix} 16 & -5 & 2 \\ 9 & -2 & 6 \end{bmatrix}$ **21.** $\begin{bmatrix} -2 & 10 \\ 0 & 8 \end{bmatrix}$

23. $\begin{bmatrix} 13 & 5 \\ 25 & 15 \end{bmatrix}$ **25.** $\begin{bmatrix} 13 \\ 29 \end{bmatrix}$ **27.** $\begin{bmatrix} 110 \\ 40 \\ -50 \end{bmatrix}$ **29.** $\begin{bmatrix} 22 & -8 \\ 11 & -4 \end{bmatrix}$

31. (a) $\begin{bmatrix} 16 & 22 \\ 7 & 19 \end{bmatrix}$ (b) $\begin{bmatrix} 5 & -5 \\ 0 & 30 \end{bmatrix}$ (c) No   (d) No

37. (a) P, P, X (b) T (c) I maintains the identity of any 2×2 matrix under multiplication.

41. (a) $\begin{array}{c} \\ S \\ C \end{array} \begin{array}{c} CC \quad MM \quad AD \\ \begin{bmatrix} .5 & .4 & .3 \\ .2 & .3 & .3 \end{bmatrix} \end{array}$ (b) $\begin{array}{c} \\ SD \\ MC \\ M \end{array} \begin{array}{c} S \quad C \\ \begin{bmatrix} 3 & 3 \\ 2 & 3 \\ 1 & 4 \end{bmatrix} \end{array}$ (c) $\begin{array}{c} \\ SD \\ MC \\ M \end{array} \begin{array}{c} CC \quad MM \quad AD \\ \begin{bmatrix} 2.1 & 2.1 & 1.8 \\ 1.6 & 1.7 & 1.5 \\ 1.3 & 1.6 & 1.5 \end{bmatrix} \end{array}$ (d) $1.60   (e) $1200 in Managua

43. (a) $\begin{bmatrix} 6 & 106 & 158 & 222 & 28 \\ 120 & 139 & 64 & 75 & 115 \\ -146 & -2 & 184 & 144 & -129 \\ 106 & 94 & 24 & 116 & 110 \end{bmatrix}$ (b) Cannot be found   (c) No   **45.** (a) $\begin{bmatrix} -1 & 5 & 9 & 13 & -1 \\ 7 & 17 & 2 & -10 & 6 \\ 18 & 9 & -12 & 12 & 22 \\ 9 & 4 & 18 & 10 & -3 \\ 1 & 6 & 10 & 28 & 5 \end{bmatrix}$

(b) $\begin{bmatrix} -2 & -9 & 90 & 77 \\ -42 & -63 & 127 & 62 \\ 413 & 76 & 180 & -56 \\ -29 & -44 & 198 & 85 \\ 137 & 20 & 162 & 103 \end{bmatrix}$ (c) $\begin{bmatrix} -56 & -1 & 1 & 45 \\ -156 & -119 & 76 & 122 \\ 315 & 86 & 118 & -91 \\ -17 & -17 & 116 & 51 \\ 118 & 19 & 125 & 77 \end{bmatrix}$ (d) $\begin{bmatrix} 54 & -8 & 89 & 32 \\ 114 & 56 & 51 & -60 \\ 98 & -10 & 62 & 35 \\ -12 & -27 & 82 & 34 \\ 19 & 1 & 37 & 26 \end{bmatrix}$

(e) $\begin{bmatrix} -2 & -9 & 90 & 77 \\ -42 & -63 & 127 & 62 \\ 413 & 76 & 180 & -56 \\ -29 & -44 & 198 & 85 \\ 137 & 20 & 162 & 103 \end{bmatrix}$ (f) Yes

■ SECTION 2.5   (PAGE 116)

1. Yes   **3.** No   **5.** No   **7.** Yes   **9.** $\begin{bmatrix} 0 & 1/2 \\ -1 & 1/2 \end{bmatrix}$ **11.** $\begin{bmatrix} 2 & 1 \\ 5 & 3 \end{bmatrix}$ **13.** No inverse   **15.** $\begin{bmatrix} 1 & 0 & 0 \\ 0 & -1 & 0 \\ -1 & 0 & 1 \end{bmatrix}$

17. $\begin{bmatrix} 15 & 4 & -5 \\ -12 & -3 & 4 \\ -4 & -1 & 1 \end{bmatrix}$ **19.** No inverse   **21.** $\begin{bmatrix} 7/4 & 5/2 & 3 \\ -1/4 & -1/2 & 0 \\ -1/4 & -1/2 & -1 \end{bmatrix}$ **23.** $\begin{bmatrix} 1/2 & 1/2 & -1/4 & 1/2 \\ -1 & 4 & -1/2 & -2 \\ -1/2 & 5/2 & -1/4 & -3/2 \\ 1/2 & -1/2 & 1/4 & 1/2 \end{bmatrix}$

25. $(-1, 4)$ **27.** $(2, 1)$ **29.** $(2, 3)$ **31.** $(a, (-a - 12)/8)$ for any real number a **33.** $(-8, 6, 1)$
35. $(15, -5, -1)$ **37.** $(-31, 24, -4)$ **39.** No inverse, no solution for system   **41.** $(-7, -34, -19, 7)$

51. (a) $\begin{bmatrix} 72 \\ 48 \\ 60 \end{bmatrix}$ (b) $\begin{bmatrix} 2 & 4 & 2 \\ 2 & 1 & 2 \\ 2 & 1 & 3 \end{bmatrix} \begin{bmatrix} x_1 \\ x_2 \\ x_3 \end{bmatrix} = \begin{bmatrix} 72 \\ 48 \\ 60 \end{bmatrix}$ (c) 8, 8, 12   **53.** (a) \$12,000 at 6%, \$7000 at 7%, and \$6000 at 10%

(b) \$10,000 at 6%, \$15,000 at 7%, and \$5000 at 10%   (c) \$20,000 at 6%, \$10,000 at 7%, and \$10,000 at 10%

55. $\begin{bmatrix} .010146 & -.011883 & .002772 & .020724 & -.012273 \\ .006353 & .014233 & -.001861 & -.029146 & .019225 \\ -.000638 & .006782 & -.004823 & -.022658 & .019344 \\ -.005261 & .003781 & .006192 & .004837 & -.006910 \\ -.012252 & -.001177 & -.006126 & .006744 & .002792 \end{bmatrix}$ **57.** No **59.** $\begin{bmatrix} 1.51482 \\ .053479 \\ -.637242 \\ .462629 \end{bmatrix}$

Entries are rounded to 6 places.

▇▇ SECTION 2.6 (PAGE 123)

1. $\begin{bmatrix} 32/3 \\ 25/3 \end{bmatrix}$ **3.** $\begin{bmatrix} 23{,}000/3579 \\ 93{,}500/3579 \end{bmatrix}$ or $\begin{bmatrix} 6.43 \\ 26.12 \end{bmatrix}$ **5.** $\begin{bmatrix} 20/3 \\ 20 \\ 10 \end{bmatrix}$ **7.** 33:47:23 **9.** 1079 metric tons of wheat and 1428 metric tons of oil

11. 1285 units of agriculture, 1455 units of manufacturing, and 1202 units of transportation **13.** 3077 units of agriculture, 2564 units of manufacturing, and 3179 units of transportation **15. (a)** 7/4 bushels of yams and 15/8 ≈ 2 pigs

(b) 167.5 bushels of yams and 153.75 ≈ 154 pigs **17.** $\begin{bmatrix} 2930 \\ 3570 \\ 2300 \\ 580 \end{bmatrix}$ **19.** $\begin{bmatrix} 1583.91 \\ 1529.54 \\ 1196.09 \end{bmatrix}$

▇▇ CHAPTER 2 REVIEW EXERCISES (PAGE 126)

1. $(-4, 6)$ **3.** $(-1, 2, 3)$ **5.** $(-9, 3)$ **7.** $(7, -9, -1)$ **9.** $((18 - 7a)/3, (3 + a)/3, a)$ for any real number a
11. 3×2; $a = 2, x = -1, y = 4, p = 5, z = 7$ **13.** 3×3 (square); $a = -12, b = 1, k = 9/2, c = 3/4, d = 3,$

$l = -3/4, m = -1, p = 3, q = 9$ **15.** $\begin{bmatrix} 8 & -6 \\ -10 & -16 \end{bmatrix}$ **17.** Not possible **19.** $\begin{bmatrix} 26 & 86 \\ -7 & -29 \\ 21 & 87 \end{bmatrix}$

21. $\begin{bmatrix} 6 & 18 & -24 \\ 1 & 3 & -4 \\ 0 & 0 & 0 \end{bmatrix}$ **23.** $\begin{bmatrix} 15 \\ 16 \\ 1 \end{bmatrix}$ **25.** $\begin{bmatrix} -7/19 & 4/19 \\ 3/19 & 1/19 \end{bmatrix}$ **27.** Does not exist **29.** $\begin{bmatrix} -\frac{1}{4} & \frac{1}{6} \\ 0 & \frac{1}{3} \end{bmatrix}$ **31.** No inverse

33. $\begin{bmatrix} \frac{1}{4} & \frac{1}{2} & \frac{1}{2} \\ \frac{1}{4} & -\frac{1}{2} & \frac{1}{2} \\ \frac{1}{8} & -\frac{1}{4} & -\frac{1}{4} \end{bmatrix}$ **35.** No inverse **37.** $X = \begin{bmatrix} 3 \\ 4 \end{bmatrix}$ **39.** $X = \begin{bmatrix} 6 \\ 15 \\ 16 \end{bmatrix}$ **41.** $(34, -9)$

43. $\begin{bmatrix} 218.1 \\ 318.3 \end{bmatrix}$ **45.** 8 thousand standard, 6 thousand extra-large **47.** 5 blankets, 3 rugs, 8 skirts

49. $\begin{bmatrix} 5 & 7 & 2532 & 52\frac{3}{8} & -\frac{1}{4} \\ 3 & 9 & 1464 & 56 & \frac{1}{8} \\ 2.50 & 5 & 4974 & 41 & -1\frac{1}{2} \\ 1.36 & 10 & 1754 & 18\frac{7}{8} & \frac{1}{2} \end{bmatrix}$ **51. (a)** $\begin{array}{c} \\ c \\ g \end{array}\begin{bmatrix} c & g \\ 0 & 1/2 \\ 2/3 & 0 \end{bmatrix}$ **(b)** 1200 units of cheese; 1600 units of goats

▇▇ EXTENDED APPLICATION (PAGE 129)

1. $A^2 = \begin{bmatrix} 9 & 2 & 3 & 2 \\ 2 & 2 & 2 & 3 \\ 3 & 2 & 6 & 4 \\ 2 & 3 & 4 & 5 \end{bmatrix}$ **(a)** 3 **(b)** 3 **(c)** 5 **(d)** 3 **2.** $A^3 = \begin{bmatrix} 12 & 12 & 22 & 21 \\ 12 & 4 & 9 & 6 \\ 22 & 9 & 12 & 12 \\ 21 & 6 & 12 & 8 \end{bmatrix}$ **(a)** 21 **(b)** 25

3. (a) $B = \begin{bmatrix} 0 & 2 & 3 \\ 2 & 0 & 4 \\ 3 & 4 & 0 \end{bmatrix}$ **(b)** $B^2 = \begin{bmatrix} 13 & 12 & 8 \\ 12 & 20 & 6 \\ 8 & 6 & 25 \end{bmatrix}$ **(c)** 12   **(d)** 14

4. (a)

	S	J	NO	H
S	0	1	2	1
J	1	0	1	0
NO	2	1	0	1
H	1	0	1	0

(b) 2   **(c)** 2   **(d)** 2

5. (a)

$C = $
	d	r	c	m
d	0	1	1	1
r	0	0	0	1
c	0	1	0	1
m	0	0	0	0

(b) $C^2 = \begin{bmatrix} 0 & 1 & 0 & 2 \\ 0 & 0 & 0 & 0 \\ 0 & 0 & 0 & 1 \\ 0 & 0 & 0 & 0 \end{bmatrix}$ C^2 gives the number of food sources once removed from the feeder. Thus, since dogs eat rats and rats eat mice, mice are an indirect as well as a direct food source.

6. See Exercises 1, 3(b), and 5(b).

▬ EXTENDED APPLICATION   (PAGE 131)

1. $PQ = \begin{bmatrix} 1 & 2 & 0 & 2 & 1 & 1 \\ 0 & 1 & 0 & 1 & 0 & 0 \\ 1 & 1 & 0 & 1 & 2 & 1 \end{bmatrix}$ **2.** None   **3.** Yes, the third person   **4.** The second and fourth persons in the third group each had four contacts in all.   **5.** See Exercise 1.

▬ EXTENDED APPLICATION   (PAGE 132)

1. (a) $A = \begin{bmatrix} .245 & .102 & .051 \\ .099 & .291 & .279 \\ .433 & .372 & .011 \end{bmatrix}$, $D = \begin{bmatrix} 2.88 \\ 31.45 \\ 30.91 \end{bmatrix}$, $X = \begin{bmatrix} x_1 \\ x_2 \\ x_3 \end{bmatrix}$ **(b)** $I - A = \begin{bmatrix} .755 & -.102 & -.051 \\ -.099 & .709 & -.279 \\ -.433 & -.372 & .989 \end{bmatrix}$ **(d)** $\begin{bmatrix} 18.2 \\ 73.2 \\ 66.8 \end{bmatrix}$

(e) \$18.2 billion of agriculture, \$73.2 billion of manufacturing, and \$66.8 billion of household would be required (rounded to three significant digits)   **2. (a)** $A = \begin{bmatrix} .293 & 0 & 0 \\ .014 & .207 & .017 \\ .044 & .010 & .216 \end{bmatrix}$, $D = \begin{bmatrix} 138,213 \\ 17,597 \\ 1,786 \end{bmatrix}$ **(b)** $I - A = \begin{bmatrix} .707 & 0 & 0 \\ -.014 & .793 & -.017 \\ -.044 & -.010 & .784 \end{bmatrix}$

(d) Agriculture, 195 million pounds; manufactured goods, 26 millions pounds; energy, 13 millions pounds

▬ CHAPTER 3   SECTION 3.1   (PAGE 144)

1.

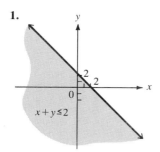

$x+y\leq2$

3.

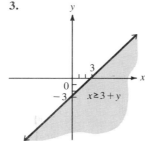

$x\geq3+y$

5.

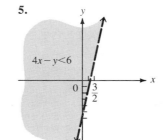

$4x-y<6$

7.

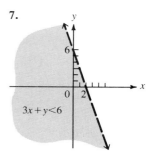

$3x+y<6$

9.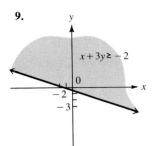

$x + 3y \ge -2$

11.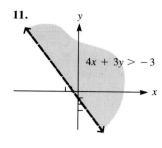

$4x + 3y > -3$

13.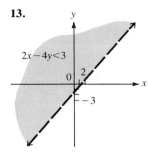

$2x - 4y < 3$

15.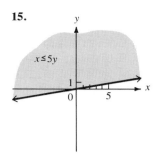

$x \le 5y$

17.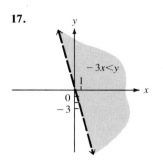

$-3x < y$

19.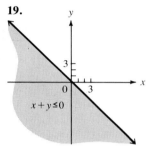

$x + y \le 0$

21.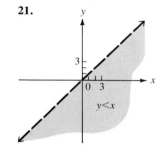

$y < x$

23.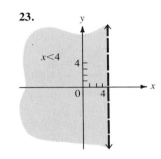

$x < 4$

25.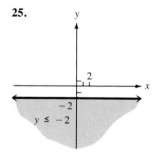

$y \le -2$

27.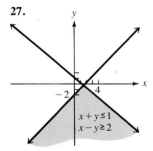

$x + y \le 1$
$x - y \ge 2$

29.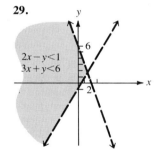

$2x - y < 1$
$3x + y < 6$

31.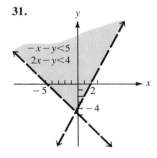

$-x - y < 5$
$2x - y < 4$

33.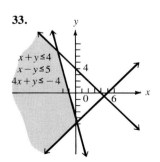

$x + y \le 4$
$x - y \le 5$
$4x + y \le -4$

35.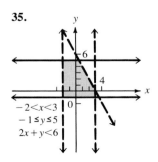

$-2 < x < 3$
$-1 \le y \le 5$
$2x + y < 6$

37.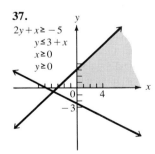

$2y + x \ge -5$
$y \le 3 + x$
$x \ge 0$
$y \ge 0$

39.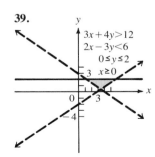

$3x + 4y > 12$
$2x - 3y < 6$
$0 \le y \le 2$
$x \ge 0$

41. (a)

	Number Made	Time on Wheel	Time in Kiln
Glazed	x	1/2	1
Unglazed	y	1	6
Maximum Time Available		8	20

(b) $(1/2)x + y \leq 8;\ x + 6y \leq 20;\ x \geq 0;\ y \geq 0$

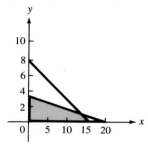

43. (a) $x \geq 3000;\ y \geq 5000;\ x + y \leq 10{,}000$
(b)

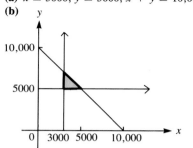

45. (a) $x \geq 1000;\ y \geq 800;\ x + y \leq 2400$
(b)

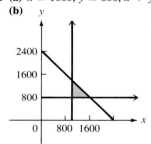

SECTION 3.2   (PAGE 150)

1. Let x be the number of product A made and y be the number of B. Then $2x + 3y \leq 45$.
3. Let x be the number of green pills and y be the number of red. Then $4x + y \geq 25$.
5. Let x be the number of pounds of \$6 coffee and y be the number of pounds of \$5 coffee. Then $x + y \geq 50$.

7. Let x be the number of engines sent to plant I and y be the number of engines sent to plant II. Then $x \geq 50$, $y \geq 27$, $x + y \leq 85$, $x \geq 0$, and $y \geq 0$. Minimize $20x + 35y$.

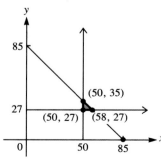

9. Let x be the number of units of policy A to purchase and y be the number of units of policy B to purchase. Then
$10x + 15y \geq 100$,
$80x + 120y \geq 1000$,
$x \geq 0$, and $y \geq 0$.
Minimize $50x + 40y$.
(Only the second constraint affects the solution.)

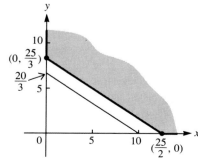

11. Let x be the number of type 1 bolts and y be the number of type 2 bolts. Then
$.1x + .1y \leq 240$,
$.1x + .4y \leq 720$,
$.1x + .5y \leq 160$, $x \geq 0$,
and $y \geq 0$. Maximize
$.10x + .12y$. (Only
$.1x + .5y \leq 160$
affects the solution.)

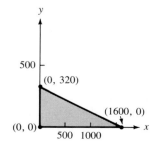

13. Let x be the number of kilograms of half-and-half mix and y be the number of kilograms of the other mix. Then $x/2 + y/3 \leq 100$, $x/2 + (2y/3) \leq 125$, $x \geq 0$, and $y \geq 0$. Maximize $6x + 4.80y$.

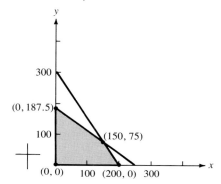

15. Let x be the number of gallons of milk from dairy II and y be the number of gallons of milk from dairy I. Then $x \leq 80$, $y \leq 50$, $x + y \leq 100$, $x \geq 0$, and $y \geq 0$. Maximize $.032x + .037y$.

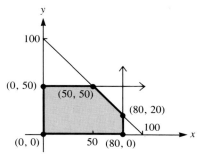

17. Let x be the amount in millions of dollars invested in treasury bonds and y be the amount in millions of dollars invested in mutual funds. Then $x + y \leq 40$, $x \geq 20$, and $y \geq 15$. Maximize $.12x + .08y$.

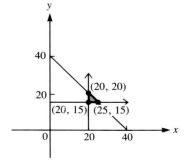

19. Let x be the number of species I prey and y be the number of species II prey. Then $5x + 3y \geq 10$, $2x + 4y \geq 8$, $x \geq 0$, and $y \geq 0$. Minimize $2x + 3y$.

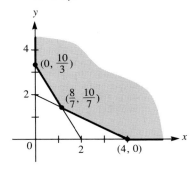

21. Let x be the number of servings of A and y be the number of servings of B. Then $3x + 2y \geq 15$, $2x + 4y \geq 15$, $x \geq 0$, and $y \geq 0$. Minimize $.25x + .40y$.

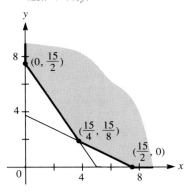

■■■ SECTION 3.3 (PAGE 159)

1. Maximum of 65 at (5, 10); minimum of 8 at (1,1)   **3.** Maximum of 9 at (0, 12); minimum of 0 at (0, 0)   **5.** No maximum; minimum of 18 at (3, 4)   **7.** Maximum of 42/5 when $x = 6/5$, $y = 6/5$ **9.** Maximum of 15 when $x = 10$, $y = 5$ **11.** Maximum of 235/4 when $x = 105/8$, $y = 25/8$ **13.** (a) Maximum of 204, when $x = 18$ and $y = 2$ **(b)** Maximum of 117 3/5 when $x = 12/5$ and $y = 39/5$ **(c)** Maximum of 102 when $x = 0$ and $y = 17/2$ **15.** 50 to plant I and 27 to plant II for a minimum cost of \$1945 (It is not surprising that costs are minimized by sending the minimum number of engines required.)   **17.** 0 units of Policy A and 25/3 or 8 1/3 units of Policy B for a minimum premium cost of \$333.33

19. 1600 Type 1 and 0 Type 2 for maximum revenue of $160   **21.** 150 kg half-and-half mix, 75 kg other mix for maximum revenue of $1260   **23.** 50 gal from dairy I and 50 gal from dairy II for maximum butterfat of 3.45 gal   **25.** $25 million in bonds and $15 million in mutual funds for maximum annual interest of $4.20 million   **27.** 8/7 units of species I and 10/7 units of species II will meet requirements with a minimum of 6.57 units of energy; however, a predator probably can catch and digest only whole numbers of prey. This problem shows that it is important to consider whether a model produces a realistic answer to a problem.   **29.** 3 3/4 servings of A and 1 7/8 servings of B for a minimum cost of $1.69   **31.** b   **33.** c

■■ CHAPTER 3   REVIEW EXERCISES   (PAGE 165)

1.
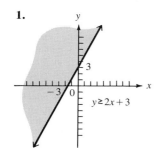

3.
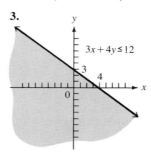

5.
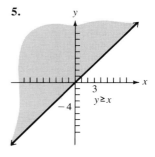

7.
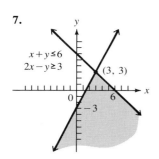

9.
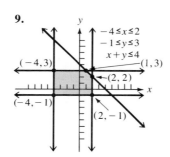

11.
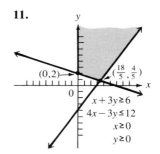

13. Minimum of 8 at (2, 1); maximum of 40 at (6, 7)   **15.** Maximum of 24 at (0, 6)

17. Minimum of 40 at any point on the segment connecting (0, 20) and (10/3, 40/3)

19. Let x = number of batches of cakes and y = number of batches of cookies. Then $x \geq 0$, $y \geq 0$, $2x + (3/2)y \leq 15$, and $3x + (2/3)y \leq 13$.

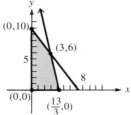

21. 3 batches of cakes and 6 of cookies for a maximum profit of $210

CHAPTER 4   SECTION 4.1   (PAGE 173)

1. $x_1 + 2x_2 + x_3 = 6$ **3.** $2x_1 + 4x_2 + 3x_3 + x_4 = 100$ **5. (a)** 3   **(b)** x_3, x_4, x_5 **(c)** $4x_1 + 2x_2 + x_3 = 20$,
$5x_1 + x_2 + x_4 = 50$, $2x_1 + 3x_2 + x_5 = 25$ **7. (a)** 2   **(b)** x_4, x_5 **(c)** $7x_1 + 6x_2 + 8x_3 + x_4 = 118$,
$4x_1 + 5x_2 + 10x_3 + x_5 = 220$ **9.** $x_1 = 0$, $x_2 = 0$, $x_3 = 20$, $x_4 = 0$, $x_5 = 15$, $z = 10$ **11.** $x_1 = 0$, $x_2 = 0$, $x_3 = 8$,
$x_4 = 0$, $x_5 = 6$, $x_6 = 7$, $z = 12$ **13.** $x_1 = 0$, $x_2 = 20$, $x_3 = 0$, $x_4 = 16$, $x_5 = 0$, $z = 60$ **15.** $x_1 = 0$, $x_2 = 0$, $x_3 = 12$,
$x_4 = 0$, $x_5 = 9$, $x_6 = 8$, $z = 36$ **17.** $x_1 = 0$, $x_2 = 0$, $x_3 = 50$, $x_4 = 10$, $x_5 = 0$, $x_6 = 50$, $z = 100$

19.

x_1	x_2	x_3	x_4	z	
2	3	1	0	0	6
4	1	0	1	0	6
−5	−1	0	0	1	0

21.

x_1	x_2	x_3	x_4	x_5	z	
1	1	1	0	0	0	10
5	2	0	1	0	0	20
1	2	0	0	1	0	36
−1	−3	0	0	0	1	0

23.

x_1	x_2	x_3	x_4	z	
3	1	1	0	0	12
1	1	0	1	0	15
−2	−1	0	0	1	0

25. If x_1 is the number of kilograms of half-and-half mix and x_2 is the number of kilograms of the other mix, find $x_1 \geq 0$, $x_2 \geq 0$, $x_3 \geq 0$, $x_4 \geq 0$ so that $(1/2)x_1 + (1/3)x_2 + x_3 = 100$, $(1/2)x_1 + (2/3)x_2 + x_4 = 125$, and $z = 6x_1 + 4.8x_2$ is maximized.

x_1	x_2	x_3	x_4	z	
1/2	1/3	1	0	0	100
1/2	2/3	0	1	0	125
−6	−4.8	0	0	1	0

27. If x_1 is the number of prams, x_2 is the number of runabouts, and x_3 is the number of trimarans, find $x_1 \geq 0$, $x_2 \geq 0$, $x_3 \geq 0$, $x_4 \geq 0$, $x_5 \geq 0$, $x_6 \geq 0$ so that $x_1 + 2x_2 + 3x_3 + x_4 = 6240$, $2x_1 + 5x_2 + 4x_3 + x_5 = 10,800$, $x_1 + x_2 + x_3 + x_6 = 3000$, and $z = 75x_1 + 90x_2 + 100x_3$ is maximized.

x_1	x_2	x_3	x_4	x_5	x_6	z	
1	2	3	1	0	0	0	6,240
2	5	4	0	1	0	0	10,800
1	1	1	0	0	1	0	3,000
−75	−90	−100	0	0	0	1	0

29. If x_1 is the number of Siamese cats and x_2 is the number of Persian cats, find $x_1 \geq 0$, $x_2 \geq 0$, $x_3 \geq 0$, $x_4 \geq 0$, $x_5 \geq 0$ so that $2x_1 + x_2 + x_3 = 90$, $x_1 + 2x_2 + x_4 = 80$, $x_1 + x_2 + x_5 = 50$, and $z = 12x_1 + 10x_2$ is maximized.

x_1	x_2	x_3	x_4	x_5	z	
2	1	1	0	0	0	90
1	2	0	1	0	0	80
1	1	0	0	1	0	50
−12	−10	0	0	0	1	0

SECTION 4.2   (PAGE 182)

1. Maximum is 20 when $x_1 = 0$, $x_2 = 4$, $x_3 = 0$, $x_4 = 0$, and $x_5 = 2$.   **3.** Maximum is 8 when $x_1 = 4$, $x_2 = 0$, $x_3 = 8$,
$x_4 = 2$, and $x_5 = 0$.   **5.** Maximum is 264 when $x_1 = 16$, $x_2 = 4$, $x_3 = 0$, $x_4 = 0$, $x_5 = 16$, and $x_6 = 0$.   **7.** Maximum is
22 when $x_1 = 5.5$, $x_2 = 0$, $x_3 = 0$, and $x_4 = .5$.   **9.** Maximum is 120 when $x_1 = 0$, $x_2 = 10$, $x_3 = 0$, $x_4 = 40$, and $x_5 = 4$.
11. Maximum is 944 when $x_1 = 118$, $x_2 = 0$, $x_3 = 0$, $x_4 = 0$, and $x_5 = 102$.   **13.** Maximum is 250 when $x_1 = 0$, $x_2 = 0$,
$x_3 = 0$, $x_4 = 50$, $x_5 = 0$, and $x_6 = 50$.   **15.** She should contact 6 churches and 2 labor unions for a maximum of $1000 per
month.   **17.** The company should produce 2 jazz albums, 3 blues albums, and 6 reggae albums for maximum weekly profit of
$10.60.   **19.** No 1-speed or 3-speed bicycles; 2700 10-speed bicycles; maximum profit is $59,400   **21. (a)** (3)   **(b)** (4)
(c) (3)   **23.** 163.6 kg of food P; none of Q; 1090.9 kg of R; 145.5 kg of S; maximum is 87,454.5   **25.** 6700 trucks and
4467 fire engines for a maximum profit of $110,997

SECTION 4.3 (PAGE 195)

1. $2x_1 + 3x_2 + x_3 = 8$, $x_1 + 4x_2 - x_4 = 7$ **3.** $x_1 + x_2 + x_3 + x_4 = 100$, $x_1 + x_2 + x_3 - x_5 = 75$, $x_1 + x_2 - x_6 = 27$
5. Change the objective function to maximize $z = -4y_1 - 3y_2 - 2y_3$.   **7.** Change the objective function to maximize
$z = -y_1 - 2y_2 - y_3 - 5y_4$.   **9.** Maximum is 480 when $x_1 = 40$ and $x_3 = 16$.   **11.** Maximum is 750 when $x_2 = 150$ and
$x_5 = 50$.   **13.** Maximum is 300 when $x_2 = 100$, $x_4 = 50$, and $x_5 = 10$.   **15.** Minimum is 40 when $x_1 = 10$ and $x_4 = 50$.
17. Minimum is 100 when $x_2 = 100$ and $x_5 = 50$.   **19.** Maximum is 133 1/3 when $x_1 = 33$ 1/3, $x_2 = 16$ 2/3, and $x_3 = $
46 2/3.   **21.** Minimum is 132 when $y_1 = 4$, $y_2 = 6$, and $y_3 = -4$ **23.** Ship 5000 barrels of oil from supplier 1 to
distributor 2; ship 3000 barrels of oil from supplier 2 to distributor 1. Minimum cost is $175,000.   **25.** 800,000 kg for whole
tomatoes and 80,000 kg for sauce for a minimum cost of $3,460,000   **27.** 1000 small and 500 large for a minimum cost of $210
29. $40,000 in government securities and $60,000 in mutual funds for maximum interest of $8800.   **31.** 3 of pill #1 and 2 of
pill #2 for a minimum cost of 70¢   **33.** 2 2/3 units of 1, none of II, and 4 of III for a minimum cost of $30.67   **35.** 1 2/3
oz of I, 6 2/3 oz of II, 1 2/3 oz of III, for a minimum cost of $1.55 per gallon

SECTION 4.4 (PAGE 207)

1. $\begin{bmatrix} 1 & 3 & 1 \\ 2 & 2 & 10 \\ 3 & 1 & 0 \end{bmatrix}$ **3.** $\begin{bmatrix} -1 & 13 & -2 \\ 4 & 25 & -1 \\ 6 & 0 & 11 \\ 12 & 4 & 3 \end{bmatrix}$ **5.** Minimize $w = 5y_1 + 4y_2 + 15y_3$ subject to $y_1 + y_2 + 2y_3 \geq 4$,

$y_1 + y_2 + y_3 \geq 3$, $y_1 + 3y_3 \geq 2$, $y_1 \geq 0$, $y_2 \geq 0$, and $y_3 \geq 0$.   **7.** Maximize $z = 50x_1 + 100x_2$ subject to $x_1 + 3x_2 \leq 1$;
$x_1 + x_2 \leq 2$; $x_1 + 2x_2 \leq 1$; $x_1 + x_2 \leq 5$; $x_1 \geq 0$; and $x_2 \geq 0$.   **9.** Minimum is 14 when $y_1 = 0$ and $y_2 = 7$.
11. Minimum is 40 when $y_1 = 10$ and $y_2 = 0$.   **13.** Minimum is 100 when $y_1 = 0$, $y_2 = 100$, and $y_3 = 0$.
15. 800,000 kg for whole tomatoes and 80,000 kg for sauce for a minimum cost of $3,460,000
17. (a) Minimize $200y_1 + 600y_2 + 90y_3 = w$ subject to $y_1 + 4y_2 \geq 1$, $2y_1 + 3y_2 + y_3 \geq 1.5$, $y_1 \geq 0$, $y_2 \geq 0$, and $y_3 \geq 0$.
(b) $y_1 = .6$, $y_2 = .1$, $y_3 = 0$, $w = 180$ **(c)** $186 ($x_1 = 114$, $x_2 = 48$)   **(d)** $179 ($x_1 = 116$, $x_2 = 42$)
19. 3 of pill #1, 2 of pill #2 for a minimum cost of 70 cents   **21.** 81 kg of the first, 382.3 kg of the second, and 286.7 kg of
the third for minimum cost of $607.24

CHAPTER 4 REVIEW EXERCISES (PAGE 210)

1. (a) $2x_1 + 5x_2 + x_3 = 50$; **(b)**
$x_1 + 3x_2 + x_4 = 25$;
$4x_1 + x_2 + x_5 = 18$;
$x_1 + x_2 + x_6 = 12$

	x_1	x_2	x_3	x_4	x_5	x_6	z	
	2	5	1	0	0	0	0	50
	1	3	0	1	0	0	0	25
	4	1	0	0	1	0	0	18
	1	1	0	0	0	1	0	12
	-5	-3	0	0	0	0	1	0

3. (a) $x_1 + x_2 + x_3 + x_4 = 90$ **(b)**
$2x_1 + 5x_2 + x_3 + x_5 = 120$
$x_1 + 3x_2 - x_6 = 80$

	x_1	x_2	x_3	x_4	x_5	x_6	z	
	1	1	1	1	0	0	0	90
	2	5	1	0	1	0	0	120
	1	3	0	0	0	-1	0	80
	-5	-8	-6	0	0	0	1	0

5. Maximum is 82.4 when $x_1 = 13.6$, $x_2 = 0$, $x_3 = 4.8$, $x_4 = 0$, and $x_5 = 0$.   **7.** Maximum is 76.67 when $x_1 = 6.67$,
$x_2 = 0$, $x_3 = 21.67$, $x_4 = 0$, $x_5 = 0$, and $x_6 = 35$.   **9.** Change the objective function to maximize $z = -10x_1 - 15x_2$.
11. Change the objective function to maximize $z = -7x_1 - 2x_2 - 3x_3$.   **13.** Minimum of 62 when $x_1 = 8$, $x_2 = 12$, $x_3 = 0$,
$x_4 = 1$, $x_5 = 0$, and $x_6 = 2$.   **15. (a)** Let $x_1 = $ number of item A, $x_2 = $ number of item B, and $x_3 = $ number of item C she
should buy. **(b)** $z = 4x_1 + 3x_2 + 3x_3$ **(c)** $5x_1 + 3x_2 + 6x_3 \leq 1200$, $x_1 + 2x_2 + 2x_3 \leq 800$, $2x_1 + x_2 + 5x_3 \leq 500$

17. (a) Let x_1 = number of gallons of fruity wine and x_2 = number of gallons of crystal wine to be made. **(b)** $z = 12x_1 + 15x_2$
(c) $2x_1 + x_2 \leq 110$, $2x_1 + 3x_2 \leq 125$, $2x_1 + x_2 \leq 90$ **19.** None of A, 400 of B, and none of C for maximum profit of $1200
21. 36.25 gallons of fruity and 17.5 gallons of crystal for maximum profit of $697.50

▰ EXTENDED APPLICATION (PAGE 213)

1. (a) $x_2 = 58.6957$, $x_8 = 2.01031$, $x_9 = 22.4895$, $x_{10} = 1$, $x_{12} = 1$, and $x_{13} = 1$ for a minimum cost of $12.55
(b) $x_2 = 71.7391$, $x_8 = 3.14433$, $x_9 = 23.1966$, $x_{10} = 1$, $x_{12} = 1$, and $x_{13} = 1$ for a minimum cost of $15.05

▰ EXTENDED APPLICATION (PAGE 215)

1. $w_1 = .95$, $w_2 = .83$, $w_3 = .75$, $w_4 = 1.00$, $w_5 = .87$, and $w_6 = .94$ **2.** $x_1 = 100$, $x_2 = 0$, $x_3 = 0$, $x_4 = 90$, $x_5 = 0$, and $x_6 = 210$

▰ CHAPTER 5 SECTION 5.1 (PAGE 225)

1. False **3.** True **5.** True **7.** True **9.** False **11.** True **13.** False **15.** True **17.** False
19. False **21.** 8 **23.** 32 **25.** 4 **27.** 0 **29.** True **31.** False **33.** False **35.** True **37.** {3, 5}
39. {7, 9} **41.** ∅ **43.** U, or {2, 3, 4, 5, 7, 9} **45.** All students in this school not taking this course **47.** All
students in this school taking accounting and zoology **49.** 6 **51.** 5 **53.** 1 **55.** B' is the set of all stocks on the list
whose price-to-earnings ratio is less than 10; $B' = $ {ATT, GE, Hershey, Mobil, RCA} **57.** $(A \cap B)'$ is the set of all stocks with
a dividend less than or equal to $3 *or* a price-to-earnings ratio less than 10}; $(A \cap B)' = $ {ATT, GE, Hershey, Mobil, RCA}
59. False **61. (a)** {s, d, c, g, i, m, h} = U **(b)** {s, d, c} **(c)** {i, m, h} **(d)** {g} **(e)** {s, d, c, g, i, m, h}
(f) {s, d, c}

▰ SECTION 5.2 (PAGE 231)

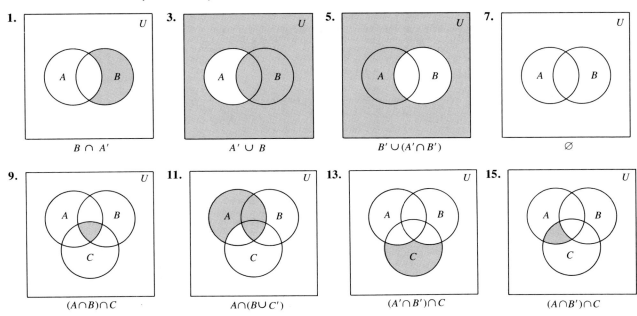

1. $B \cap A'$ **3.** $A' \cup B$ **5.** $B' \cup (A' \cap B')$ **7.** ∅

9. $(A \cap B) \cap C$ **11.** $A \cap (B \cup C')$ **13.** $(A' \cap B') \cap C$ **15.** $(A \cap B') \cap C$

17. 9   **19.** 16

21.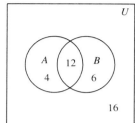

23.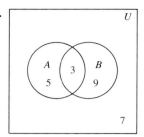

25.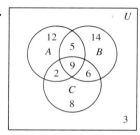

27.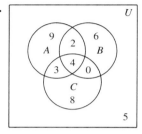

35. **(a)** 75 **(b)** 39 **(c)** 14 **(d)** 40 **(e)** 61   **37.** **(a)** 37 **(b)** 22 **(c)** 50 **(d)** 11 **(e)** 25 **(f)** 11
39. **(a)** 50 **(b)** 2 **(c)** 14   **41.** **(a)** 54 **(b)** 17 **(c)** 10 **(d)** 7 **(e)** 15 **(f)** 3 **(g)** 12 **(h)** 1   **43.** **(a)** 40 **(b)** 30
(c) 95 **(d)** 110 **(e)** 160 **(f)** 65

▰▰ SECTION 5.3   (PAGE 241)

1. 12   **3.** 720   **5.** 1   **7.** 24   **9.** 156   **11.** 2.490952×10^{15} **13.** 240,240   **15.** 30   **17.** **(a)** 27,600
(b) 35,152   **(c)** 1104   **19.** 15   **21.** **(a)** 17,576,000   **(b)** 11,232,000   **(c)** 22,464,000   **23.** 720   **25.** 120
27. 2730   **29.** 1,256,640   **31.** 1440   **33.** **(a)** 840 **(b)** 180 **(c)** 420   **35.** **(a)** 362,880 **(b)** 1728 **(c)** 1260
37. 55,440

▰▰ SECTION 5.4   (PAGE 250)

1. 56   **3.** 792   **5.** 1   **7.** 352,716   **9.** 2,042,975   **11.** 19,448   **13.** 1326   **15.** 65,780   **17.** **(a)** 10
(b) 7   **19.** 50   **21.** **(a)** 9 **(b)** 6 **(c)** 3; yes from both   **23.** Combinations; 6   **25.** Combinations; **(a)** 56 **(b)** 462
(c) 3080   **27.** Combinations; **(a)** 220 **(b)** 220   **29.** Permutations; 479,001,600   **31.** Combinations; **(a)** 84 **(b)** 10
(c) 40 **(d)** 28   **33.** Combinations; **(a)** 21 **(b)** 6 **(c)** 11   **35.** Permutations; 5040   **37.** Combinations **(a)** 10 **(b)** 0
(c) 1 **(d)** 10 **(e)** 30 **(f)** 15 **(g)** 0   **39.** Permutations; **(a)** 720 **(b)** 360   **41.** 2,598,960   **43.** 6.3501356×10^{11}

▰▰ SECTION 5.5   (PAGE 256)

1. $m^4 + 4m^3n + 6m^2n^2 + 4mn^3 + n^4$ **3.** $729x^6 - 2916x^5y + 4860x^4y^2 - 4320x^3y^3 + 2160x^2y^4 - 576xy^5 + 64y^6$
5. $m^5/32 - 15m^4n/16 + 45m^3n^2/4 - 135m^2n^3/2 + 405mn^4/2 - 243n^5$ **7.** $p^{10} + 10p^9q + 45p^8q^2 + 120p^7q^3$
9. $a^{15} + 30a^{14}b + 420a^{13}b^2 + 3640a^{12}b^3$ **11.** 1024   **13.** 64   (This result includes a pizza made with *none* of the six
toppings.)   **15.** 60,439   **17.** 64

▤ CHAPTER 5 REVIEW EXERCISES (PAGE 257)

1. False **3.** False **5.** True **7.** False **9.** False **11.** {6, 7}; 2 **13.** {1, 2, 3, 4}; 4 **15.** 16
17. {a, b, e} **19.** {c, d, g} **21.** {a, b, e, f} **23.** *U* **25.** All employees in the accounting department who also have at least 10 years with the company **27.** All employees who are in the accounting department or who have MBA degrees
29. All employees who are not in the sales department and who have worked less than 10 years with the company
31.

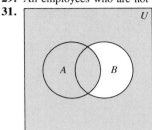

$A \cup B'$

33.

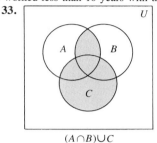

$(A \cap B) \cup C$

35. 52 **37.** 12 **39.** 720 **41. (a)** 90 **(b)** 10 **(c)** 120 **(d)** 220 **43.** 24
45. (a) 840 **(b)** 2045 (This assumes that a dinner must include at least one item from either A or B.)
47. $32m^5 + 80m^4n + 80m^3n^2 + 40m^2n^3 + 10mn^4 + n^5$
49. $x^{20} - 10x^{19}y + 95x^{18}y^2/2 - 285x^{17}y^3/2$ **51.** 247

▤ CHAPTER 6 SECTION 6.1 (PAGE 267)

1. {January, February, March, . . . , December} **3.** {0, 1, 2, . . . , 80} **5.** {go ahead, cancel}
7. {0, 1, 2, 3, . . . , 5000} **9.** {*hhhh, hhht, hhth, hthh, thhh, hhtt, htth, tthh, thth, thht, htht, httt, thtt, ttht, ttth, tttt*}
11. {*h, th, tth, ttth, tttth,* . . .} **13.** {(1, 1), (1, 2), (1, 3), (1, 4), (1, 5), (1, 6), (2, 1), (2, 2), (2, 3), (2, 4), (2, 5), (2, 6), (3, 1), (3, 2), (3, 3), (3, 4), (3, 5), (3, 6), (4, 1), (4, 2), (4, 3), (4, 4), (4, 5), (4, 6), (5, 1), (5, 2), (5, 3), (5, 4), (5, 5), (5, 6), (6, 1), (6, 2), (6, 3), (6, 4), (6, 5), (6, 6)} **(a)** {(3, 1), (3, 2), (3, 3), (3, 4), (3, 5), (3, 6)} **(b)** {(2, 6), (3, 5), (4, 4), (5, 3), (6, 2)} **(c)** ∅ **15.** {*hh, hth, thh, htth, thth, tthh, tthh, tthh, tthth, thtth, htth, htttt, thttt, tthtt, tttht, ttth, tttt*}; **(a)** {*hh*}
(b) {*hth, thh*} **(c)** {*tttt, ttth, ttht, tthtt, thtt, htttt*} **17. (a)** Worker is male **(b)** Worker has worked 5 yr or more.
(c) Worker is female and has worked less than 5 yr. **(d)** Worker has worked less than 5 yr or has contributed to a voluntary retirement plan. **(e)** Worker is female or does not contribute to a voluntary retirement plan. **(f)** Worker has worked 5 yr or more and does not contribute to a voluntary retirement plan. **19.** 1/6 **21.** 2/3 **23.** 1/13 **25.** 1/26 **27.** 1/52
29. {2}, {4}, {6} **31.** {ace of spades} **33.** {profit goes up}, {profit does down}, {profit stays the same} **35.** {(1, 1)},
{(2, 2)}, {(3, 3)}, {(4, 4)}, {(5, 5)}, {(6, 6)} **37.** Possible **39.** Not possible **41.** Not possible

▤ SECTION 6.2 (PAGE 278)

1. No **3.** No **5.** Yes **7. (a)** 1/36 **(b)** 1/12 **(c)** 1/9 **(d)** 5/36 **9. (a)** 5/18 **(b)** 5/12 **(c)** 11/36
11. (a) 2/13 **(b)** 7/13 **(c)** 3/26 **(d)** 3/4 **13. (a)** 1/2 **(b)** 2/5 **(c)** 3/10 **15. (a)** 1/10 **(b)** 2/5 **(c)** 7/20
17. (a) 7/18 **(b)** 13/18 **(c)** 5/9 **(d)** 2/3 **(e)** 2/9 **19.** Not empirical **21.** Empirical **23.** Empirical **25.** Not empirical **27.** Not empirical **29. (a)** .12 **(b)** .06 **(c)** .24 **31. (a)** .25 **(b)** .75 **(c)** .36 **33. (a)** .50 **(b)** .24
(c) .09 **(d)** .83 **35.** 1 to 5 **37.** 2 to 1 **39. (a)** 6 to 19 **(b)** 1 to 1 **41.** 11 to 4 **43.** 4/11 **45.** No; "odds *in favor* of a direct hit are very low . . ." **47.** Yes **49.** No **51. (a)** .951 **(b)** .527 **(c)** .473 **(d)** .466 **(e)** .007
(f) .515 **53. (a)** 1/4 **(b)** 1/2 **(c)** 1/4
Theoretical probabilities are given for Exercises 55 and 57. Answers using the Monte Carlo method will vary.
55. (a) 5/32 **(b)** .03125 **57.** .632

▄▄ SECTION 6.3 (PAGE 292)

1. 0 **3.** 1 **5.** 1/6 **7.** 4/17 **9.** 25/51 **11.** .00050 **13.** .1875 **15.** .00198 **17.** .296 **19.** 1/10
21. 0 **23.** 2/7 **25.** 2/7 **27.** .05 **29.** .25 **31.** Not very realistic **33.** .999985; fairly realistic **35.** True
37. True **39.** True **41.** False **43.** 0 **47.** The probability of a customer cashing a check, given that the customer
made a deposit, is 5/7. **49.** The probability of a customer not cashing a check, given that the customer did not make a deposit,
is 1/4. **51.** The probability of a customer not both cashing a check and making a deposit is 6/11. **53.** 1/6 **55.** 0
57. 2/3 **59.** .06 **61.** 1/4 **63.** 1/4 **65.** 1/7 **67.** .049 **69.** .534 **71.** 42/527 or .080 **73.** Yes
75. 1/2000 = .0005 **77.** $(1999/2000)^a$ **79.** $(1999/2000)^{Nc}$ **81.** .259

▄▄ SECTION 6.4 (PAGE 302)

1. 1/3 **3.** 2/41 **5.** 21/41 **7.** 8/17 **11.** .146 **13.** $119/131 \approx .908$ **15.** .824 **17.** $1/176 \approx .006$
19. 1/11 **21.** 5/9 **23.** 5/26 **25.** 72/73 **27.** 165/343 **29.** .082

▄▄ SECTION 6.5 (PAGE 310)

1. 7/9 **3.** 5/12 **5.** .424 **7.** 1/6 **9.** 3/10 **11.** 1326 **13.** $33/221 \approx .149$ **15.** $52/221 \approx .235$
17. $130/221 \approx .588$ **19.** $(1/26)^5 \approx 8.42 \times 10^{-8}$ **21.** $1 \cdot 25 \cdot 24 \cdot 23 \cdot 22/(26)^4 = 18,975/28,561 \approx .664$
23. $4/\binom{52}{5} = 1/649,740 \approx 1.54 \times 10^{-6}$ **25.** $13 \cdot \binom{4}{4}\binom{48}{1}/\binom{52}{5} = 1/4165 \approx 2.40 \times 10^{-4}$ **27.** $1/\binom{52}{13}$
29. $\binom{4}{3}\binom{4}{3}\binom{44}{7}/\binom{52}{13}$ **31. (a)** .054 **(b)** .011 **(c)** .046 **(d)** .183 **(e)** .387 **(f)** .885 **33.** $1 - P(365, 41)/(365)^{41} \approx .9032$
35. 1 **37.** $120/343 \approx .3499$ **39.** 6 **41.** 362,880 **43.** 1/3

▄▄ SECTION 6.6 (PAGE 318)

1. $5/16 \approx .313$ **3.** $1/32 \approx .031$ **5.** $3/16 \approx .188$ **7.** $13/16 \approx .813$ **9.** .0000000005 **11.** .269 **13.** .875
15. $1/32 \approx .031$ **17.** $13/16 \approx .813$ **19.** $5/32 \approx .156$ **21.** $9/256 \approx .035$ **23.** .0217 **25.** .0394 **29.** .026
31. .999 **33.** .016 **35.** .442 **37.** 9 teams **39.** .246 **41.** .017 **43.** .243 **45.** .925 **47.** .072
49. .035 **51.** .096 **53.** .104 **55.** .185 **57.** .256 **59.** .0025 **61.** .0055 **63.** .000078 **65. (a)** .199
(b) .568 **(c)** .133 **(d)** .794 **67. (a)** .000036 **(b)** .000054 **(c)** 0 **69. (a)** .148774 **(b)** .021344 **(c)** .978656

▄▄ CHAPTER 6 REVIEW EXERCISES (PAGE 322)

1. {1, 2, 3, 4, 5, 6} **3.** {0, .5, 1, 1.5, 2, · · · , 299.5, 300} **5.** {(3, R), (3, G), (5, R), (5, G), (7, R), (7, G), (9, R),
(9, G), (11, R), (11, G)} **7.** {(3, G), (5, G), (7, G), (9, G), (11, G)} **9.** $E' \cap F'$ or $(E \cup F)'$ **11.** 1/4 **13.** 3/13
15. 1/2 **17.** 1 **19.** 1 to 25 **21.** .86 **23.** $5/36 \approx .139$ **25.** $1/6 \approx .167$ **27.** $1/6 \approx .167$
29. $2/11 \approx .182$ **31.** .66 **33.** .71 **35.** $4/165 \approx .024$ **37.** $2/11 \approx .182$ **39.** $4/33 \approx .121$ **41.** $25/102 \approx$
.245 **43.** $15/34 \approx .441$ **45.** $4/17 \approx .235$ **47.** 3/10 **49.** 2/5 or .4 **51.** 3/4 or .75 **53.** $19/22 \approx .864$
55. 1/3 **57.** $1/7 \approx .143$ **59.** $5/16 \approx .313$ **61.** $11/32 \approx .344$ **63.** .00004 **65.** 1.0000 **67. (a)** 8 wells
(b) 10 wells **69.** 1/4 **71.** 1/4

▄▄ EXTENDED APPLICATION (PAGE 326)

1. .715; .569; .410; .321; .271 **2.** There are 94 other possible values for *n*, namely, 6, 7, 8, . . . , 99. Many of these are
quite unlikely, but, for instance, 3rd and 10 is not uncommon. Hence, the outcomes listed do not exhaust the sample space.

▄▄ EXTENDED APPLICATION (PAGE 326)

1. .076 **2.** .542 **3.** .051

▦ CHAPTER 7   SECTION 7.1   (PAGE 334)

1. (a)

Number	0	1	2	3	4	5
Probability	0	0	.1	.3	.4	.2

(b)

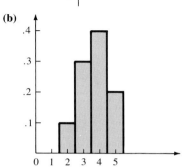

3. (a)

Number	0	1	2	3	4	5	6
Probability	0	.04	0	.16	.40	.32	.08

(b)

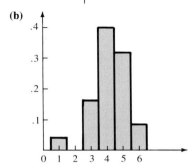

5. (a)

Number	0	1	2	3	4	5
Probability	.15	.25	.3	.15	.1	.05

(b)

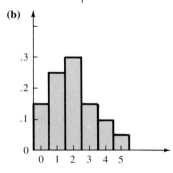

7.

Number of heads	0	1	2	3	4
Probability	1/16	1/4	3/8	1/4	1/16

9.

Number of aces	0	1	2	3
Probability	$4324/5525 \approx .783$	$1128/5525 \approx .204$	$72/5525 \approx .013$	$1/5525 \approx .0002$

11.

Number of hits	0	1	2	3	4
Probability	.254	.415	.254	.069	.007

13.

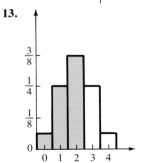

15.

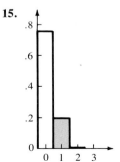

17.

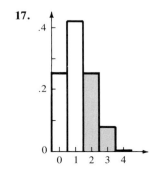

19.

E:	17.0%	T:	9.1%	N:	7.1%	R:	7.1%	I:	6.6%	A:	6.2%	H:	5.8%
S:	5.8%	L:	5.0%	C:	4.6%	O:	4.1%	F:	3.7%	G:	3.7%	U:	3.7%
D:	2.1%	Y:	2.1%	M:	1.7%	P:	1.7%	W:	1.7%	Q:	1.2%	V:	.4%
B:	0%	J:	0%	K:	0%	X:	0%	Z:	0%				

21. Five of them are, but H would be a better choice than L.

▬ SECTION 7.2 (PAGE 343)

1. 3.6 **3.** 14.64 **5.** 2.7 **7.** 18 **9.** 51 **11.** 130 **13.** 49 **15.** 9 **17.** 64 **19.** 6.1, 6.3
21. $-64¢$; no **23.** $9/7 \approx 1.3$ **25.** (a) $5/3 \approx 1.67$ (b) $4/3 \approx 1.33$ **27.** 1 **29.** No, the expected value is about
$-21¢$. **31.** $-2.7¢$ **33.** $-20¢$ **35.** (a) Yes, the probability of a match is still 1/2. (b) 40¢ (c) $-40¢$ **37.** 2.5
39. 10¢; no **41.** Job A, which offers the higher expected value of $45,000 **43.** 1.58 **45.** $4500

47.

Account	EV	Total	Class
3	$2000	$22,000	C
4	$1000	$51,000	B
5	$25,000	$30,000	C
6	$60,000	$60,000	A
7	$16,000	$46,000	B

49. (a) $94.0 million for seeding; $116.0 million for not seeding
(b) Seed

▬ SECTION 7.3 (PAGE 354)

1. $\sqrt{407.4} \approx 20.2$ **3.** $\sqrt{215.3} \approx 14.7$ **5.** $\sqrt{59.4} \approx 7.7$ **7.** Var $= .81$; $\sigma = .9$ **9.** Var $= .000124$; $\sigma = .011$
11. .700 **13.** .745 **15.** (a) **17.** 2.41 **19.** (a) At least 3/4 (b) At least 15/16 (c) At least 24/25
21. (a) Mean $= \$320$; $\sigma = \$170.81$ (b) 6 (c) 6 (d) At least 3/4 **23.** (a) Mean $= 25.5$; $\sigma = 6.9$ (b) Forever Power
(c) Forever Power **25.** (a) 1/3, 2, $-1/3$, 0, 5/3, 7/3, 1, 4/3, 7/3, 2/3 (b) 2.1, 2.6, 1.5, 2.6, 2.5, .6, 1.0, 2.1, .6, 1.2
(c) 1.13 (d) 1.68 (e) 4.41, -2.15 (f) 4.31, 0 **27.** Mean $= 46.647$; $\sigma = .932$ **29.** Mean $= 5.0876$; $\sigma = .1065$

▬ SECTION 7.4 (PAGE 367)

1. 49.38% **3.** 17.36% **5.** 45.64% **7.** 49.91% **9.** 7.7% **11.** 47.35% **13.** 92.42% **15.** 32.56%
17. -1.64 or -1.65 **19.** 1.04 **21.** .0062 **23.** .4325 **25.** $38.62 and $25.88 **27.** .0823 **29.** 5000
31. 4332 **33.** 642 **35.** 9918 **37.** 19 **39.** .1587 **41.** .0062 **43.** 84.13% **45.** 37.79% **47.** 2.28%
49. 99.38% **51.** 189 **53.** 6.68% **55.** 38.3% **57.** 82 **59.** 70
Answers to Exercises 61–69 may vary slightly because of different computers or software.
61. .0796 **63.** .9933 **65.** .3172 **67.** .1527 **69.** .0051

▬ SECTION 7.5 (PAGE 376)

1. (a)

x	0	1	2	3	4	5	6
$P(x)$	.335	.402	.201	.054	.008	.001	.000

(b) 1.00 **(c)** .91

3. (a)

x	0	1	2	3
$P(x)$	.941	.058	.001	.000

(b) .06 **(c)** .24

5. (a)

x	0	1	2	3	4
$P(x)$	.0081	.0756	.2646	.4116	.2401

(b) 2.8 **(c)** .92

7. 12.5; 3.49   **9.** 51.2; 3.2   **11.** .1974   **13.** .1210   **15.** .0240   **17.** .9032   **19.** .8665   **21.** .0956
23. .0760   **25.** .6443   **27.** .9945   **29.** .0146   **31.** .1974   **33.** .0092   **35.** .0001   **37.** For Exercise 67:
(a) .0001,   **(b)** .0002,   **(c)** .0000; for Exercise 68: **(a)** .0472,   **(b)** .9875,   **(c)** .8814; for Exercise 69: **(a)** .1686,   **(b)** .0304,
(c) .9575; for Exercise 70: **(a)** .0000,   **(b)** .0018

▆▆ CHAPTER 7   REVIEW EXERCISES   (PAGE 378)

1. (a)

Number	1	2	3	4	5
Probability	.125	.292	.375	.125	.083

(b)

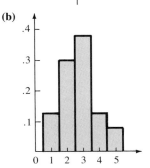

3. (a)

Number	0	1	2	3
Probability	1/8 = .125	3/8 = .375	3/8 = .375	1/8 = .125

(b)

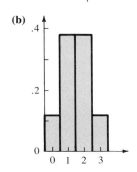

5. (a)

Number	0	1	2	3	4	5
Probability	.1	.1	.2	.3	.3	0

(b)

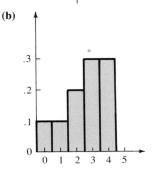

7. .6   **9.** −28¢   **11.** \$8500   **13.** 11.87   **15.** 2.6   **17.** \$1.29
19. Variance = 1.64; σ = 1.28   **21. (a)** 98.76% **(b)** Chebyshev's
inequality would give "at least 84%."   **23.** z = 1.41   **25. (a)** Stock 1:
8%; 6.5%; Stock 2: 8%; 2.2%   **(b)** Stock 2

27.

x	0	1	2	3	4
Probability	.9801	.0197	.0001485	.0000004975	6×10^{-10}

Mean = .02; σ = .141

29. (a) .0959   **(b)** .0027
31. (a) 25.14%   **(b)** 28.10%   **(c)** 22.92%   **(d)** 56.25%

▆▆ EXTENDED APPLICATION   (PAGE 381)

1. (a) \$69.01   **(b)** \$79.64   **(c)** \$58.96   **(d)** \$82.54   **(e)** \$62.88   **(f)** \$64.00   **2.** Stock only part 3 on the truck
3. The events of needing parts 1, 2, and 3 are not the only events in the sample space.   **4.** 2^n

▆▆ EXTENDED APPLICATION   (PAGE 382)

1. .9886 million dollars   **2.** −.9714 million dollars

CHAPTER 8   SECTION 8.1   (PAGE 392)

1. No   **3.** Yes   **5.** No   **7.** Yes   **9.** No   **11.** Yes

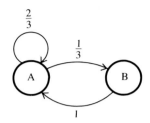

13. No   **15.** No   **17.** Yes; $\begin{bmatrix} .9 & .1 & 0 \\ .1 & .6 & .3 \\ 0 & .3 & .7 \end{bmatrix}$

19. $A = \begin{bmatrix} 1 & 0 \\ .8 & .2 \end{bmatrix}$; $A^2 = \begin{bmatrix} 1 & 0 \\ .96 & .04 \end{bmatrix}$; $A^3 = \begin{bmatrix} 1 & 0 \\ .992 & .008 \end{bmatrix}$; 0

21. $C = \begin{bmatrix} .5 & .5 \\ .72 & .28 \end{bmatrix}$; $C^2 = \begin{bmatrix} .61 & .39 \\ .5616 & .4384 \end{bmatrix}$; $C^3 = \begin{bmatrix} .5858 & .4142 \\ .596448 & .403552 \end{bmatrix}$; .4142

23. $E = \begin{bmatrix} .8 & .1 & .1 \\ .3 & .6 & .1 \\ 0 & 1 & 0 \end{bmatrix}$; $E^2 = \begin{bmatrix} .67 & .24 & .09 \\ .42 & .49 & .09 \\ .3 & .6 & .1 \end{bmatrix}$; $E^3 = \begin{bmatrix} .608 & .301 & .091 \\ .483 & .426 & .091 \\ .42 & .49 & .09 \end{bmatrix}$; .301

25. (a) $\begin{array}{c} \\ \text{Red} \\ \text{White} \end{array} \begin{array}{c} \text{Red} \quad \text{White} \\ \begin{bmatrix} .75 & .25 \\ .50 & .50 \end{bmatrix} \end{array}$ **(b)** [.75 .25]   **(c)** [.6875 .3125]   **(d)** [.6719 .3281]   **(e)** [.6680 .3320]   **(f)** [.6670 .3330]

27. (a) [.53 .47]   **(b)** [.5885 .4115]   **(c)** [.6148 .3852]   **(d)** [.62666 .37334]   **29. (a)** 42,500; 5000; 2500
(b) 36,125; 8250; 5625   **(c)** 30,706; 10,213; 9081   **(d)** 26,100; 11,241; 12,659   **31. (a)** [.257 .597 .146]
(b) [.255 .594 .151]   **(c)** [.254 .590 .156]

33. (a) $\begin{array}{c} \\ \text{Single} \\ \text{Multiple} \end{array} \begin{array}{c} \text{Single} \quad \text{Multiple} \\ \begin{bmatrix} .90 & .10 \\ .05 & .95 \end{bmatrix} \end{array}$ **(b)** [.75 .25]   **(c)** 68.8% in single-family dwellings, 31.3% in multiple-family dwellings
(d) 63.4% in single-family dwellings, 36.6% in multiple-family dwellings

35. The first power is the given transition matrix; $\begin{bmatrix} .2 & .15 & .17 & .19 & .29 \\ .16 & .2 & .15 & .18 & .31 \\ .19 & .14 & .24 & .21 & .22 \\ .16 & .19 & .16 & .2 & .29 \\ .16 & .19 & .14 & .17 & .34 \end{bmatrix}$; $\begin{bmatrix} .17 & .178 & .171 & .191 & .29 \\ .171 & .178 & .161 & .185 & .305 \\ .18 & .163 & .191 & .197 & .269 \\ .175 & .174 & .164 & .187 & .3 \\ .167 & .184 & .158 & .182 & .309 \end{bmatrix}$;

$\begin{bmatrix} .1731 & .175 & .1683 & .188 & .2956 \\ .1709 & .1781 & .1654 & .1866 & .299 \\ .1748 & .1718 & .1753 & .1911 & .287 \\ .1712 & .1775 & .1667 & .1875 & .2971 \\ .1706 & .1785 & .1641 & .1858 & .301 \end{bmatrix}$; $\begin{bmatrix} .17193 & .17643 & .1678 & .18775 & .29609 \\ .17167 & .17689 & .16671 & .18719 & .29754 \\ .17293 & .17488 & .17007 & .18878 & .29334 \\ .17192 & .17654 & .16713 & .18741 & .297 \\ .17142 & .17726 & .16629 & .18696 & .29807 \end{bmatrix}$; .18719

37. (a) .847423 or about 85%   **(b)** [.0128 .0513 .0962 .8397]

SECTION 8.2 (PAGE 402)

1. Regular   **3.** Not regular   **5.** Regular   **7.** [2/5 3/5]   **9.** [4/11 7/11]   **11.** [14/83 19/83 50/83]
13. [170/563 197/563 196/563]   **15.** [2/3 1/3]   **17.** [0 0 1]   **19.** [2/17 11/17 4/17]   **21.** [.244 .529 .227]

25.

$$\begin{array}{c} \\ \text{Works} \\ \text{Doesn't Work} \end{array} \begin{array}{cc} \text{Works} & \text{Doesn't Work} \\ \begin{bmatrix} .9 & .1 \\ .7 & .3 \end{bmatrix} \end{array}; \quad \text{long-range probability of line working correctly is 7/8}$$

27. [51/209 88/209 70/209]   **29.** [1/4 1/2 1/4]   **31.** [.348 .286 .366]
33. [.171898 .176519 .167414 .187526 .296644]

SECTION 8.3 (PAGE 411)

1. State 2 is absorbing; matrix is that of an absorbing Markov chain.   **3.** State 2 is absorbing; matrix is not that of an absorbing Markov chain.   **5.** States 2 and 4 are absorbing; matrix is that of an absorbing Markov chain.   **7.** States 2 and 4 are absorbing; matrix is that of an absorbing Markov chain.   **9.** $F = [2]$; $FR = [.4 \quad .6]$ **11.** $F = [5]$; $FR = [3/4 \quad 1/4]$
13. $F = [3/2]$; $FR = [1/2 \quad 1/2]$

15. $F = \begin{bmatrix} 1 & 0 \\ 1/3 & 4/3 \end{bmatrix}$; $FR = \begin{bmatrix} 1/3 & 2/3 \\ 4/9 & 5/9 \end{bmatrix}$ **17.** $F = \begin{bmatrix} 25/17 & 5/17 \\ 5/34 & 35/34 \end{bmatrix}$; $FR = \begin{bmatrix} 4/17 & 15/34 & 11/34 \\ 11/34 & 37/68 & 9/68 \end{bmatrix}$

19. **(a)** $F = \begin{bmatrix} 3/2 & 1 & 1/2 \\ 1 & 2 & 1 \\ 1/2 & 1 & 3/2 \end{bmatrix}$; $FR = \begin{bmatrix} 3/4 & 1/4 \\ 1/2 & 1/2 \\ 1/4 & 3/4 \end{bmatrix}$ **(b)** 3/4   **(c)** 1/4   **21.** .8756

23.

p	.1	.2	.3	.4	.5	.6	.7	.8	.9
x_a	.9999999997	.99999905	.99979	.98295	.5	.017046	.000209	.00000095	.0000000003

25. **(a)**

$$\begin{array}{c} \\ \text{Automobile} \\ \text{Light Rail} \\ \text{Stay Out} \end{array} \begin{array}{ccc} \text{Automobile} & \begin{array}{c}\text{Light}\\\text{Rail}\end{array} & \begin{array}{c}\text{Stay}\\\text{Out}\end{array} \\ \begin{bmatrix} .80 & .15 & .05 \\ .05 & .80 & .15 \\ 0 & 0 & 1 \end{bmatrix} \end{array};\quad F = \begin{bmatrix} 6.154 & 4.615 \\ 1.538 & 6.154 \end{bmatrix}; FR = \begin{bmatrix} 1.000 \\ 1.000 \end{bmatrix}$$ **(b)** 1.000

27. **(a)** $F = \begin{bmatrix} 1.0625 & .1875 \\ .0625 & 1.1875 \end{bmatrix}$; $FR = \begin{bmatrix} 1 \\ 1 \end{bmatrix}$ **(b)** 0   **29.** **(a)** .6   **(b)** .5

CHAPTER 8 REVIEW EXERCISES (PAGE 414)

1. Yes   **3.** Yes   **5.** **(a)** $C = \begin{bmatrix} .6 & .4 \\ 1 & 0 \end{bmatrix}$; $C^2 = \begin{bmatrix} .76 & .24 \\ .6 & .4 \end{bmatrix}$; $C^3 = \begin{bmatrix} .696 & .304 \\ .76 & .24 \end{bmatrix}$ **(b)** .76

7. **(a)** $E = \begin{bmatrix} .2 & .5 & .3 \\ .1 & .8 & .1 \\ 0 & 1 & 0 \end{bmatrix}$; $E^2 = \begin{bmatrix} .09 & .8 & .11 \\ .1 & .79 & .11 \\ .1 & .8 & .1 \end{bmatrix}$; $E^3 = \begin{bmatrix} .098 & .795 & .107 \\ .099 & .792 & .109 \\ .1 & .79 & .11 \end{bmatrix}$ **(b)** .099   **9.** [.453 .547];

[5/11 6/11] or [.455 .545]   **11.** [.48 .28 .24]; [47/95 26/95 22/95] or [.495 .274 .232]   **13.** Regular
15. Not regular   **17.** State 2 is absorbing; matrix is not that of an absorbing Markov chain.   **19.** $F = [5/4]$; $FR = [5/8 \quad 3/8]$
21. $F = \begin{bmatrix} 7/4 & 4/5 \\ 1 & 8/5 \end{bmatrix}$; $FR = \begin{bmatrix} .55 & .45 \\ .6 & .4 \end{bmatrix}$ **23.** **(a)** [.54 .46]   **(b)** [.6464 .3536]   **25.** [.428 .322 .25]

27. [.431 .284 .285] **29.** .2 **31.** .196 **33.** .28 **35.** [.195 .555 .25] **37.** [.194 .556 .25]

39. States 1, 3, and 6 **41.** $F = \begin{bmatrix} 5/2 & 1 & 1/2 \\ 1 & 2 & 1 \\ 1/2 & 1 & 5/2 \end{bmatrix}$; $FR = \begin{bmatrix} 11/16 & 1/8 & 3/16 \\ 3/8 & 1/4 & 3/8 \\ 3/16 & 1/8 & 11/16 \end{bmatrix}$

CHAPTER 9   SECTION 9.1   (PAGE 420)

1. (a) Buy speculative **(b)** Buy blue-chip **(c)** Buy speculative; $24,300 **(d)** Buy blue-chip **3. (a)** Set up in the stadium **(b)** Set up in the gym **(c)** Set up in both; $1010

5. (a)
$$\begin{array}{c} \\ \text{Market} \\ \text{Don't Market} \end{array} \begin{array}{cc} \text{Better} & \text{Not Better} \\ \begin{bmatrix} 50,000 & -25,000 \\ -40,000 & -10,000 \end{bmatrix} \end{array}$$
(b) $5000 if they market new product, −$22,000 if they don't; market the new product

7. (a)
$$\begin{array}{c} \\ \text{Bid } \$30,000 \\ \text{Bid } \$40,000 \end{array} \begin{array}{cc} \text{Strike} & \text{No Strike} \\ \begin{bmatrix} -5500 & -4500 \\ -4500 & 0 \end{bmatrix} \end{array}$$
(b) $40,000

9. Emphasize environment; 14.25

SECTION 9.2   (PAGE 427)

1. $6 from B to A **3.** $2 from A to B **5.** $1 from A to B **7.** Yes, column 2 dominates column 3.

9. $\begin{bmatrix} -2 & 8 \\ -1 & -9 \end{bmatrix}$ **11.** $\begin{bmatrix} 4 & -1 \\ 3 & 5 \end{bmatrix}$ **13.** $\begin{bmatrix} 8 & -7 \\ -2 & 4 \end{bmatrix}$ **15.** (1, 1); 3; strictly determined

17. No saddle point; not strictly determined **19.** (3, 1); 3; strictly determined **21.** (1, 3); 1; strictly determined

23. No saddle point; not strictly determined **25.**
$$\begin{array}{c} \\ \text{Stone} \\ \text{Scissors} \\ \text{Paper} \end{array} \begin{array}{ccc} \text{Stone} & \text{Scissors} & \text{Paper} \\ \begin{bmatrix} 0 & 1 & -1 \\ -1 & 0 & 1 \\ 1 & -1 & 0 \end{bmatrix} \end{array}; \text{no}$$
29. (2, 3); 6

31.
$$\begin{array}{cc} & \begin{array}{ccc} & B & \\ 1 & 2 & 3 \end{array} \\ A \begin{array}{c} 1 \\ 2 \\ 3 \end{array} & \begin{bmatrix} 15 & -2 & 6 \\ 7 & 15 & 9 \\ 3 & -3 & 15 \end{bmatrix} \end{array}; \text{no}$$

SECTION 9.3   (PAGE 437)

1. (a) −1 **(b)** −.28 **(c)** −1.54 **(d)** −.46 **3.** Player A: 1: 1/5, 2: 4/5; player B: 1: 3/5, 2: 2/5; value 17/5
5. Player A: 1: 7/9, 2: 2/9; player B: 1: 4/9, 2: 5/9; value −8/9 **7.** Player A: 1: 8/15, 2: 7/15; player B: 1: 2/3, 2: 1/3; value 5/3 **9.** Player A: 1: 6/11, 2: 5/11; player B: 1: 7/11, 2: 4/11; value −12/11 **11.** Strictly determined game; saddle point at (2, 2); value −5/12 **13.** Player A: 1: 2/5, 2: 3/5; player B: 1: 1/5, 2: 4/5; value 7/5 **15.** Player A: 1: 1/14, 2: 0, 3: 13/14; player B: 1: 1/7, 2: 6/7; value 50/7 **17.** Player A: 1: 2/3, 2: 1/3; player B: 1: 0, 2: 1/9, 3: 8/9; value 10/3 **19.** Player A: 1: 0, 2: 3/4, 3: 1/4; player B: 1: 0, 2: 1/12, 3: 11/12; value 33/4 **21.** Player A: 1: 1/2, 2: 1/2; Player B: 1: 1/2, 2: 1/2; value 0; fair game

23. (a)

		Number of Fingers	
		0	2
Number of Fingers	0	0	−2
	2	−2	4

(b) For both players A and B: choose 0 with probability 3/4 and 2 with probability 1/4. The value of the game is − 1/2.

25. Allied should use T.V. with probability 10/27 and use radio with probability 17/27. The value of the game is 1/18, which represents increased sales of $55,556. **27.** He should invest in rainy day goods about 5/9 of the time and sunny day goods about 4/9 of the time for a steady profit of $72.22.

CHAPTER 9 REVIEW EXERCISES (PAGE 441)

1. Hostile **3.** Hostile; $785 **5.** Oppose **7.** Oppose; 2700 **9.** $2 from A to B **11.** $7 from B to A

13. Row 3 dominates row 1; column 1 dominates column 4 **15.** $\begin{bmatrix} -11 & 6 \\ -10 & -12 \end{bmatrix}$ **17.** $\begin{bmatrix} -2 & 4 \\ 3 & 2 \\ 0 & 3 \end{bmatrix}$

19. (1, 1); value − 2 **21.** (2, 2); value 0; fair game **23.** (2, 3); value − 3 **25.** Player A: 1: 5/6, 2: 1/6; player B: 1: 1/2, 2: 1/2; value 1/2 **27.** Player A: 1: 1/9, 2: 8/9; player B: 1: 5/9, 2: 4/9; value 5/9 **29.** Player A: 1: 1/5, 2: 4/5; player B: 1: 3/5, 2: 0, 3: 2/5; value − 12/5 **31.** Player A: 1: 3/4, 2: 1/4, 3: 0; player B: 1: 1/2, 2: 1/2; value 1/2

EXTENDED APPLICATION (PAGE 443)

1. $E_1 = 18.61M$, $E_2 = 2.898M − 42.055$, $E_3 = .56M − 48.6$ **2.** $E_1 = 19.7325M$, $E_2 = .556M − 48.475$, $E_3 = .108M − 49.73$ **3.** $E_1 = 19M$, $E_2 = 1.7M − 45$, $E_3 = .6M − 48.5$

EXTENDED APPLICATION (PAGE 446)

1.

		Japanese	
		North	South
Kenney	North	2	2
	South	1	3

2. (North, North); Kenney should concentrate reconnaissance on the northern route and the Japanese should sail the northern route

CHAPTER 10 SECTION 10.1 (PAGE 455)

1. $50 **3.** $1312.50 **5.** $72.54 **7.** $168 **9.** $83.76 **11.** $292.60 **13.** $258.75 **15.** $109.89 **17.** $14,423.08 **19.** $5205.72 **21.** $15,072.29 **23.** $6363.50 **25.** $336.89 **27.** $27,894.30 **29.** $1732.90 **31.** $5827.29 **33.** $3773.00; 13.58% **35.** $7251.52

Answers in the remainder of this chapter may differ by a few cents depending on the table or calculator used.

SECTION 10.2 (PAGE 463)

1. $1593.85 **3.** $1515.80 **5.** $13,213.14 **7.** $5105.58 **9.** $60,484.66 **11.** $1167.56 **13.** $2551.12 **15.** $8265.24 **17.** $988.94 **19.** $5370.38 **21.** 4.04% **23.** 8.16% **25.** 12.36% **27.** $8763.47 **29.** $16,288.95 **31.** $30,611.30 **33.** $3951.57 **35.** $1000 now **37.** About 18 yr **39.** About 12 yr **41.** $142,886.40 **43.** $123,506.50 **45.** $63,685.27 **47.** $904.02

▬ SECTION 10.3   (PAGE 471)

1. 11, 17, 23, 29, 35; arithmetic   **3.** $-4, -1, 2, 5, 8$; arithmetic   **5.** $-2, -8, -14, -20, -26$; arithmetic   **7.** 2, 4, 8, 16, 32; geometric   **9.** $-2, 4, -8, 16, -32$; geometric   **11.** 6, 12, 24, 48, 96; geometric   **13.** 1/3, 3/7, 1/2, 5/9, 3/5; neither   **15.** 1/2, 1/3, 1/4, 1/5, 1/6; neither   **17.** Arithmetic; $d = 8$ **19.** Arithmetic; $d = 3$ **21.** Geometric; $r = 3$ **23.** Neither   **25.** Arithmetic; $d = -3$ **27.** Arithmetic; $d = 3$ **29.** Geometric; $r = -2$ **31.** Neither   **33.** 70   **35.** 65   **37.** 78   **39.** -65 **41.** 63   **43.** 678   **45.** 183   **47.** -18 **49.** 48   **51.** -648 **53.** 81   **55.** 64   **57.** 15   **59.** 156/25 = 6.24   **61.** -208 **63.** -15 **65.** 80   **67.** 125   **69.** 134   **71.** 70   **73.** -56 **75.** 125,250   **77.** 90   **79.** 170   **81.** 160   **83.** \$15,600   **85. (a)** \$1680.70 **(b)** \$576.48

▬ SECTION 10.4   (PAGE 478)

1. 15.91713   **3.** 21.82453   **5.** 22.01900   **7.** 20.02359   **9.** \$1318.08   **11.** \$305,390.00   **13.** \$671,994.62   **15.** \$4,180,929.88   **17.** \$180,307.41   **19.** \$27,541.18   **21.** \$516,397.05   **23.** \$6294.79   **25.** \$158,456.00   **27.** \$13,486.56   **29.** \$6517.42   **31.** \$628.25   **33.** \$952.33   **35.** \$354.79   **37.** \$691.08   **39.** \$4566.33   **41.** \$139.29   **43.** \$137,895.84   **45.** \$67,940.98   **47.** \$226.11   **49.** \$522.85   **51.** \$112,796.87   **53.** \$84,579.40   **55.** \$23,023.98

57. (a) \$1200 **(b)** \$3511.58 **(c)**

Payment Number	Amount of Deposit	Interest Earned	Total
1	\$3511.58	\$0	\$3511.58
2	\$3511.58	\$105.35	\$7128.51
3	\$3511.58	\$213.86	\$10,853.95
4	\$3511.58	\$325.62	\$14,691.15
5	\$3511.58	\$440.73	\$18,643.46
6	\$3511.58	\$559.30	\$22,714.34
7	\$3511.58	\$681.43	\$26,907.35
8	\$3511.58	\$807.22	\$31,226.15
9	\$3511.58	\$936.78	\$35,674.51
10	\$3511.58	\$1070.24	\$40,256.33
11	\$3511.58	\$1207.69	\$44,975.60
12	\$3511.58	\$1349.27	\$49,836.45
13	\$3511.58	\$1495.09	\$54,843.12
14	\$3511.59	\$1645.29	\$60,000.00

▬ SECTION 10.5   (PAGE 485)

1. 9.71225   **3.** 12.65930   **5.** 14.71787   **7.** 5.69719   **9.** \$6246.89   **11.** \$7877.72   **13.** \$153,724.50   **15.** \$160,188.18   **17.** \$111,183.90   **19.** \$97,122.50   **21.** \$76,060.80   **23.** \$160.08   **25.** \$5309.69   **27.** \$11,727.32   **29.** \$258.90   **31.** \$476.81   **33.** \$530.35   **35.** \$6698.98   **37.** \$12,493.78   **39.** \$158.00

41.

Payment Number	Amount of Payment	Interest for Period	Portion to Principal	Principal at End of Period
0	—	—	—	\$4000
1	\$1207.68	\$320.00	\$887.68	\$3112.32
2	\$1207.68	\$248.99	\$958.69	\$2153.63
3	\$1207.68	\$172.29	\$1035.39	\$1118.24
4	\$1207.70	\$ 89.46	\$1118.24	\$0

43.

Payment Number	Amount of Payment	Interest for Period	Portion to Principal	Principal at End of Period
0	—	—	—	$7184
1	$189.18	$71.84	$117.34	$7066.66
2	$189.18	$70.67	$118.51	$6948.15
3	$189.18	$69.48	$119.70	$6828.45
4	$189.18	$68.28	$120.90	$6707.55
5	$189.18	$67.08	$122.10	$6585.45
6	$189.18	$65.85	$123.33	$6462.12

45. (a) $4025.90   **(b)** $2981.93

47.

End of Year	To Interest	To Principal	Total Interest	Total Principal	Balance
1	$3225.54	$2557.95	$ 3225.54	$ 2557.95	$35389.55
2	$3008.11	$2775.38	$ 6233.65	$ 5333.33	$32614.17
3	$2772.20	$3011.29	$ 9005.85	$ 8344.62	$29602.88
4	$2516.24	$3267.25	$11522.09	$11611.87	$26335.63
5	$2238.53	$3544.96	$13760.62	$15156.83	$22790.67
6	$1937.21	$3846.28	$15697.83	$19003.11	$18944.39
7	$1610.27	$4173.22	$17308.10	$23176.33	$14771.17
8	$1255.55	$4527.94	$18563.65	$27704.27	$10243.23
9	$ 870.67	$4912.82	$19434.32	$32617.09	$ 5330.41
10	$ 453.09	$5330.41	$19887.41	$37947.50	$0

▬ CHAPTER 10   REVIEW EXERCISES   (PAGE 488)

1. $848.16   **3.** $921.50   **5.** $48.99   **7.** $24,000   **9.** $78,717.19   **11.** $739.09   **13.** 12.36%
15. $1999.00   **17.** $43,988.32   **19.** $3247.37   **21.** $3576.92   **23.** $6002.84   **25.** $6289.04   **27.** $14,202.12
29. $2501.24   **31.** $18,207.65   **33.** $1067.71   **35.** $-2, -6, -10, -14, -18$; arithmetic   **37.** 1/2, 4/7, 5/8, 2/3,
7/10   **39.** 61   **41.** -24 **43.** $10,078.44   **45.** $137,925.91   **47.** $25,396.38   **49.** $886.05   **51.** $5132.48
53. $2815.31   **55.** $47,988.11   **57.** $12,806.37   **59.** $2305.07   **61.** $546.93

63.

Payment Number	Amount of Payment	Interest for Period	Portion to Principal	Principal at End of Period
0	—	—	—	$5000
1	$985.09	$250.00	$735.09	$4264.91
2	$985.09	$213.25	$771.84	$3493.07
3	$985.09	$174.65	$810.44	$2682.63
4	$985.09	$134.13	$850.96	$1831.67
5	$985.09	$ 91.58	$893.51	$ 938.16
6	$985.07	$ 46.91	$938.16	$0

65. $6072.05   **67.** $3653.45   **69.** $27,320.71   **71.** $3598.01

▬ EXTENDED APPLICATION   (PAGE 495)

1. $14,038   **2.** $9511   **3.** $8837   **5.** $3968

CHAPTER 11   SECTION 11.1   (PAGE 500)

1.
$$
\begin{array}{c}
 & \begin{matrix} R & S & T \end{matrix} \\
\begin{matrix} R \\ S \\ T \end{matrix} &
\begin{bmatrix} 0 & 1 & 0 \\ 0 & 0 & 0 \\ 2 & 1 & 0 \end{bmatrix}
\end{array}
$$

3.
$$
\begin{array}{c}
 & \begin{matrix} X & Y & Z \end{matrix} \\
\begin{matrix} X \\ Y \\ Z \end{matrix} &
\begin{bmatrix} 0 & 1 & 0 \\ 1 & 0 & 0 \\ 0 & 0 & 0 \end{bmatrix}
\end{array}
$$
Adjacency matrix

5.
$$
\begin{array}{c}
 & \begin{matrix} P & R & S & T \end{matrix} \\
\begin{matrix} P \\ R \\ S \\ T \end{matrix} &
\begin{bmatrix} 0 & 2 & 0 & 1 \\ 0 & 0 & 1 & 1 \\ 0 & 0 & 0 & 0 \\ 1 & 0 & 1 & 0 \end{bmatrix}
\end{array}
$$

7.
$$
\begin{array}{c}
 & \begin{matrix} P & Q & R & S \end{matrix} \\
\begin{matrix} P \\ Q \\ R \\ S \end{matrix} &
\begin{bmatrix} 0 & 1 & 1 & 0 \\ 1 & 0 & 1 & 1 \\ 0 & 0 & 0 & 1 \\ 1 & 0 & 1 & 0 \end{bmatrix}
\end{array}
$$
Adjacency matrix

9.
$$
\begin{array}{c}
 & \begin{matrix} E & F & G & H & K \end{matrix} \\
\begin{matrix} E \\ F \\ G \\ H \\ K \end{matrix} &
\begin{bmatrix} 0 & 2 & 1 & 1 & 1 \\ 1 & 0 & 1 & 1 & 0 \\ 1 & 0 & 0 & 1 & 1 \\ 1 & 1 & 0 & 0 & 1 \\ 0 & 0 & 1 & 0 & 0 \end{bmatrix}
\end{array}
$$

11.
$$
\begin{array}{c}
 & \begin{matrix} A & B & C & D & E & F \end{matrix} \\
\begin{matrix} A \\ B \\ C \\ D \\ E \\ F \end{matrix} &
\begin{bmatrix} 0 & 1 & 1 & 1 & 1 & 0 \\ 1 & 0 & 1 & 1 & 1 & 1 \\ 0 & 0 & 0 & 1 & 1 & 1 \\ 0 & 1 & 1 & 0 & 1 & 0 \\ 1 & 0 & 0 & 1 & 0 & 1 \\ 1 & 0 & 0 & 0 & 0 & 0 \end{bmatrix}
\end{array}
$$
Adjacency matrix

13. A ⇄ B

15.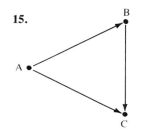

17.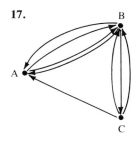

19.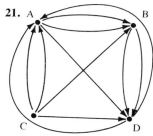

21. A graph with vertices A, B, C, D

23. Al, Bill, Cynthia, Donna graph
$$
\begin{array}{c}
 & \begin{matrix} A & B & C & D \end{matrix} \\
\begin{matrix} A \\ B \\ C \\ D \end{matrix} &
\begin{bmatrix} 0 & 1 & 1 & 0 \\ 1 & 0 & 0 & 0 \\ 0 & 1 & 0 & 1 \\ 1 & 1 & 1 & 0 \end{bmatrix}
\end{array}
$$

SECTION 11.2   (PAGE 505)

1. $\begin{bmatrix} 0 & 0 \\ 0 & 0 \end{bmatrix}$; A dominates 1 person, B dominates no one   **3.** $\begin{bmatrix} 0 & 0 & 1 \\ 1 & 0 & 0 \\ 0 & 1 & 0 \end{bmatrix}$; each dominates 2 people   **5.** Not the matrix of a dominance digraph

7. $\begin{bmatrix} 0 & 1 & 2 & 0 \\ 1 & 0 & 0 & 0 \\ 0 & 1 & 0 & 1 \\ 1 & 0 & 1 & 0 \end{bmatrix}$; R dominates 5 people, S dominates 2 people, T dominates 3 people, and V dominates 4 people   **9.** $\begin{bmatrix} 0 & 0 & 1 & 1 & 2 \\ 1 & 0 & 0 & 2 & 1 \\ 2 & 1 & 0 & 1 & 0 \\ 0 & 1 & 2 & 0 & 1 \\ 1 & 2 & 1 & 0 & 0 \end{bmatrix}$; each dominates 6 people

11.
$\begin{array}{c} \\ A \\ B \\ C \\ D \end{array}\begin{array}{cccc} A & B & C & D \\ \left[\begin{array}{cccc} 0 & 1 & 1 & 1 \\ 0 & 0 & 1 & 0 \\ 0 & 0 & 0 & 0 \\ 0 & 1 & 1 & 0 \end{array}\right. \end{array}$; A has 3 two-stage dominances (B through D, and C through both B and D), and D has 1 two-stage dominance (C through B)

13.
$\begin{array}{c} \\ A \\ B \\ C \\ D \end{array}\begin{array}{cccc} A & B & C & D \\ \left[\begin{array}{cccc} 0 & 0 & 0 & 1 \\ 1 & 0 & 1 & 0 \\ 1 & 0 & 0 & 1 \\ 0 & 1 & 0 & 0 \end{array}\right. \end{array}$; A has 1 two-stage dominance (of B through D), B has 3 two-stage dominances (of A through C and D through both A and C), C has 2 two-stage dominances (of B through D and of D through A), and D has 2 two-stage dominances (of A through B and C through B)

15. C won 8 games in one- or two-stage dominance, while A won 7. Declare C the winner.   **17.** Green has 5 one- or two-stage dominances, and red has 4. Make the cereal green.   **19.** B has 8 and E has 7, so wine B wins the taste test.
21. A dominates B in 6 ways, C in 4 ways, D in 1 way, and E in 1 way. B dominates A in 1 way, C in 6 ways, D in 1 way, and E in 4 ways. C dominates A in 1 way, B in 1 way, D in 4 ways, and E in 6 ways. D dominates A in 6 ways, B in 4 ways, C in 1 way, and E in 1 way. E dominates A in 4 ways, B in 1 way, C in 1 way, and D in 6 ways.

SECTION 11.3   (PAGE 511)

1.
$\begin{array}{c} \\ A \\ B \\ C \\ D \end{array}\begin{array}{cccc} A & B & C & D \\ \left[\begin{array}{cccc} 0 & 0 & 1 & 1 \\ 0 & 0 & 1 & 1 \\ 1 & 1 & 0 & 2 \\ 1 & 1 & 2 & 0 \end{array}\right. \end{array}$
Communication digraph

3.
$\begin{array}{c} \\ P \\ Q \\ R \\ S \\ T \end{array}\begin{array}{ccccc} P & Q & R & S & T \\ \left[\begin{array}{ccccc} 0 & 0 & 1 & 0 & 0 \\ 0 & 0 & 0 & 1 & 1 \\ 1 & 0 & 0 & 1 & 0 \\ 0 & 1 & 1 & 0 & 1 \\ 0 & 1 & 0 & 1 & 0 \end{array}\right. \end{array}$
Communication digraph
and organizational
communication digraph

5.
$\begin{array}{c} \\ M \\ N \\ P \\ Q \\ R \end{array}\begin{array}{ccccc} M & N & P & Q & R \\ \left[\begin{array}{ccccc} 0 & 1 & 0 & 0 & 2 \\ 1 & 0 & 1 & 0 & 1 \\ 0 & 1 & 0 & 1 & 0 \\ 0 & 0 & 1 & 0 & 1 \\ 2 & 1 & 0 & 1 & 0 \end{array}\right. \end{array}$
Communication digraph

7.
$\begin{array}{c} \\ T \\ W \\ X \\ Y \\ Z \end{array}\begin{array}{ccccc} T & W & X & Y & Z \\ \left[\begin{array}{ccccc} 0 & 1 & 0 & 1 & 0 \\ 1 & 0 & 0 & 0 & 1 \\ 0 & 0 & 0 & 0 & 1 \\ 1 & 0 & 0 & 0 & 1 \\ 0 & 1 & 1 & 1 & 0 \end{array}\right. \end{array}$
Communication digraph
and organizational
communication digraph

9.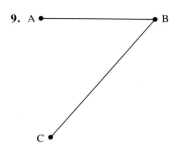

Two-stage paths are A to C through B and C to A through B.

11.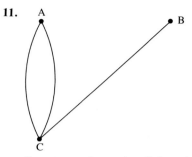

Two-stage paths are A to B through C in 2 ways and B to A through C in 2 ways.

13.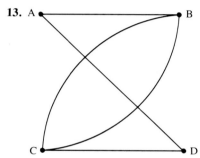

Two-stage paths are A to C in 3 ways (2 through B and 1 through D), B to D in 3 ways (2 through C and 1 through A), C to A in 3 ways (2 through B, 1 through D), and D to B in 3 ways (2 through C, 1 through A)

15.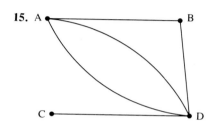

Two-stage paths are A to B through D in 2 ways, A to C through D in 2 ways, A to D through B in 1 way, B to A through D in 2 ways, B to D through A in 2 ways, B to C through D in 1 way, C to A through D in 2 ways, C to B through D in 1 way, D to A through B in 1 way, and D to B through A in 2 ways.

17. AED, ACED, ABED   **19.** ACBED, ABCED, ABED, ACED

━━━ SECTION 11.4   (PAGE 517)

1. ADCBA, ACBA, ADCA   **3.** RTSWVR, RSWVR, RTSR

5.

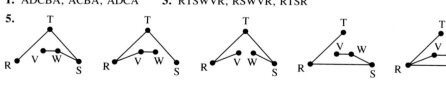

7.

  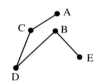

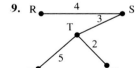 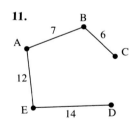

9.

11.

13. 18

15.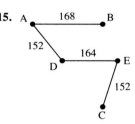

![] CHAPTER 11 REVIEW EXERCISES (PAGE 520)

1.

	M	N	P	Q
M	0	1	0	1
N	1	0	0	1
P	0	0	0	0
Q	0	0	1	0

Adjacency matrix

3.

	T	W	X	Y	Z
T	0	0	0	1	1
W	1	0	0	1	0
X	0	1	0	1	0
Y	0	0	0	0	0
Z	0	0	1	1	0

Adjacency matrix

5.

7.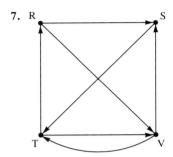

9.
$$\begin{array}{c} \\ A \\ B \\ C \end{array} \begin{array}{ccc} A & B & C \\ \left[\begin{array}{ccc} 0 & 0 & 1 \\ 1 & 0 & 0 \\ 0 & 1 & 0 \end{array}\right] \end{array}$$
Total dominances in one- or two-stage dominances: A dominates B and C, B dominates A and C, C dominates A and B

11. Not the matrix of a dominance digraph

13.
$$\begin{array}{c} \\ P \\ Q \\ R \\ S \end{array} \begin{array}{cccc} P & Q & R & S \\ \left[\begin{array}{cccc} 0 & 0 & 1 & 1 \\ 1 & 0 & 0 & 1 \\ 0 & 1 & 0 & 1 \\ 0 & 0 & 0 & 0 \end{array}\right] \end{array}$$
Total dominances in one- or two-stage dominances: P dominates Q and R once each and S twice; Q dominates P and R once each and S twice; R dominates P and Q once each and S twice; S dominates no one

15.
$$\begin{array}{c} \\ A \\ B \\ C \\ D \end{array} \begin{array}{cccc} A & B & C & D \\ \left[\begin{array}{cccc} 0 & 1 & 1 & 0 \\ 0 & 0 & 1 & 1 \\ 0 & 0 & 0 & 0 \\ 1 & 0 & 1 & 0 \end{array}\right] \end{array}$$
Two-stage dominances: A dominates C and D (1 way each), B dominates A and C in 1 way each, C has no two-stage dominances, D dominates B and C in 1 way each

17. B wins with $3 + 5 = 8$ one- or two-stage dominances, while D has only $3 + 4 = 7$.

19.
$$\begin{array}{c} \\ A \\ B \\ C \\ D \end{array} \begin{array}{cccc} A & B & C & D \\ \left[\begin{array}{cccc} 0 & 1 & 2 & 1 \\ 1 & 0 & 0 & 0 \\ 2 & 0 & 0 & 1 \\ 1 & 0 & 1 & 0 \end{array}\right] \end{array}$$
Communication digraph

21.
$$\begin{array}{c} \\ R \\ S \\ T \\ V \\ W \end{array} \begin{array}{ccccc} R & S & T & V & W \\ \left[\begin{array}{ccccc} 0 & 1 & 1 & 1 & 1 \\ 1 & 0 & 0 & 1 & 1 \\ 1 & 0 & 0 & 1 & 0 \\ 1 & 1 & 1 & 0 & 1 \\ 1 & 1 & 0 & 1 & 0 \end{array}\right] \end{array}$$
Communication digraph and organizational communication digraph

23.

Two-stage paths are A to B through C in 2 ways and B to A through C in 2 ways.

25.

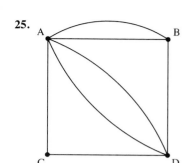

Two-stage paths are A to B in 2 ways, A to C in 2 ways, A to D in 3 ways, B to A in 2 ways, B to C in 3 ways, B to D in 4 ways, C to A in 2 ways, C to B in 3 ways, C to D in 2 ways, D to A in 3 ways, D to B in 4 ways and D to C in 2 ways.

27. ABED, AED   **31.** XYZWX, XYZX, XZWX

33.

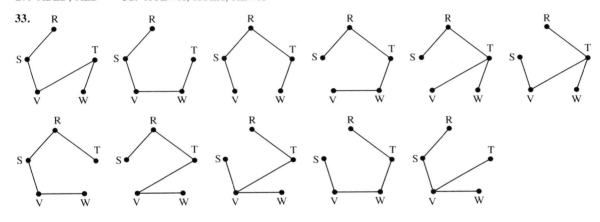

35.

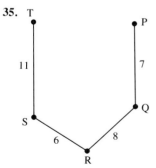

INDEX

Absorbing Markov chain, 404
   properties of, 410
Addition of matrices, 93
Addition principle, for simple
   events, 266
Addition property
   of equality, 2
   of inequality, 5
Additive identity of a matrix, 94
Additive inverse of a matrix, 94
Adjacency matrix, 500
Amortization, 483
Annuity, 473
   future value of, 473
   ordinary, 473
   present value of, 481
Annuity due, 476
Area under a normal curve, 363
Arithmetic mean, 337
Arithmetic sequence, 466
   sum of terms of, 468
Artificial variable, 193
Asymmetric matrix, 502
Athletic contracts, 491
Augmented matrix, 79
Average, 337
   weighted, 338
Average rate of change, 34

Basic probability principle, 265
Basic variable, 171
Bayes' formula, 297, 299
   special case of, 298
Bernoulli trials, 305, 312
   probability in, 314
Binomial distribution, 370
Binomial expansion, 253
Binomial theorem, 254

Booz, Allen & Hamilton, 62
Boundary, 139
Bounded region, 156
Break-even analysis, 36
Break-even point, 31

Cardinal number of a set, 220
Cartesian coordinate system, 11
Certain event, 264
Change in x, 21
Change in y, 21
Chebyshev's theorem, 352
Closed model, 122
Coefficient of correlation, 50
Coefficient matrix, 112
Column matrix, 92
Column vector, 92
Combination, 244
Common difference, 466
Common ratio, 468
Communication digraph, 507
Complement
   of an event, 271
   of a set, 222
Compound amount, 457
Compound interest, 457
Conditional probability, 283
   definition of, 284
Connected digraph, 509
Constraints
   of inequalities, 187
   of linear inequalities, 146
Constructing mathematical
   models, 44
Contagion, 131
Continuous compounding, 460
Continuous distribution, 359

Continuous probability
   function, 332
Coordinate system, 11
Corner point, 148
Corner point theorem, 156
Correlation, 50
Cost analysis, 34
Cost function, 35
Curve fitting, 47
Cycle, 513

Decision making, 418
Decision theory, 418
Degeneracies, 195
Delta x, 21
Demand curve, 15
Demand matrix, 120
Dependent events, 286
Dependent system, 67
Dependent variable, 8
Depreciation, 38
   straight-line, 38
   sum-of-the-years'-digits, 44
Deterministic phenomena, 261
Difference of matrices, 94
Digraph, 498
   communication, 507
   connected, 509
   dominance, 501
   matrix representation of,
     499
   organizational
     communication, 508
   path of, 509
Dimension of a matrix, 91
Directed edge, 498
Directed graph, 498

Discount, 453, 454
Discrete probability function, 332
Disjoint sets, 223
Distinguishable permutation, 240
Distribution
  binomial, 370
  continuous, 359
  frequency, 329
  normal, 358
  of states, 391
  probability, 277, 329
  skewed, 359
Domain, 9
Dominance, 501
  four-stage, 504
  one-stage, 502
  three-stage, 504
  two-stage, 503
Dominant strategy, 424
Duality, 200
  theorem of, 204

Echelon method, 68
  solution by, 73
Edge, 497
  directed, 498
Effective rate
  compound interest, 460
  simple interest, 454
Element
  of a matrix, 79
  of a set, 218
Empty set, 220
Entry of a matrix, 79
Equal matrices, 92
Equal sets, 218
Equilibrium demand, 17
Equilibrium price, 17
Equilibrium supply, 17
Equilibrium vector, 398
Equivalent system, 68
Event, 264
  certain, 264
  complement of, 271
  dependent, 286
  impossible, 264
  independent, 291
  mutually exclusive, 271
  odds favoring, 275
  probability of, 265
  simple, 264
Exact interest, 452
Expected value, 336
Experiment, 261

Factorial notation, 238
Fair game, 340, 426
Feasible region, 142
  bounded, 156
  corner point of, 148
  unbounded, 150
First-degree equation, 66
First down, 326
Fixed cost, 35
Fixed vector, 398
Four-stage dominance, 504
Frequency distribution, 329
Function
  continuous, 332
  cost, 35
  definition of, 9
  discrete, 332
  domain of, 9
  linear, 8
  objective, 146
  probability distribution, 331
  range of, 9
  sequence, 465
  vertical line test for, 12
Function notation, 9
Fundamental matrix, 409
Future value, 451
  of an annuity, 473, 475

Gambler's ruin problem, 406
Game, 423
  fair, 426
  strictly determined, 423,
    425, 426
  two-person, 423
  value of, 425
  zero-sum, 424
Game theory, 423
Gauss-Jordan method, 81
General term of a sequence, 466
Geometric sequence, 468
  nth term of, 469
  sum of terms of, 470
Graph, 11, 497
  directed, 498
Graph theory, 497

Half-plane, 141
Histogram, 332

Ice cream, 213
Identity matrix, 108

Impossible event, 264
Inconsistent system, 67
Independent events, 291
Independent variable, 8
Indicators, 169
Inequality
  linear, 5, 138
  system, 142
Initial simplex tableau, 169
Input-output matrix, 119
Intercept, 13
Interest, 451
  compound, 457
  continuous compounding of, 460
  discount, 453
  exact, 452
  nominal rate of, 460
  ordinary, 452
  rate of, 451
  simple, 451
  stated rate of, 460
Intersection of sets, 222

Least squares line, 49
Least squares method, 47
Leontief's model, 132
Linear depreciation, 38
Linear equation, 2
  standard form of, 28
Linear function, 8
  definition of, 12
  graphing, 12
Linear inequality, 5, 138
  definition of, 138
  graphing, 138
Linear model, 32
Linear programming, 146
  duality, 200
  graphical method, 154
  maximum value, 154
  minimum value, 154
  nonstandard problem, 192
  simplex method, 168, 182
  standard maximum, 168
  standard minimum, 190

Marginal cost, 35
Markov chain, 386
  absorbing, 404
  equilibrium vector of, 398
  fixed vector of, 398
  properties of, 400, 410
  regular, 397

Mathematical model, 2, 146
  constructing, 44
Matrix, 79
  addition of, 93
  additive identity of, 94
  additive inverse, 94
  adjacency, 500
  asymmetric, 502
  augmented, 79
  coefficient, 112
  column, 92
  demand, 120
  difference of, 94
  dimension of, 91
  element of, 79
  entry of, 79
  equal, 92
  fundamental, 409
  identity, 108
  input-output, 119
  multiplicative inverse of, 108
  negative of, 94
  of digraphs, 499
  order, 91
  payoff, 418, 424
  product, 99
  production, 119
  regular transition, 387
  row, 92
  row operations, 80
  square, 92
  subtraction of, 94
  sum of, 93
  symmetric, 507
  technological, 119
  transition, 386
  transpose, 202
  zero, 94
Matrix operations, 90
Maturity value, 451
Maximization problems, 176
Mean, arithmetic, 337
Measure of central tendency, 341
Median, 341
Medical diagnosis, 326
Member of a set, 218
Minimization problem, 190
Minimum spanning tree, 514
Mixed constraints, 187
Mixed strategy, 430, 434
Mode, 342
Most negative indicator, 178
Multiplication principle, 236
Multiplication property
  of equality, 2
  of inequality, 5

Multiplicative inverse of a
  matrix, 108
Mutually exclusive events, 271

n-factorial, 238
Negative of a matrix, 94
Net cost, 38
Network, 513
Nominal rate, 460
Normal curve, 359, 361
Normal distribution, 358
  probabilities for, 361
nth term
  of an arithmetic sequence, 466
  of a geometric sequence, 469

Objective function, 146
  maximum value of, 154
  minimum value of, 154
Odds, 275
One-stage dominance, 503
Open model, 122
Operations on sets, 224
Optimum strategy, 425, 433
Order of a matrix, 91
Ordered pair, 11
Ordinary annuity, 473
Ordinary interest, 452
Organizational communication
  digraph, 508
Origin, 11
Outcome, 261

Parameters, 73
Pascal's triangle, 255
Path of a digraph, 509
Payment period, 473
Payoff, 418, 424
Payoff matrix, 418, 424
Permutation, 236, 238
  distinguishable, 240
Phases of a solution, 188
Pivot, 172
Point-slope form, 26
Present value
  compound interest, 461
  of an annuity, 481
  simple interest, 452
Principal, 451
Probability, 261
  addition principle, 266
  Bayes' formula, 297, 299
  Bernoulli trials, 314

Probability *(continued)*
  conditional, 283
  empirical, 276
  of an event, 265
  k trials for m successes, 317
  odds, 275
  product rule, 286, 291
  properties of, 267
  union rule, 271
Probability distribution, 277, 329
  normal, 358
  variance of, 350
Probability distribution
  function, 331
Probability function, 331
  continuous, 332
  discrete, 332
  standard deviation for, 351
  variance of, 351
Probability principle, 265
Probability vector, 392
Proceeds, 453
Product
  of a matrix and a scalar, 98
  of matrices, 99
Production matrix, 119
Product rule
  for dependent events, 286
  for independent events, 291
Pure strategy, 425, 430

Quadrant, 11

Random variable, 329
Range, 9
Rate of interest, 451
Region of feasible solutions, 142
Regular Markov chain, 397
  properties of, 400, 410
Regular transition matrix, 397
Relative frequency, 261, 330
Repeated trials, 312
Routing, 129
Row matrix, 92
Row operations, 80
Row vector, 92

Saddle point, 426
Sales analysis, 33
Sample space, 262
Scalar of a matrix, 98
Scrap value, 38

Sequence, 465
  arithmetic, 466
  general term of, 466
  geometric, 468
  *n*th term of, 466
  term of, 465
Set, 218
  cardinal number of, 220
  complement of, 222
  disjoint, 223
  element of, 218
  empty, 220
  equal, 218
  intersection of, 222
  member of, 218
  subset of, 219
  union of, 223
  universal 219
  universe of discourse, 219
Set operations, 224
Set-builder notation, 219
Shadow cost, 207
Signal Oil, 382
Simple discount notes, 453
Simple event, 264
Simple interest, 451
Simple interest notes, 453
Simplex method, 168, 182
Simplex tableau, 169
Sinking fund, 477
Skewed distribution, 359
Slack variable, 169
Slope, 21
  definition of, 22
Slope-intercept form, 24
Solution of a first-degree
    equation, 67
Southern Pacific
    Transportation, 493
Spanning tree, 514
  minimum, 514
Square matrix, 92
Standard deviation, 350
Standard form of a linear
    equation, 28
Standard maximum form, 168
Standard minimum form, 190
Standard normal curve, 361
Stated rate, 460
State of nature, 418

Statistical process control, 353
Stochastic process, 289
Straight-line depreciation, 38
Strategy, 418, 423
  dominant, 424
  mixed, 430, 434
  optimum, 425, 433
  pure, 425, 430
Strictly determined game,
    423, 425, 426
Subset, 219
Subtraction of matrices, 94
Summation notation, 337
Sum of matrices, 93
Sum-of-the-years' digits, 44
Supply and demand, 15
Supply curve, 15
Surplus variable, 187
Symmetric matrix, 507
System of equations, 66
  dependent, 67
  echelon method for solving, 68
  equivalent, 68
  Gauss-Jordan method for
      solving, 81
  inconsistent, 67
  linear, 66
  linear applications, 75
  matrix solutions, 112
  parameters, 73
  transformations, 68
System of inequalities, 142
  applications, 143
  solution of, 142

Tableau, 169
Technological matrix, 119
Term
  of an annuity, 473
  of a sequence, 465
Theorem of duality, 204
Three-stage dominance, 504
Transformations of a system, 68
Transition diagram, 386
Transition matrix, 386
  regular, 387
Transpose of a matrix, 202
Tree, 513
  spanning, 514

Tree diagram, 221
Trial, 261
  repeated, 312
Two-person game, 423
Two-stage dominance, 503

Unbounded region, 150
Union of sets, 223
Union rule for probability, 271
Universal set, 219
Universe of discourse, 219
Upjohn Company, 61, 215
Utility, 420

Value, 425
Variable
  artificial, 193
  basic, 171
  dependent, 8
  independent, 8
  random, 329
  slack, 169
  surplus, 187
Variable cost, 35
Vector
  equilibrium, 398
  fixed, 398
  probability, 392
Venn diagram, 220
  applications of, 227
Vertex of a graph, 497
Vertical line test, 12

Weighted average, 338

x-axis, 11
x-coordinate, 11
x-intercept, 13

y-axis, 11
y-coordinate, 11
y-intercept, 13

Zero matrix, 94
Zero-sum game, 424
z-score, 363